Metallurgical Process Simulation and Optimization

Metallurgical Process Simulation and Optimization

Editors

Qing Liu
Jiangshan Zhang

MDPI • Basel • Beijing • Wuhan • Barcelona • Belgrade • Manchester • Tokyo • Cluj • Tianjin

Editors
Qing Liu
University of Science and Technology Beijing
China

Jiangshan Zhang
University of Science and Technology Beijing
China

Editorial Office
MDPI
St. Alban-Anlage 66
4052 Basel, Switzerland

This is a reprint of articles from the Special Issue published online in the open access journal *Materials* (ISSN 1996-1944) (available at: https://www.mdpi.com/journal/materials/special_issues/metallurgical_process).

For citation purposes, cite each article independently as indicated on the article page online and as indicated below:

LastName, A.A.; LastName, B.B.; LastName, C.C. Article Title. *Journal Name* **Year**, *Volume Number*, Page Range.

ISBN 978-3-0365-6430-2 (Hbk)
ISBN 978-3-0365-6431-9 (PDF)

Cover image courtesy of Jiangshan Zhang and Wuhu Xinxing Ductile Iron Pipes Co., Ltd.

Contents

About the Editors

Qing Liu

Qing Liu is a Professor at the University of Science and Technology Beijing. His research interests include the fine control of the solidification process in continuous casting, metallurgical process engineering, simulation and optimization of metallurgical process, and highly efficient utilization of metallic resources. He has published around 200 peer-reviewed papers.

Jiangshan Zhang

Jiangshan Zhang is an Associate Professor at the University of Science and Technology Beijing. His research interests include advanced material manufacturing and process metallurgy simulation. He has published around 50 peer-reviewed papers.

Preface to "Metallurgical Process Simulation and Optimization"

Metallurgy involves the art and science of extracting metals from their ores and modifying the metals for use. With thousands of years of development, many interdisciplinary technologies have been introduced into this traditional and large-scale industry. In modern metallurgical practices, modelling and simulation are widely used to provide solutions in the areas of design, control, optimization, and visualization, and are becoming increasingly significant in the progress of digital transformation and intelligent metallurgy.

This Special Issue (SI), entitled "Metallurgical Process Simulation and Optimization", has been organized as a platform to present the recent advances in the field of modelling and optimization of metallurgical processes, which covers the processes of electric/oxygen steel-making, secondary metallurgy, (continuous) casting, and processing. Both fundamental insights and practical foresights are presented, including the topics of thermodynamics, kinetics, physical modelling, numerical simulation, CFD, 3D visualization, microstructural evolution, quality control, artificial intelligence, big data, and cloud computation. Eighteen articles have been included that cover various aspects of the topic.

We would like to take this opportunity to express our most profound appreciation to the MDPI Book staff; the editorial team of the Materials journal, especially Mr. Felix Guo; the Assistant Editor of this Special Issue; the talented authors; and the hardworking and professional reviewers.

Qing Liu and Jiangshan Zhang
Editors

materials

Editorial

Metallurgical Process Simulation and Optimization

Jiangshan Zhang *, Yuhong Liu and Qing Liu *

State Key Laboratory of Advanced Metallurgy, University of Science and Technology Beijing, Beijing 100083, China
* Correspondence: jszhang@ustb.edu.cn (J.Z.); qliu@ustb.edu.cn (Q.L.); Tel.: +86-10-8237-5255 (Q.L.)

Citation: Zhang, J.; Liu, Y.; Liu, Q. Metallurgical Process Simulation and Optimization. *Materials* **2022**, *15*, 8421. https://doi.org/10.3390/ma15238421

Received: 16 November 2022
Accepted: 22 November 2022
Published: 26 November 2022

Metallurgy involves the art and science of extracting metals from their ores and modifying them for use. Over thousands of years of development, many interdisciplinary technologies have been introduced into this large-scale, traditional industry. In modern metallurgical practices, modeling and simulation have been widely used to provide solutions for design, control, optimization, and visualization, and they are increasingly significant in the progress of digital transformation and intelligent metallurgy. This Special Issue (SI), "Metallurgical Process Simulation and Optimization", presents recent advances in the modeling and optimization of metallurgical process, and includes nearly twenty articles covering various aspects of the topic. A second volume of this successful SI is being organized in which researchers from both academia and industry are invited to publish their new work. The purpose of the current editorial is to briefly summarize the publications included in this SI.

Advanced modeling methods have been widely used to simulate the fluid flow, heat transfer, and mass transfer behavior of metallurgical vessels, and are usually coupled with species, multiphase, and electromagnetic fields to optimize their design/operation. The effect of interfacial tension on alumina inclusion motion behavior was studied in a continuous casting mold using a commercial CFD software package, as reported by Siddiqui et al. [1]. It was found that inclusions were vulnerable to engulfment by the solidification front under a high-surface-tension gradient. Tie et al. [2] shed light on the fluid flow inside an asymmetric multi-strand tundish using both numerical and physical simulation, and the flow field was optimized to improve the overall quality of bloom castings and con-rod products. In Zhao et al.'s [3] numerical study of mesoscopic fluid-particle flow, a bifurcated pool-type SEN under steady operating conditions was employed using the lattice Boltzmann method (LBM) coupled with the large eddy simulation (LES) model with validation. Liu et al. [4] investigated the influence of M-EMS on fluid flow and initial solidification in slab continuous casting; a model coupled with an electromagnetic field was developed to simulate the transient turbulence flow and initial solidification, with special consideration of the effects of different electromagnetic stirring (EMS) currents and casting speeds. The function of an Electric Arc Furnace is essentially to melt metals, and Moskal et al. [5] optimized the melting process using statistical-thermodynamic modeling based on, among other things, multiple linear regression (MLR). In the work of Duan et al. [6], the gas–solid heat transfer and decomposition processes of limestone calcined with blast furnace gas were studied in a parallel-flow regenerative lime kiln; a Porous Medium Model (PMM) and a Shrinking Core Model (SCM) were used to examine the feasibility of calcining limestone with low-calorific fuel gas. The software ANSYS was used in Deng et al.'s [7] study to simulate the temperature field and stress–strain distribution on the working layer of a four-strand tundish under three preheating stages through the indirect coupling method.

State-of-the-art characterization techniques and experimental methods enable more detailed and accurate insights into the metallurgical process, and provide enriched information for modeling and process optimization. Li et al. [8] examined the nucleation process of zirconium oxide inclusions in steel using classical nucleation theory and first principles, accompanied by SEM, TEM, and XRD characterization methods. An optimized nucleation pathway of ZrO2 was achieved at a high temperature. Molecular dynamics simulations (MD) were combined with experiments in Du et al.'s work [9], which revealed the existence of the form and functional mechanism of CaF2 in phosphosilicate systems. The hot deformation characteristics of Ti alloys were analyzed using the strain hardening exponent, strain rate sensitivity, a processing map, and microstructure observation in Piao et al.'s study [10], and an optimum processing temperature and strain rate were proposed. Wang et al. [11] developed an in situ observation method to characterize the solidification of duplex steel during solidification, and phase-field simulation was also carried out to study its solidification and heat transfer behavior. The influence of the mineralogical structure of mold flux film on heat transfer was studied during continuous casting of peritectic steel in the work of Liu et al. [12], which covers the layered structure, crystallization ratio, mineralogical species, and morphology features of flux films. Optimized mineral phase structures of flux film were put forward to eliminate longitudinal cracks. An electro-gas welding experiment on thickness E36 steel plates was conducted in Fu et al.'s study [13]. A semi-ellipsoid heat source model was developed using linear, sinusoidal, or oscillate-stop paths. Different heating paths were recommended for corresponding thicknesses of steel plate. Malinowski et al. [14] investigated the beneficial effect of rare-earth metal oxides on the wear resistance of surface layers applied to castings, which are intended for structural elements of machinery and equipment in mining and recycling. Exploratory data analysis was carried out for the evaluation of tribological properties, and improvements were made to the modified surface wear resistance. The effects of another rare-earth metal, Cerium (Ce), were studied on the casting slab quality, microstructure and inclusions of cryogenic vessel steel in Wu et al.'s work [15]. Moreover, Zaba et al. [16] used full-field image correlation and infrared thermography techniques to study the effects of strain rate, specimen orientation, and plastic strain on the value and distribution of temperature in dog-bone stainless-steel specimen deformation in uniaxial tensile tests. In addition, Liu et al. [17] used positron annihilation lifetime spectroscopy (PALS), high-resolution transmission electron microscopy (HRTEM), and positron annihilation Coincidence Doppler broadening (CDB) techniques synergistically to study the microstructural evolution of alloys in their early aging stages and at low temperatures. Moreover, an optimized measurement temperature was proposed.

In addition, thermodynamic analysis of H_2 behavior was conducted in an iron-making blast furnace at different stages of the process of integrating top-gas recycling and CO_2 electrolysis for H_2-rich gas injection, as discussed in Hu et al.'s study [18]; this is of fundamental importance for better performance in H_2 metallurgy.

Funding: This research was funded by the National Natural Science Foundation of China (No. U21A20112, 52004024).

Conflicts of Interest: The authors declare no conflict of interest.

References

1. Siddiqui, M.I.H.; Albaqami, A.; Arifudin, L.; Alluhydan, K.; Alnaser, I.A. Simulation of Inclusion Particle Motion Behavior under Interfacial Tension in Continuous Casting Mold. *Materials* **2022**, *15*, 7458. [CrossRef] [PubMed]
2. Tie, Z.; Tang, H.; Wang, K.; Miao, H.; Cai, S.; Xian, F.; Zhang, J. Effect of Flow Field Optimization of an Asymmetric Multi-Strand Tundish on the Quality Consistency of Cracking Con-Rod Steel. *Materials* **2022**, *15*, 3698. [CrossRef] [PubMed]
3. Zhao, P.; Piao, R.; Zou, Z. Mesoscopic Fluid-Particle Flow and Vortex Structural Transmission in a Submerged Entry Nozzle of Continuous Caster. *Materials* **2022**, *15*, 2510. [CrossRef] [PubMed]
4. Liu, G.; Lu, H.; Li, B.; Ji, C.; Zhang, J.; Liu, Q.; Lei, Z. Influence of M-EMS on Fluid Flow and Initial Solidification in Slab Continuous Casting. *Materials* **2021**, *14*, 3681. [CrossRef] [PubMed]
5. Moskal, M.; Migas, P.; Karbowniczek, M. Multi-Parameter Characteristics of Electric Arc Furnace Melting. *Materials* **2022**, *15*, 1601. [CrossRef] [PubMed]
6. Duan, S.; Li, B.; Rong, W. Numerical Simulation Study of Gas-Solid Heat Transfer and Decomposition Processes of Limestone Calcined with Blast Furnace Gas in a Parallel Flow Regenerative Lime Kiln. *Materials* **2022**, *15*, 4024. [CrossRef] [PubMed]

7. Deng, X.; Li, J.; Xie, X. Effect of Preheating Temperature on Thermal-Mechanical Properties of Dry Vibrating MgO-Based Material Lining in the Tundish. *Materials* **2022**, *15*, 7699. [CrossRef] [PubMed]
8. Li, Y.; Wang, L.; Chen, C.; Yang, S.; Li, X. New Insights into the Mechanism of Nucleation of ZrO_2 Inclusions at High Temperature. *Materials* **2022**, *15*, 7960. [CrossRef]
9. Du, Y.; Huang, Z.; Long, M.; Duan, H.; Chen, D. Unveiling the Effect of CaF_2 on the Microstructure and Transport Properties of Phosphosilicate Systems. *Materials* **2022**, *15*, 7916. [CrossRef]
10. Piao, R.; Zhu, W.; Ma, L.; Zhao, P.; Hu, B. Characterization of Hot Deformation of near Alpha Titanium Alloy Prepared by TiH_2-Based Powder Metallurgy. *Materials* **2022**, *15*, 5932. [CrossRef] [PubMed]
11. Wang, T.; Wexler, D.; Guo, L.; Wang, Y.; Li, H. In Situ Observation and Phase-Field Simulation Framework of Duplex Stainless-Steel Slab during Solidification. *Materials* **2022**, *15*, 5517. [CrossRef] [PubMed]
12. Liu, L.; Han, X.; Li, M.; Zhang, D. Influence of Mineralogical Structure of Mold Flux Film on Heat Transfer in Mold during Continuous Casting of Peritectic Steel. *Materials* **2022**, *15*, 2980. [CrossRef] [PubMed]
13. Fu, J.; Tao, Q.; Yang, X.; Nenchev, B.; Li, M.; Tao, B.; Dong, H. The Effect of Heat Source Path on Thermal Evolution during Electro-Gas Welding of Thick Steel Plates. *Materials* **2022**, *15*, 2215. [CrossRef] [PubMed]
14. Malinowski, P.; Kasińska, J.; Rutkowski, S.; Madej, M. Exploratory Data Analysis for the Evaluation of Tribological Properties of Wear-Resistant Surface Layers Modified with Rare-Earth Metals. *Materials* **2022**, *15*, 2032. [CrossRef] [PubMed]
15. Wu, L.; Zhi, J.; Zhang, J.; Zhao, B.; Liu, Q. Effect of Cerium on the Microstructure and Inclusion Evolution of C-Mn Cryogenic Vessel Steels. *Materials* **2021**, *14*, 5262. [CrossRef] [PubMed]
16. Żaba, K.; Trzepieciński, T.; Rusz, S.; Puchlerska, S.; Balcerzak, M. Full-Field Temperature Measurement of Stainless Steel Specimens Subjected to Uniaxial Tensile Loading at Various Strain Rates. *Materials* **2021**, *14*, 5259. [CrossRef] [PubMed]
17. Liu, X.; Zhang, P.; Wang, B.; Cao, X.; Jin, S.; Yu, R. Study of Interaction Mechanism between Positrons and Ag Clusters in Dilute Al–Ag Alloys at Low Temperature. *Materials* **2021**, *14*, 1451. [CrossRef] [PubMed]
18. Hu, Y.; Qiu, Y.; Chen, J.; Hao, L.; Rufford, T.E.; Rudolph, V.; Wang, G. Integrating a Top-Gas Recycling and CO_2 Electrolysis Process for H2-Rich Gas Injection and Reduce CO_2 Emissions from an Ironmaking Blast Furnace. *Materials* **2022**, *15*, 2008. [CrossRef] [PubMed]

materials

MDPI

Article

Simulation of Inclusion Particle Motion Behavior under Interfacial Tension in Continuous Casting Mold

Md Irfanul Haque Siddiqui [1], Ayidh Albaqami [1,*], Latif Arifudin [1], Khalid Alluhydan [1] and Ibrahim Abdullah Alnaser [1,2]

1 Department of Mechanical Engineering, King Saud University, Riyadh 11451, Saudi Arabia
2 Center of Excellence for Research in Engineering Material (CEREM), King Saud University, Riyadh 11451, Saudi Arabia
* Correspondence: 441106631@student.ksu.sa

Abstract: Inclusions entrapped by the solidifying front during continuous casting adversely affect the properties of the final steel products. In this study, we investigated the effect of the interfacial tension due to surfactant concentration, particularly sulfur, on alumina inclusion motion behavior during molten steel solidification in a continuous casting mold. A two-dimensional numerical model was developed in Ansys Fluent software to simulate the inclusion motion in a continuous casting mold. Further, the impacts of different values of the alumina inclusion diameter, sulfur concentration, and melt temperature were studied to understand the inclusion motion behavior. The inclusion diameter affected the inclusion distribution throughout the domain. The alumina inclusion entrapment percentage varied in the case of sulfur mixing (using an empirical relationship for modeling). It was found that the removal percentage varied according to the sulfur concentration. The addition of sulfur at concentrations from 10 ppm to 70 ppm resulted in a 4% increase in the removal of alumina inclusions (trapped in the solidifying shell), except for the 100-ppm case. Smaller-sized inclusion particles had a 25% higher chance of entrapment at the top level of the mold. Under the effect of a higher surface tension gradient between inclusions and the melt, the predicted findings show that inclusions were vulnerable to engulfment by the solidification front.

Keywords: alumina inclusion; interfacial tension; steelmaking; continuous casting mold; surfactant; sulfur; simulation

Citation: Siddiqui, M.I.H.; Albaqami, A.; Arifudin, L.; Alluhydan, K.; Alnaser, I.A. Simulation of Inclusion Particle Motion Behavior under Interfacial Tension in Continuous Casting Mold. *Materials* **2022**, *15*, 7458. https://doi.org/10.3390/ma15217458

Academic Editors: Qing Liu and Jiangshan Zhang

Received: 25 July 2022
Accepted: 30 August 2022
Published: 24 October 2022

1. Introduction

Steel's mechanical properties and chemical composition are continually changing, and worries about cost, energy, and the environment in the manufacturing process are becoming increasingly relevant. To fulfill demand, steel's strength, ductility, durability, and corrosion resistance have all increased over time. A variety of processes, strongly influenced by the equipment, modify the final product's shape, appearance, and properties [1]. Due to its superior features, steel has been widely used for various applications since it was commercialized at the beginning of the 1900s. The need for clean steel is an important aspect for steel producers in the context of greenhouse emissions, competition, customer requirements, and sustainability. It has been reported widely that nonmetallic inclusions significantly influence the properties of steel products. The mold remains the last stage in the continuous casting process, where the proper implementation of processes can lead to significant inclusion removal. To properly control nonmetallic inclusions during the continuous casting process, it is of the utmost importance to understand the inclusion motion behavior in the continuous casting mold [2–4].

1.1. Inclusions in the Continuous Casting Mold

There are two kinds of nonmetallic inclusions in steel, each with its unique mode of formation. Indigenous oxide inclusions, which form when the steel melt is deoxidized, are one kind. The bulk of these oxides is removed from the melt during refining and degassing in the ladle, but some nonmetallic oxide inclusions remain. When the deoxidized steel melt is reoxidized by air or entrained slag is introduced into the melt during the melt transfer from the ladle to the mold, exogenous inclusions occur. Exogenous inclusions are generally much larger than native inclusions and hence more harmful. Inclusions cause problems during the casting, rolling, and heat treatment processes, and they may cause the steel to break during use. The size and nature of nonmetallic inclusions that decrease steel properties are not constant but rather fluctuate according to the application [5–9].

Typically, in the secondary steelmaking process, oxygen impurities are removed by aluminum deoxidation. However, alumina inclusions are formed during the deoxidation process. The steel components and structure require control of alumina inclusions to ensure their design performance. Inadequate removal of alumina inclusions during the steelmaking process can induce micro-surface cracks during continuous casting and hot rolling [10,11]. Generally, solid nonmetallic inclusions are not well-wetted with liquid steel. Inclusions distributed in liquid steel are sometimes engulfed in the solidification interface in the mold and become defects in steel products. Inclusions entrapped by the solidifying front during continuous casting will adversely affect the properties of the final steel products [1,12–14]. Therefore, the behavior of inclusions at the interface should be investigated in detail.

1.2. Inclusion Motion Characteristics

The interfacial tension, which is also referred to as the pushing and engulfment phenomenon, has an impact on the distribution of inclusion particles trapped in the solidifying slab of the continuous casting mold. The interfacial characteristics are connected to the pushing and engulfment of inclusion particles [15–17]. Particles move from one position with higher interfacial tension to another position with lower interfacial tension when there is a gradient of interfacial tension around a gas bubble or a small liquid inclusion. The Marangoni force's impact on inclusions at the solid–liquid border is seen in Figure 1. As is the case for pure metals, surface tension typically decreases as temperature rises. The presence of a surface-active element in a metal is not necessarily a given, though. To regulate Marangoni convection in the molten metal and the subsequent solidification in the mold, the concentration of minor elements in steels is important [18]. Some researchers have used bubbles to conduct extensive research on this phenomenon in water model experiments. Mukai et al. [15–17] reported that the pushing and engulfment of small bubbles occur in the solid–liquid interface due to the presence of a gradient of interfacial tension in the boundary layer. Further, Shibata et al. [19] studied the inclusion/bubble motion behavior at the solid–liquid interface of the solidifying metal surface. Some researchers emphasize that the Marangoni effect will affect the entire process of microgravity experiments due to temperature gradients or concentration gradients of surfactant elements [18,20]. Yin and Emi [20] reported that Marangoni flow was vigorously strong in molten steel, even with low concentrations of oxygen and sulfur. J. Jeong et al. (2020) [11] carried out comprehensive experimental work to find the correlation of interfacial tension between alumina inclusions and molten automotive steel. They studied the effect of surfactant concentration (sulfur) and temperature on interfacial tension.

1.3. Background of the Research

As a structural material, steel has benefits such as strength, ductility, and durability. Durability relates to resistance to wear, fatigue, hydrogen-induced cracking, and stress corrosion cracking, whereas ductility is exemplified by properties such as deep drawability, cold formability, and low-temperature toughness. Large nonmetallic impurities significantly reduce steel's ductility and durability. These inclusions are known as "dirty" inclusions, and they are seen in steel. The increasing need for enhanced ductility and durability in demanding applications can only be met by "clean" steels, which have a smaller number of evenly distributed, small-size impurities.

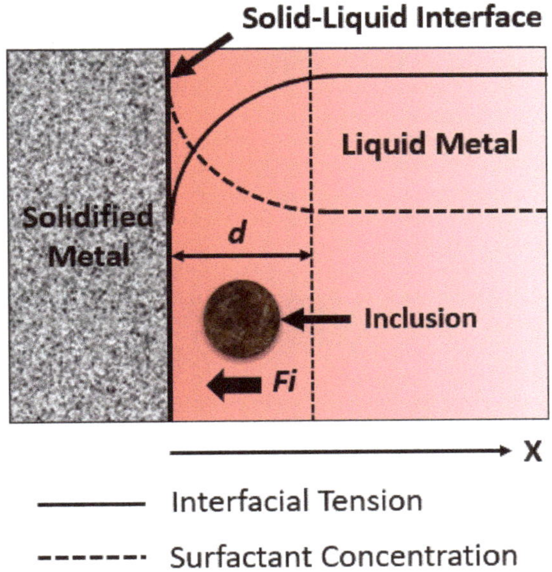

Figure 1. The effect of Marangoni force on an inclusion near the solidifying front (d = distance between solidifying boundary and saturation-level interfacial tension; Fi = interfacial tension).

Oxides, sulfides, nitrides, carbides, and their compounds or composites are examples of nonmetallic inclusions. During typical cooling of steel below the solidus temperature, sulfides, carbides, and nitrides precipitate. To improve steel qualities, small particles of certain oxide inclusions, sulfides, carbides, and nitrides have been used to manipulate the microstructure. Most big oxide inclusions and certain sulfide inclusions, on the other hand, occur when the steel is still liquid. They may create errors in casting goods, cause processing problems and failures, lower productivity, impair product quality, and diminish premium yield if they are not eliminated from the steel melt before solidification.

Improvement of the performance of continuous casting processes has been a subject of extensive research in the metallurgical industry. High-speed casting in the continuous casting operation is an essential objective of many of these studies, where the goal is to improve productivity without sacrificing the quality of the casts. In the continuous casting process, the fluid flow phenomenon greatly affects the quality of steel and defects in the final steel product [21]. It is part of a long-term effort to develop and apply comprehensive models of continuous casting and is supported by computational models and experimental measurements.

The particles in the liquid melt can move toward or away from the freezing front during the solidification of a liquid containing scattered second-phase particles; particles close to the freezing front will either be consumed or rejected. During solidification, these two events cause the particle distribution to change. The molten steel flows in the continuous casting process include several interacting phenomena that interact with each other in a complex form, such as heat transfer, solidification, multiphase turbulent flow, clogging, electromagnetic effects, complex interfacial behavior, particle entrapment, thermal-mechanical distortion, stress, cracks, segregation, and microstructure formation [22–24]. In the continuous casting mold, studies on fundamental phenomena, such as temperature, solidified shell growth, turbulent fluid flow along with multiphase physical processes, electromagnetic effects and particle transport, microstructure and grain structure, thermal properties, distortion, and stress, have all made significant strides toward accurately predicting these factors by computational models. However, the accurate prediction of the inclusion motion behavior near the solidifying boundary layer in response to the surfactant concentration needs much further study [25]. Mukai and Masafumi (2003) [15] studied the motion of inclusion particles at the solidifying boundary layer. They reported that the movement of inclusion particle motion due to the surfactant gradient is inextricably linked to some of the phenomena in the casting process, such as bubble- or inclusion-related defects in steel. This effect may be noticed due to the presence of a surface tension gradient around the solidifying boundary induced by the surfactant concentration. Yin et al. (2018) [26] studied inclusion motion and entrapment during full solidification in a curved billet caster. They created a numerical model in three dimensions linking flow, solidification, and inclusion motion. It was noted that the distribution of inclusions inside the mold's liquid pool is not exactly symmetrical. Additionally, the flow pattern of the molten steel and the inclusions' buoyancy force have a significant impact on the mobility and trapping of tiny inclusions in the mold.

Z. Liu and B. Li (2018) [27] developed a model to simulate the transient fluid flow, heat transfer, and solidification processes in various vertical-bending continuous casting casters. The enthalpy-porosity approach was used to simulate the heat transfer and solidification of steel in the caster. The vertical length has less effect on the removal ratio of inclusions from the top surface but determines the entrapment positions of inclusions. More inclusions can be carried and captured deep in the mold cavity with a larger vertical length. J. Jeong et al. (2020) [11] predicted the behavior of an alumina inclusion in front of the solid–liquid interface during solidification, and the interfacial tension between SPFH590 microalloyed steel and alumina was experimentally determined. According to the research results, the surface stresses of steel samples decreased as the temperature rose. Additionally, the contact angles of samples containing 11 to 72 ppm sulfur decreased with rising temperature, but they rose for the sample containing 94 ppm sulfur. The estimated interfacial tension values were used to anticipate the behavior of an alumina inclusion in front of the solid–liquid interface in SPFH590 steel. According to estimates, inclusions transitioned from being trapped to being driven away from the solid–liquid interface as the sulfur concentration rose from 5 to 10 ppm.

Using the empirical results established by Ref. [11], Siddiqui et al. (2020) [28] used a novel correlation of the interfacial tension of alumina inclusions in molten steel under different conditions of sulfur concentration to study inclusion motion at the solid–liquid boundary. It was reported that inclusions were prone to engulfment by the solidification front under the influence of higher interfacial tension between the inclusions and the melt.

H. Yin and T. Emi (2003) [20] investigated the direction and velocity of the surface flow of the steel melt in the vicinity of the solid–melt interface. It was found that the solutal Marangoni flow was very strong during the solidification of the steel melt with very low O and S contents. Further, many fine inclusion particles were brought to the free surface near the interface by this flow. Fei et al. (2019) [29] investigated the surface flaws in interstitial-free (IF) steel and carried out quantitative metallographic assessments of near-surface inclusions as well as surface liquid flow detection using the nail-board tipping method. The findings demonstrate that at casting rates of 0.8 and 1.0 m/min, a thin liquid mold flux layer develops, non-uniform argon bubble floating causes the entrainment and subsequent entrapment of the liquid flux, and small inclusion particles of Al_2O_3 may also collect at the solidification front.

Q. Wang and Zhang (2016) [30] studied the entrapment and final distribution of inclusions in the solidified shell. Their results indicated the distribution of oxide inclusions along the radial direction. Further, with a larger diameter, inclusions tended to be entrapped toward the center area of the billet. Chen, Ren, and Zhang (2018) [31] investigated the fluid flow, solidification, motion, and entrapment of inclusions in an actual vertical-bending continuous casting (CC) strand. It was reported that the inclusion removal fraction increased significantly with the increase in inclusion diameter, while the percentage of large-size inclusions in the entrapped inclusions was decreased. Kölbl and Harmuth (2019) [32] reported that the quality of steels is mainly influenced by the infiltration of mold slag into the gap between the strand and the mold and the temperature-dependent solidification behavior of the slag. You et al. (2017) [2] reported that the formation of nonmetallic inclusions in the solidification process can essentially influence the properties of steels. They stated that microsegregation and inclusion formation are fundamental to simulating the phenomenon in the solidification process. Wein Chen et al. (2019) [33] carried out work in which a coupled three-dimensional model, the Lagrangian Discrete Phase Model, and a VOF multiphase model were developed to investigate the transient two-phase flow and bubble distribution in continuous casting strands. The influence of the lift force, bubble diameter, and argon flow rate on the transient two-phase flow and bubble distribution were investigated. Soumava Chakraborty et al. (2020) [34] studied the efficiency of the removal of solid alumina inclusions by filtration and the distribution of inclusions in a stainless steel casting. The study documented that inclusion floatation inside the mold cavity plays a role in reducing the inclusion concentration in the casting.

It is noted from the literature review that several techniques have been implemented to prevent inclusions in steel, particularly in a continuous casting mold. It is also evident that the inclusion motion behavior and entrapment in the mold are dependent upon several factors, such as thermo-physical properties and molten steel flow characteristics. Amongst these variables, the surfactant concentration plays an important role in inclusion motion near the solid–liquid boundary layer. The concentration level affects the interfacial tension, and thus, Marangoni forces play an important role in the pushing and engulfment phenomenon. The interfacial tension gradient along the interface between liquid and inclusion particles will propel the particle in the direction of decreasing interfacial tension. It is established in studies in the literature that inclusion particles are driven in the mold by the interfacial tension gradient, especially in front of a solidifying interface and in the vicinity of the interface between liquid steel [35]. Further, it is also evident that minor concentrations of certain surfactants, such as oxygen, nitrogen, and sulfur, play an important role in inducing interfacial tension at the solid–liquid boundary.

In most of the previous studies, inclusion motion has been characterized in terms of fluid flow movement in a continuous casting mold. Very few experimental studies have reported on the inclusion motion behavior under Marangoni forces induced by the surfactant concentration in the continuous casting mold. None of the experimental or numerical research works has reported on inclusion motion using experimental correlations between interfacial tension, surfactant concentration, and temperature. Amongst the few available numerical works, no research has been reported to identify the effect of sulfur concentration on the pushing and engulfment of alumina inclusions near the solidifying front of automotive steel in a continuous casting mold.

In this work, we investigated the inclusion motion behavior in a continuous casting mold using two-dimensional turbulent modeling. The inclusion motion characteristic has been studied with different parameters, such as the input temperature of molten steel in the mold and the size of the inclusion particles. Further, we focused on inclusion motion in the solidifying zone. This research study focused on alumina inclusion motion behavior in molten automotive steel (SPFH590) in a continuous casting mold. Firstly, a full-scale model of the mold was utilized to simulate the melt flow and inclusion flow characteristics depending on some fundamental parameters. Secondly, interfacial correlation and thermo-physical property data from the literature were used for the simulation. In this work, the numerical simulation was simplified by utilizing only important phenomena, such as heat transfer and the interfacial tension distribution in a two-dimensional domain, assuming low to negligible velocity at the solid–liquid interface.

2. Numerical Modeling

A computational fluid dynamic-based numerical model was developed by using Ansys Fluent to simulate the molten steel flow and alumina inclusion motion in the continuous casting mold. Table 1 shows the alloying elements of the steel modeled in this study. The simulation model considers the molten steel flow, heat transfer, solidification of molten steel, and inclusion motion. Therefore, a two-dimensional model was used for the simulation. The study was carried out using different parameters, such as the surfactant concentration and the temperature of molten metal. Tables 2 and 3 show the details of thermo-physical properties for numerical modeling.

Table 1. Chemical composition of selected steel (SPFH590) (wt%).

Element	C	Mn	Si	P	Al	Nb
Concentration	≤0.1	≤3.0	≤0.5	≤0.1	≤0.1	≤0.1

Table 2. Thermo-physical properties of SPFH590 steel.

Parameters	Values
Density of molten steel [36]	ρ (kg m^{-3}) = 8621.17 − 0.88T
* Viscosity of molten steel [36]	μ (mPa s) = 0.2388 ∗ exp(47.44/(RT))
Specific heat [37]	750 J kg^{-1} K^{-1}
Thermal conductivity [37]	41 W m^{-1} K^{-1}
Surface tension (σL) and interfacial tension (σPL)	Equations (9) and (10) {Ref: [10,11]}
Solidus temperature	1781 K
Liquidus temperature	1798 K
Alumina inclusion size	5, 100, and 300 μm

* where μ is the viscosity, R is the molar gas constant, and T is the absolute temperature (K).

Table 3. Process Parameters.

Process Parameters	Values
Mold width	1500 mm
Mold length	3600 mm
SEN submergence depth	160 mm
Nozzle port downward angle	15 degree
Inlet velocity	2 m/s
Outlet	Pressure outlet condition
Alumina inclusion density	2500 kg/m^3
Shell surface temperature	1273 K
Mold conductivity	315 W/mk
Latent heat	272,000 (J/kg)

The experimental data for the interfacial tension of alumina inclusions and other thermo-physical properties were from the research works of J. Jeong et al. (2020) [10] and Siddiqui et al. [28].

The Governing Equations

The numerical model was developed to simulate the melt flow and predict the inclusion motion behavior in a continuous casting mold. In the first stage of the research work, we used a full-scale model of the continuous casting mold for the simulation. A transient two-dimensional numerical model was developed using Ansys Fluent (Canonsburg, PA, USA). The inclusion particles were incorporated using the discrete phase modeling method. In addition, the computational fluid dynamics (CFD) model was based on a realizable k-epsilon model and was initially treated as a single-phase model. A major advantage of the realizable k-epsilon model is its ability to accurately predict the spreading rate of both planar and round jets, which is highly related to this work.

In the second stage of the work, we used small geometry to understand the alumina inclusion motion near the solid boundary. In order to investigate the impact of weld parameters and other interfacial phenomena, a two-dimensional multiphase model was built for this example. The governing equations for mass and momentum for the current simulation model, which was created using a mathematical model based on CFD, are as follows:

$$\frac{\partial \rho}{\partial t} + \nabla \cdot \left(\rho \vec{v}\right) = S_m \tag{1}$$

$$\frac{\partial}{\partial t}\left(\rho \vec{v}\right) + \nabla \cdot \left(\rho \vec{v} \vec{v}\right) = -\nabla p + \nabla \cdot \left(\overline{\overline{\tau}}\right) + \rho \vec{g} + \vec{F} \tag{2}$$

where p denotes the static pressure, g denotes the gravitational constant, S_m denotes the mass source term, and F is the external gravitational force. The velocity field, pressure, and temperature of the fluid in the specified domain are determined using the solutions of the aforementioned Equation contains the stress tensor $\overline{\overline{\tau}}$ information (3).

$$\overline{\overline{\tau}} = \mu\left[\left\{\nabla \vec{v} + \nabla \vec{v}^{T}\right\} - \frac{2}{3}\nabla \cdot \vec{v} I\right] \tag{3}$$

where μ is the molecular viscosity, I is the unit tensor, and the second term on the right-hand side is the effect of volume dilation. The energy equation is expressed in Equation (4):

$$\frac{\partial}{\partial t}(\rho E) + \nabla \cdot \left\{\vec{v}(\rho E + p)\right\} = \nabla \cdot \left\{k_{eff} \nabla T - \sum_{j} h_j \vec{J}_j + \left(\overline{\overline{\tau}}_{eff} \cdot \vec{v}\right)\right\} + S_h \tag{4}$$

where k_{eff} is the effective conductivity ($k + k_t$, where k_t is the turbulent thermal conductivity, defined according to the turbulence model being used), and $\vec{J}_j$ is the diffusion flux of species j. S_h is a volumetric heat source term.

Melt pool (molten metal) solidification was quantified using the enthalpy-porosity approach. The mushy zone, or the liquid portion of molten metal in the range of 0 to 1, was regarded as a porous medium. The liquid fraction of each cell was determined after taking the porosity into account. Additionally, the solidified cells were handled as if they had a porosity percentage of one and were non-porous. As a result, it was assumed that the velocities of fully solidified cells were zero. Additionally, a "pseudo"-porous medium was developed to represent the mushy zone. This indicates that the porosity percentage in the mushy zone ranges from 0 to 1. The sensible enthalpy, h, and the latent gas were combined to compute the metal enthalpy in this model:

$$H = h + \Delta H \tag{5}$$

where

$$h = h_{ref} + \int_{T_{ref}}^{T} C_p dT$$

Additionally, h_{ref} is the reference enthalpy, T_{ref} is the reference temperature, and C_p is the specific heat at constant pressure.

Further, the liquid fraction, £, can be defined as:

$$£ = 0 \; if \; Temperature \; (To) < Solidus \; Temperature (Ts)$$

$$£ = 1 \; if \; Temperature \; (To) > Solidus \; Temperature (Ts)$$

$$£ = \frac{To - Ts}{Tl - Ts} \; if \; Ts < T < Tl$$

The molten metal's latent heat content is given as L, ΔH = £. For both solids and liquids, the latent heat content ranges from 0 to 1. The value of latent heat can be obtained from. Furthermore, the energy equation for solidification/melting problems is written as:

$$\frac{\partial}{\partial t}(\rho H) + \nabla \cdot \left(\rho \vec{v} H\right) = \nabla \cdot (k \nabla T) + S \tag{6}$$

where S is the source term, and H is the enthalpy. The fundamental strategy used to handle melting and solidification is to modify the thermal energy equation by including a phenomenological heat source factor. The domain started with a certain temperature of molten steel at the start of the calculation. The initial circumstances were used to determine the heat source term.

The local mass fraction of molten steel, sulfur material, Y_i, is predicted by the solution of a convection–diffusion equation for the ith species. The conservation formula for all liquid phases is as follows:

$$\frac{\partial}{\partial t}(\rho Y_i) + \nabla \cdot \left(\rho \vec{v} Y_i\right) = -\nabla \cdot \vec{J}_j \tag{7}$$

Therefore, one of the components of microalloyed steel is sulfur, and one goal of this work is to comprehend how alumina inclusions move when there is interfacial tension. The following equation can be used to determine the diffusion coefficient in this situation since sulfur diffuses in steel:

$$D = \frac{kT}{2\pi\mu d}\left[\frac{m_1 + m_2}{2m_2}\right]^{\frac{1}{2}} \tag{8}$$

where d is the diameter of atoms, and m_1 and m_2 are the atomic mass of the solute and solvent, respectively. T is the temperature of the melt, μ is the viscosity of the molten metal, and k is Boltzmann's constant (1.38×10^{-23} J/K).

3. Numerical Details

The goal of the current effort was to create a numerical model that can simulate melt flow and forecast inclusion motion behavior in a continuous casting mold. Using a full-scale model of a continuous casting mold, CFD simulation was performed in the first part of the research project. Utilizing Ansys Fluent, a transient two-dimensional numerical model was developed (Academic version: 21.0, ANSYS, Inc., Canonsburg, PA, USA). With the use of the discrete phase modeling technique, inclusion particles (1000 particles at t = 0 s) were then injected. In this study, the aggregation phenomenon of inclusion particles was ignored in favor of the spherical form assumption.

A schematic representation of the domain, information on the phases, and the meshed zone are shown in Figure 2a, along with the dimensions of a full-scale mold. The mold's exterior sidewalls were regarded as stiff walls because of their heat transfer conductivity of 315 W/mk. The simulation's thermo-mechanical characteristics were derived from earlier work by Siddiqui et al. [28]. The domain's exterior walls were regarded as convective barriers with appropriate heat transfer rates. We determined the mean velocity along the vertical length of the mold for the grid independence test.

The specifics of the grid independence test are shown in Figure 2b (the approximate number of elements is illustrated). It was determined that, in comparison to a greater number of elements, about 200,000 elements produced outcomes that were satisfactory (see Figure 3). According to the grid independence test, the domain's mid-vertical plane had a mean velocity change of no more than 5%. However, adding more items resulted in a noticeable increase in computing time. By taking an average temperature reading at the mid-vertical plane of the domain, the convergence of the solution was investigated. From Ref. [28], it was possible to determine the thermo-physical characteristics of molten steel and alumina inclusions.

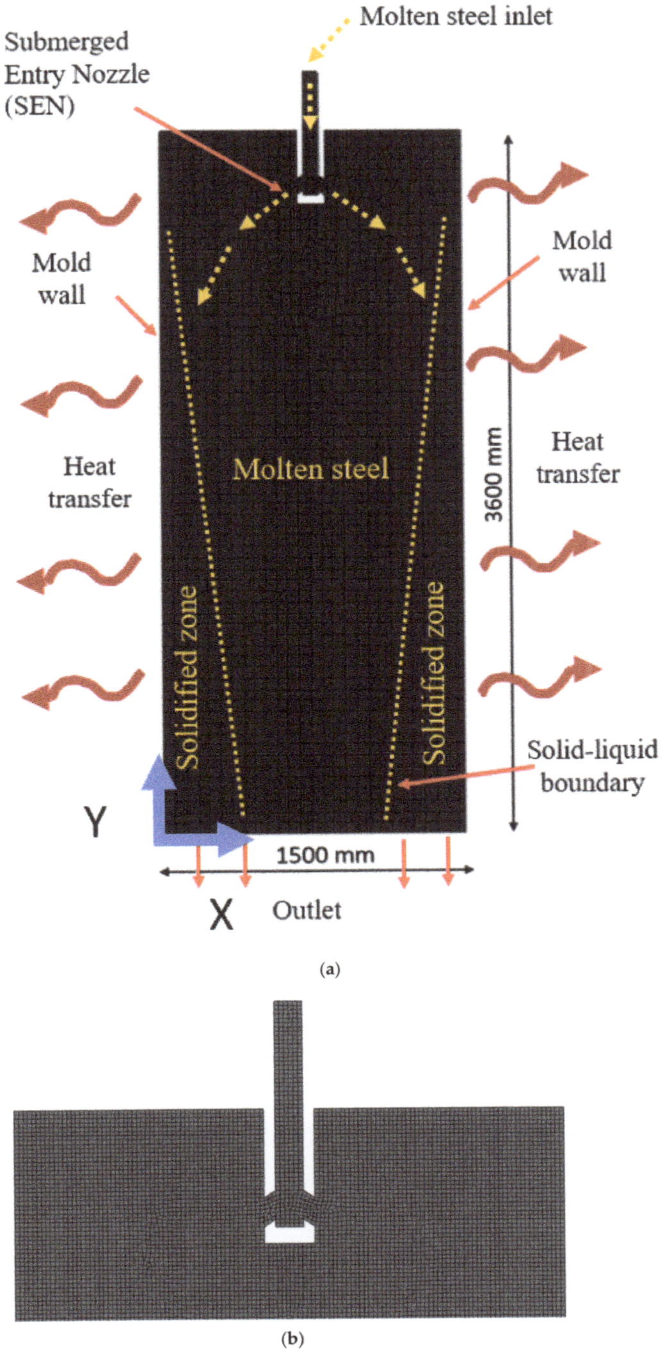

Figure 2. (**a**) Schematic of the domain and (**b**) domain meshing.

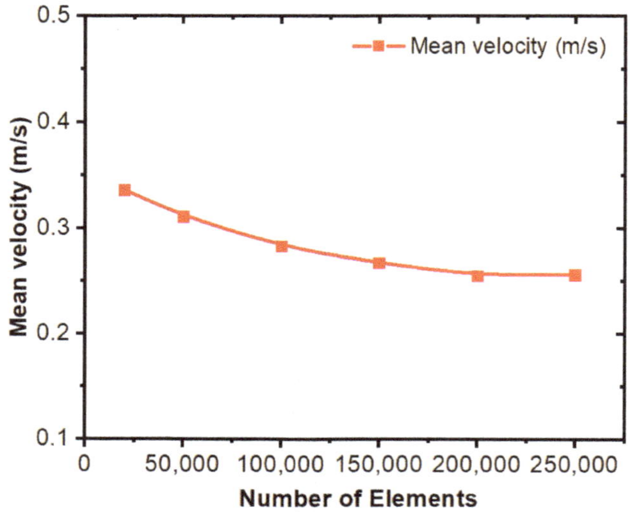

Figure 3. Grid independence test.

Equation (9) was employed in the simulation to account for the surface tension correlation between molten steel and alumina. Equation (10) was also used in a computer model to measure the interfacial tension between molten steel and alumina inclusions. We used empirical relationships from Jeong et al. [11]. They conducted experiments to determine the surface tension of SPFH590 steel and the interfacial tension of SPFH590 steel and an alumina inclusion. The following formulas can be used to calculate the surface tension of SPFH590 steel:

$$\begin{aligned}\sigma_L = {} & (1511 + 0.08277\,T) \\ & -(1041 - 0.5156\,T) \\ & \times\{\ln[1 + exp(-3.583 + 19846/T)(wt.\%\,S)]\}\end{aligned} \tag{9}$$

where σ_L is the surface tension, T is the temperature in K, and S is the sulfur concentration in ppm.

The following equation describes the interfacial stress between SPFH590 steel and an alumina inclusion.

$$\begin{aligned}\sigma_{PL} = {} & \Big\{3050.51 + 131437.97 \times (wt.\%\,S) - 1.544 \times 10^7 (wt.\%\,S)^2 \\ & -3.378 \times 10^9\,(wt.\%\,S)^3\Big\} \\ & +\Big\{-0.8498 - 79.739 \times (wt.\%\,S) + 7655.06 \times (wt.\%\,S)^2 \\ & +1.962 \times 10^6 \times (wt.\%\,S)^3\Big\}T\end{aligned} \tag{10}$$

where σ_{PL} is the interfacial tension between SPFH590 steel and an alumina inclusion.

Furthermore, sulfur concentration was achieved across the domain by using the species model. In this numerical model, the solidification model was also taken into account. The species continuity equation was solved at each time step during the simulation's two transient periods. The numerical model for molten steel solidification was validated with the numerical results of Cho et al. [38], as shown in Figure 4. The shell thickness profile was compared with predicted numerical results. The predicted results seem to have good agreement with the previous work of Cho et al. [38].

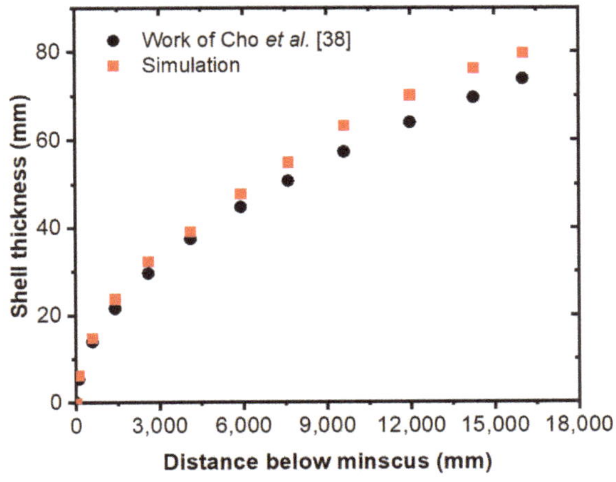

Figure 4. Comparison of solidifying steel shell thickness in the mold (Cho et al. [38]).

4. Results and Discussion

Mold and Inclusion Characteristics

In the first part of our work, we investigated the inclusion distribution in a continuous casting mold. A two-dimensional discrete phase model (DPM) was used to analyze and track the movement of inclusion particles. In addition to this solidification, a model was employed to simulate the solidified zone in the mold. A suitable turbulence model, the k-ϵ realizable model, was selected to properly configure the turbulence flow of molten steel. Further, inclusions were assumed to be spherical in shape and non-conglomerating to simplify the numerical modeling procedure.

Figure 5a,b shows the distribution of 5-micron-diameter inclusions at the mid-section of the mold at different periods. It can be observed that inclusion concentrations (calculated by the discrete phase model, DPM) increase over time at a given section. Secondly, it is also observed that a major part of the inclusion distribution is available on the left and right sides of the mold, which are solidifying boundary layers. Further, the midzone of the mold has a lower concentration of inclusions. The inclusion motions are affected by several parameters, such as the velocity of the local zone, buoyancy force, natural convection, and interfacial tension. It is expected that inclusions move to the solidifying zone due to an increase in the thermal gradient and a lower local velocity field.

Figure 5c,d show the distribution of 100-micron-diameter inclusions at the mid-section of the mold at different periods. In this case, inclusion concentrations have different distribution characteristics as compared with small (5-micron size) inclusions. It can be noticed here that inclusions have a lump-sum uniform distribution over the x-axis length. However, it cannot be denied that some part of the inclusion population is concentrated towards the solidifying boundary. A significant portion of the inclusion distribution is noticed on the left and right sides of the mold, which are hardening border layers. We later study the larger-sized inclusion particles and their distribution in the mold zone. It was found that inclusions are more evenly concentrated in the mold zone as compared to the smaller sizes. This implies that larger inclusions can be more influenced by melt flow velocity and buoyancy forces. Furthermore, the effect of temperature on the inclusion distribution was investigated to analyze the impact of natural convection and interfacial tension. Figure 6 shows the inclusion concentration on the same plane but with elevated temperature conditions. It can be seen that the inclusion concentration is more evenly distributed in the mold as compared to lower-temperature conditions.

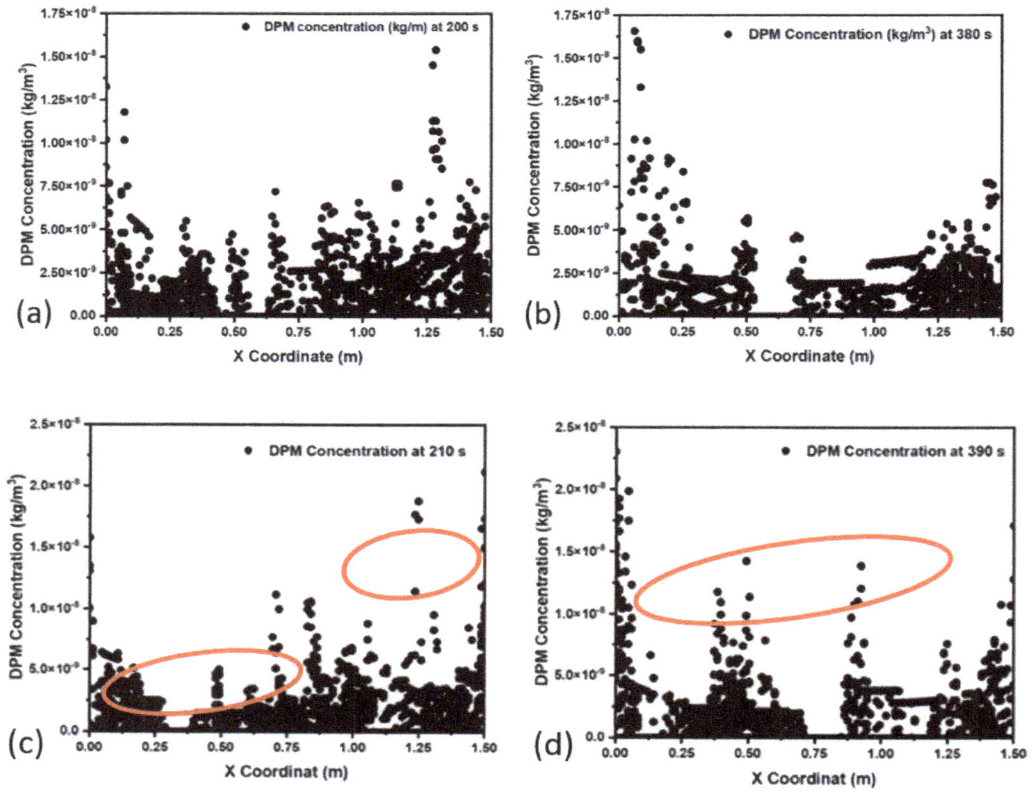

Figure 5. Inclusion (DPM) concentration in the mold at 1853 K; sizes: (**a**,**b**) 5-micron diameter and (**c**,**d**) 100-micron diameter.

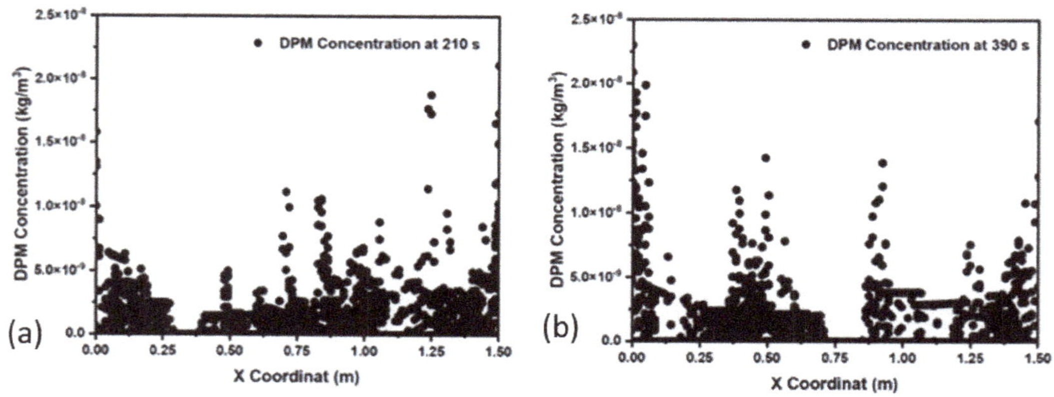

Figure 6. *Cont.*

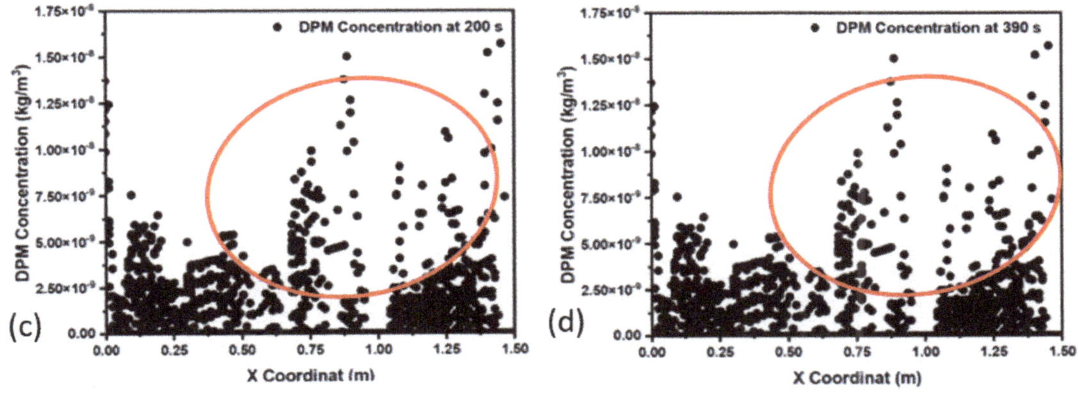

Figure 6. Inclusion (DPM) concentration with a 100-micron diameter in the mold at (**a**,**b**) 1853 K and (**c**,**d**) 1893 K.

Figure 7 shows the velocity profiles of 5-, 100-, and 300-micron inclusion particles at different sections of the mold. On the top side of the mold, we can see that inclusions have a higher flow velocity especially at the left and right sides of the solidifying boundary. Similarly, for the midzone, we can see that the velocity of inclusions has been reduced due to a reduction in flow field velocity. The flow field velocity is reduced due to a reduction in temperature and an increase in the viscosity of the melt. In the lower zone (Y = 0 m), we can observe the minimal velocities of inclusion particles. The velocity profile of 100-micron inclusion particles at different sections of the mold is shown in Figure 7. We can see that inclusions have a larger flow velocity on the top side of the mold, especially on the left and right sides of the solidifying border. Similarly, we can see that the velocity of inclusions decreases at the midzone (Y = 1.95 m) due to a decrease in flow field velocity. Because of the lower temperature and higher viscosity of the melt, the flow field velocity is lower. We can see the lowest velocities of inclusion particles in the bottom zone (Y = 0 m).

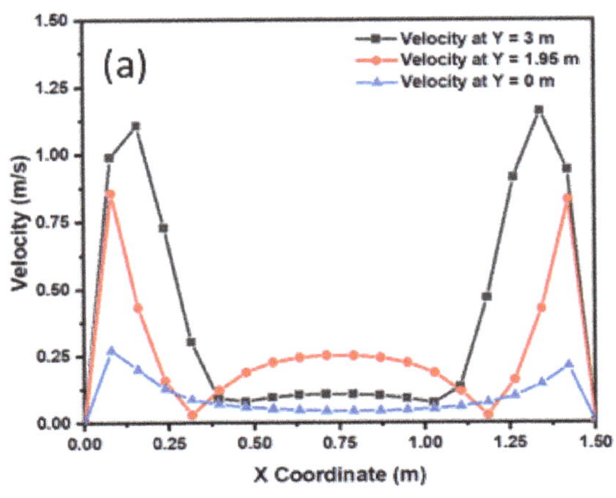

Figure 7. *Cont.*

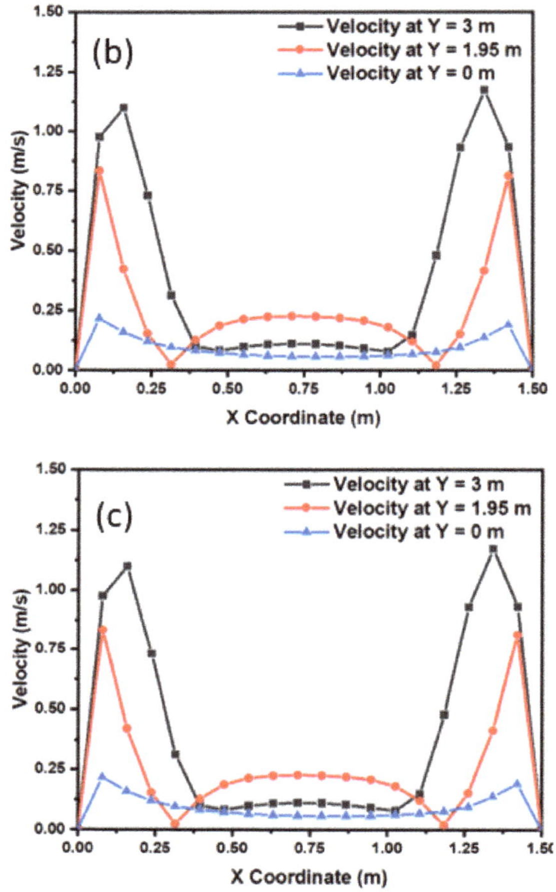

Figure 7. The inclusion velocity profile in the mold at 1853 K; size: (**a**) 5 microns, (**b**) 100 microns, and (**c**) 300 microns.

Further, it depicts the velocity curve of 300-micron inclusion particles at various mold sections. On the top side of the mold, especially on the left and right sides of the solidifying border, inclusions have a higher flow velocity. Similarly, due to a drop in flow field velocity, the velocity of inclusions is reduced at the midzone (Y = 1.95 m). The flow field velocity is reduced because of the lower temperature and increased viscosity of the melt. The bottom zone (Y = 0 m) has the slowest inclusion particle velocities.

The average inclusion concentration profile on the *x*-axis of the domain is shown in Figure 8 with respect to inclusion size and time. Figure 8a represents the distribution of small inclusions (5-micron diameter) over different timescales, namely, 200 s, 380 s, and 560 s. It is observed that the average inclusion concentration remains unvaryingly distributed throughout the domain. However, during a certain period of the simulation, two peaks can be observed in the domain. This suggests that the inclusion distribution increases near the middle zone of the domain. Further, when we changed the inclusion size in the same conditions, it was observed that inclusions are concentrated in less time (390 s), as shown in Figure 8b. A similar pattern is also observed in Figure 8c, where the inclusion diameter is significantly increased to 300 microns. In another case, we tried to see the effect of the inlet temperature on the average distribution of inclusions under the same conditions. Figure 8d shows the effect of the inlet temperature change. It was observed that the peak pattern of the average distribution has an impact on temperature, and inclusion concentrations are formed on a larger timescale.

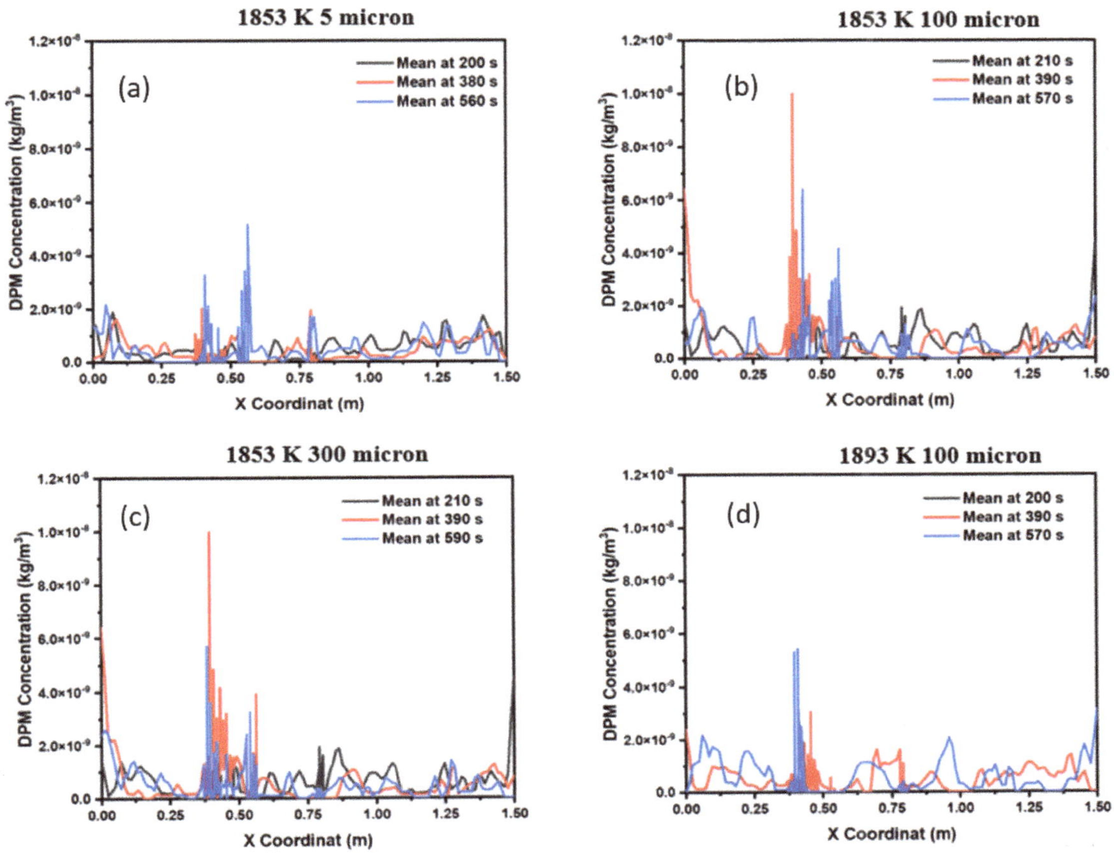

Figure 8. Inclusion concentration profile in the mold: (**a**) 5 microns, (**b**) 100 microns, (**c**) 300 microns, and (**d**) 100 microns (at 1893 K).

In Figure 9, the inclusion concentration profile is shown in the whole domain of the simulation by using different sizes of inclusions. It can be noted that the inclusion distribution is mostly on the middle-left side of the mold. On the other hand, we can see almost the same level of inclusion distribution. In addition to this, inclusions more or less have the same concentration at six different time lengths. The inclusion distribution density is depicted in Figure 9a–c throughout the domain for different cases. This density concentration further confirms the inclusion aggregation towards the solidifying boundary zone. Further, the inclusion motion with streamlines is shown in Figure 10. The position and motions of inclusions are shown at a well-established flow time (1000 s) in the mold.

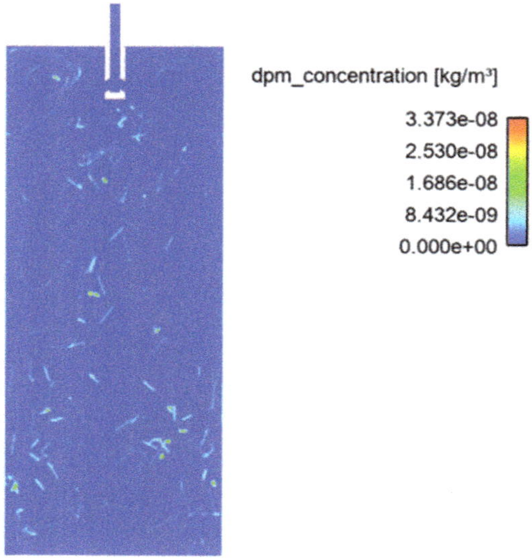

(**a**)

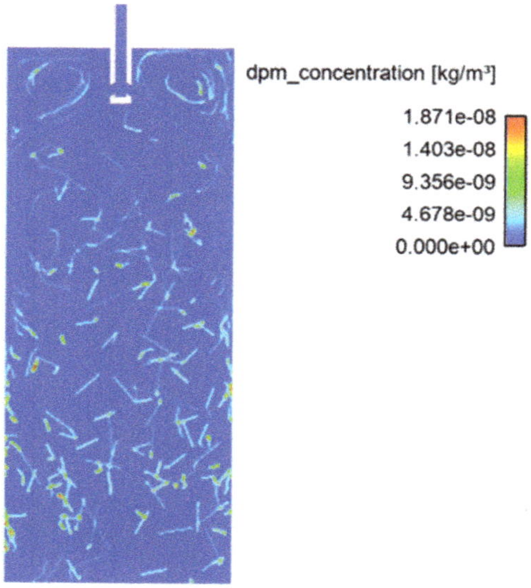

(**b**)

Figure 9. *Cont.*

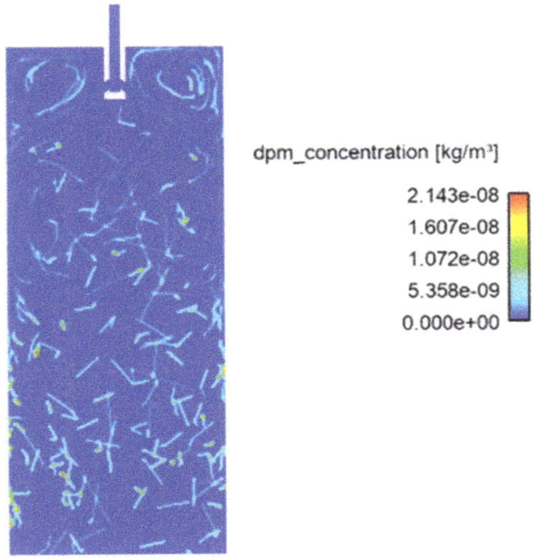

(c)

Figure 9. Inclusion concentration profile in the mold at 1853 K; diameter size: (**a**) 5 microns, (**b**) 100 microns, and (**c**) 300 microns.

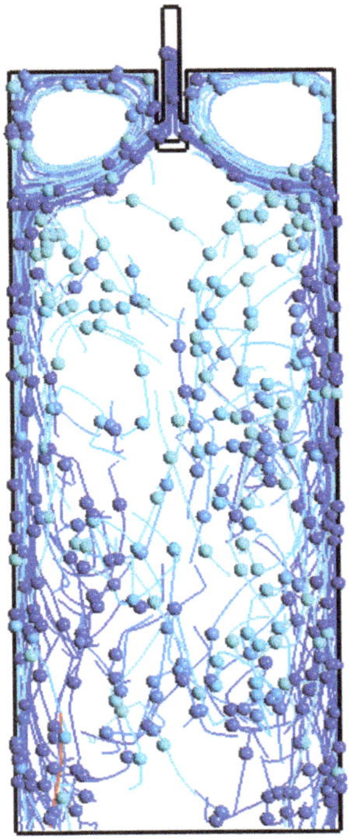

Figure 10. Alumina inclusion motion in the mold.

Figure 11 shows the temperature contour of the domain at different time intervals. The initial temperature of the melt inflow was 1853 K. The temperature of molten steel decreases after some time, and the boundary layer of solidifying steel is visible in the domain. Figure 12 shows the velocity contour in the domain at different time intervals. Figure 13 shows the total number of inclusion particles available in the domain as a function of injection time using the discrete phase model. We studied the effect of sulfur concentration on the entrapment of alumina inclusions near the solid–liquid boundary layer in the mold. The sulfur concentration affects the interfacial tension at the solid–liquid boundary, and thus, it impacts the motion of alumina inclusions due to interfacial tension. Figure 14 shows the trapped inclusions near the solid–liquid boundary in the mold at different sulfur concentrations. It can be noted here that when there was no mixing of sulfur, the entrapment percentage was low with respect to time. Because of that condition, we did not utilize the empirical relationship in the simulation. However, the entrapment percentage varied in the case of sulfur mixing (using the empirical relationship in modeling). It can be noticed that a 10 ppm sulfur concentration does not have much impact. However, as we increase the sulfur concentration up to 70 ppm, we can see that there is a significant rise in the entrapment of alumina inclusions in the solidifying boundary layer. Another important outcome that we notice here is that the maximum entrapment of alumina inclusions occurs at 70 ppm, but the entrapment percentage decreases at 100 ppm sulfur. The variation in entrapment can be directly correlated with the experimental correlation between the interfacial tension and sulfur concentration, as stated in Equations (9) and (10).

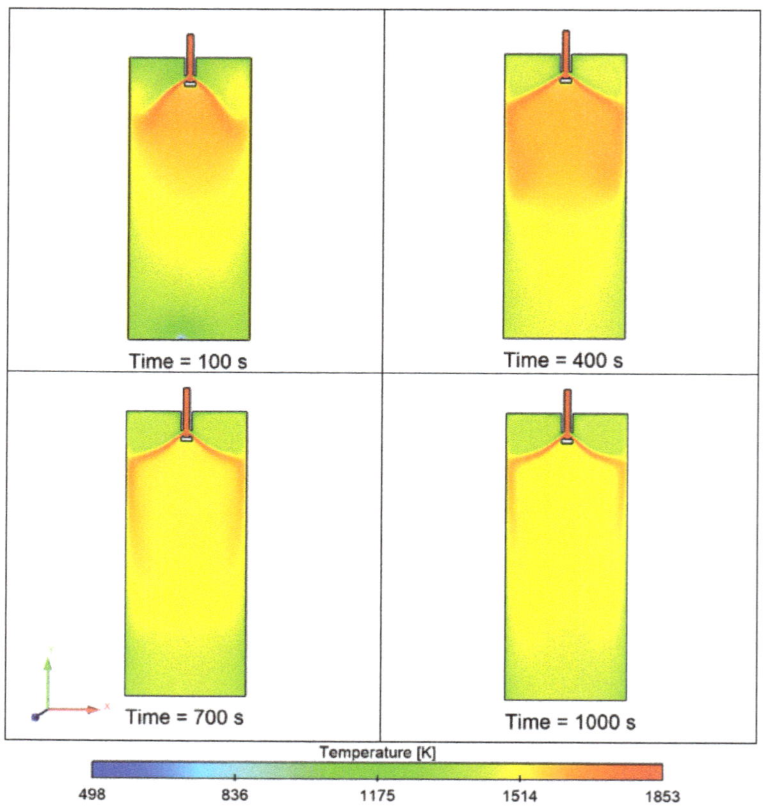

Figure 11. Temperature contour of the mold domain (initial melt temperature: 1853 K).

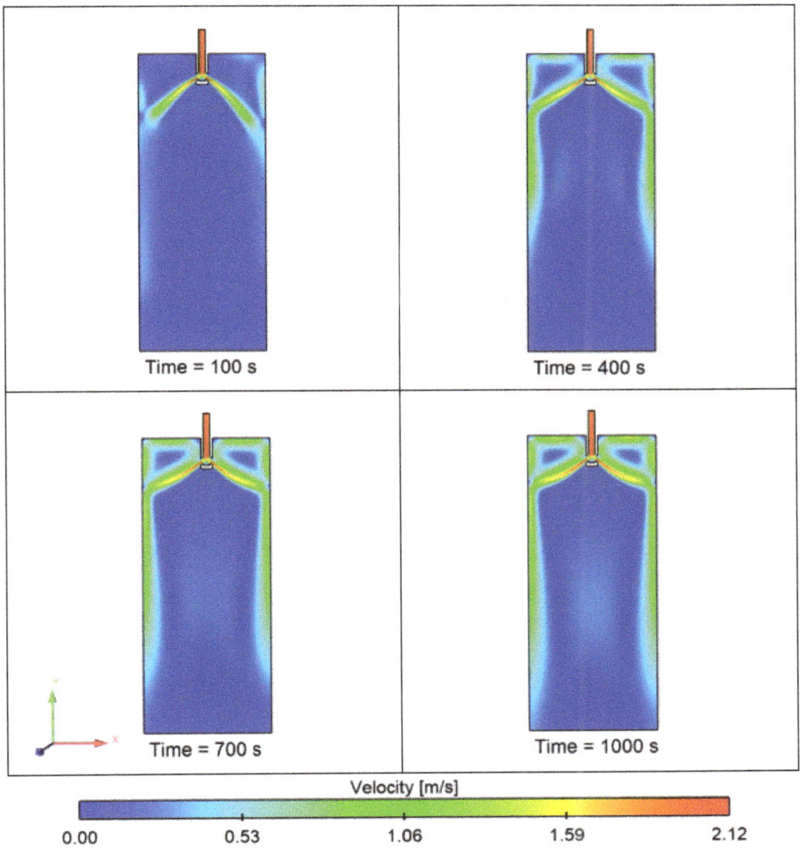

Figure 12. Velocity contour of the mold domain (initial melt temperature: 1853 K).

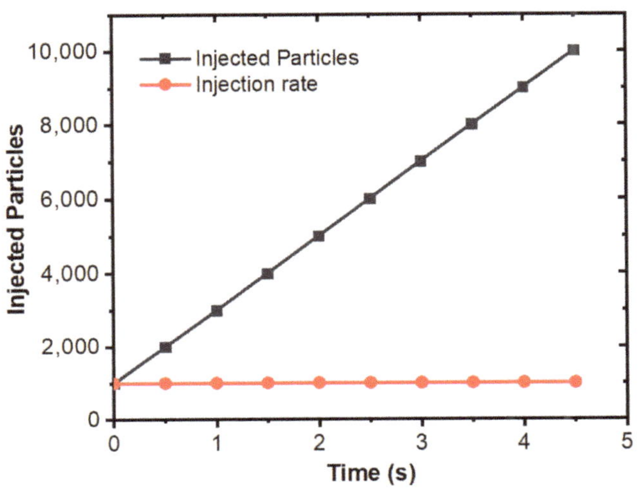

Figure 13. Injection rate and total particles in the mold using DPM modeling.

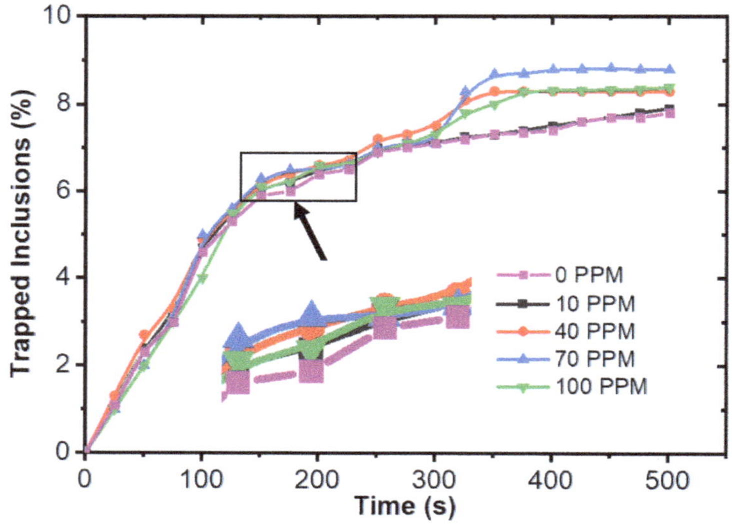

Figure 14. Trapped inclusions near the solid–liquid boundary in the mold at different sulfur concentrations.

Figure 15 shows the removal percentage of inclusions from the outlet (i.e., outlet inclusion particles are assumed to have moved out or absorbed in the solidified shell). It can be noted here that the removal percentage varies according to the sulfur concentration (around 4%). We can observe that an increase in sulfur ppm up to 70 ppm leads to higher removal percentages, but the results are the opposite at 100 ppm. Moreover, larger particles are more prone to move out of solidified shells as compared to smaller particles.

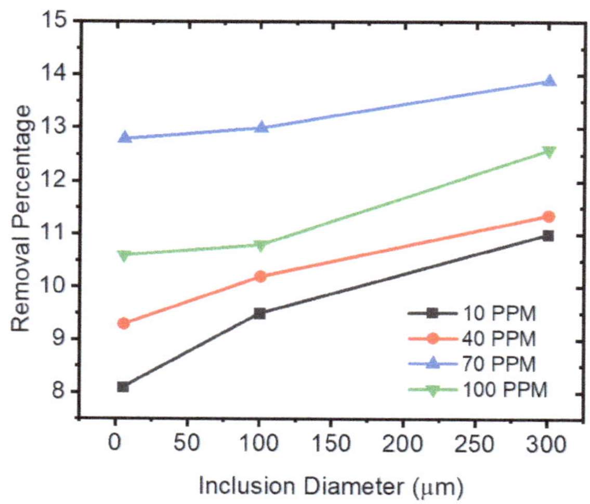

Figure 15. Removal percentage of alumina inclusions in different conditions of sulfur concentration and inclusion diameters.

We also studied the entrapment of inclusion particles at the top surface of the mold. In a typical continuous casting mold, the slag layer is artificially created on the top surface for many reasons. One of the advantages is that inclusion particles become stuck in the slag layer. Thus, in this simulation model, we considered the entrapment of particles when they reach the top surface. Hence, a particular inclusion particle's velocity is frozen at zero, and the discrete phase model stops tracking it. Figure 16 shows the entrapment of inclusion particles on the top surface of the mold. We observed that the smaller-sized inclusion particles have higher chances of entrapment (more than 25%) and are thus removed from the molten metal. However, the larger-sized inclusion particles sustain their motion in the melt for a longer period, and their percentage entrapment is lower. The other conclusion that we made is that interfacial tension plays an important role in entrapment, as can be seen in Figure 16. An increase in sulfur concentration up to 70 ppm leads to more entrapment; however, an increase to 100 ppm does not make a more significant difference.

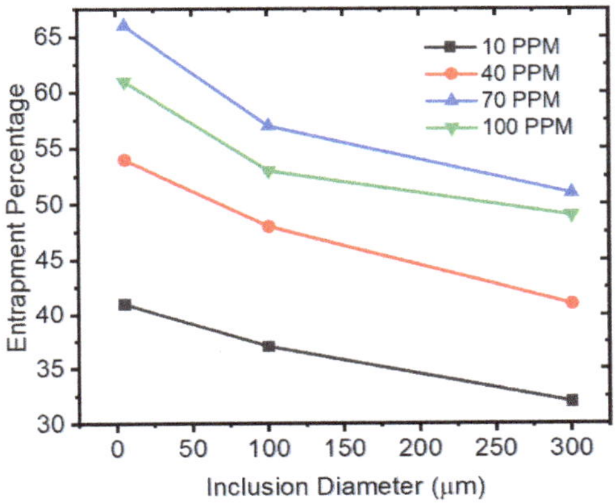

Figure 16. Entrapment (at the top surface) of alumina inclusions under different conditions of sulfur concentration and inclusion diameters.

Further, we also studied the effect of melt inlet temperature and its effect on inclusion entrapment and removal (from the outlet/solidified shell). Figure 17 shows the entrapment and removal (from the outlet) percentage of alumina inclusions. It can be noted that the inclusion entrapment percentage is increased as the inlet temperature is increased. The temperature at the inlet and top surface zone affects the surface tension and melt flow characteristics. Secondly, we can also observe that the removal percentage does not vary significantly.

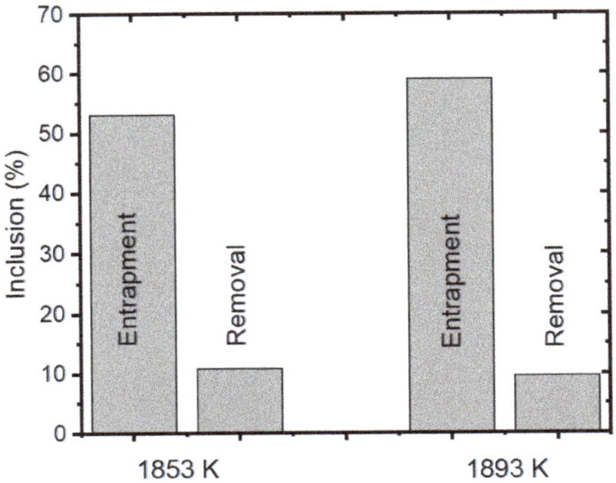

Figure 17. Entrapment and removal (from outlet) percentages of alumina inclusion.

5. Conclusions

A numerical model was developed to study alumina inclusion motion behavior in a continuous casting mold. Initially, we investigated the inclusion motion behavior in a continuous casting mold using a two-dimensional CFD model. The inclusion motion characteristic was studied with different parameters, such as the input temperature of molten steel in the mold and the size of the inclusion particles. Secondly, we focused on inclusion motion under the influence of interfacial tension. The effect of surfactant concentration (sulfur in this work) on the motion of alumina inclusions was studied. The following are the main conclusions obtained from this research work:

- A transient, two-dimensional model of a continuous casting mold was developed using Ansys Fluent software. Alumina inclusions in microalloyed steel were tracked under the condition of solidification at the boundary side. Further, the empirical relationship of interfacial tension was used to study the effects of temperature and surfactant concentration (sulfur) on inclusion motion near the solidifying boundary layer in the mold.
- The simulation findings indicate that the size of the inclusions and the inlet temperature have significant effects on the distribution of inclusions. It was also observed that a major part of the inclusion distribution was on the left and right sides of the mold, which are solidifying boundary layers. Further, the midzone of the mold had a lower concentration of inclusions. The inclusion motions were affected by several parameters, such as the velocity of the local zone, buoyancy force, natural convection, and interfacial tension.
- The inclusion diameter affected the inclusion distribution throughout the domain. Larger-sized inclusions had different distribution characteristics as compared to small (5-micron size) inclusions. It was found that larger inclusions were more evenly concentrated in the mold zone as compared to the smaller sizes. This implies that larger inclusions can be more influenced by the melt flow velocity and buoyancy forces.
- The effect of temperature on the inclusion distribution was investigated to analyze the impact of natural convection and interfacial tension. It was noticed that the inclusion concentration was more evenly distributed in the mold as compared to lower-temperature conditions.
- The entrapment percentage of alumina inclusions was low over time when the interfacial tension equation was not used. However, the entrapment percentage varied in the case of sulfur mixing (using the empirical relationship in modeling). It was found that a 10 ppm sulfur concentration did not have much impact. However, we observed that there was a significant rise in the entrapment of alumina inclusions in the solidifying boundary layer at sulfur concentrations up to 70 ppm. Another important outcome was that the maximum entrapment of alumina inclusions occurred at 70 ppm, but the entrapment percentage was lower at 100 ppm sulfur.
- It was found that the removal percentage varied according to the sulfur concentration. The addition of sulfur at concentrations from 10 ppm to 70 ppm resulted in around a 4% increase in the removal of alumina inclusions (trapped in the solidifying shell), except for the 100-ppm case.
- The smaller-sized inclusion particles had a 25% higher chance of entrapment at the top level of the mold (slag layer). However, the larger-sized inclusion particles sustained their motion in the melt for a longer period, and their percentage entrapment was lower.
- It can be noted that the inclusion entrapment percentage increased as the inlet temperature was increased. The temperature at the inlet and top surface zone affected the surface tension and melt flow characteristics. Secondly, we also observed that the removal percentage did not vary significantly.

Author Contributions: Conceptualization, M.I.H.S. and A.A.; methodology, A.A.; software, L.A. and A.A.; formal analysis, K.A., I.A.A. and L.A.; writing—original draft preparation, M.I.H.S. and A.A.; writing—review and editing, K.A. and I.A.A. All authors have read and agreed to the published version of the manuscript.

Funding: This research received no external funding.

Institutional Review Board Statement: Not applicable.

Informed Consent Statement: Not applicable.

Data Availability Statement: Available on request.

Acknowledgments: The authors would like to thank Deanship of scientific research in King Saud University (Riyadh, Saudi Arabia) for funding and supporting this research through the initiative of DSR Graduate Students Research Support (GSR).

Conflicts of Interest: The authors declare no conflict of interest.

References

1. Kovačič, M.; Župerl, U. Genetic programming in the steelmaking industry. *Genet. Program. Evolvable Mach.* **2020**, *21*, 99–128. [CrossRef]
2. You, D.; Michelic, S.K.; Presoly, P.; Liu, J.; Bernhard, C. Modeling Inclusion Formation during Solidification of Steel: A Review. *Metals* **2017**, *7*, 460. [CrossRef]
3. Qin, X.; Cheng, C.; Li, Y.; Zhang, C.; Zhang, J.; Jin, Y. A simulation study on the flow behavior of liquid steel in tundish with annular argon blowing in the upper nozzle. *Metals* **2019**, *9*, 225. [CrossRef]
4. Wang, Y.; Yang, S.; Wang, F.; Li, J. Optimization on reducing slag entrapment in 150 × 1270 mm slab continuous casting mold. *Materials* **2019**, *12*, 1774. [CrossRef] [PubMed]
5. Merder, T.; Pieprzyca, J.; Warzecha, M.; Warzecha, P.; Hutny, A. Evolution of the numerical model describing the distribution of non-metallic inclusions in the tundish. *Materials* **2021**, *14*, 2229. [CrossRef] [PubMed]
6. Yu, Q.; Yang, X.; Lai, C.; Tong, Z. Study on mns inclusion aggregation along continuous casting slab thickness of medium carbon structural steel. *Metals* **2022**, *12*, 56. [CrossRef]
7. Lu, X.; Zhang, Z.; Lv, M.; Li, X.; Song, B.; Fang, M. Evolution of Non-Metallic Inclusions in 27SiMn Steel. *Metals* **2022**, *12*, 718. [CrossRef]
8. Tie, Z.; Tang, H.; Wang, K.; Miao, H.; Cai, S.; Xian, F.; Zhang, J. Effect of Flow Field Optimization of an Asymmetric Multi-Strand Tundish on the Quality Consistency of Cracking Con-Rod Steel. *Materials* **2022**, *15*, 3698. [CrossRef]
9. Migas, P.; Ślęzak, M.; Karbowniczek, M.; Szczęch, S.; Hornik, A. Numerical Investigation of Outflow of Non-Metallic Inclusions during Steel Refining in the Ladle. *Materials* **2022**, *15*, 3039. [CrossRef]
10. Jeong, J.; Park, D.; Shim, S.; Na, H.; Bae, G.; Seo, S.-J.; Lee, J. Interfacial tension between SPFH590 microalloyed steel and alumina. *Metall. Mater. Trans. B* **2019**, *51*, 690–696. [CrossRef]
11. Jeong, J.; Park, D.; Shim, S.; Na, H.; Bae, G.; Seo, S.J.; Lee, J. Prediction of Behavior of Alumina Inclusion in Front of Solid–Liquid Interface in SPFH590 Steel. *Metall. Mater. Trans. B Process Metall. Mater. Process. Sci.* **2020**, *51*, 690–696. [CrossRef]
12. Yin, Y.; Zhang, J.; Dong, Q.; Zhou, Q.H. Fluid flow and heat transfer with nail dipping method in mould during continuous casting process. *Ironmak. Steelmak.* **2019**, *46*, 855–864. [CrossRef]
13. Yin, Y.; Zhang, J.; Dong, Q.; Zhou, Q.H. Challenges in Special Steel Making. *Ironmak. Steelmak.* **2019**, *46*, 855–864. [CrossRef]
14. Siddiqui, M.I.H.; Maurya, A.; Ashraf, M.; Asiri, F. Modeling of Inclusion Motion Under Interfacial Tension in a Flash Welding Process. In *Advanced Computational Methods in Mechanical and Materials Engineering*; CRC Press: Boca Raton, FL, USA, 2021; pp. 91–110.
15. Mukai, K.; Zeze, M. Motion of fine particles under interfacial tension gradient in relation to continuous casting process. *Steel Res.* **2003**, *74*, 131–138. [CrossRef]
16. Wang, Z.; Mukai, K.; Lee, J. Behavior of fine bubbles in front of the solidifying interface. *ISIJ Int.* **1999**, *39*, 553–562. [CrossRef]
17. Matsushita, T.; Mukai, K.; Zeze, M. Correspondence between surface tension estimated by a surface thermodynamic model and number of bubbles in the vicinity of the surface of steel products in continuous casting process. *ISIJ Int.* **2013**, *53*, 18–26. [CrossRef]
18. Scheller, P.R.; Lee, J.; Yoshikwa, T.; Tanaka, T. Treatise on Process Metallurgy Volume 2: Process Phenomena. In *Treatise on Process Metallurgy*; Seetharaman, S., Ed.; Elsevier: Oxford, UK, 2013; Volume 2, pp. 119–139.
19. Shibata, H.; Yin, H.; Yoshinaga, S.; Emi, T.; Suzuki, M. In Situ Observation of Engulfment and Pushing of Nonmetallic Inclusions in Steel Melt by Advancing Melt/Solid Interface. *ISIJ Int.* **1998**, *38*, 49–56. [CrossRef]
20. Yin, H.; Emi, T. Marangoni flow at the gas/melt interface of steel. *Metall. Mater. Trans. B* **2003**, *34*, 483–493. [CrossRef]
21. Thomas, B.G. Modeling of the continuous casting of steel—Past, present, and future. *Metall. Mater. Trans. B Process Metall. Mater. Process. Sci.* **2002**, *33*, 795–812. [CrossRef]
22. Siddiqui, M.I.H.; Jha, P.K. Assessment of turbulence models for prediction of intermixed amount with free surface variation using CLSVOF method. *ISIJ Int.* **2014**, *54*, 2578–2587. [CrossRef]
23. Mazumdar, D.; Evans, J.W. *Modeling of Steelmaking Processes*; CRC Press: Boca Raton, FL, USA, 2009.
24. Sahai, Y.; Emi, T. *Tundish Technology for Clean Steel Production*; World Scientific: Singapore, 2008.

25. Thomas, B.G. Review on Modeling and Simulation of Continuous Casting. *Steel Res. Int.* **2018**, *89*, 1700312. [CrossRef]
26. Yin, Y.; Zhang, J.; Dong, Q.; Zhou, Q.H. Modelling on inclusion motion and entrapment during the full solidification in curved billet caster. *Metals* **2018**, *5*, 320. [CrossRef]
27. Liu, Z.; Li, B. Effect of vertical length on asymmetric flow and inclusion transport in vertical-bending continuous caster. *Powder Technol.* **2018**, *323*, 403–415. [CrossRef]
28. Siddiqui, M.I.H.; Geleta, D.D.; Bae, G.; Lee, J. Numerical Modeling of the Inclusion Behavior during AC Flash Butt Welding. *ISIJ Int.* **2020**, *60*, 2503–2511. [CrossRef]
29. Fei, P.; Min, Y.; Liu, C.; Jiang, M. Effect of continuous casting speed on mold surface flow and the related near-surface distribution of non-metallic inclusions. *Int. J. Miner. Metall. Mater.* **2019**, *46*, 855–864. [CrossRef]
30. Wang, Q.; Zhang, L. Determination for the Entrapment Criterion of Non-metallic Inclusions by the Solidification Front during Steel Centrifugal Continuous Casting. *Ironmak. Steelmak.* **2016**, *46*, 855–864. [CrossRef]
31. Chen, W.; Ren, Y.; Zhang, L. Large Eddy Simulation on the Fluid Flow, Solidification and Entrapment of Inclusions in the Steel Along the Full Continuous Casting Slab Strand. *JOM* **2018**, *46*, 855–864. [CrossRef]
32. Kölbl, N.; Harmuth, H. Automated break temperature determination of mould slags for the continuous casting of steel based on temperature-dependent viscosity data. *Ironmak. Steelmak.* **2020**, *47*, 899–902. [CrossRef]
33. Chen, W.; Ren, Y.; Zhang, L.; Scheller, P.R. Numerical Simulation of Steel and Argon Gas Two-Phase Flow in Continuous Casting Using LES + VOF + DPM Model. *Ironmak. Steelmak.* **2019**, *71*, 1158–1168. [CrossRef]
34. Chakraborty, S.; O'Malley, R.J.; Bartlett, L. Removal of Alumina Inclusions from Molten Steel by Ceramic Foam Filtration. *Int. J. Met.* **2021**, *15*, 1006–1020. [CrossRef]
35. Mukai, K.; Zeze, M.; Morohoshi, T. Motion of fine particles under interfacial tension gradient in relation to solidification of steel. *Mater. Sci. Forum* **2006**, *508*, 211–220. [CrossRef]
36. Iida, T.G. *RIL: The Physical Properties of Liquid Metals*; Clarendon Press: Oxford, UK, 1993.
37. Maurya, A.; Jha, P.K. Influence of electromagnetic stirrer position on fluid flow and solidification in continuous casting mold. *Appl. Math. Model.* **2017**, *48*, 736–748. [CrossRef]
38. Cho, S.M.; Thomas, B.G.; Hwang, J.Y.; Bang, J.G.; Bae, I.S. Modeling of inclusion capture in a steel slab caster with vertical section and bending. *Metals* **2021**, *11*, 654. [CrossRef]

materials

Article

Effect of Flow Field Optimization of an Asymmetric Multi-Strand Tundish on the Quality Consistency of Cracking Con-Rod Steel

Zhanpeng Tie [1,2], Haiyan Tang [1,*], Kaimin Wang [1], Hongsheng Miao [2], Sen Cai [1], Fenqiang Xian [2] and Jiaquan Zhang [1,*]

1 School of Metallurgical and Ecological Engineering, University of Science and Technology Beijing, No. 30 Xueyuan Road, Haidian District, Beijing 100083, China; tzp15003685525@163.com (Z.T.); wangkaimin0906@163.com (K.W.); 15901131365@163.com (S.C.)

2 Technology & Quality Center, Xining Special Steel Co., Ltd., 52 Chaidamu West Road, Xining 810005, China; mhs1972@126.com (H.M.); xfq1456322319@163.com (F.X.)

* Correspondence: tanghaiyan@metall.ustb.edu.cn (H.T.); jqzhang@metall.ustb.edu.cn (J.Z.)

Abstract: Cracking con-rod is an advanced high-precision connecting structure based on brittle expansion, breaking and reconnection of steel, to solve the problem of assembly circle missing. High carbon micro-alloyed steel C70S6, as a dominant material for the production of cracking con-rod, has extremely strict requirements on non-metallic inclusions in steel and microstructure stability. Continuous casting tundish plays an important role in removing large-sized inclusions and stabilizing casting quality. Aiming at the inconsistent casting quality of C70S6 steel produced by a three-strand asymmetric tundish and the frequent occurrence of slag entrapment problems in Xining Special Steel, the tundish structure was optimized by means of physical modelling combined with numerical simulation, and the quality of the bloom castings and subsequent hot-rolled products before and after optimization were compared based on volume production. The results show that a new flow control design to the tundish can effectively improve the consistency of its metallurgical effect for each of the three strands and the following overall product quality, in which the flow field and temperature field in the tundish are more uniform. This is due to the adoption of a vortex inhibitor and an optimized wall structure according to the measured RTD curve, ink trajectory and numerical simulation on the 3-D streamline contours and temperature distribution in the tundish. The peak concentration of outlet 1 is decreased from 6.5 before optimization to less than 2.0 after optimization, which means the elimination or alleviation of the local short-circuit flow. The maximum temperature difference of C70S6 molten steel measured at the outlets of the tundish three strands is decreased from 2–5 °C to 1–3 °C, which is in good agreement with the numerical simulation results. The difference in columnar crystal ratio of the corresponding bloom castings is decreased from 2.27–3.17% to 1.26–1.85%, and the consistency of central carbon segregation index is also significantly improved. In addition, the difference in oxygen content among the three strand blooms is decreased from 1.7–3.5 ppm to 0.8–1.9 ppm. As a result, the overall mechanical properties and microstructure stability of the hot-rolled products are improved statistically, in which the hardness fluctuation is decreased from 84 HBW to 60 HBW, the inclusion grade of types B + C + D + Ds is reduced to 1.105, and the occurrence rate of Ds dropped to 0.118%. Accordingly, the failure rate of the cracking con-rod is controlled stably within 4‰, and the fracture is generally smoother than that before tundish optimization. In summary, the flow field optimization to a multi-strand asymmetric tundish has a clear effect on improving the overall quality of its bloom castings and rolled products, which should be paid more attention industrially. Meanwhile, the present study provides a reliable theoretical and experimental reference for the improvement of metallurgical effects of an asymmetric-typed tundish commonly used in special steel production.

Keywords: high carbon steel; cracking con-rod; C70S6; tundish; flow field; water modelling; mathematical simulation; temperature field; bloom quality consistency

Citation: Tie, Z.; Tang, H.; Wang, K.; Miao, H.; Cai, S.; Xian, F.; Zhang, J. Effect of Flow Field Optimization of an Asymmetric Multi-Strand Tundish on the Quality Consistency of Cracking Con-Rod Steel. *Materials* **2022**, *15*, 3698. https://doi.org/10.3390/ma15103698

Academic Editor: Antonio Lamura

Received: 11 April 2022
Accepted: 11 May 2022
Published: 21 May 2022

1. Introduction

Cracking con-rods are important transmission components connecting crankshafts and their pistons, which bear high requirements on load capacity, installation and service precision. As an advanced connecting structure, a cracking con-rod is made based on brittle expansion, breaking and reconnection, which is expected to solve the frequent issue of assembly circle missing. To produce a precise cleavage fracture for tight joints and avoid big head distortion of the reconnecting rod due to plastic deformation, very strict demands for a uniform microstructure of steels needs to be met [1]. The non-quenched and tempered free cutting steel C70S6 is a near eutectoid micro-alloyed steel invented in Germany for cracking con-rod manufacture [2]. It is also a new type of energy-saving and efficient steel, with features of high-carbon, micro-alloyed, high-sulfur content and narrow chemical composition range, thus it is relatively difficult to produce by continuous casting. In addition, non-metallic inclusions in steel destroy the continuity of the steel matrix, causing stress concentration and inducing crack formation, thus improving the cleanliness of molten steel is one of the key demands to solve this problem.

From the perspective of structure and performance requirements on finished products, on the one hand, the high-carbon and high-sulfur steel has a large tendency to shrink and crack in the process of continuous casting, which will easily result in the failure of ultrasonic detection. Additionally, its large segregation tendency due to its inherent chemistry can easily cause poor stability of the microstructures and mechanical properties of the subsequent rolling products [3–5]. Therefore, an in-depth understanding of the as-cast steel microstructure and its defect characteristics are of great significance on developing a continuous casting practice of this type of high-class steel. For this reason, we systematically characterized its solidification microstructures and composition segregation characteristics in the present continuous casting condition based on industrial development tests [6]. In this study, the as-cast genetic defects which may affect the requirements of the expanding cleavage fracture during the processing of steel products were revealed, and the technological approaches of optimizing continuous casting process were put forward. As a result, the quality of the bloom castings has been improved to an applicable degree of the users. However, due to the asymmetric three-strand tundish used in this steel plant, the temperature difference is large among different strands in tundish caused by an uneven flow field, in which the maximum temperature difference of molten steel measured at the outlets near the stopper reached 5 °C. Moreover, the bloom qualities of C70S6 steel and subsequent rolled products corresponding to each strand are not at a consistent level. Therefore, it is of great necessity to control the consistency and stability of casting conditions from tundish, at the very beginning of continuous casting.

Tundish, a transition container connecting ladle and mold in the casting process of molten steel, not only plays the role of receiving liquid steel and distributing it to the molds, but also has the metallurgical functions of homogenizing the composition and temperature of liquid steel and removal of non-metallic inclusions [7–9]. For the continuous casting of high-quality special steel, the consistency of the metallurgical effect of each strand in the tundish is key to improving and stabilizing the overall quality of subsequent products, representing a popular issue in recent years [10,11]. A reasonable tundish structure can ensure a good flow-state of liquid steel in the tundish, which is beneficial to improving the consistency of each strand, reducing the temperature difference between tundish outlets, and prolonging the residence time of liquid steel in the tundish, thus promoting the removal of inclusions.

In view of the high temperature characteristics of continuous casting tundish, the optimization of its structure is mainly carried out by water modelling [12–14] and numerical simulation [15–17]. For example, Tang et al. [18] optimized a 4-strand tundish in a steel mill through the above method. After application in the steel production, it was found that macroinclusions and microinclusions in the billet castings were reduced by 44.9% and 2.7%, respectively, compared with the use of the prototype tundish, and the total oxygen and inclusion contents in the four-strand billet castings were very consistent. Chen et al. [19]

used the same method to optimize the flow field of a two-strand symmetrical tundish for slab production. By increasing the distance between dam and retaining wall, the average residence time of water in the simulated tundish and volume fraction of the piston zone were increased, and the fraction of dead zone was reduced. The total oxygen content of billet was reduced by 30.32% after applying the optimized tundish in production. Merder et al. [20] studied the motion behavior of non-metallic inclusions in a two-strand tundish by using different simulation models and showed that the process of microparticle distribution can be well described by the Euler–Lagrange method according to the DPM model. In addition, their distribution characteristics and local concentrations can be reflected with a high accuracy. Sheng et al. [21] applied a CFD model to study the effect of thermal buoyancy on fluid flow and residence time distribution in a single-strand tundish. The results showed that thermal buoyancy has a significant impact on the flow pattern and temperature distributions of molten steel in the tundish. The increase of heat loss through the top surface shortens the mean residence time of molten steel in the tundish, and thus leads to an increase in dead volume fraction and a decrease in plug-flow volume fraction. Sheng [22] also investigated certain key aspects regarding the CFD simulation of a tundish, such as the choice of turbulence model, meshing, boundary conditions and a discretization scheme. In his study, a mesh size of 0.006 m, realizable k-ε turbulence model and second-order upwind discretization scheme are the recommended modeling settings for tundish flow simulation.

In the present paper, a three-strand continuous casting tundish was studied for continuous casting of cracking con-rod steel in Xining Special Steel. The tundish has an asymmetric geometric structure with a casting position of the ladle long nozzle close to its No.1 and 2 strand, but far from the No. 3 strand. Usually, an asymmetric tundish is more difficult to design than a symmetric tundish, and the uniformity of the flow field is also poorer. The industrial big data found that under the prototype flow control device of the tundish, the molten steel temperature at the first strand outlet and the occurrence rate of inclusions in the final cast products were higher, while its equiaxed crystal ratio measured lower than the other strands. Although there are many reports on the optimization of tundishes, most studies focus on symmetric multi-strand tundishes, which are evidently different from their asymmetric counterparts.

To meet the strict requirements of the cracking con-rod regarding the consistencies of the microstructure and properties of the rolled products, the tundish structure is optimized by physical and numerical simulation methods in the present study, and an overall quality comparison of the bloom castings and the following hot-rolled products before and after optimization is given based on a large number of industrial tests. These on-site volume data will be of great value in evaluating the casting practice for the quality production of special steel. Meanwhile, this study intends to provide a roadmap or benchmark for the optimization of other asymmetric multi-strand tundishes.

2. Production Process and Basic Parameters of C70S6 Steel

The manufacturing process of C70S6 steel is as follows: 110 t Consteel EAF (electric arc furnace) → 75 t LF (ladle furnace) + 75 t VD (Vacuum degas) → three-strand 250 mm × 280 mm bloom continuous caster → hot rolling → magnetic plus ultrasonic flaw detection. The chemical composition is shown in Table 1, indicating that this steel has the characteristics of high C, S and N content, and low Si and Al content.

Table 1. Chemical composition of steel C70S6, wt%.

C	Si	Mn	P	S	Cr	Ni	Mo	Al	V	N
0.67–0.73	0.15–0.25	0.53–0.60	≤0.045	0.060–0.070	0.10–0.15	0.04–0.08	≤0.03	≤0.010	0.030–0.040	0.012–0.016

Steel is cast by a three-strand tundish whose structure and original flow control device is shown in Figure 1. The flow of molten steel is mainly controlled through a retaining wall with a diversion hole measuring 140 mm in diameter and angled 8° upwards on every side wall. After the molten steel is poured into the impact zone from a long nozzle, it is transferred through the diversion hole in the side wall to three submerged entry nozzles (SEN). The process parameters of the continuous casting of C70S6 steel are shown in Table 2.

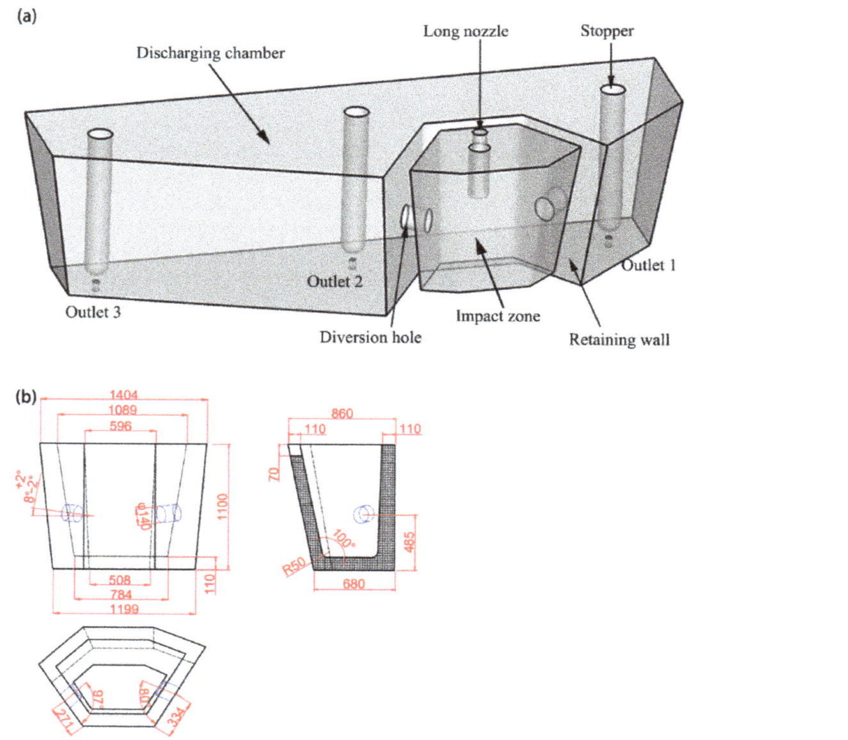

Figure 1. Schematic diagram of the prototype tundish (**a**) and flow control device (**b**), unit: mm.

Table 2. Process parameters for the continuous casting of C70S6 steel.

Parameter	Value
Tundish capacity	21 t
SEN inner diameter	40 mm
Inner diameter of ladle nozzle	85 mm
Depth of the molten bath in tundish	950 mm
Section size of the bloom caster	250 mm × 280 mm
Casting speed	0.75 m/min

3. Research Methods

3.1. Physical Modelling

Taking the industrial tundish shown in Figure 1 as the prototype, a model tundish with a geometric similarity scale of 1:2 was designed based on the similarity principle, and water was used to simulate the flow of molten steel in the tundish. According to the similarity principle, the Froude (*Fr*) numbers of the prototype and model should be equal [13], which can be expressed as:

$$Fr_{\mathrm{m}} = Fr_{\mathrm{p}} \tag{1}$$

where m denotes the tundish model, and p is the prototype.

The Fr is defined as:

$$Fr = \frac{u^2}{gL} \tag{2}$$

where u represents flow velocity, L is characteristic length of tundish and g is gravity acceleration.

Thus, the main geometrical and operation parameters between the model and prototype can be expressed with similarity ratio λ as follows:

$$L_m/L_p = \lambda,\ V_m/V_p = \lambda^3,\ u_m/u_p = \lambda^{0.5},\ Q_m/Q_p = \lambda^{2.5} \tag{3}$$

where V is tundish volume and Q is the volumetric flowrate of molten steel.

Substituting in the data from Table 2, Q_m and V_m can be obtained, respectively.

The stimulation–response method [13,23] was adopted in this experiment. When the liquid level of tundish reached a given height, 100 mL saturated KCl solution was injected as a pulse tracer from a side port mounted at the upper position of the ladle shroud. The conductivity of KCl at the three outlets of tundish was monitored by three conductivity probes. The conductivity curve with time (RTD curve) was obtained by DJ800 (IWHR, Beijing, China) system.

The stagnation time t_{min} and peak time t_{peak} of the fluid in the tundish was directly read from the RTD curve. Other flow characteristic parameters were obtained according to the modified mixing model proposed by Sahai Y [24], as shown in Table 3, where $C(t)$ is the measured tracer concentration at the time t; j is a specific strand, j = 1, 2, 3; n is total strand number, n = 3. The optimization effect of flow control device was evaluated by analyzing the flow characteristic parameters of molten steel in tundish and the RTD curve.

Table 3. Calculation formula of flow characteristic parameters of molten steel in tundish.

Parameters	Symbol and Unit	Formula
Theoretical mean residence time	t_f, s	V_m/Q_m
Actual average residence time	t_{av}, s	$\int_0^\infty tC(t)dt / \int_0^\infty C(t)dt$
Volume fraction of plug flow	θ_p, %	$\frac{1}{2}(t_{min} + t_{peak})/t_f$
Volume fraction of dead zone	θ_d, %	$1 - t_{av}/t_f$
Volume fraction of well-mixed zone	θ_m, %	$1 - \theta_p - \theta_d$
Standard deviation between strands	S, s	$\sqrt{[\sum_{j=1}^{n} (x_j - \overline{x})^2]/(n-1)}$

To visually display the flowing trajectory of molten steel from ladle to mold, ink was added to the water in the tundish from the same position where KCl was injected, and its dispersion was recorded by a camera.

3.2. Mathematical Model

A mathematical model including four main governing equations was developed to calculate the flow field and temperature profiles of molten steel in the prototype and optimized tundishes, with the casting speed set as 0.75 m/min for a bloom section of 250 mm × 280 mm. These equations and their designations are given in Table 4, and details of the model are described in refs. [10,13,25].

Table 4. Governing equations of the mathematical model.

Governing Equations	Symbols
1. Continuity equation $\frac{\partial(\rho u_i)}{\partial x_i} = 0$	ρ is density of molten steel, u_i is fluid velocity, μ_{eff} is effective viscosity, P is pressure.
2. Momentum conservation equation $\frac{\partial(\rho u_i u_j)}{\partial x_j} = -\frac{\partial p}{\partial x_i} + \frac{\partial}{\partial x_i}[\mu_{eff}(\frac{\partial u_i}{\partial x_j} + \frac{\partial u_j}{\partial x_i})] + \rho g_i$	x is the spatial coordinates at i and j directions.
3. Turbulent equations $\frac{\partial(\rho k u_i)}{\partial x_i} = \frac{\partial}{\partial x_i}\left[(\mu_{eff} + \frac{\mu_t}{\sigma_k})\frac{\partial k}{\partial x_i}\right] + G - \rho\varepsilon$ $\frac{\partial(\rho \varepsilon u_i)}{\partial x_i} = \frac{\partial}{\partial x_i}\left[(\mu_{eff} + \frac{\mu_t}{\sigma_\varepsilon})\frac{\partial \varepsilon}{\partial x_i}\right] + C_1\frac{\varepsilon}{k}G - C_2\rho\frac{\varepsilon^2}{k}$ $G = \mu_t\frac{\partial u_j}{\partial x_i}(\frac{\partial u_i}{\partial x_j} + \frac{\partial u_j}{\partial x_i})$ $\mu_{eff} = \mu + \mu_t = \mu + \rho C_\mu\frac{k^2}{\varepsilon}$ $C_1 = 1.44,\ C_2 = 1.92,\ C_\mu = 0.09,\ \sigma_k = 1.0,\ \sigma_\varepsilon = 1.3$	k is turbulence kinetic energy, ε is the turbulent kinetic energy dissipation rate, σ_k and σ_ε are the Schmidt numbers for k and ε, G is generation rate of turbulence energy and μ_{eff} is effective viscosity.
4. Energy equation $\frac{\partial}{\partial x_i}(\rho T u_i) = \frac{\partial}{\partial x_i}\left(K_{eff}\frac{\partial T}{\partial x_i}\right)$	T is temperature, K_{eff} is effective thermal conductivity.

3.3. *Assumptions and Boundary Conditions*

3.3.1. Main Assumptions

The following assumptions were made for simulation calculation: liquid steel in tundish was incompressible single-phase turbulent flow; influence of the slag layer on flow was ignored; the slag-steel surface was treated as a free surface.

3.3.2. Boundary Conditions and Solution Methods

The inlet of the molten steel in tundish adopted a velocity inlet. The inlet temperature was set as 1773 K based on the casting temperature of steel C60S6. The direction of speed was perpendicular to the section of the long nozzle inlet, and its value was obtained by the size of the casting section and the casting speed according to the mass conservation, which is 0.486 m/s in the present study. The tundish outlet adopted outflow boundary conditions, the same casting speed for each strand; the melt surface was considered to be flat, and the vertical gradient of all variables on the free surface was zero; the tundish wall surface was regarded as a non-slipping condition, and the standard wall function was adopted to deal with the near wall; the heat dissipation of each wall of the tundish adopted the second type of boundary condition. The material parameters and boundary conditions of the model are shown in Table 5, in which the turbulence kinetic energy and turbulence energy dissipation rate were obtained by semi-empirical formulas $k = 0.01v^2$ and $\varepsilon = 2 \times k^{1.5}/D$, respectively, ($v$ represents the flow rate of molten steel and D represents the diameter of the long nozzle), and the others originated from reference studies [10,16].

Table 5. Boundary conditions and material parameters.

Boundary Condition and Material Parameter	Value	Boundary Condition and Material Parameter	Value
Longitudinal wall heat flux, W/m^2	−4600	Density of liquid steel, kg/m^3	7014
Transversal wall heat flux, W/m^2	−4000	Specific heat of liquid steel, J/(kg·K)	750
Free surface heat flux, W/m^2	−15,000	Thermal conductivity of liquid steel, W/(m·K)	41
Bottom heat flux, W/m^2	−1800	Viscosity of liquid steel, kg/(m·s)	0.0061
k, m^2/s^2	0.00238	ε, m^2/s^2	0.00274

The flow and temperature distribution of molten steel in the tundish were simulated by commercial CFD software FLUENT 14.5 (ANSYS, Pittsburgh, PA, USA, 2012). Choosing the pressure-based solver for calculation, and the pressure and velocity coupling were based on the SIMPLE algorithm. The computational convergence residual was set to less than 10^{-6} for the energy equation and 10^{-4} for the other equations.

4. Experimental Results and Discussion

4.1. Results of Water Modelling

In total, 30 groups of experimental schemes were designed and carried out, but only the tundish prototype and the final optimization scheme are relevant to the comparative quality analysis here, thus this paper only lists the two schemes for comparison. The retaining wall structure of prototype case A was shown in Figure 1. The optimized case B is shown in Figure 2, in which a diversion hole measuring 90 mm in diameter and upward angle of 10° is displayed on the right side of the wall near outlet 1, and two diversion holes are displayed on the left side near outlet 2, with each hole at the same diameter and upward angle as on the right side but a different horizontal inclination of 3° and 5°, respectively. Additionally, a vortex inhibitor is designed as shown in Figure 2b. The flow characteristic parameters of the two schemes are shown in Table 6.

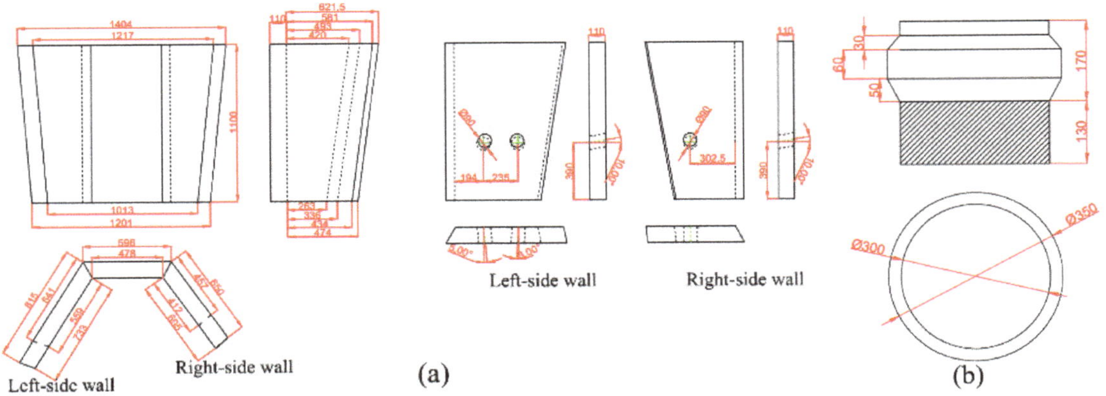

Figure 2. Retaining wall structure (**a**) and vortex inhibitor (**b**) in optimized case B and their geometry sizes, unit: mm.

Table 6. Flow characteristic parameters of the two experimental schemes.

Case	Outlet No.	t_{min}/s	t_{peak}/s	t_{av}/s	θ_d/%	θ_p/%	θ_m/%	S_{tmin}/s	S_{tav}/s
	1	17	20	267					
A	2	16	19	369	53.08	2.38	44.53	19.20	25.56
	3	50	64	434					
	1	30	41	651					
B	2	31	71	637	21.10	5.59	73.31	11.12	15.22
	3	50	60	591					

As can be seen from Table 6, the proportion of dead zone (θ_d) in the tundish prototype case A reaches as high as 53.08%, while the piston zone θ_p is only 2.38%. A large proportion of dead zone indicates that molten steel flows much slowly in some areas of tundish, which renders present non-metallic inclusions difficult to collide and grow, and the local molten steel temperature will also be relatively low due to weak heat exchange. The stagnation times t_{min} of outlets 1 and 2 are 17 and 16 s, respectively, and the peak time t_{peak} is 20 and 19 s, respectively. The difference between t_{min} and t_{peak} is only 3 s, suggesting that the short-circuit flow is possibly formed at the two outlets. The stagnation time of outlet 3 is about 2.5 times that of outlets 1 and 2, and its average residence time is much longer than that of the other two, indicating poor consistency among strands. This is because the diversion holes are far from the No. 3 strand, while pointing to Nos. 1 and 2 in direction, and the fluid flows out from outlets 1 and 2 first. Its kinetic energy is small when flowing to outlet 3, resulting in larger standard deviations of the stagnation time and average residence time among the three strands, S_{tmin} and S_{tav}, are 19.20 s and 25.56 s, respectively.

Compared with case A, the dead zone proportion of case B is only 21.1%, and the stagnation times of outlets 1 and 2 are extended to 30 and 31 s, respectively. The standard deviations of stagnation time and average residence time are reduced to 11.12 and 15.22 s, respectively, indicating a clear improvement of the consistency among strands.

Figure 3 shows the RTD curves of the two schemes, where the ordinate is dimensionless concentration and the abscissa is dimensionless time. As can be observed, after the tracer KCl is added into the tundish in case A, the concentrations of outlets 1 and 2 rapidly reach the peak value, and the peak concentration of outlet 1 is as high as 6.5, indicating that a short-circuit flow has formed in the tundish. The short-circuit flow will make the molten steel flow out of the tundish rapidly. In addition, its residence time in the tundish is short, and there is inadequate time for inclusions to grow or float. This may be related to the lower pass rate of ultrasonic detection of the rolled products corresponding to tundish outlet 1 in industrial production.

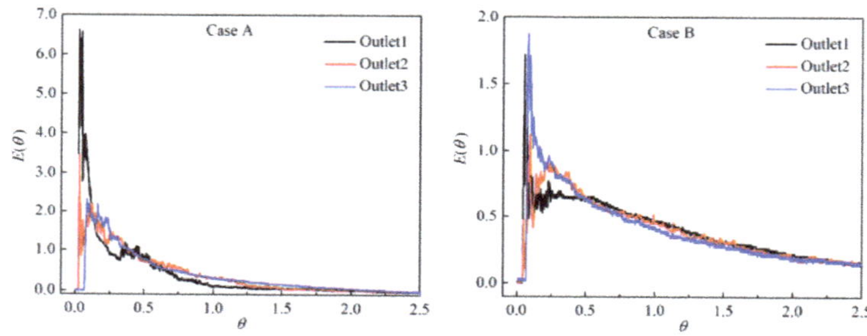

Figure 3. Comparison of RTD curves between cases A and B.

Although the RTD curve of case B is not very smooth, the peak concentration is greatly reduced to less than 2.0, and the consistency among different strands and the short-circuit flow are also improved as compared with case A. In addition, the average residence time of the three strands ranges from 591 to 651 s, much longer than the 267 to 434 s in the original case. This will provide enough time for inclusions in the molten steel to float and be removed, and thus lead to an improvement in the cleanliness of the steel.

Figure 4 compares the ink trajectory of the two cases. The ink tracer consists of 50 mL black ink mixed into 150 mL water. From Figure 4a, it can be seen that the ink from diversion holes in case A flows downwards and reaches outlets 1 and 2 at about 16 s, then reaches outlet 3 at about 60 s. This is basically in agreement with the experimental measurement. The ink moves to the side wall of the tundish and then flows back to the middle area. The top region of outlet 2 mixes slowest, possibly representing the main dead zone in the tundish. The local temperature of molten steel will possibly also be low. The ink fills the whole tundish at about 180 s in this case. By comparison, the ink in case B (Figure 4b) flows upwards due clearly to the increased upward inclination angle of the diversion holes from 8° to 10°, with which the fluid will be delayed in arriving at outlets 1 and 2. However, because of the reduced hole size (from 140 mm to 90 mm) in the wall, the fluid velocity will increase, which accordingly shortens the stagnation time of outlet 3 to 51 s. As a result, the flow of molten steel will become more uniform in case B than in case A. The ink is completely mixed at 124 s, 56 s shorter than that in case A.

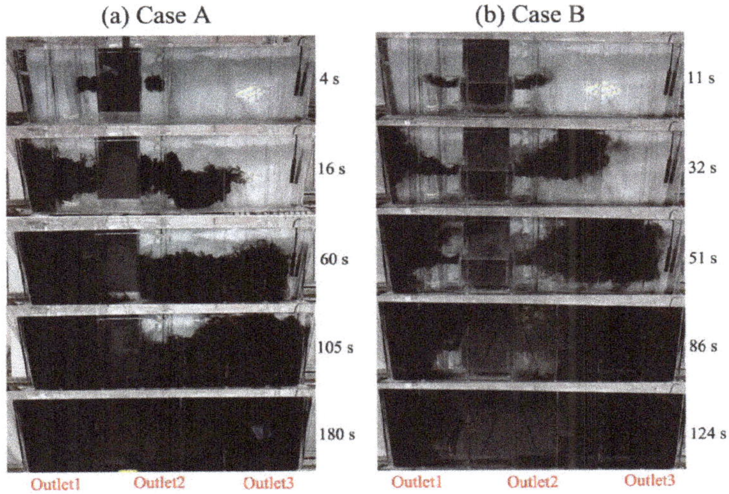

Figure 4. Ink trajectory comparison of cases A (**a**) and B (**b**).

4.2. Validation of CFD Modelling and Mesh Independence Test

The numerical simulation method has been widely used in the structural optimization of continuous casting tundish, and the authors have also conducted an abundance of research works on this topic [10,13,18] with its reliability confirmed by water simulation experiments and industrial application results [10,18]. This paper adopts the same mathematical modeling and calculation method, and the reliability of the simulation will be further verified through actual temperature measurements.

Meanwhile, a mesh independence test was carried out because of its importance to the simulation results. As shown in Table 7, different numbers of grids were designed for case A and case B, respectively. By applying the same boundary conditions and parameters, the temperature of each tundish outlet was calculated and compared with industrial measurement data. Data from the table indicate that the temperature of each outlet is slightly changed and the calculation accuracy can be met when the number of grids is about 300,000~600,000. Therefore, 394,471 tetrahedral grids for case A and 559,800 tetrahedral grids for case B are chosen for the following studies.

Table 7. Mesh independence study and model validation.

Case	Number of Grids	$T_{outlet1}$/K	$T_{outlet2}$/K	$T_{outlet3}$/K	Maximum Temperature Difference among Three Strands for Industrial Measurement/°C
A	309,506	1768	1765	1767	2–5, average 3.2
	394,471	1768	1765	1767	
	489,901	1768	1765	1767	
	740,025	1769	1765	1766	
	1,093,481	1768	1767	1765	
B	321,107	1766	1766	1767	1–3, average 1.8
	431,562	1766	1766	1767	
	559,800	1766	1766	1767	
	736,498	1766	1766	1767	
	1,031,898	1766	1767	1767	

4.3. Results of Numerical Simulation

Figure 5 shows the 3D streamline contours of cases A and B. It is seen that in case A the velocity of molten steel in impact zone is great with most at 0.05 m/s. In addition, the flow is active in the area near outlet 1 as the diversion hole on one side of retaining wall points to it adjacently. The local velocity reaches over 0.02 m/s. However, the molten steel at outlets 2 and 3 flows slowly at under 0.01 m/s. This is due to deviation of the diversion hole on the right wall away from outlet 2 and further still from outlet 3. By comparison the flow field of case B is more uniform, and the flow velocity and streamline of molten steel near outlets 2 and 3 are clearly intensified, due to the adoption of an optimized wall structure. The area with flow velocity over 0.02 m/s increases, and the velocity of the impact zone is greatly reduced with the help of a vortex inhibitor. The flow field in the tundish is crucial to the temperature uniformity of the molten steel within.

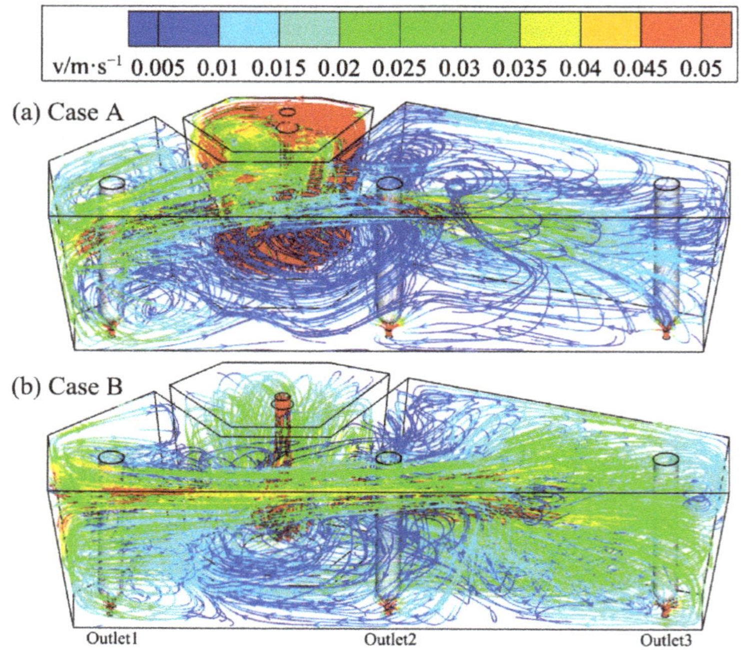

Figure 5. Three-dimensional streamline contours of tundish in cases A (**a**) and B (**b**).

Figure 6 compares the temperature distribution of cases A and B in the vertical section of the outlet center. Similarly, the temperature distribution in case B is more uniform than in case A. The maximum temperature difference at three outlets of tundish is only 1 K, 2 K lower than case A. The maximum temperature difference in the whole section is decreased from 8 K (1769 minus 1761 K) to 6 K (1769 minus 1763 K), and the high (1768~1769 K) and low temperature areas (1761~1763 K) are both also reduced in case B.

From the results given above, the consistency of flow and temperature distribution of case B are both significantly improved compared with case A, which is expected to improve the stability and consistency of steel product quality among different strands.

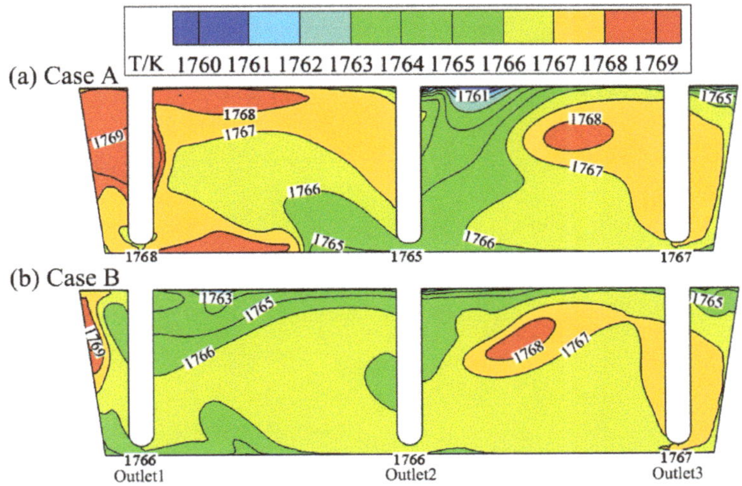

Figure 6. Comparison of temperature distribution of cases A (**a**) and B (**b**) in the vertical section of outlet center.

5. Comparison of Industrial Application Effect before and after Tundish Optimization

Based on the results of the physical and numerical simulation, the optimized case B was successfully adopted to industrial production of C70S6 steel. The temperature difference of molten steel at the three outlets of the tundish, and the qualities of bloom castings and subsequent rolled products before and after optimization were compared under the same given refractory material and operating conditions.

5.1. Comparison of Temperature Difference of Molten Steel

The molten steel temperature in the tundish was measured comparatively before and after tundish optimization, in which the 2nd, 3rd, 4th, 5th and 7th heats were selected for each casting test, three times for each heat during the casting process. A temperature detector (TC1200-S-AL, Heraeus, Hanau, Germany) was inserted into the molten steel from the stopper holes between strands, and the maximum temperature difference among the three strands was recorded. For better comparison, measurement of the three strands was carried out at nearly the same casting time. There are 13 groups of effective data obtained for the comparison, which are shown in the Figure 7.

As is seen, the maximum temperature difference between strands measures 2–5 °C before tundish optimization, and the average temperature difference is 3.2 °C. After optimization, the maximum temperature difference decreases to 1–3 °C, and the average temperature difference drops to 1.8 °C, indicating an improved temperature field uniformity of the tundish. Additionally, the maximum temperature difference calculated by numerical simulation was 3 °C before optimization and 1 °C after optimization, which also verifies the reliability and accuracy of the numerical simulation from an industrial perspective. Meanwhile, the results also show that the flow field optimization has significantly improved regarding temperature consistency in the tundish. As we know, uniformity of the temperature field is helpful in stabilizing and unifying the superheat for the three-strand castings, and thus stabilizing the as-cast solidification structure of the strands, along with the identical quality of the as-rolled and subsequent products.

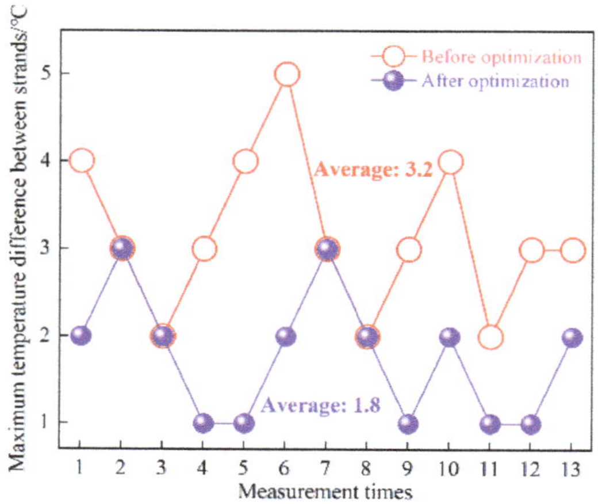

Figure 7. Maximum temperature difference among strands before and after tundish optimization.

5.2. Comparison of Erosion of Molten Steel on Refractory Material in Tundish

The change of flow field of the tundish may affect the erosion of its lining, and thus affect the cleanliness of blooms and the following rolled products. For this reason, the eroded appearance of the tundish working layer at three positions near the slag line was compared at the end of the casting. The test positions (A, B, C) are illustrated in Figure 8. The eroded appearances of the three tundishes were examined before optimization, and five tundishes were examined at the same locations after optimization. To ensure the comparability and effectiveness of erosion data before and after optimization, the material of the working layer, amount and method of cover slag added, level height of molten steel in tundish, fluctuation of casting speed and temperature measurement times were strictly controlled identical under the two working conditions. The erosion width and depth are shown in Table 8.

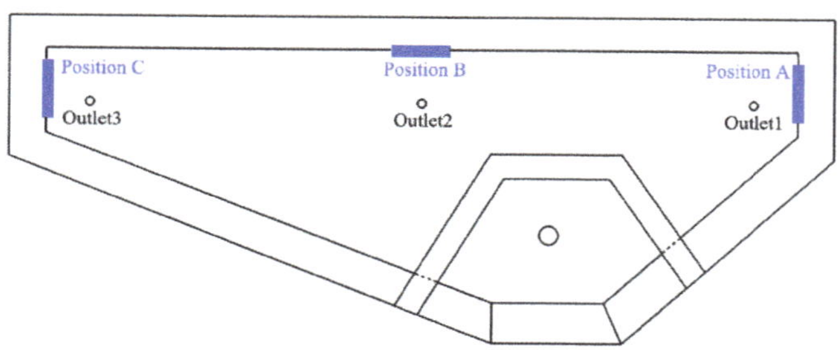

Figure 8. Schematics of erosion examination locations.

Table 8. Erosion state of the working layer in tundish.

Item		Before Optimization	After Optimization
Measured number of tundish, #		3	5
Erosion width, mm	A	90–120	90–100
	B	90–110	90–110
	C	90–110	90–110
Erosion depth, mm	A	10–12	9–11
	B	10–12	9–11
	C	10–12	9–11

As shown in the Table 8, the erosion width of the working layer at position A close to strand No. 1 decreases from 90–120 mm before optimization to 90–100 mm after optimization, and the depth decreases from 10–12 mm to 9–11 mm. This is because a diversion hole in the retaining wall before optimization directly points to zone A, imparting significant washing to the local lining material. After optimization, the diversion hole is provided with an upward and horizontal deflection angle to avoid direct impingement on the tundish wall. This comparison indicates that a good flow field can reduce erosion of the lining, and thus hopefully reduce exogenous inclusions from erosion. The erosion widths of positions B (near No. 2 strand) and C (No. 3 strand) are basically the same before and after optimization, while the erosion depth is slightly reduced after optimization.

5.3. Comparison of Bloom Quality Consistency

The effect of tundish flow field optimization on bloom quality should be embodied mainly by the improvement of bloom quality and that of its consistency among different strands. For this reason, 10 batches of C70S6 steel blooms in total were compared, with 5 batches (1#–5#), respectively, before and after tundish optimization. For each batch, No. 1 to No. 3 strand blooms at the same casting time were taken and machined, then acid etched with 1:1 HCl solution at 70 °C for 20 min. The columnar, equiaxed crystal rates and the center shrinkage were evaluated and their differences among the three strands were revealed. Moreover, the rod samples were taken from the same position of each strand bloom, and then the total oxygen content was measured with an oxygen–nitrogen analyzer (TCH600, St Joseph, MI, USA). The results are shown in Table 9.

Table 9 shows that before flow field optimization, the columnar crystal percentage of three-strand billets distributes from 43.86–58.19%, and the difference among different strands ranges between 2.27–3.17%. After optimization, the difference of columnar crystal rates among strands is reduced to 1.26–1.85%, and the difference of equiaxed crystal rates decreases from 2.25–2.87% to 0.48–2.10%. There is no significant change in center porosity before and after optimization. It is well known that equiaxed and columnar crystal proportions are closely related to casting superheat. A high superheat means a greater temperature gradient between the inner liquid core and the outer, along with a lower solidification nucleation rate, and thus resulting in a more developed columnar crystal. Meanwhile, the excessive growth of columnar crystals causes the enrichment of solute elements in the center, generating center segregation. Before tundish optimization, the temperature of molten steel near No.1 strand is higher than that of other two strands, thus the superheat is relatively high, and accordingly the proportion of columnar crystals is slightly higher than that of other strands from statistics. After optimization, the temperature field is more homogeneous, thus the microstructure difference among strands is reduced.

Moreover, Table 9 shows that the oxygen content difference between three strands of blooms is 0.8–1.9 ppm after flow field optimization, while it is 1.7–3.5 ppm before that, indicating that the consistency of the bloom cleanliness has been effectively improved. Meanwhile, the oxygen content in the No. 1, No. 2 and No. 3 strands measures 11.2–14.3 ppm, 8.8–11.8 ppm and 8.5–13.1 ppm, respectively, before optimization, while it is 8.8–11.8 ppm, 9.2–10.8 ppm and 7.9–11.6 ppm, respectively, after optimization. The oxygen content after optimization is lower than that before optimization, especially in the first and third strands. This is because, as mentioned above, the stagnation time and average residence time of molten steel in tundish after optimization are both improved, which provides sufficient time for the inclusions in molten steel to float. In addition, the consistency among strands significantly improves due to the more uniform flow field after optimization.

Table 9. Comparison of C70S6 bloom quality before and after tundish optimization.

Item	Working Condition		Before Optimization					After Optimization				
	Detected Sample No.		1#	2#	3#	4#	5#	1#	2#	3#	4#	5#
As-cast macrostructure	Columnar crystal rate, %	No. 1	49.25	46.27	56.1	48.75	53.81	47.80	52.56	51.00	49.96	53.88
		No. 2	48.56	43.86	58.19	46.36	51.50	49.06	53.50	52.85	48.75	55.17
		No. 3	46.08	46.48	55.92	46.87	54.15	48.70	54.25	51.35	48.54	54.86
		Δ_{max}	3.17	2.62	2.27	2.39	2.65	1.26	1.69	1.85	1.42	1.29
	Equiaxed crystal rate, %	No. 1	28.32	25.04	22.21	28.71	27.96	30.36	25.40	24.38	26.21	20.06
		No. 2	28.41	26.84	24.55	27.91	28.13	30.17	26.36	23.28	26.43	21.72
		No. 3	30.87	27.74	23.66	30.78	30.21	29.88	27.24	24.49	27.81	19.62
		Δ_{max}	2.55	2.70	2.34	2.87	2.25	0.48	1.84	1.21	1.60	2.10
	Central shrinkage	No. 1	1.5	1.0	1.5	1.0	1.5	1.5	1.5	1.5	1.5	1.0
		No. 2	1.5	1.0	1.5	1.0	1.5	1.5	1.5	2.0	1.0	1.0
		No. 3	1.5	1.0	1.5	1.5	2.0	1.5	1.5	1.5	1.5	1.5
		Δ_{max}	0.0	0.0	0.0	0.5	0.5	0.0	0.0	0.5	0.5	0.5
C segregation index at different regions of bloom	Columnar crystal zone	No. 1	0.97–1.03	0.98–1.03	0.98–1.08	-	-	0.97–1.03	0.96–1.04	-	-	0.98–1.03
		No. 2	0.96–1.04	0.95–1.01	0.95–1.03	-	-	0.96–1.02	0.96–1.02	-	-	0.96–1.03
		No. 3	0.95–1.02	0.94–1.03	0.99–1.02	-	-	0.99–1.02	0.98–1.01	-	-	0.96–1.02
	Equiaxed crystal zone	No. 1	0.95–1.04	0.93–1.02	0.96–1.03	-	-	0.95–1.03	0.96–1.04	-	-	0.95–1.03
		No. 2	0.97–1.05	0.96–1.03	0.95–1.01	-	-	0.97–1.01	0.98–1.03	-	-	0.97–1.02
		No. 3	0.95–1.03	0.98–1.03	0.97–1.03	-	-	0.96–1.03	0.95–1.02	-	-	0.96–1.3
	Center point	No. 1	1.11	1.05	1.07	-	-	1.10	1.09	-	-	1.08
		No. 2	1.03	1.12	1.12	-	-	1.09	1.03	-	-	1.09
		No. 3	1.05	1.02	1.03	-	-	1.10	1.08	-	-	1.04
T. O.	ppm	No. 1	12.3	14.3	13.2	11.8	11.2	10.8	8.8	11.2	9.9	11.8
		No. 2	8.8	11.8	10.4	9.4	9.9	9.2	9.2	10.8	10.2	10.7
		No. 3	10.2	10.9	13.1	8.5	9.5	8.9	7.9	11.6	10.8	10.5
		Δ_{max}	3.50	3.40	2.80	3.30	1.70	1.90	1.30	0.80	0.90	1.30

The as-cast pieces of blooms after acid etching were sampled by drilling with a φ5 mm drill for chemical analysis as shown in Figure 9. A total of 25 points were drilled in each bloom from No. 1 to No. 25, and their carbon contents were measured by an HCS-140 infrared carbon and sulfur meter. The local carbon segregation index was calculated based on the ratio of its carbon content to the average carbon content at all points, namely: $C_i/(\sum C_i/25)$, where $i = 1, 2, \ldots \ldots 25$. Three heats were taken respectively before and after optimization, and each heat took three strands of blooms. Therefore, there are 18 pieces (3 heats × 3 strands × 2 working conditions = 18) in total for the statistics. The C segregation indexes of all points are shown in Table 9, and those along the cross-section center line (from points 1 to 9) are shown in Figure 10.

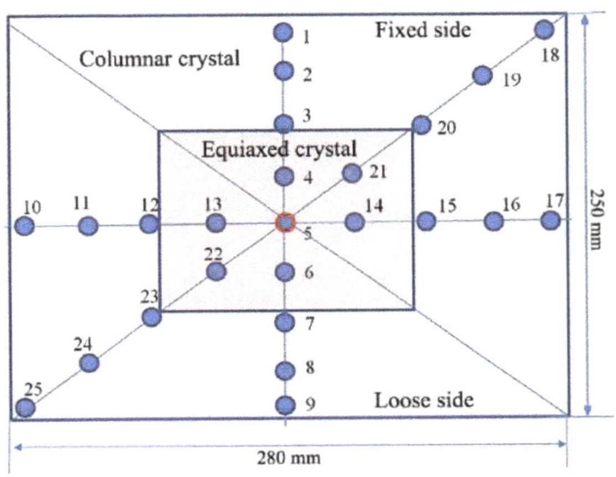

Figure 9. Schematic diagram of sampling location for C segregation measurement.

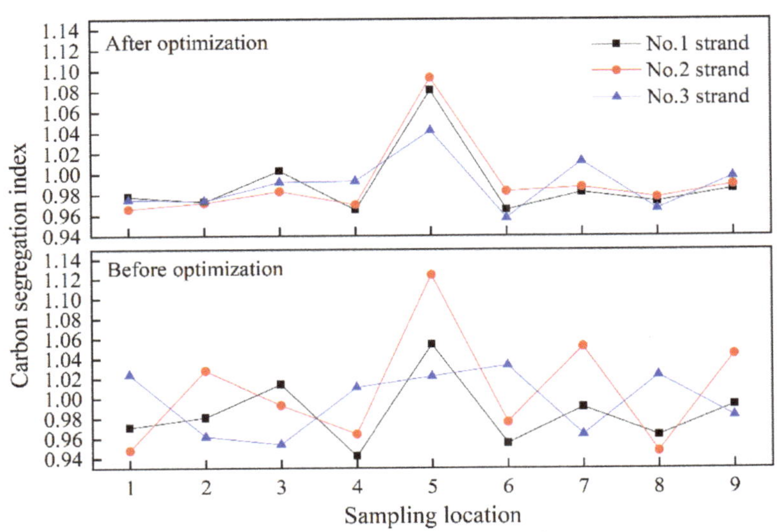

Figure 10. Carbon segregation index before and after optimization.

As can be observed in Table 9, the carbon segregation indexes in equiaxed crystal and columnar crystal zones for different batches of bloom pieces fluctuates between 0.95 and 1.05. There seems to be no clear change before and after optimization, but the consistency among different strands is significantly improved as shown in Figure 10, and the segregation index of the center point 5 in No. 2 strand is also reduced from 1.13 to 1.10. This is because the element segregation is relevant to the as-cast structure, which is closely related to the superheat of molten steel [26]. After optimization, the temperature distribution of molten steel in tundish is more uniform as shown in Figure 6. The superheat for each strand casting is at the same level, and thus their solidification and segregation consistency are significantly improved accordingly, together with a greatly reduced center carbon segregation index.

5.4. Comparison on the Quality of Rolled Products by Overall Stactistical Data

The mechanical properties of hot-rolled products before and after tundish optimization, including tensile strength Rm, yield strength Rel, elongation A and reduction of area Z, were measured using a universal material testing machine WAW-Y500C, and the hardness measured with an HB-3000B Brinell hardness tester. Additionally, the ferritic phases and band structures were observed using a Carl Zeiss optical microscope after samples were etched with 2% nitric acid alcohol solution. In addition, the grain sizes were evaluated by the standard method with water-quenched samples after insulating at 860 °C for one hour. Finally, the product quality indexes measured above are comparatively listed in Table 10.

According to the big data analysis, after the flow field of the tundish is optimized, the pass rate of tensile strength of rolled products is 98.9%, which is higher than 98.3% before optimization, while the yield strength, elongation and section shrinkage show no clear change. The hardness fluctuation (maximum value minus minimum value) decreases from 84 HBW before optimization to 60 HBW after optimization. The ferrite content is stabilized from 3–9% to 3–8%. The band structure decreases from 2.08 to 1.87 grade. Inclusions B + C + D + Ds decrease from 1.143 to 1.105 grade, and the occurrence rate of Ds inclusions decrease from 0.192% to 0.118%. Ds refers to single particle inclusion, which are large in size and harmful to steel service property, thus requiring strict control in many special steels. In this study, the occurrence rate of Ds-type inclusions in C60S6 steel is reduced by the optimization of tundish, since the stagnation time and average residence time of strands 1 and 2 are prolonged, and the short-circuit flow is eliminated after optimization, which gives the inclusions adequate time to float and be removed. In addition, as mentioned above, the optimized tundish structure can alleviate the lining erosion in zone A and thus reduce large-sized exogenous inclusions into the molten steel. The A-type inclusion was not counted in this study because the steel contains S and no conspicuous related service defects in its rod products from A-type inclusions. Moreover, no clear grain-size index variation was observed in this study.

5.5. Products Customer Feedback

According to user feedback regarding the cracking con-rod products, the failure rate of products was controlled within 4‰ during their forming processes, and the fracture of connecting rod was relatively smooth compared with that before casting practice optimization.

Table 10. Comparison of the quality of rolled products before and after tundish optimization.

Item		Before Optimization				After Optimization			
Mechanical property	Number of samples, #	1153				912			
	Measured item	Rm, MPa	Rel, MPa	A, %	Z, %	Rm, MPa	Rel, MPa	A, %	Z, %
	Standard requirement	900–1050	>550	>10	20	900–1050	>550	>10	20
	Measured value	878–1050	485–599	8.5–16	13–29	849–1018	439–692	6.5–19.5	8–48
	Average value	943	555	12.7	21.5	941	552	11.9	20.5
	Pass rate, %	98.3	99.6	97.8	98.5	98.9	99.6	97.9	98.4
Hardness	Number of samples, #	1095				875			
	Measured value, HBW	229–313				225–285			
	Average value, HBW	258				256			
	Hardness difference (Maximum minus minimum), HBW	84				60			
Microstructure	Number of samples, #	70				93			
	Microstructure	F + P				F + P			
	Ferritic content, %	3–9				3–8			
Grain size	Number of samples, #	551				448			
	Measured value	7.0–8.0				7.0–8.0			
Band structure index	Number of samples, #	54				56			
	Measured value	1.0–2.5				1.0–2.5			
	Average value	2.08				1.87			
Inclusions	Number of samples, #	1044				848			
	Grade of (B + C + D + Ds)	1.143				1.105			
	Ds occurrence rate, %	0.192				0.118			

6. Conclusions

The flow field optimization to a multi-strand asymmetric tundish has been conducted through water modelling combined with a numerical calculation method for improved metallurgical behavior, and the beneficial results were revealed accordingly through volume production. The conclusions are drawn as follows:

(1) The water modeling experiments show that dead zone proportion of the tundish is reduced from 53.08% originally to 21.10% after optimization, and the stagnation time of its outlets 1 and 2 are extended to about twice that of before, indicating clear improvement in the consistency among strands. The average residence time of the three strands measures 591 to 651 s, much longer than the 267 to 434 s in the original case. This will provide sufficient time for inclusions in the molten steel to float and be removed, and thus improve the cleanliness of steel.

(2) The numerical simulation results show that the flow field in the tundish is more uniform due to the adoption of a vortex inhibitor and an optimized wall structure, which is beneficial for temperature-field uniformity. The maximum temperature difference at the three outlets of the tundish is only 1 K, 2 K lower than the original case. The maximum temperature difference in the whole outlet vertical section is decreased from 8 K to 6 K and both the high and low temperature areas are reduced by the optimized case.

(3) The maximum temperature difference of molten steel measured among the outlets around stoppers of the three strands is decreased from an initial 2–5 °C to 1–3 °C after optimization; the lining erosion in region A near the No. 1 strand of the tundish is clearly alleviated due to the disappearance of the local short-circuit flow.

(4) The differences in columnar crystal rate among different strands of the bloom castings are decreased from 2.27–3.17% to 1.26–1.85% after tundish optimization; the consistency of the strand central line carbon segregation index is significantly improved. The total oxygen content difference among strands is reduced from 1.7–3.5 ppm to 0.8–1.9 ppm, with the No. 1 strand showing a more significant effect than the other two strands.

(5) The main mechanical properties and microstructure stability of the rolled products are statistically well improved owing to the generally identical as-cast primary quality, in which the pass rate of tensile strength reaches 98.9%, the hardness dispersion is decreased from 84 HBW to 60 HBW, the non-metallic inclusion grade of B + C + D + Ds is decreased from 1.143 to 1.105, and the occurrence rate of Ds type inclusion is decreased from 0.192% to 0.118%. Additionally, the band-structure level of the hot-rolled products drops from 2.08 to 1.87.

In summary, flow-field optimization of the tundish reveals a clear beneficial effect to its subsequent castings and hot-rolled products, especially for an improved identical quality index while cast by different strands.

Author Contributions: Conceptualization, Z.T. and H.T.; methodology, Z.T and F.X.; software, K.W. and S.C.; validation, Z.T., H.T. and K.W.; formal analysis, Z.T. and F.X.; investigation, Z.T. and F.X.; resources, H.M.; data curation, H.M.; writing—original draft preparation, Z.T. and H.T.; writing—review and editing, H.T. and J.Z.; visualization, K.W. and S.C.; supervision, H.T. and J.Z.; project administration, H.T., J.Z. and H.M.; funding acquisition, H.T. All authors have read and agreed to the published version of the manuscript.

Funding: This research was funded by National Natural Science Foundation of China (grant number 51874033) and Beijing Natural Science Foundation (grant number 2182038) to Haiyan Tang.

Institutional Review Board Statement: Not applicable.

Informed Consent Statement: Not applicable.

Data Availability Statement: The data presented in this study are available on request from the corresponding author.

Conflicts of Interest: The authors declare no conflict of interest.

References

1. Wang, Z.H.; Li, S.; Ge, Y.; Zhou, L.; Zhao, X.M.; Mao, X.Y.; Hu, Q.X. Research on Microstructure and Mechanical Properties and Fracture Splitting Properties of Forged Connecting Rod of C70S6 Non-quenched and Tempered Steel. *Hot Work. Technol.* **2015**, *44*, 39–43.
2. Deng, X.Y.; Li, J.; Xie, J.B.; Hu, D.L.; Wang, Y.; Fu, J.X. Comparison of Quality between C70S6 Steels Produced by Chinese and Overseas Steelmakers for Fracture Splitting Connection Rod. *Iron Steel Vanadium Titan.* **2018**, *39*, 148–154.
3. Zhang, Z.; Li, H.Y.; Zhou, L.; Liu, H.S.; Tang, H.Y.; Zhang, J.Q. As-Cast Spot Segregation of Gear Steel and Its Evolution in the Rolled Products. *Acta Metall. Sin.* **2021**, *57*, 1281–1290.
4. Li, L.; Zhang, Z.H.; Luo, M.; Li, B.; Lan, P.; Zhang, J.Q. Control of Shrinkage Porosity and Spot Segregation in Ø195 mm Continuously Cast Round Bloom of Oil Pipe Steel by Soft Reduction. *Metals* **2021**, *11*, 9. [CrossRef]
5. Yang, F.F.; Zhang, Z.H.; Liu, H.S.; Luo, M.; Li, G.; Zhang, J.Q. Banded segregation and its related band-typed mixed grain structure in high strength oil well pipe steel. *J. Iron Steel Res.* **2021**, *33*, 979–986.
6. Xia, Y.; Li, L.; Wang, P.; Tie, Z.P.; Lan, P.; Tang, H.Y.; Zhang, J.Q. Characteristics of the as-cast high-carbon microalloyed continuous casting bloom steel for expansion-break connecting rods. *Chin. J. Eng.* **2022**, *44*, 189–197.
7. Merder, T.; Warzecha, M. Optimization of a Six-Strand Continuous Casting Tundish: Industrial Measurements and Numerical Investigation of the Tundish Modifications. *Metall. Trans. B* **2012**, *43*, 856–868. [CrossRef]
8. Li, S.X.; Tie, Z.P.; Zheng, B.A.; Kou, Y.S.; Yan, Q.Z.; Tang, H.Y.; Zhang, W.; Zhang, J.Q. Numerical simulation for an optimized flow field in a three–strand tundish of bloom casting. *Contin. Cast.* **2018**, *44*, 47–53.
9. Ai, X.G.; Dong, H.; Li, S.L.; Zeng, H.B.; Li, H.Y. Optimization of flow uniformity control device for six–stream continuous casting tundish. *J. Iron Steel Res. Int.* **2020**, *27*, 1035–1044. [CrossRef]
10. Tang, H.Y.; Li, X.S.; Zhang, S.; Zhang, J.Q. Fluid Flow and Heat Transfer in a Tundish with Channel Induction Heating for Sequence Casting with a Constant Superheat Control. *Acta Metall. Sin.* **2020**, *56*, 1629–1642.
11. Deng, W.; Niu, S.; Yang, E.J.; Wang, D.J.; Wang, K.M.; Tang, H.Y. Study on an identical metallurgical effect for muti-strand tundish during special steel bloom casting. *J. Iron Steel Res.* **2021**, *33*, 1144–1153.
12. Manish, M.; Tripathi, K.P.; Hosahali, T.; Kumar, D.S.; Konangi, R. Reduction in skull loss in billet caster tundish through water modelling studies. *Ironmak. Steelmak.* **2021**, *48*, 803–810. [CrossRef]
13. Tang, H.Y.; Guo, L.Z.; Wu, G.H.; Xiao, H.; Yao, H.Y.; Zhang, J.Q. Hydrodynamic Modeling and Mathematical Simulation on Flow Field and Inclusion Removal in a Seven–Strand Continuous Casting Tundish with Channel Type Induction Heating. *Metals* **2018**, *8*, 374. [CrossRef]
14. Ranjan, S.; Mukherjee, S.; Agnihotri, A.; Mazumder, D. Physical Modelling of Nitrogen Variation in Continuously Cast Blooms Resulting from Atmospheric Exposure of Steel in Tundish during Initial Stages of Teeming. *ISIJ Int.* **2022**, *62*, 609–612. [CrossRef]
15. Sheng, D.Y.; Chen, D.F. Comparison of Fluid Flow and Temperature Distribution in a Single-Strand Tundish with Different Flow Control Devices. *Metals* **2021**, *11*, 796. [CrossRef]
16. Yue, Q.; Zhang, C.B.; Pei, X.H. Magnetohydrodynamic flows and heat transfer in a twin-channel induction heating tundish. *Ironmak. Steelmak.* **2017**, *44*, 227–236. [CrossRef]
17. Zhang, B.L.; Liu, F.H.; Zhu, R.; Zhu, J.F. Effects of Multiple-Hole Baffle Arrangements on Flow Fields in a Five-Strand Asymmetric Tundish. *Materials* **2020**, *13*, 5129. [CrossRef]
18. Tang, H.Y.; Yu, M.; Li, J.S.; Bao, Y.P.; Jiang, J.; Ji, C.B.; Li, T. Numerical and physical simulation on inner structure optimization of a continuous casting tundish and its metallurgical effect. *J. Univ. Sci. Technol. Beijing* **2009**, *31*, 38–42.
19. Chen, Y.; Ou, X.D.; Wei, H.R.; Xue, L.Q.; Yuan, P.F. Physical simulation and application of flow control devices optimization in two-strand tundish. *Contin. Cast.* **2019**, *44*, 75–78.
20. Kumar, A.; Mazumdar, D.; Koria, S.C. Modeling of Fluid Flow and Residence Time Distribution in a Four-strand Tundish for Enhancing Inclusion Removal. *ISIJ Int.* **2008**, *48*, 38–47. [CrossRef]
21. Merder, T.; Pieprzyca, J.; Warzecha, M.; Warzecha, P.; Hutny, A. Evolution of the Numerical Model Describing the Distributionof Non-Metallic Inclusions in the Tundish. *Materials* **2021**, *14*, 2229. [CrossRef] [PubMed]
22. Sheng, D.Y.; Jönsson, P. Effect of Thermal Buoyancy on Fluid Flow and Residence-Time Distribution in a Single-Strand Tundish. *Materials* **2021**, *14*, 1906. [CrossRef] [PubMed]
23. Sheng, D.Y. Synthesis of a CFD Benchmark Exercise: Examining Fluid Flow and Residence-Time Distribution in a Water Model of Tundish. *Materials* **2021**, *14*, 5453. [CrossRef] [PubMed]
24. Sahai, Y.; Emi, T. Melt Flow Characterization in Continuous Casting Tundishes. *ISIJ Int.* **1996**, *36*, 667–672. [CrossRef]
25. Launder, B.E.; Spalding, D.B. The numerical computation of turbulent flows. *Comput. Methods Appl. Mech. Eng.* **1974**, *3*, 269–289. [CrossRef]
26. Wang, Y.; Song, J.X.; Cheng, N.L.; Guo, Z.H.; Li, J.S.; Yang, S.F.; Zhao, M.J.; Wang, C. Application of Graphite Electrode Plasma Heating Technology in Continuous Casting. *Materials* **2022**, *15*, 2590. [CrossRef]

materials MDPI

Article

Mesoscopic Fluid-Particle Flow and Vortex Structural Transmission in a Submerged Entry Nozzle of Continuous Caster

Peng Zhao [1], Rongxun Piao [2,*] and Zongshu Zou [3]

[1] School of Materials Science and Engineering, Shandong Jianzhu University, Jinan 250101, China; zhaopeng3@hotmail.com
[2] Department of Mechanical Engineering, Anhui University of Science and Technology, Huainan 232001, China
[3] School of Metallurgy, Northeastern University, Shenyang 110819, China; zouzs@mail.neu.edu.cn
* Correspondence: yhpark617@163.com

Abstract: Understanding the essence of the flow oscillations within a submerged-entry nozzle (SEN) is essential to control flow patterns in the continuous casting mold and consequently increase the superficial quality of steel products. A numerical study of the mesoscopic fluid-particle flow in a bifurcated pool-type SEN under steady operating conditions is conducted using the lattice Boltzmann method (LBM) coupled with the large eddy simulation (LES) model. The accuracy of the model has been verified by comparing vortex structures and simulated velocities with published experimental values. The LBM modeling is also verified by comparing the "stair-step" jet patterns observed in the experiment. The geometrical parameters and operational conditions of physical experiments are reproduced in the simulations. By comparing the time-averaged velocities of Reynolds-averaged Navier–Stokes equations (RANS) with LBM models, transient mesoscopic fluid-particles and related vortex structures can be better reproduced within the SEN. The visualization of internal flow within the SEN is illustrated through the mass-less Discrete Phase Model (DPM) model. The trajectories show that the LBM–LES–DPM coupled model is good at predicting the transient vortical flow within the SEN. A large vortex is found inside the exit port and continuously changes in shape and size therein. The monitoring points and lines within the SEN are selected to illustrate the velocity variations and effective viscosity, which can reflect the oscillating characteristics even under stable operating conditions without changes at the exit from the SEN. Furthermore, the formation, development, diffusion, and dissipation of the vortex structures from the exit port of the SEN are also investigated using the *Q* criteria. The comparison of the power spectrum with high-frequency components along the exit port indicates that the flow oscillations must originate from within the SEN and are intensified in the exit port. The mesoscopic LBM model can replicate the fluid-particle flow and vortex structure transmission as well as their turbulence effects inside the SEN in detail.

Keywords: fluid-particle flow; submerged entry nozzle; continuous casting; lattice Boltzmann method; large eddy simulation; vortex structures

Citation: Zhao, P.; Piao, R.; Zou, Z. Mesoscopic Fluid-Particle Flow and Vortex Structural Transmission in a Submerged Entry Nozzle of Continuous Caster. *Materials* **2022**, *15*, 2510. https://doi.org/10.3390/ma15072510

Academic Editors: Qing Liu and Jiangshan Zhang

Received: 31 January 2022
Accepted: 23 March 2022
Published: 29 March 2022

1. Introduction

A common element in flow networks (industrial piping systems) is a T-junction that splits the flow into two nearly symmetric streams. What these systems often have in common is the presence of low-density particles or air bubbles, which get trapped at the junction, accumulate, and ultimately change the flow distribution. The unanticipated trapping mechanism can create malfunctions, and unexpected dangers in industrial environments [1]. The flow behavior presented inside the SEN has similarities to the fluid flow behavior in other engineering fields. One of the most significant challenges in continuous casting is to obtain clean steel. The quality of steel produced by continuous casting depends mainly on the characteristics of the liquid steel flow pattern within the mold. This pattern depends on the flow dynamics of the nozzle that is immersed in liquid steel. The behavior of the transient flow within the SEN has a determinant influence on the flow pattern, which was believed to be associated with the internal and superficial steel quality, such as its effects on asymmetrical defects such as slip cracking, pinholes, and blisters [2–7]. Consequently, these results show the importance of an accurate description of the behavior of the flow. Therefore, a better understanding of the flow structures and development of vortices within the SEN can be beneficial to quality control and improvements of the process such as reducing energy consumption and the cost of high-quality slabs and foundry alloy (silicon aluminum alloys) [8].

Since the effect of SEN flow on mold flow is difficult to detect under real operating conditions, experimental models have been applied to investigate the flow of steel in practice. As a supplement to experimental methods of investigation, the flow within the SEN also has been subject to numerical simulations. To reproduce the flow pattern in an SEN, the behavior of the fluid inside the SEN has been studied using the models based on the Reynolds-averaged Navier–Stokes equations (RANS), such as the k-ε turbulence model [9–12]: RANS models can reproduce time-averaged average flow patterns. The symmetrical flow inside the SEN is obtained when the velocity inlet is aligned perfectly with the SEN bore, however, they do not allow replication of transient dynamic flow behaviors. The observations in steel-making and physical simulations show that the transient behavior inside the SEN ports evolves with time. If gas bubbles or inclusions are introduced into SEN flow, the simulations will become significantly more complicated. Common to these works is the investigation of fundamental aspects that modify the flow pattern inside the mold. For example, Calderon-Ramos et al. [13] studied the effect of both the shape of the exit ports and the angle of inclination of the ports on symmetric jets. Similarly, Zhang et al. [14] studied the influence of the port angle and immersion depth of the SEN exit ports on the liquid steel flow pattern, which can reduce mold-level fluctuations.

Compared with the RANS model that provides time-averaged variables of turbulent flows, the LES model can provide more detailed instantaneous turbulence to capture small vortices better. In the LES model, large-scale eddies are solved directly with the filtered Navier–Stokes equations, while small-scale eddies are modeled as sub-grid scale grids. SEN flow and mold flow have been further investigated through the LES model on regular grids. Similarly, the presence of unsteady flow inside the SEN has also been described in previous work, allowing comparison of their results with physical experiments. For example, Real et al. [15] studied the periodic flow behavior inside the SEN using the LES model. Flow separation occurs in the upper region of SEN, where the area suddenly increases in channels. There are differences in the influence that each of these geometric properties has on the flow pattern of liquid steel inside the nozzle. Liu et al. [16] further developed computational fluid dynamics (CFD) methods by coupling them with LES models to simulate asymmetrical phenomena at both sides of the mold. Gonzalez-Trejo et al. [17] characterized the fluid dynamics of two separate SENs without exit ports, one with a well and the other without a bottom well, through LES with dynamic k-equation filtering. The model reproduced the dynamic nature of the internal flow pattern seen in physical experiments, however, only the bifurcation of the flow of steel in nozzle interior was examined and the effects of the port shape were not considered in their research.

Recently, the mesoscopic LBM model has been used to demonstrate micro-fluidics and complicated turbulent flows: this might be of interest to engineers running continuous casting operations. Pirker et al. [18] predicted the horizontal secondary vortices and bubble aggregation through the embedded lattice Boltzmann sub-region positioned in the bottom SEN region. The author [19,20] predicted the complicated flow pattern and coherent vortices inside a continuous casting mold. These studies have produced significant results, reflecting the potential advantages that predict local complicated flow behaviors. Flow structures play an essential role in complex flow regimes inside the SEN; however, previous work has paid little attention to the exploration of interior asymmetrical flow structure and their turbulent interaction mechanisms using the LBM–LES model inside the SEN.

The present work explores the fluid-particle flow and vortex structure transmission in jets that emerge from the nozzle ports as well as their turbulent effects. The effects of the tundish sliding nozzle and argon bubbles on liquid steel flow or any change in the operational parameters are not included. The bifurcated pool-type SEN and rectangular geometry of the SEN ports should be considered in this simulation. The fluid-particle flow within the SEN can be reproduced using the LBM model. The accuracy of the model is verified by comparing vortex structures and simulated velocities with results reported in the literature. The transmission of the vortex structures along the exit port of the SEN is also explored by the *Q*-criterion. The monitoring points and lines are finally selected to study asymmetrical turbulent flow at different positions of the exit port inside the SEN.

2. Model Formulation

2.1. Lattice Boltzmann Method

Lattice Boltzmann Method is given by [21]:

$$f_i(x+c_i,t+\delta t)-f_i(x,t) = \frac{1}{\tau_{\mathrm{LB}}}(f_i(x,t)-f_i^{\mathrm{eq}}(x,t))+\vec{F}_i \tag{1}$$

where c_i represents the discrete lattice velocities; δ denotes the time step; $f_i(x,t)$ is the discretized distribution functions; τ_{LB} represents the dimensionless relaxation time; $\vec{F}$ is the gravitational force adopted into the LBM model; $f_i^{\mathrm{eq}}(x,t)$ is Maxwell distribution equilibrium function given by:

$$f_i^{\mathrm{eq}}(x,t)=\rho w_i\left[1+\frac{c_i\cdot u}{C_s^2}+\frac{(c_i\cdot u)^2}{2C_s^4}-\frac{u^2}{2C_s^2}\right] \tag{2}$$

where C_s is the sound speed, $C_s=1/\sqrt{3}$; w_i is weight according to the lattice discretization, and i range from 0 to 18, as follows:

$$w_i=\left\{\begin{array}{cc}\frac{1}{3} & i=0\\ \frac{1}{18} & i=1\text{–}6\\ \frac{1}{36} & i=7\text{–}8\end{array}\right\} \tag{3}$$

The model D3Q19 is applied for the LBM model, and the collision operator employs the Bhatnagar-Gross-Krook (BGK) model with a single relaxation time [22].

The macroscopic quantities are calculated by taking the moments of distribution functions:

$$\rho=\sum_i f_i \tag{4}$$

$$\rho u=\sum_i c_i f_i \tag{5}$$

where ρ is the macroscopic density; u is the macroscopic velocity.

2.2. Large Eddy Simulation

The subgrid-scale stresses τ_{ij} is defined by Smagorinsky subgrid model [23]:

$$\tau_{ij}-\frac{1}{3}\delta_{ij}\tau_{kk}=-2v\overline{S}_{ij} \tag{6}$$

where δ_{ij} is the Kronecker delta function, the isotropic part of the Reynolds stress term τ_{kk} is included in the pressure term; the subscripts i and j represent three Cartesian directions.

$\overline{S}_{ij}$ is expressed as the rate of large-scale strain tensor given by

$$\overline{S}_{ij} = \frac{1}{2}\left(\frac{\partial \overline{u}_i}{\partial x_j} + \frac{\partial \overline{u}_j}{\partial x_i}\right) \tag{7}$$

The viscosity v in the Smagorinsky model is defined as:

$$v = v_0 + v_{\text{eddy}} \tag{8}$$

where v_0 is the kinematic viscosity; v_{eddy} is eddy viscosity given by:

$$v_{\text{eddy}} = C\Delta^2|\overline{S}| \tag{9}$$

where Δ is expressed as the minimum size; C is the Smagorinsky constant, which depends on the size of the grid, and the value ranges between 0.1 and 0.2.

Relaxation time τ_{LB} can be expressed as:

$$\tau_{\text{LB}} = 3\left(v_0 + C\Delta^2|\overline{S}|\right) + \frac{1}{2} \tag{10}$$

where $|\overline{S}|$ is the intensity of the local filtered stress tensor given by:

$$|\overline{S}| = \frac{\sqrt{v_0^2 + 18C\Delta^2\sqrt{\overline{\Pi}_{i,j}\overline{\Pi}_{i,j}}} - v_0}{6C\Delta^2} \tag{11}$$

where $\overline{\Pi}_{i,j} = \sum_{\alpha} c_{\alpha i} c_{\alpha j}\left|\overline{f}_i - \overline{f}_i^{\text{eq}}\right|$ is the local non-equilibrium stress tensor.

3. Model Setup and Validation

3.1. Calculation Details

The computational domain includes a bifurcated SEN nozzle with the bottom pool. The global computational domain comprises the incoming duct, the flow deflection region, the two opposing port openings, and part of the emerging confined jets. Table 1 lists the geometrical SEN, simulation conditions, and operating parameters considered in the simulation.The lattice spacing is set to 2 mm. This computational domain is discretized by approximately 500,000 elements in the LBM model. The time step for these simulations is 0.02 s. The resolved scale is 0.002 m. The starting time step for these simulations is 0.001 s. The inlet and outlet boundary conditions in the model are based on the bounce-back condition for the non-equilibrium portion of the distribution. The constant velocity inlet boundary condition is based on the casting speed at the inlet of SEN. The pressure outlet condition is applied in the exit outlet at both sides of the SEN. Figure 1 shows a schematic diagram of the geometry and boundary conditions as well as monitoring points and lines within the SEN considered in the study.

Table 1. Simulation conditions considered in the model.

Parameter	Value
Mold width (m)	1.2
Mold thickness (m)	0.23
Casting speed ($m{\cdot}min^{-1}$)	1.3
SEN submergence depth (mm)	200
SEN port angle (deg)	15°
SEN port shape	Rectangle
SEN port height (mm)	45
SEN port width (mm)	35
Molten steel density ($kg{\cdot}m^{-3}$)	7000
Molten steel viscosity (Pa·s)	0.006

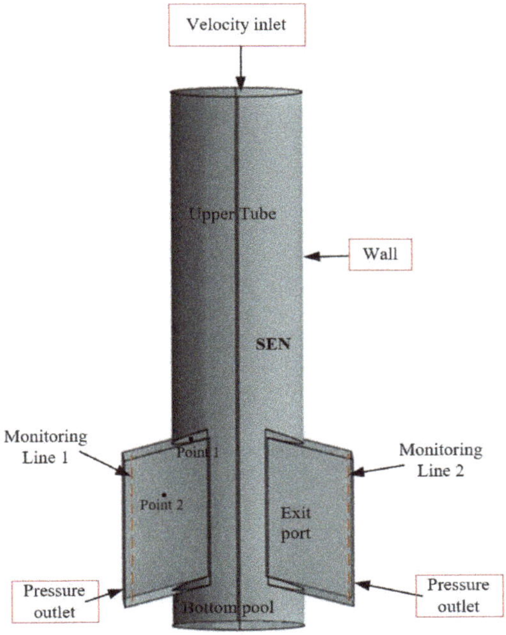

Figure 1. SEN geometry and two monitoring points.

3.2. Validation

The complex flow consists of multiple vortices in the mold. The behaviors of the jet emerging from the exit port of the SEN play an essential role in the flow pattern in the continuous casting mold. The jet vortices induced by the Q-criterion are investigated using the LBM model. The Q-criterion is given by [24]:

$$Q = -\frac{1}{2}\frac{\partial u_i}{\partial x_j}\frac{\partial u_j}{\partial x_i} = \frac{1}{2}(\Omega_{ij}\Omega_{ij} - \Psi_{ij}\Psi_{ij}) > 0 \tag{12}$$

where Ω_{ij} and Ψ_{ij} are the symmetric and anti-symmetric components of ∇u, respectively. Thus, the evolution of structures flow is deduced from the Q-criterion.

The models have been applied to a one-half scale water model for comparison, and the details of geometry and operating conditions are treated as being under the same conditions. Simulated velocity profiles are compared with previous experimental measurements. Figure 2a–c illustrates a comparison of the "stair-step" jet outside the SEN through (a) water experiment, (b) predicted velocity, and (c) vortex structures. The simulation results of the patterns and vortex structures of the jets also agree with the observations from water experiments, which reproduce the "stair-step" jets at two sides of the SEN when black ink is injected into the upper SEN tube. By comparing the simulation and the water experiment, the LBM model shows that it can be used to model the behavior of the oscillating jets within the SEN.

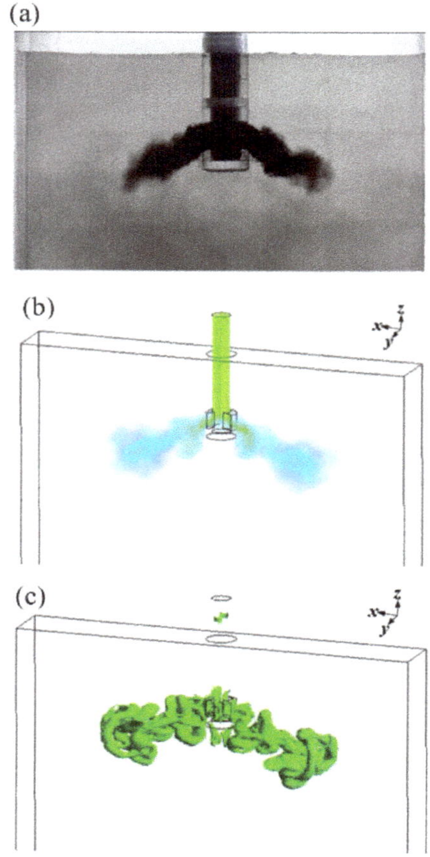

Figure 2. Comparison of the "stair-step" jet inside the mold under the same conditions through (**a**) the water experiment, (**b**) predicted velocity, and (**c**) vortex structures.

To verify the accuracy of the modeling the interior flow inside the SEN, the LBM model also has been applied to the water experiments for a comparison of experimental velocities and simulation results. Simulated velocity profiles are compared with those measured results from the water experiment under the same conditions, and the details of geometry and operating conditions are available in the literature [17]. Figure 3 illustrates a comparison of the turbulent fluid flow (left) through the LBM model with the vortex structures outlined by the fluid velocity vectors for Case A (right) inside the nozzle internal prototype (NIP). The NIP is used to evaluate the vortex distribution. As can be seen from the figures, there is a dominant vortex that occupies a large fraction of the volume of the nozzle. Figure 3a shows the vortex structure (isosurface of v = 1.2 m/s) in this simulation close to the exit port of the SEN. Vortex tube outlined by the velocity vectors is present in the same projection in Figure 3b. This surface outlines the shape of the vortex that can be produced inside the NIP. The comparison shows the vortex structures in this model are consistent with the vortex surface along with the velocity vectors in the previous simulation for Case A with the well pool. The LBM model can be validated to obtain internal flow and structure inside the SEN.

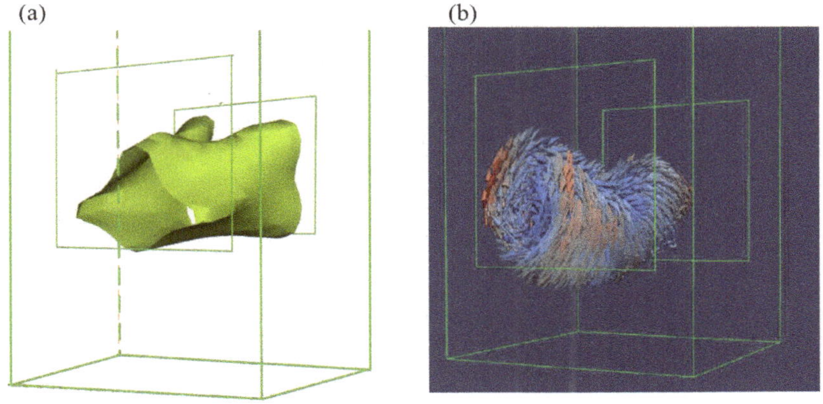

Figure 3. Comparison of (**a**) turbulent fluid flow pattern through the LBM model (left) with (**b**) vortex structure outlined by the fluid velocity vectors for Case A [17] (right) inside the NIP.

To compare the velocity of the flow inside the NIP, the velocity magnitudes along two straight lines from the physical model are available in the literature [17]. Figure 4 illustrates a comparison of experimental results along the path in Case A and simulated time-averaged velocities that measured three times under the same conditions. The measurements are undertaken along vertical lines close to the left side of the exit nozzle: these show the shape of the velocity profiles changes at different heights of the SEN. The velocity fluctuates significantly at the nozzle exit. The predicted velocity increases with the height of the exit port of the nozzle until it reaches a maximum value and then decreases with the height thereof. In the right-hand figure, the agreement between both methods is not as good as the velocity profile across the height of the exit port; however, the result reproduces the velocity profile as a function of SEN height, qualitative agreement is obtained.

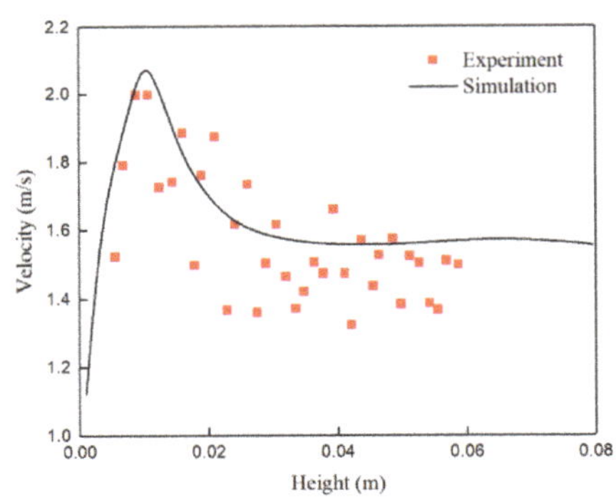

Figure 4. Comparison of experimental results for Case A [17] and predicted velocities prediction at the left-hand line of the SEN.

4. Results and Discussion

4.1. Fluid-Particle Flow

The LBM model for the fluid-particle flow is also conducted in this simulation. To show the transient process of fluid-particle flow in the SEN under the steady-state conditions, Figure 5a–d show the fluid-particle flow patterns at 0.1, 0.2, 0.25, 0.3, and 0.35 s, respectively. The transient process of fluid-particles can be described as follows: transient fluid-particles emerge from the inlet of the SEN bore. The upper SEN pipe is filled with the fluid-particles;

thereafter, the particle-based jet impinges onto the bottom of the SEN pool at a high velocity, and flow towards two sides of the SEN port. Meanwhile, the fluid-particles at the SEN pool interact with those from the upper tube; finally, the fluid-particles almost fill the entire the SEN port, meanwhile, backflow is also found in the upper region of the SEN port, as shown in Figure 5e. The fluid-particles at the SEN port interact with those within the backflow, resulting in more complex transient flow within the SEN. The results show that the mesoscopic LBM model can obtain the transient behavior of fluid-particle flow within the SEN in detail.

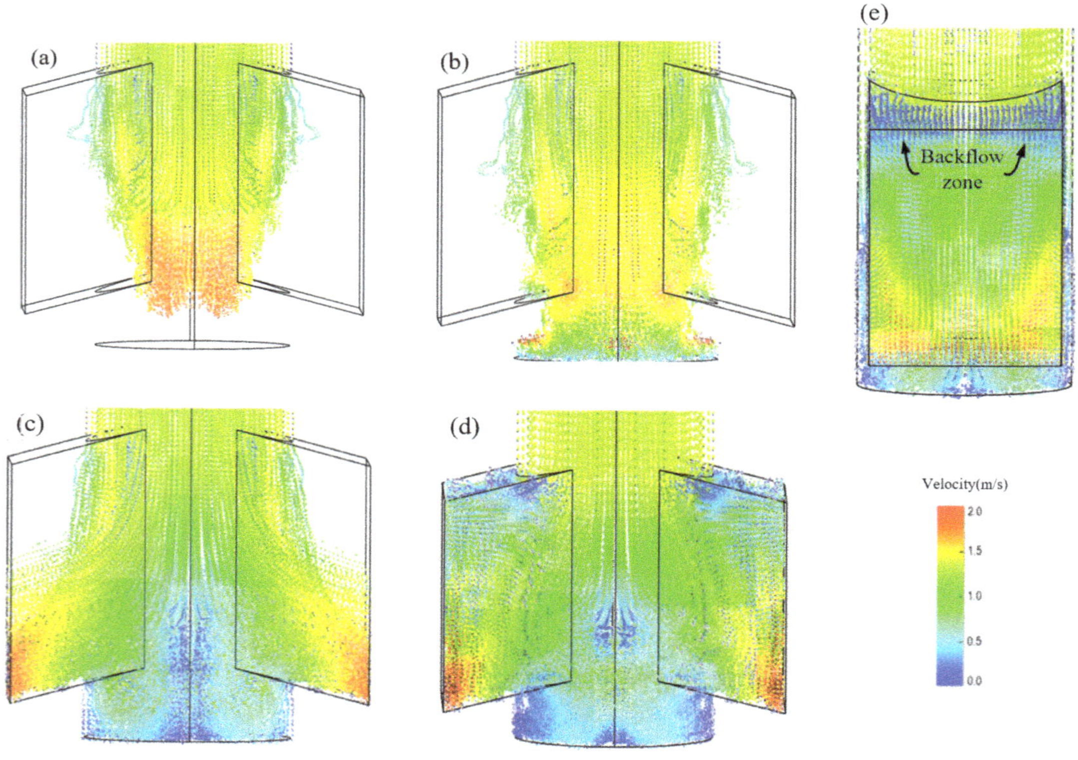

Figure 5. Transient evolution of mesoscopic fluid-particle flow inside the SEN at different times of (**a**) 0.1, (**b**) 0.2, (**c**) 0.25, (**d**) 0.3, and (**e**) 0.35 s.

Passive stream-tracers allow the simulation to track Lagrangian particles (mass-less) along the fluid, which is thus useful to better observe the streamlines lines or flow trajectories. The mass-less Discrete Phase Model (DPM) is used here to describe the vortical flow within the SEN. The visualization of internal flow within the SEN is illustrated in the following process (Figure 6a–d): Mass-less particles start to be separated at the corners between the upper tube and the exit port of the SEN. Then, some particles flow directly towards the exit port, while others move into the lower region until they impinge on the bottom wall of the SEN. These mass-less particles finally flow along the bottom region toward the exit port, accompanied by swirling particles-flowlines close to port of the SEN (see enlarged area Figure 6d). The trajectories show that the LBM–LES–DPM coupled model is good at predicting the transient vortical flow within the SEN because the inclusions are found dispersed at a fairly low volume fraction (usually less than 20%) in the steel through actual industrial experiment.

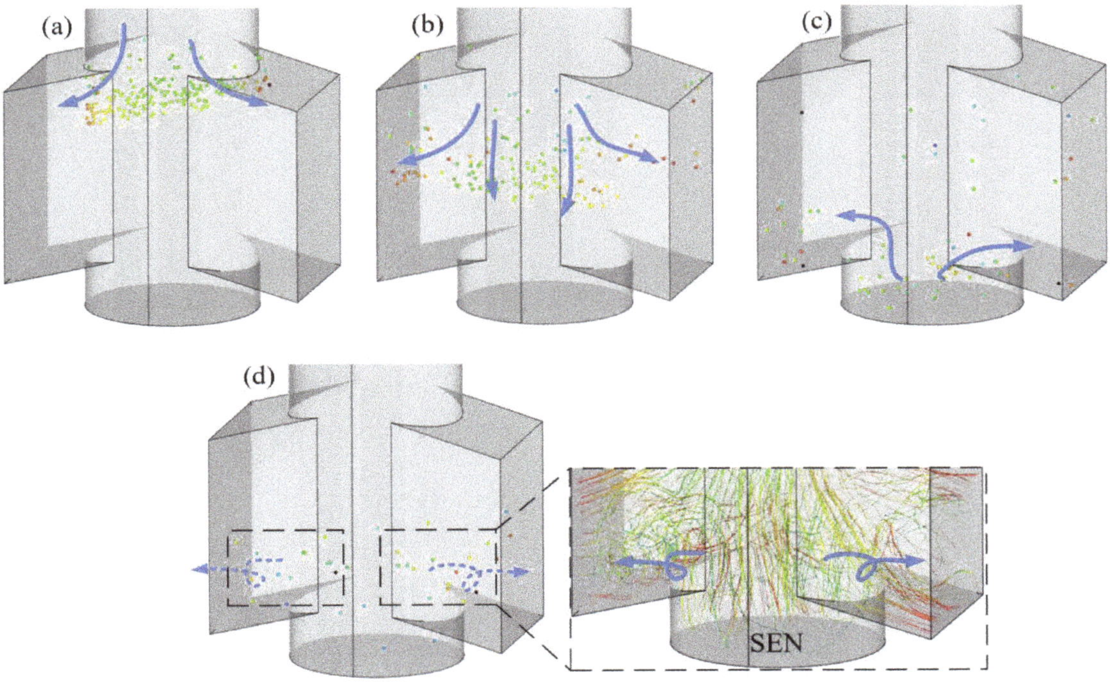

Figure 6. The evolution of passive particles flow and stream-tracers within the SEN at different stages: (**a**) corner separation, (**b**) central diffusion, (**c**) bottom diffusion, and (**d**) vortical flow.

In order to better explain the mechanism of internal vortices structures inside SEN, Figure 7a–c schematically shows the formation of internal flow corresponding coherent vortices inside the SEN. The main processes are as follows: (a) Vortex structures are formed at upper region, meanwhile, they interact with the backflow zone at the upper region of the SEN; (b) vortex structures are concentrated near the concave shape, and they also interact with the incoming flow from the upper tube; (c) vortices structures from the upper region interact with those at the bottom concave, which can cause a more complex transient flow inside the SEN. In short, fluid separation occurs at the upper separation and the lower concave of the SEN. Vortex vortices split into small vortices and interact with each other, leading to the accumulation in these regions and intensifying the oscillating jets from the SEN port thereon.

To quantify the time-averaged velocity of this mesoscopic fluid, Figure 8a illustrates the time-averaged velocity contour on the middle plane (XZ). The simulation results are measured three times with an interval of 10 s. These maps indicate that the flow patterns in both exit ports are identical. The symmetric flows on both sides of the SEN ports are present. Additionally, the flow velocity contour and vectors at the exit port of the SEN are also included in Figure 8b. The corresponding value of X for the plane is -0.03 m. Note that the area has a very low velocity along the upper edges of the exit ports, while fluid flow with high velocity is uniformly distributed at the bottom of the exit ports. A close inspection of the low-velocity vectors directed towards the interior at the upper SEN shows the existence of reverse flow (Figure 8c), which promotes the oscillation of the jet as it emerges from the exit port of the SEN.

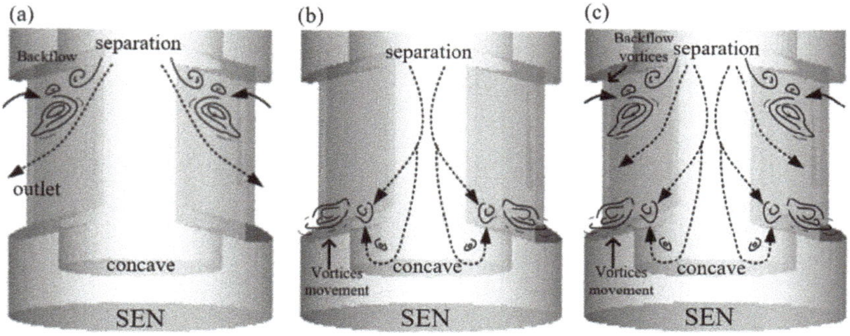

Figure 7. Schematic streamlines of the formation and development of internal flow corresponding coherent vortices inside the SEN at different stages: (**a**) corner separation, (**b**) bottom diffusion, and (**c**) vortical flow.

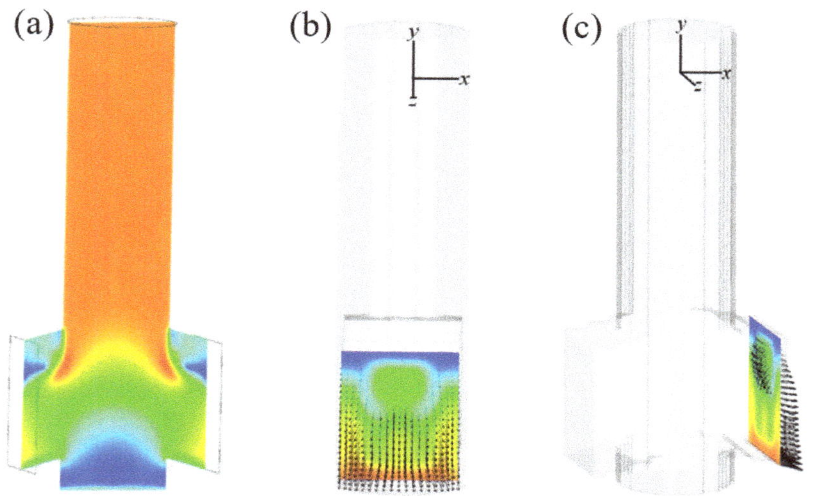

Figure 8. The time-averaged velocity contour and vectors at the exit port of the SEN from different views: (**a**) the middle plane *XZ*, (**b**) the plane *XY*, and (**c**) the exit port.

The better to show the evolution of flow structure at the exit port, Figure 9a–c shows transient vortex contour and trace streams perpendicular to the exit port of the SEN under the same condition. The simulation takes, as an example, the situation where a large vortex is generated inside the SEN with a well pool. The analysis of the simulation led us to the following observations: the big vortex located close to the bottom of the SEN is always present and is cylindrical, however, there is always a bias of the vortex axis with respect to the central plane of the SEN. The rotation axis of the large vortex is not substantially parallel to that of the exit ports. This large vortex oscillates continuously, changing in both shape and size.

To quantify the turbulent flow close to the exit port, the effective viscosities (on the monitoring lines) at both sides of the exit points are investigated. The locations of the two reference lines are illustrated in Figure 1. Figures 10 and 11 show the effective viscosities of the turbulent flow along the vertical monitoring lines close to exit port of the SEN at different times (Figure 1). From the figures, the turbulent viscosity of flow near the wall of the mold always exhibits fluctuation characteristics at both sides of the exit port. Figure 10 demonstrates the asymmetric viscosity at both sides, where the viscosity on the left upper port is higher than that on the right. Figure 11 shows a similar mirror image of the viscosity distribution, where the viscosity on the right upper port is higher than that on the left. It can be inferred that the viscosity change is related to the wobbling jet behavior, reflecting asymmetric changes in flow patterns from the exit port.

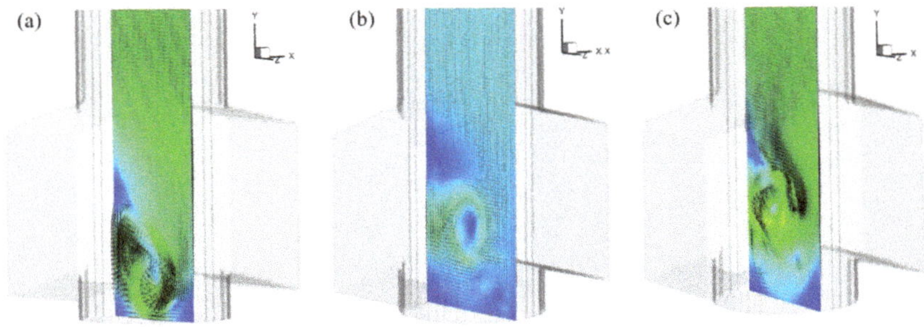

Figure 9. Transient vortex contours and trace streams on the center plane perpendicular to the exit port of the SEN at different times: (**a**) the large vortex appears at the bottom, and (**b**) the vortex gradually flows upward, (**c**) the vortex moves towards the middle output.

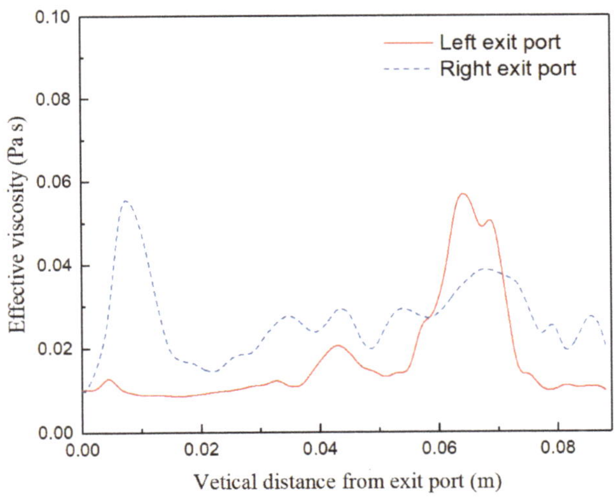

Figure 10. Asymmetric effective viscosities of the turbulent flow along the vertical monitoring lines close to the exit port of the SEN.

Transient fluid-particle flow within the SEN is reproduced, the oscillation of fluid-particle flow is quantitatively analysed in this section. The periodic behavior within the SEN will be characterised mathematically here. The velocity magnitude fluctuations at monitoring Points 1 and 2, located at the upper corner and the middle along the exit port of the SEN (Figure 1), respectively. Figure 12 shows the comparison between the velocity time series at Point 1 using different models. Numerical simulations using the RANS (k-ε turbulent) model are conducted to compare the transient flow within the SEN through LBM model. The solid bold lines correspond to simulations using the LBM model, and dotted lines represent the RANS model. The RANS results illustrate that the velocity fluctuations at the exit port are relatively stable, which are not as severe as those obtained from the

LBM results. The behavior of mesoscopic fluid-particles inside the SEN exhibits periodic changes.

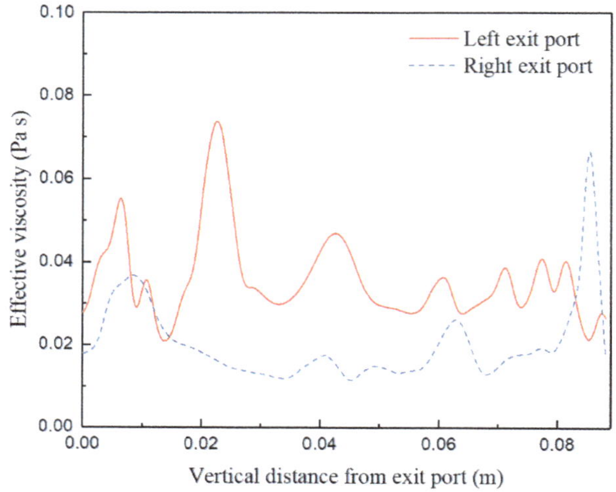

Figure 11. Mirror image of effective viscosities of the turbulent flow along the vertical monitoring lines close to the exit port of the SEN.

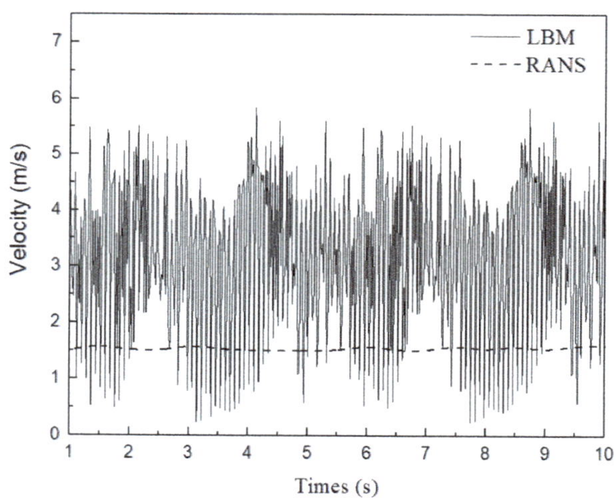

Figure 12. Comparison between velocity time series at Point 1 using the LBM and RANS models.

All numerical simulations presented are performed under the same the conditions. Figure 13 demonstrates the velocity change for Point 2 through the LBM and RANS models. In the simulation of the RANS model, the dynamic change in the SEN is not evident; however, the LBM simulation also demonstrates that the flow behavior in periodic variation predicts a greater velocity oscillation close to the port (Point 2) than one at the corner (Point 1) within the SEN, which promotes the oscillations of the jet thereon. From the above comparison, the results show that the flow oscillations initially occur on the interior bifurcation tip due to flow separation, where the area suddenly changes in the shape of SEN geometry. The oscillations of the jet are intensified when the internal flow develop along the exit port of the SEN.

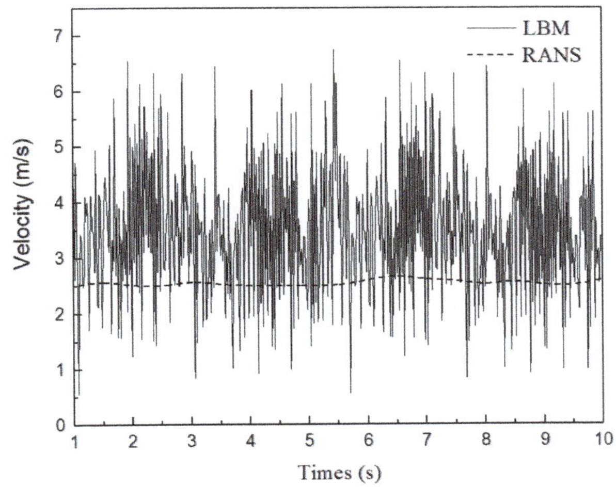

Figure 13. Comparison between velocity time series at Point 2 using LBM and RANS models.

4.2. Flow Structures

The vortex structures play a significant role in the flow pattern on both sides of the SEN. To clarify this, the process of vortices development within the SEN is schematically divided into four stages, as shown in Figure 14a–c:

(I) Formation of the vortices: the vortices emerge at the upper corner of the SEN. The area of a flow suddenly increases in channels, and separate at the boundary layer;
(II) Development of the vortices: the size of the vortices increases gradually. The vortices develop along with the exit port and produce more vortices inside the SEN;
(III) Dissipation of vortices: the vortices interact with each other and disappear inside the SEN. The vortices persist for a relatively short period.

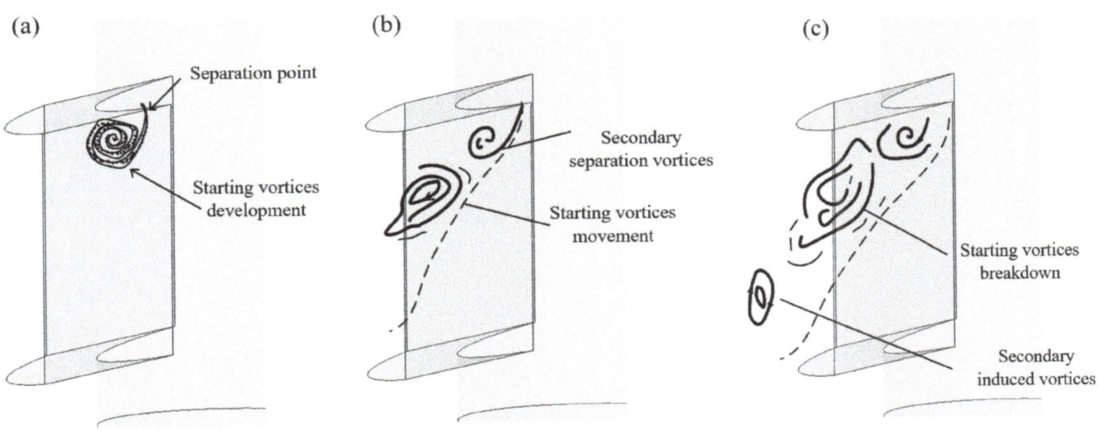

Figure 14. Formation, development, and dissipation of the vortices at the channel of the SEN at different stages: (**a**) formation of the vortices, (**b**) development of the vortices, and (**c**) dissipation of vortices.

Differing from those studies, in this simulation, we propose that the evolution of the vortex structures is induced in terms of satisfaction of the *Q*-criterion. Figures 15–18 show the vortices induced by the *Q*-criterion and corresponding streamlines. To illustrate the evolution of the vortices, Figure 15 illustrates that the vortices with the greatest strength are mainly concentrated at the upper corner of the SEN. The intensities (driving force) of the vortices are relatively large, which are of great importance to the evolution of vortex structures. Next, the corresponding vortices propagate from the upper corner to the exit port of the SEN, as shown in Figure 8. After that, the vortices then spread and deform along with the exit port, as shown in Figures 16 and 17. The small vortices are finally separated from their larger counterparts close to the exit port, increasing the instability of the jet flow from the exit port. In addition, the separated vortices are also found (Figure 18) accompanied by small swirling flows, which are consistent with the phenomenon of the "stepped" jets that are observed in a water experiment.

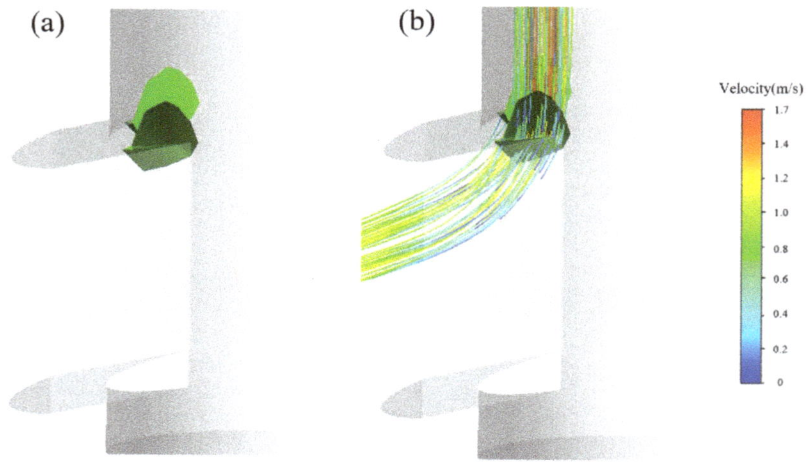

Figure 15. Formation of vortex structures (**a**) at the interior corner of the SEN port and (**b**) corresponding streamlines.

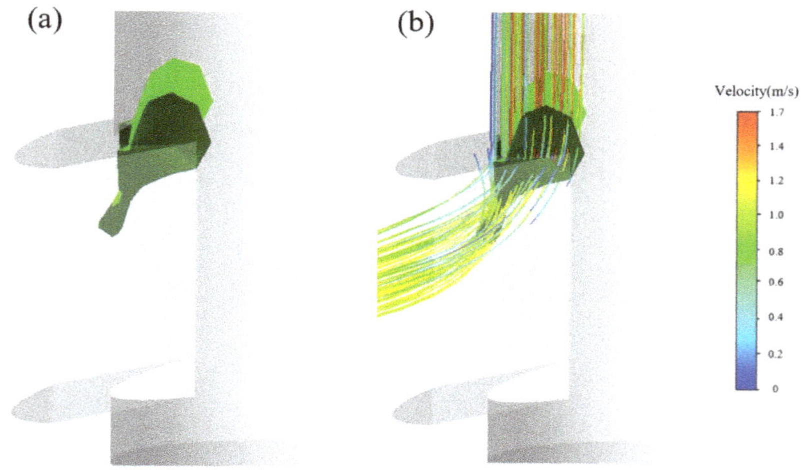

Figure 16. Development of vortex structures (**a**) at the upper edge of the SEN port (**b**) corresponding streamlines.

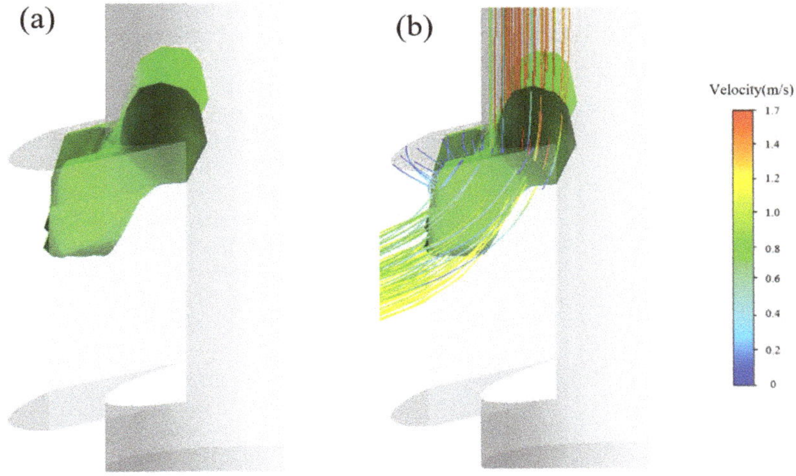

Figure 17. Diffusion of vortices (**a**) along with the channel of SEN port and (**b**) corresponding streamlines.

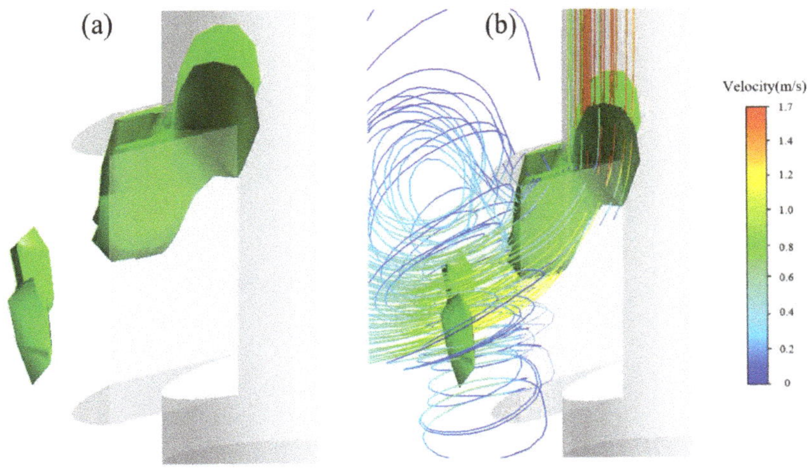

Figure 18. Separation of vortex structures (**a**) nearby the SEN port and (**b**) corresponding streamlines as well as accompanying swirling flow.

To investigate the oscillating flow, the velocity time series obtained from the LBM model is processed into a power spectrum to estimate the oscillation period of the circulation. These samples are then processed into a power spectrum by using a fast Fourier transform (FFT) routine as defined by [25]:

$$P(f_k) = \frac{1}{N^2}|f_k|^2\ k = 1,\ 2, \ldots,\ N/2 \quad (13)$$

$$U_k = \sum^{N} ue^{2\pi ijk/N}\ k = 1,\ 2, \ldots,\ N \quad (14)$$

$$f_k = \frac{k}{N\Delta t_N}\ k = 1,\ 2, \ldots,\ N/2 \quad (15)$$

where u is the velocity in the time series, t_N denotes the times at the start and the end of the sampling period; N represents the number of sampling points. Thus, a frequency analysis of the velocity magnitude time series is conducted. In order to study the oscillating flow along the exit port of the SEN, frequencies of velocity fluctuation are analyzed at monitoring points P1, and P2 (Figure 1). Figures 19 and 20 show the power spectrum of velocity magnitude with the SEN port at the corner (Point 1) and the exit port (Point 2), respectively. The dominant peak represents the oscillation frequency. There are many frequencies in the flow at the SEN exit port. The oscillations are not a simple combination of modes inside the SEN. The maximum amplitude is 0.03 at frequency 1 Hz at Point 1, however, this exceeds 0.07 at Point 2. It can be concluded that the flow oscillations have been intensified close to the exit port of the SEN.

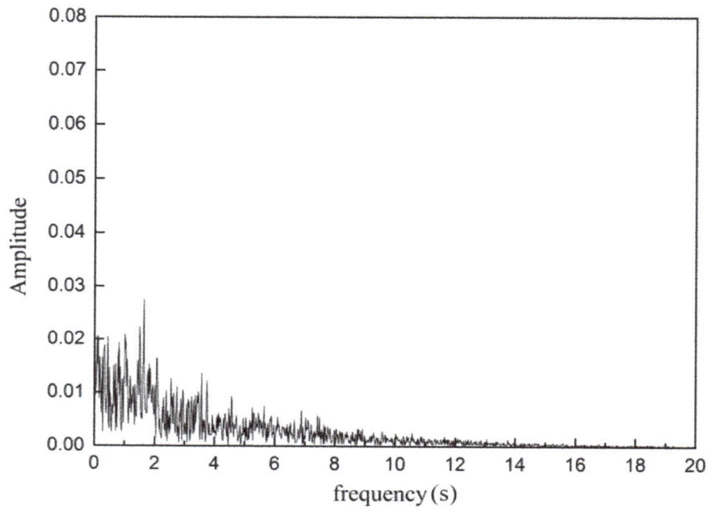

Figure 19. Power spectrum of velocity magnitude inside the exit point at Point 1.

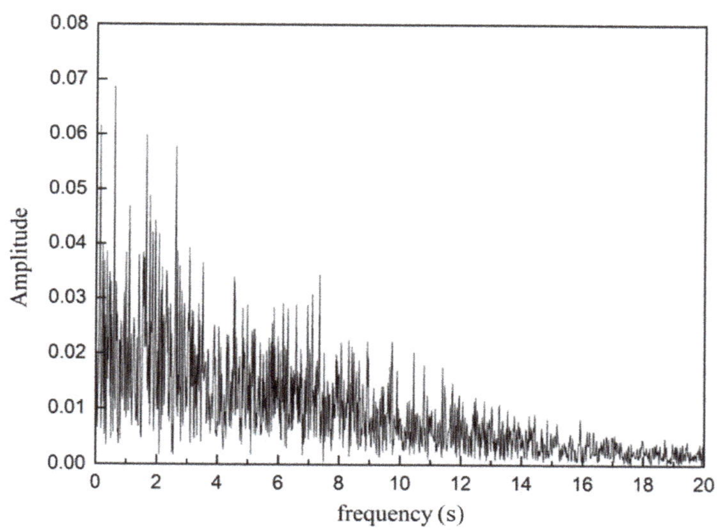

Figure 20. Power spectrum of velocity magnitude inside the exit point at Point 2.

5. Conclusions

The fluid-particle flow inside a bifurcated SEN is studied to evaluate flow oscillations through the LBM model. From the results, the following conclusions are drawn:

(1) The accuracy of the model has been verified by comparing vortex structures and simulated velocities with published experimental values inside and outside the SEN. Comparing simulations with RANS and LBM models, the results show that the mesoscopic LBM can predict the transient fluid-particle flow along the exit port of the SEN in detail, which can better reproduce the behavior of the oscillating jet within the SEN.

(2) The transient flow behavior results from the dynamic interaction of fluid-particle flow within the SEN. The velocities always exhibit strong high-frequency components. The periodic changes of mesoscopic fluid-particles flow within the SEN can be reproduced even under stable operating conditions. The visualization of internal flow within the SEN show that the LBM–LES–DPM coupled model is good at predicting the transient vortical flow within the SEN.

(3) Vortex structures are of great importance to the development of turbulent flow along the exit port of the SEN. The turbulent viscosity of flow near exit output of the SEN always fluctuates at both sides of the exit port. The asymmetric viscosities at both sides become similarly mirror-like at different times. The formation, development, and dissipation of detached vortices induced by the *Q*-criterion that identifies vortex structures can cause oscillation of jet within the SEN.

(4) There are many frequencies (between 1 Hz and 3 Hz) seen in flow oscillations along the channel of the SEN port. The oscillations are not a simple combination of modes and flow oscillations have been intensified simultaneously. The jet oscillations must originate from within the SEN interior and are intensified along the exit port.

This work represents our further efforts towards fluid-particles flow, vortices structures and their effects on turbulence inside bifurcated SEN flow. Further studies are needed to study the behaviour of introduced particles or bubbles under the framework of LBM model, the interaction between vortex structure and bubbles as well as their interactions (aggregation and breakdown).

Author Contributions: Software, P.Z.; validation, P.Z.; investigation, Z.Z.; writing—original draft preparation, P.Z.; writing—review and editing, R.P.; funding acquisition, R.P. All authors have read and agreed to the published version of the manuscript.

Funding: This research was funded by the National Natural Science Foundation of China (No. 51174122), and Anhui University of science and technology introduction of talent research start-up fund project (No. 13200456).

Institutional Review Board Statement: Not applicable.

Informed Consent Statement: Not applicable.

Data Availability Statement: Not applicable.

Acknowledgments: National Engineering Laboratory for Vacuum Metallurgy, Kunming University of Science and Technology for the supports.

Conflicts of Interest: The authors declare no conflict of interest.

References

1. Vigolo, D.; Radl, S.; Stone, H.A. Unexpected trapping of particles at a T junction. *Proc. Natl. Acad. Sci. USA* **2014**, *111*, 4770–4775. [CrossRef] [PubMed]
2. Yang, J.P.; Liu, Q.; Guo, W.D.; Zhang, J.G. Quantitative evaluation of multi-process collaborative operation in steelmaking-continuous casting sections. *Int. J. Miner. Metall. Mater.* **2021**, *28*, 1353–1366. [CrossRef]
3. Wu, Y.; Liu, Z.; Wang, F.; Li, B.; Gan, Y. Experimental investigation of trajectories, velocities and size distributions of bubbles in a continuous-casting mold. *Powder Technol.* **2021**, *387*, 325–335. [CrossRef]

4. Liu, G.; Lu, H.; Li, B.; Ji, C.; Zhang, J.; Liu, Q.; Lei, Z. Influence of M-EMS on Fluid Flow and Initial Solidification in Slab Continuous Casting. *Materials* **2021**, *14*, 3681. [CrossRef]
5. Hibbeler, L.C.; Thomas, B.G. Mold slag entrainment mechanisms in continuous casting molds. *Iron Steel Technol.* **2013**, *10*, 121–136.
6. Li, X.; Li, B.; Liu, Z.; Niu, R.; Liu, Y.; Zhao, C.; Yuan, T. Large eddy simulation of multi-phase flow and slag entrapment in a continuous casting mold. *Metals* **2019**, *9*, 7. [CrossRef]
7. Zhao, P.; Li, Q.; Kuang, S.B.; Zou, Z. Mathematical modeling of liquid slag layer fluctuation and slag droplets entrainment in a continuous casting mold based on VOF-LES method. *High Temp. Mater. Process.* **2017**, *36*, 551–565. [CrossRef]
8. Kuz'min, M.P.; Chu, P.K.; Qasim, A.M.; Larionov, L.M.; Kuz'mina, M.Y.; Kuz'min, P.B. Obtaining of AleSi foundry alloys using amorphous microsilica e crystalline silicon production waste. *J. Aalloys Compd.* **2019**, *806*, 806–813. [CrossRef]
9. Thomas, B.G.; Yuan, Q.; Mahmood, S.; Liu, R.; Chaudhary, R. Transport and entrapment of particles in steel continuous casting *Metall. Mater. Trans. B* **2014**, *45*, 22–35. [CrossRef]
10. Dash, S.K.; Mondal, S.S.; Ajmani, S.K. Mathematical simulation of surface wave created in a mold due to submerged entry nozzle *Int. J. Numer. Methods Heat Fluid Flow* **2004**, *14*, 606–632. [CrossRef]
11. Wondrak, T.; Eckert, S.; Gerbeth, G.; Klotsche, K.; Stefani, F.; Timmel, K.; Yin, W. Combined electromagnetic tomography for determining two-phase flow characteristics in the submerged entry nozzle and in the mold of a continuous casting model. *Metall. Mater. Trans. B* **2011**, *42*, 1201–1210. [CrossRef]
12. Torres-Alonso, E.; Morales, R.D.; Garcia-Hernandez, S. Cyclic turbulent instabilities in a thin slab mold. Part II: Mathematical model. *Metall. Mater. Trans. B* **2010**, *41*, 675–690. [CrossRef]
13. Calderón-Ramos, I.; Morales, R.D.; Servín-Castañeda, R.; Pérez-Alvarado, A.; García-Hernández, S.; de Jesús Barreto, J.; Arreola-Villa, S.A. Modeling study of turbulent flow in a continuous casting slab mold comparing three ports SEN designs. *ISIJ Int.* **2019**, *59*, 76–85. [CrossRef]
14. Zhang, H.; Fang, Q.; Xiao, T.; Ni, H.; Liu, C. Optimization of the flow in a slab mold with argon blowing by divergent bifurcated SEN. *ISIJ Int.* **2019**, *59*, 86–92. [CrossRef]
15. Real, C.; Miranda, R.; Vilchis, C.; Barron, M.; Hoyos, L.; Gonzalez, J. Transient internal flow characterization of a bifurcated submerged entry nozzle. *ISIJ Int.* **2006**, *46*, 1183–1191. [CrossRef]
16. Liu, Z.Q.; Li, B.K.; Jiang, M.F. Asymmetric flow and bubble transport inside a slab continuous-casting mold. *Metall. Mater. Trans. B* **2014**, *45*, 675–697. [CrossRef]
17. Gonzalez-Trejo, J.; Real-Ramirez, C.A.; Carvajal-Mariscal, I.; Sanchez-Silva, F.; Cervantes-De-La-Torre, F.; Miranda-Tello, R.; Gabbasov, R. Hydrodynamic analysis of the flow inside the submerged entry nozzle. *Math. Probl. Eng.* **2020**, *2020*, 6267472. [CrossRef]
18. Pirker, S.; Kahrimanovic, D.; Schneiderbauer, S. Secondary vortex formation in bifurcated submerged entry nozzles: Numerical simulation of gas bubble entrapment. *Metall. Mater. Trans. B* **2015**, *46*, 953–960. [CrossRef]
19. Zhao, P.; Li, Q.; Kuang, S.B.; Zou, Z. LBM-LES simulation of the transient asymmetric flow and free surface fluctuations under steady operating conditions of slab continuous casting process. *Metall. Mater. Trans. B* **2017**, *48*, 456–470. [CrossRef]
20. Zhao, P.; Yang, B.; Piao, R. Lattice Boltzmann Method Modeling of Flow Structures and Level Fluctuations in a Continuous Casting Process. *ACS Omega* **2019**, *4*, 13131–13142. [CrossRef]
21. He, X.; Luo, L.S. Theory of the lattice Boltzmann method: From the Boltzmann equation to the lattice Boltzmann equation. *Phys. Rev. E* **1997**, *56*, 6811–6817. [CrossRef]
22. Qian, Y.; D'Humières, D.; Lallemand, P. Lattice BGK models for Navier-Stokes equation. *EPL Eur. Lett.* **1992**, *17*, 479. [CrossRef]
23. Smagorinsky, J. General circulation experiments with the primitive equations: I. The basic experiment. *Weather Rev.* **1963**, *91*, 99–164. [CrossRef]
24. Hunt, J.C.; Wray, A.A.; Moin, P. Eddies, stream, and convergence zones in turbulent flows. In *Studying Turbulence Using Numerical Simulation Databases—II*; Center for Turbulence Research: Stanford, CA, USA, 1988; pp. 193–208.
25. Lawson, N.J.; Davidson, M.R. Oscillatory flow in a physical model of a thin slab casting mold with a bifurcated submerged entry nozzle. *J. Fluids Eng.* **2002**, *124*, 535–543. [CrossRef]

materials

MDPI

Article

Influence of M-EMS on Fluid Flow and Initial Solidification in Slab Continuous Casting

Guoliang Liu [1,2], Haibiao Lu [3], Bin Li [3], Chenxi Ji [2], Jiangshan Zhang [1], Qing Liu [1,*] and Zuosheng Lei [3,*]

1 State Key Laboratory of Advanced Metallurgy, University of Science and Technology Beijing, Beijing 100083, China; liuguoliang1983@163.com (G.L.); zjsustb@163.com (J.Z.)
2 Shougang Group Co., Ltd., Research Institute of Technology, Beijing 100043, China; jicx5617@shougang.com.cn
3 State Key Laboratory of Advanced Special Steel & Shanghai Key Laboratory of Advanced Ferrometallurgy, Shanghai University, Shanghai 200444, China; luhaibiao@shu.edu.cn (H.L.); gumufeng1023@i.shu.edu.cn (B.L.)
* Correspondence: qliu@ustb.edu.cn (Q.L.); lei_zsh@staff.shu.edu.cn (Z.L.)

Abstract: A mathematical model coupled with electromagnetic field has been developed to simulate the transient turbulence flow and initial solidification in a slab continuous casting mold under different electromagnetic stirring (EMS) currents and casting speeds. Through comparing the magnetic flux density, flow field with measured results, the reliability of the mathematical model is proved. The uniform index of solidified shell thickness has been introduced to judge the uniformity of the solidified shell. The results show that a horizonal recirculation flow has been generated when EMS is applied, and either accelerated or decelerated regions of flow field are formed in the liquid pool. Large EMS current and low casting speed may cause the plug flow near the mold narrow face and a suitable EMS current can benefit to the uniform growth of solidified shell. Meanwhile, an industrial test exhibits that EMS can weaken the level fluctuation and number density of inclusion. Overall, a rational EMS current range is gained, when the casting speed is 1.2 m/min, the rational EMS current is 500–600 A.

Keywords: continuous casting; electromagnetic stirring; flow behavior; uniform index of solidified shell

Citation: Liu, G.; Lu, H.; Li, B.; Ji, C.; Zhang, J.; Liu, Q.; Lei, Z. Influence of M-EMS on Fluid Flow and Initial Solidification in Slab Continuous Casting. *Materials* **2021**, *14*, 3681. https://doi.org/10.3390/ma14133681

Academic Editor: Francesco Iacoviello

Received: 3 June 2021
Accepted: 26 June 2021
Published: 1 July 2021

1. Introduction

Continuous casting has been widely developed as the most important production process in the steelmaking industry. During this process, molten steel flows into the mold through a submerged entry nozzle (SEN), solidifies against the water-cooled copper mold walls in the presence of many complex metallurgical phenomena, including multiphase flow, heat transfer, solidification and solute transport, and it finally forms a solidified shell [1]. Flow of molten steel in the mold and solidification in this process is of great importance because it is responsible for many surface and internal defects [2], such as hooks and longitudinal cracks. Therefore, it is important to control the flow and solidification within an acceptable process to avoid defects.

To enhance or even control the liquid steel flow, many techniques have been developed and used, including argon blowing [3], electromagnetic brake [4,5] (EMBr), electromagnetic stirring [6–12] (EMS), etc. Together with these techniques, EMS has been shown to be one of the most effective countermeasures to improve the molten steel flow and initial solidification in the mold and most works focused mainly on the rotary stirring in billet or round bloom.

Li et al. [9] developed a mathematical model coupling the electromagnetic and flow fields; it was developed to investigate the influence of the SEN clogging rate on the flow field and the influence of electromagnetic stirring (EMS) on the asymmetric mold flow.

Fang et al. [13] and Ren et al. [14] simulated the flow, temperature and solidification field in billet mold, and the results showed that EMS can reduce the impinging effect of jet flow and eliminate molten steel superheat. Wang et al. [15] developed a 3-D coupled model considering electromagnetic field, flow field, heat transfer and level fluctuation under different EMS positions in continuous billet mold, and found that the velocity and wave height at steel/slag interface decreases under lower stirrer position. Maurya et al. [16–18] analyzed the influence of EMS on flow and solidified shell under different EMS currents and frequencies. However, little research considered the influence of EMS on the flow and initial solidification on slab.

In a series of works from Fujisaki [7,19,20] and his coworkers, a 3-D magnetohydrodynamic model was developed to evaluate the flow, heat transfer as well as free surface in mold with EMS separately, and the results showed that EMS makes the solidified shell uniform and the dynamic deviation of temperature stable. Li et al. [21] investigated the influence of EMS position on flow field, possibility of slag entrapment and inclusion removal in slab mold, the flow pattern in the mold changes greatly under different EMS positions and the low stirrer position is favorable to inclusion removal, but the solidified behavior of shells are not considered in his study. In addition, most of them mainly considered the influence of EMS current on the flow and heat transfer in the mold, while the influence of EMS under other process parameters, such as casting speed, SEN depth, etc. has not yet been investigated.

In this paper, the fluid flow and solidification in a continuous casting slab mold under EMS were numerically simulated by a multi-physics model. By comparison with the magnetic flux density data and flow field, the reliability of the mathematical model is proved. The uniform index of solidified shell thickness has been introduced to judge the uniformity or solidified shell. Afterwards, the metallurgical behaviors under different EMS currents and casting speeds were investigated and compared to obtain optimal process parameters with a relatively reasonable flow pattern and solidified shell. According to the simulation results, the industry test was carried out to estimate the influence of EMS on level fluctuation and number density of inclusion.

2. Mathematical Modeling

2.1. Assumption

In order to simplify the numerical simulation, the present work includes the following assumptions and simplifications:

(1) The influence of flow field on the electromagnetic field is ignored due to the small magnetic Reynolds number [22], and the electromagnetic field is assumed to be quasi-static.
(2) The influence of Joule heat generated by currents is ignored in simulation of heat transfer and solidification due to its low frequency.
(3) The liquid steel and the liquid slag behave as incompressible Newtonian fluids.
(4) The effects of mold oscillation and mold curvature are not taken into account [23].

2.2. Governing Equation

A transient three-dimensional mathematical model for slab continuous casting has been developed which was coupled with electromagnetic field, flow and heat transfer. The corresponding governing equations are written as follows:

2.2.1. Electromagnetic Model

Electromagnetic field was determined by solving Maxwell's equations:

$$\nabla \cdot \vec{B} = 0 \tag{1}$$

$$\nabla \times \vec{H} = \vec{J} \tag{2}$$

$$\nabla \times \vec{E} = -\frac{\partial \vec{B}}{\partial t} \tag{3}$$

$$\vec{J} = \sigma\left(\vec{E}\right) \tag{4}$$

where $\vec{B}$ is the magnetic flux density, σ is the electric conductivity, $\vec{E}$ is the electric field strength, V/m; $\vec{J}$ is the induced current density and $\vec{H}$ is the magnetic field strength.

The time-averaged electromagnetic force can be calculated by:

$$\vec{F} = \frac{1}{2}\mathrm{Re}(\vec{J} \times \vec{B}^{*}) \tag{5}$$

where $\vec{F}$ is the time-average electromagnetic volume force, $\vec{B}^{*}$ is the complex conjugate of $\vec{B}$ and Re denotes the real part of the complex quantity.

2.2.2. Fluid Flow and Solidification Model

Continuity equation:

$$\nabla \cdot (\rho \vec{u}) = 0 \tag{6}$$

Momentum equation (N-S):

$$\frac{\partial(\rho \vec{u})}{\partial t} + \nabla \cdot (\rho \vec{u} \vec{u}) = -\nabla p + \nabla \cdot (\mu_{eff} \nabla \vec{u}) + \rho g + S_m + \vec{F} \tag{7}$$

where $\vec{u}$ is the fluid velocity, p is the pressure, ρ is the fluid density and μ_{eff} is the effective viscosity, the standard $\kappa - \varepsilon$ turbulent model is applied to calculate the effective viscosity, μ_{eff}.

Energy equation,

$$\frac{\partial}{\partial t}(\rho H) + \nabla \cdot (\rho \vec{u} H) = \nabla \cdot \left[(\lambda + C_p \frac{\mu_t}{\mathrm{Pr}_t}) \nabla T\right] \tag{8}$$

where,

$$H = h_{ref} + \int_{T_{ref}}^{T} c_p dT + f_l L_s \tag{9}$$

where λ is the thermal conductivity of the fluid, C_p is the specific heat capacity of the fluid, Pr_t is the turbulence Prandtl number (0.85), H is enthalpy, h_{ref} is the reference enthalpy relative to temperature T_{ref}, f_l is liquid fraction at mushy zone and L_s is latent heat of solidification.

The enthalpy-porous model is used to simulate the solidification of steel in a continuous slab casting mold, the liquid-solid mushy zone is treated as a porous zone. The sink (S_m) is added to the momentum equations as a source term:

$$S_m = \frac{(1 - f_l)^2}{(f_l^3 + x)} A_{mush}(\vec{u} - u_c) \tag{10}$$

where, A_{mush} is a mushy zone constant, 10^8 [24] and u_c is the casting speed.

2.3. Geometry Model and Boundary Conditions

2.3.1. Geometry Model

A three-dimensional mathematical model was created that consists of submerged entry nozzle (SEN), Liu et al. [25] as the authors studied the chamfered slab mold in studied metals, a chamfered slab mold with the section of 1600 mm × 230 mm, length of 800 mm; considering the influence of lower recirculation flow on the flow field of molten steel in the mold, we extended the mold model to 3000 mm. Figure 1a shows the geometry model of an electromagnetic simulation, a pair of traveling-wave electromagnetic stirrers is designed on both sides of the mold wide face. The schematic of fluid simulation is presented in Figure 1b, and the other process parameters for numerical simulation are given in Table 1

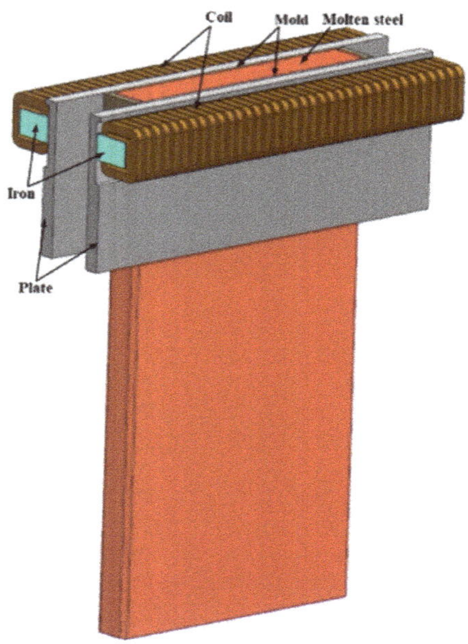

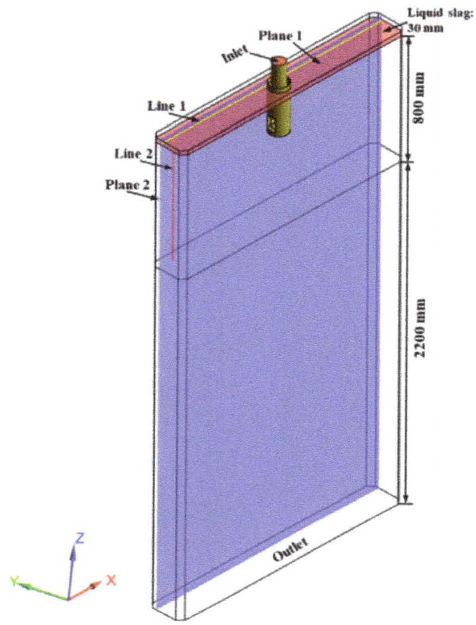

Figure 1. Schematics of the calculation model: (**a**) model used for electromagnetic simulation (surrounding air cuboid is not shown), (**b**) model used for fluid simulation.

Table 1. Process parameters used in mathematical simulation.

Parameters	Value	Parameters	Value
Section size of mold (mm^2)	1600 × 230	Casting speed (m/min)	0.8, 1.0, 1.2, 1.4
Size of SEN port (mm^2)	80 × 60	Coil number of each stirrer	36
Outer diameter of SEN (mm)	120	Turn number of each coil	20
Inner diameter of SEN (mm)	80	Stirrer center from meniscus (mm)	75
Inclination angle (°)	15	EMS frequency (Hz)	4
Submergence entry depth (mm)	170	EMS current (A)	400, 500, 600, 700

2.3.2. Boundary Conditions for Electromagnetic Simulation

The whole geometry model of the electromagnetic field was taken to be surrounded by an air cuboid (2.95 m × 1.2 m × 3.6 m) in which most of the magnetic flux lines are closed. Boundary conditions are applied on the external surface of this cuboid with magnetically flux parallel boundary [14].

2.3.3. Boundary Conditions for Fluid Field Simulation

(1) The inlet velocity of the SEN was calculated based on the mass conservation, and turbulent kinetic energy and the energy dissipation rate are estimated by the semi-empirical equations [26]. The casting temperature is set as 1827 K.

(2) The outlet boundary at the bottom of the calculation domain is a fully developed outflow condition.
(3) The mold wall is treated with the no-slip boundary condition and the heat flux on the wide and narrow faces is a function of distance toward the mold bottom, as shown in Equation (11), which is similar to the form proposed by Savage [27]. The convective heat transfer boundary condition is imposed on the extended region of the continuous caster, and the average heat transfer coefficient for wide and narrow faces is 320 W/(m^2·K) and 360 W/(m^2·K), respectively.

$$q = 2.68 - \varphi\sqrt{\frac{60L}{u_c}} \tag{11}$$

where q is the heat flux density, φ represents constants for wide and narrow faces calculated according to the imported and exported temperature difference and the cooling water flow rate in the mold, which is 0.288 and 0.292, respectively, and L is the distance from the meniscus, m.
(4) The top surface is treated as a free-slipped boundary condition and considering the heat insulation of mold flux, adiabatic condition is applied to it.

2.4. Numerical Solution Procedure

In this study, the numerical simulation is divided into two sections: first, the magnetic field and Lorentz force are computed using the commercial software, ANSYS EMAG, by solving the Maxwell equations, and then the time-averaged electromagnetic force is interpolated into the momentum equation as a source term of ANSYS FLUENT. The calculation domain of fluid field model is divided into 2 million finite volumes and grid-independent verification was performed; the structured mesh is used to simulate the fluid flow and heat transfer in the slab mold. The pressure-implicit with splitting of operators (PISO) algorithm was used for the pressure-velocity coupling and first-order upwind is used for the discretization of momentum and energy equations and the time step size is 0.004 s. The physical parameters for numerical simulation are given in Table 2.

Table 2. Physical parameters used in mathematical simulation [28,29].

Molten steel Parameters	Value
Density (kg/m^3)	7000
Specific heat (J/(kg·K))	720
Thermal conductivity (W/(m·K))	31
Viscosity (Pa·s)	0.0065
Latent heat (J/kg)	275,000
Solidus temperature (K)	1802
Liquidus temperature (K)	1812

3. Results and Discussion

3.1. Validation of Electromagnetic Field and Flow Field

Figure 2 displays the distribution of measured and numerical magnetic flux density at 15 mm from the mold's fixed side of the stirrer mid-plane with operating condition of 700 A/4 Hz, where magnetic flux density data was measured using a CT-3 Teslameter within an empty mold. It can be observed that the numerical results matched well with the measured data; both the predicted data and distribution tendency are similar, the biggest numerical magnetic flux density is about 98 mT, which was located near the left side of the mold.

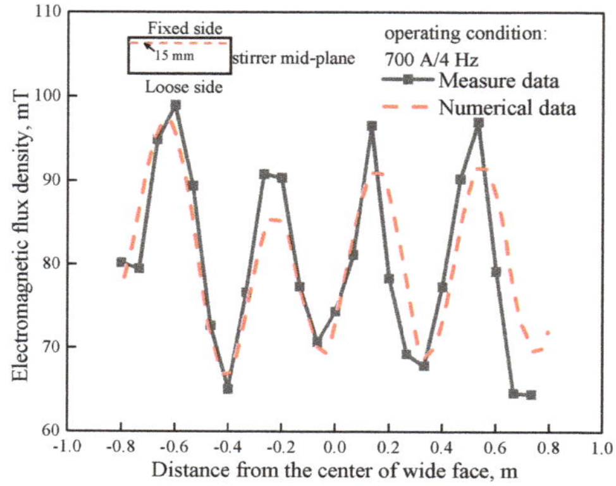

Figure 2. Distribution of measured and numerical magnetic flux density at stirrer mid-plane (Z = −0.075 m).

Figure 3a depicts the distribution of time-averaged electromagnetic force with the operating condition of 700 A/4 Hz, and it can be seen that the distribution of time-averaged electromagnetic force is centrosymmetric, which resulted from the similar distribution of magnetic flux density [9]. The tangential components of electromagnetic force are in the same direction at each edge, but their directions are opposite to each other, and thus, produce a horizontal recirculation. Four transverse swirls of electromagnetic force exist in the interior of the cross-section, which correspond to the pole number acting on the molten steel. Figure 3b illustrates the variation of electromagnetic force at 15 mm from the mold's fixed side of the stirrer mid-plane under different EMS currents. It can be seen that electromagnetic force increases with the EMS current, as the EMS current increases from 400 to 800 A, the maximum electromagnetic force is 2133 N/m^3 and 7950 N/m^3.

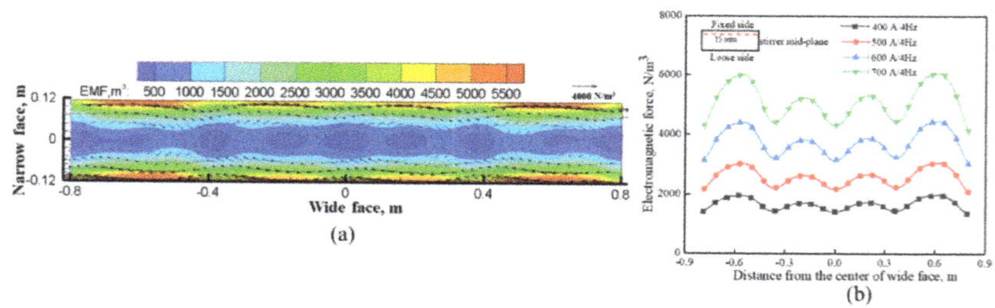

Figure 3. Distribution of (**a**) time-average electromagnetic force at the center of stirrer and (**b**) electromagnetic force under different EMS currents.

In order to validate the mathematical model, a 1/5th scaled physical model using mercury when EMS is not applied was established; the fluid velocity in the mold was measured by means of ultrasound Doppler velocimetry (UDV), and Li et al. [30] as the authors published this apparatus' description in *ISIJ International*. Meanwhile, a mathematical model was set corresponding to the physical model both on the geometry and material properties, etc.

Figure 4a exhibits the flow pattern obtained from the physical model and Figure 4b exhibits the flow pattern obtained from the mathematical model at a quarter longitudinal plane near the fixed side of the mold. When EMS is not applied, a typical roll-flow pattern is observed, both the measured flow pattern and numerical flow pattern are similar not only in flow structure but also in characteristic points, for example, the impinging point at mold narrow face. It can be seen that the numerical results matched well with the experimental results. Overall, by comparison of the magnetic flux density and flow field between simulation and experimental results, the mathematical model is proven to be reliable.

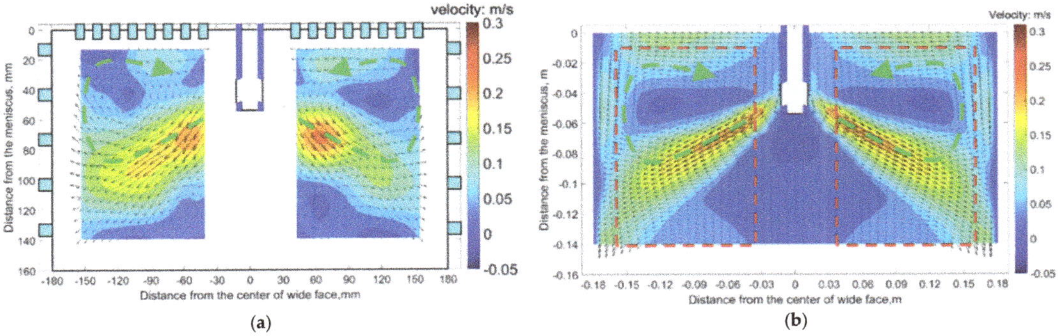

Figure 4. Vector and contour of fluid flow obtained at quarter plane of mold fixed side under the condition of 1.2 m/min-0 A: (**a**) physical model and (**b**) numerical model.

3.2. Effect of EMS Current on Fluid Flow in the Mold

In order to analyze the flow field inside the liquid pool, two planes and corresponding characteristic lines were selected, as indicated in Figure 1b. Plane 1 is a transverse plane which −5 mm distance from the meniscus, and line 1 is located at Y = 57.5 mm of plane 1, while plane 2 is a longitudinal plane at Y = 57.5 mm iso-surface, and line 2 is located at X = −700 mm of plane 2. The velocity distribution at plane 1 and plane 2 under different EMS currents when the casting speed is 1.2 m/min are shown in Figure 5.

As shown in Figure 5a, when EMS is off, the flow pattern at plane 2 is a classical double-roll structure as the jet travels to the narrow face and then splits into an upward (to the free surface and back towards the SEN) and downward (to carry molten steel deep into the mold) flow. For plane 1, molten steel impinges the free surface near the mold narrow face and results in a large velocity at this region, whereas the flow pattern near the SEN is almost stagnant. This nonuniform velocity distribution may affect the temperature distribution near the meniscus to some extent. EMS can improve this nonuniform distribution of velocity and temperature to some extent.

When EMS is applied (shown in Figure 5b,c), for plane 1, a horizonal recirculating flow has been generated due to the effect of electromagnetic force, this trend is similar to Yin' work in [31]. Meanwhile, it can be seen that the velocity at both side of SEN increases significantly compared with that without EMS. For plane 2, the upper and lower recirculation flow disappears gradually, instead, molten steel at the left side tends to move towards the right side of mold. In fact, the flow pattern in the mold is the interaction of inertial flow caused by jet flow and driving flow induced by electromagnetic force; Figure 5d shows the schematic of flow pattern with EMS, two different accelerated and decelerated regions exist both at the longitudinal plane and transverse plane which are related to the direction of electromagnetic force and initial flow pattern without EMS. With the increase of EMS current, it can be seen that an integrated recirculation flow has been generated at plane 1, and the velocity magnitude also increases, while for plane 2, the jet flow has obviously been inhibited at the left side of the SEN, meanwhile, the jet angle at the right side of the SEN decreases due to large electromagnetic force.

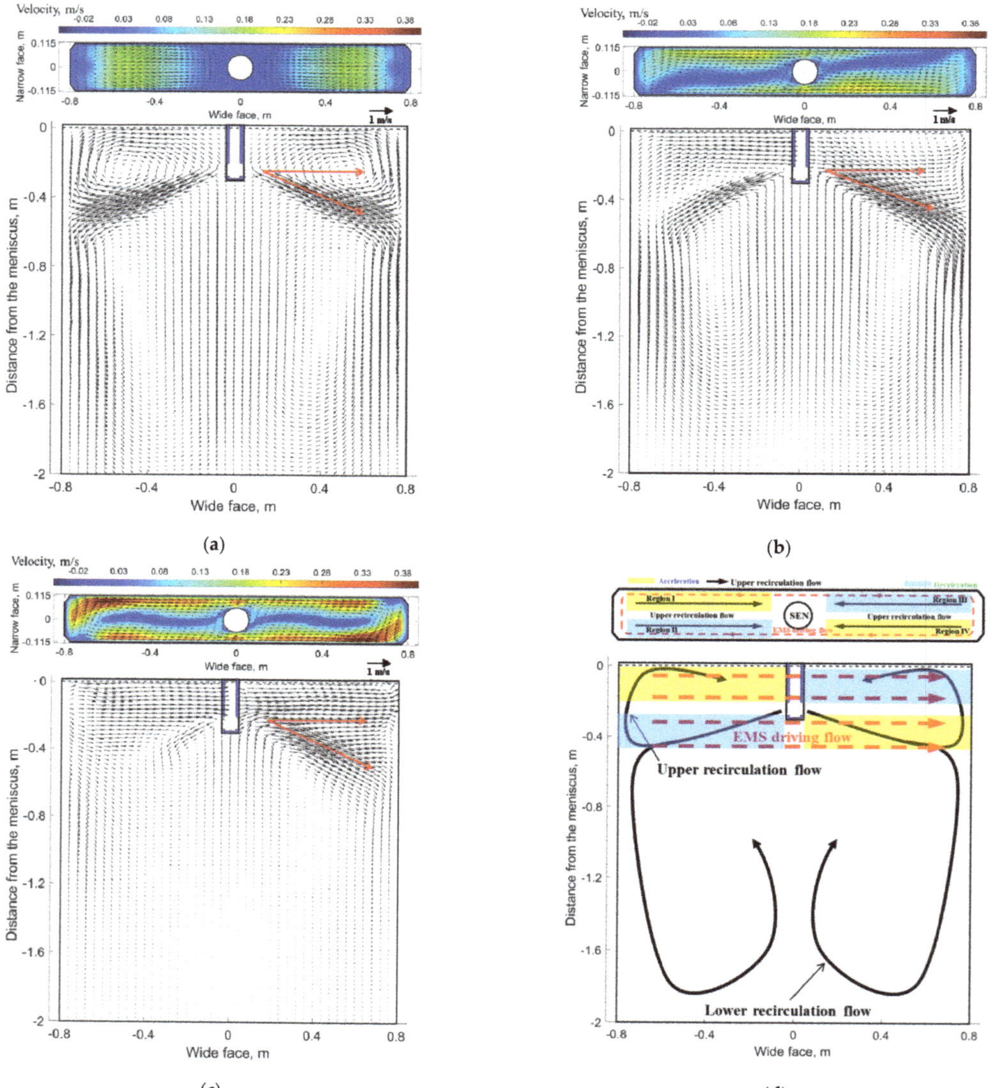

Figure 5. Velocity distribution at transverse plane (plane 1) and a quarter longitudinal plane (plane 2): (**a**) without-EMS, (**b**) 400 A-EMS, (**c**) 700 A-EMS (**d**) schematic of flow pattern with EMS.

Further justification of velocity under the influence of EMS is plotted in Figure 6. As shown in Figure 6a, when EMS is off, the X-velocity has the same magnitude at both sides of the SEN but with opposite direction; the maximum value is 0.18 m/s which corresponds to the impinging region of upper recirculation flow of Figure 5a. When EMS is on, with the increase of EMS current, the value of X-velocity at the left side of the SEN increases due to the accelerated region of Figure 5d, while for the right side of SEN, the direction of X-velocity changed firstly because of the horizonal swirling flow, and then the value of X-velocity increases with the increase of EMS current. For the small EMS current (400–600 A), it can be observed that the maximum value of X-velocity is less than that without EMS

(0.18 m/s), it means that small parts of upper recirculation flow still exist which hinders the horizonal recirculation flow, while when EMS current is greater than 600 A, the maximum value of X-velocity is 0.21 m/s which is larger than 0.18 m/s, it indicates that the upper recirculation flow has been broken completely in Figure 6b, with the increase of EMS current, the velocity in the mold decreases, as the EMS current increases to 700 A, the Z-velocity all changes to negative, which indicates that a plug flow has generated near the mold narrow face. This flow pattern may not benefit to the removal of inclusion.

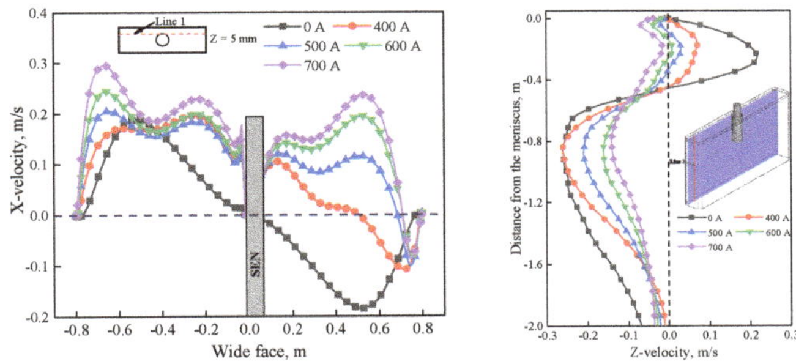

(a) (b)

Figure 6. Variation of velocity under different EMS currents: (**a**) X-velocity at line 1, (**b**) Z-velocity at line 2.

3.3. *Effect of EMS Current on Initial Solidification*

The 3D variation of solidified shell thickness along the casting direction under different EMS currents at the casting speed of 1.2 m/min are shown in Figure 7.

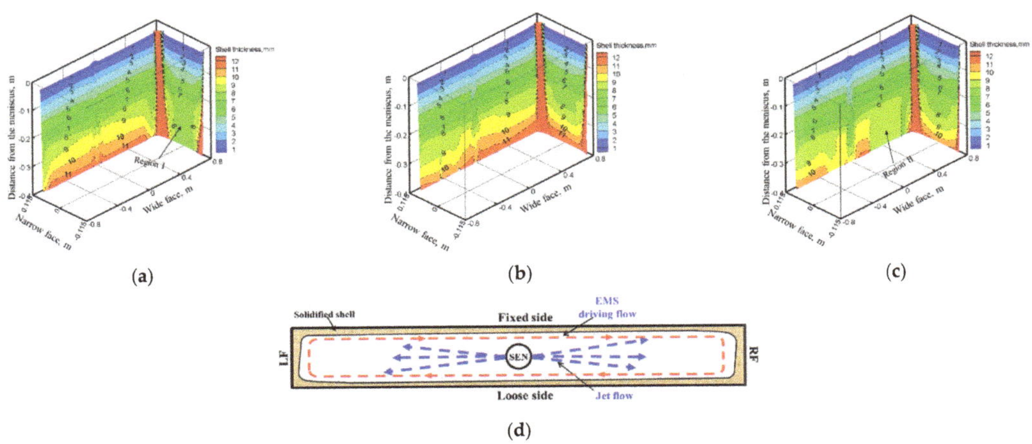

Figure 7. 3D variation of solidified shell thickness along the casting direction under different EMS currents: (**a**) 0 A, (**b**) 400 A, (**c**) 700 A, (**d**) schematic of transverse flow near jet flow with EMS.

As shown in Figure 7, the initial solidified shell was generated at −0.042 m distance from the meniscus, and its thickness increases along the casting direction. When EMS is off, the jet flow pouring from the SEN impinges to the solidified shell at the narrow face of the mold, which may result in the thin and uneven distribution of solidified shell at the impact region. When EMS is applied, the solidified shell thickness at the mold narrow face increases because that the direct impact of jet flow has been weakened under the influence of electromagnetic force. While for the mold wide face, due to the washing effect of transverse flow, the uniformity of solidified shell thickness changes even for a suitable EMS current (Figure 7b), however, for the higher EMS current (700 A), the solidified shell thickness at the right side of wide face decreases The main reason is that as shown in Figure 7d, an electromagnetic force gradient exists from the edge to the center of the slab, the jet flow may deviate towards the mold wide face and lead to the remelting of solidified shell; this phenomenon may not benefit the uniformity of the solidified shell to any extent.

Figure 8 shows the variation of solidified shell thickness at the center of the mold wide face and narrow face under different EMS currents, respectively. It can be seen that the solidified shell becomes thinner as the EMS current increases for the mold wide face. This phenomenon can be attributed to the superheat dissipation effect of EMS. While for the mold narrow face, when the EMS is not applied, it can be observed that the solidified shell stops growing due to the remelting phenomenon of high temperature liquid steel, which may increase the risk of break out. With the increase of EMS, the solidified shell thickness at the mold exit increases. When the EMS current increases from 0 to 700 A, the shell thickness of wide face and narrow face at the mold exit changes from 19.51 and 10.2 mm to 18.85 and 18.5 mm, respectively.

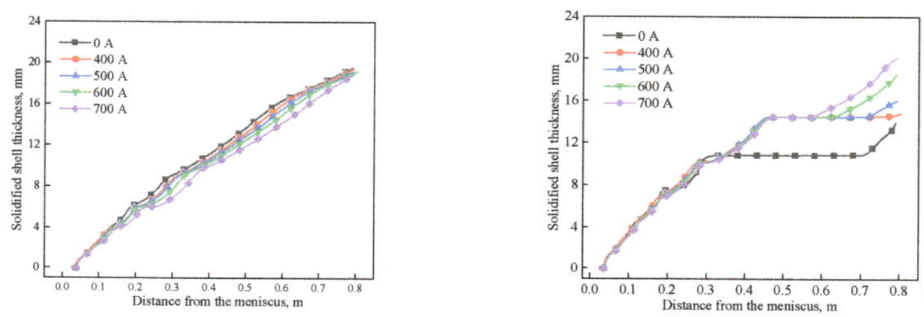

(a) (b)

Figure 8. Variation of solidified shell at the center of: (**a**) wide face, (**b**) narrow face under different EMS currents.

In the continuous casting process, the initial solidified shell near the meniscus around the perimeter of the mold is of great importance because it is associated with crack formation [1]. Therefore, in order to describe the uniformity of solidified shell thickness quantitatively, the uniform index of solidified shell thickness is set up, U_i, and the relevant definitions are depicted in Equations (12)–(14):

$$U_i = 1 - \frac{\sigma_i}{h_{i-ave}} \quad (12)$$

$$h_{i-ave} = \left(\sum_{j=1}^{m} h_{i-j} \right) / m_i \quad (13)$$

$$\sigma_i = \sqrt{\frac{\sum_{j=1}^{m} \left(h_{i-j} - h_{i-ave} \right)^2}{m_i}} \quad (14)$$

where, i represents the wide face (WF) and narrow face (NF) at mold transverse plane, respectively. h_{i-j}, h_{i-ave} and m_i are solidified shell thickness in mesh element j, averaged solidified shell thickness and total mesh element number, respectively, and σ_i is the standard deviation of the solidified shell. The larger the uniform index is, the more uniformly solidified the shell will be. What is more, because the influence of the electromagnetic force is mainly concentrated at −0.4 m distance from the meniscus for the current stirrer position, therefore, in the current study, we use the z = −0.4 m transverse plane to analyze the uniformity. Figure 9 and Table 3 show the uniform index at the mold wide face and narrow face under different EMS currents.

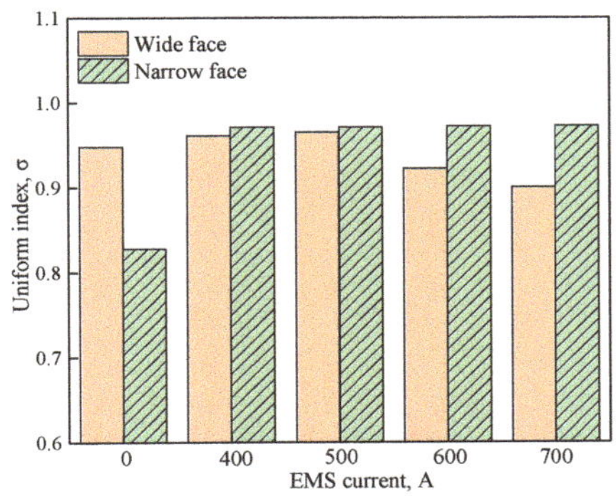

Figure 9. Uniform index at z = −0.4 m under different EMS currents.

Table 3. Uniform index at z = −0.4 m under different EMS currents.

EMS Current (A)	Uniform Index	
	Wide Face	Narrow Face
0	0.948	0.828
400	0.961	0.971
500	0.965	0.97
600	0.923	0.972
700	0.9	0.973

As shown in Figure 9, compared with that without EMS, EMS can effectively improve the uniformity of the solidified shell, especially for the mold narrow face. For the mold narrow face, the uniform index changes little as the EMS current increases from 400 to 700 A, the main reason is that the electromagnetic driving flow can suppress the direct impact of jet flow on the mold narrow face, and benefit to the uniform growth of solidified shell. For the mold wide face, it can be observed that the uniform index increases at first and then decreases accompanied with the increase of EMS current, this can be attributed that for the high EMS current, the remelt phenomenon at region II intensified, so the uniform index decreased.

3.4. *Effect of Casting Speeds on Fluid Flow in the Mold*

Considering that the flow velocity uniformity on the left and right sides of the mold is better at 600 A, as shown in Figure 6a, and the thickness of the shell with narrow surface is thicker at 600 A, as shown in Figure 8b, the thickness of the shell with narrow surface is conducive to improving the speed. According to Figures 6a and 8b, it is better at 700 A, but considering the uniform index of the solid shell as shown in Figure 9, the uniformity index of the wide face shell is worse at 700 A, therefore, 600 A was selected for further study.

The velocity distribution at plane 1 and plane 2 under different casting speeds when the EMS current is 600 A and frequency is 4 Hz are shown in Figure 10.

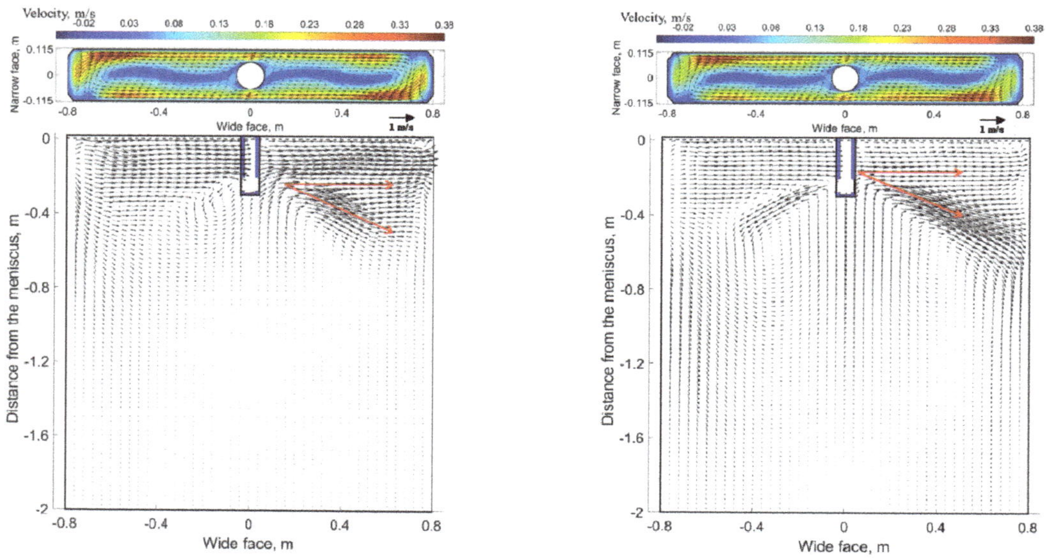

(a) (b)

Figure 10. Velocity distribution at transverse plane (plane 1) and a quarter longitudinal plane (plane 2) at (600 A, 4 Hz): (**a**) 0.8 m/min, (**b**) 1.4 m/min.

As shown in Figure 10, for plane 2, molten steel tends to flow from the left side of the SEN to the right side of the SEN due to the effect of electromagnetic force. When the casting speed is 0.8 m/min, it can be seen that the jet flow at the left side of SEN has nearly disappeared by comparing it with that of 1.4 m/min. While at the right side of the SEN, the jet angle increases significantly as the casting speed increase from 0.8 to 1.4 m/min. The main reason is that under the same EMS current, for the low casting speed (0.8 m/min), the jet velocity poured from the SEN is relatively small. Therefore, under this condition, the jet flow at the left side of the SEN is constrained more significantly, while for the right side of the SEN, the jet flow can also be easily dragged under the low casting speed, and results in a small jet angle. For plane 1, although a horizonal recirculating flow has been generated for these two cases, the velocity at the decelerated region is small when the casting speed is 1.4 m/min due to the hindering effect of stronger upper recirculation flow. Figure 11a shows the variation of velocity under different casting speeds at line 1, the velocity at the left side of SEN is greater than that of the right side; because of the accelerated region, with the increase of casting speed, although a transverse flow was generated for all casting speeds, the difference of maximum velocity at both sides of the SEN increases, as casting speed increases from 0.8 to 1.4 m/min, the difference increases from 0.0188 to 0.072 m/s. For the longitudinal line (Figure 11b), for the low casting speed, a plug flow has generated near the mold wall, it means that a critical casting speed exists to avoid the plug flow.

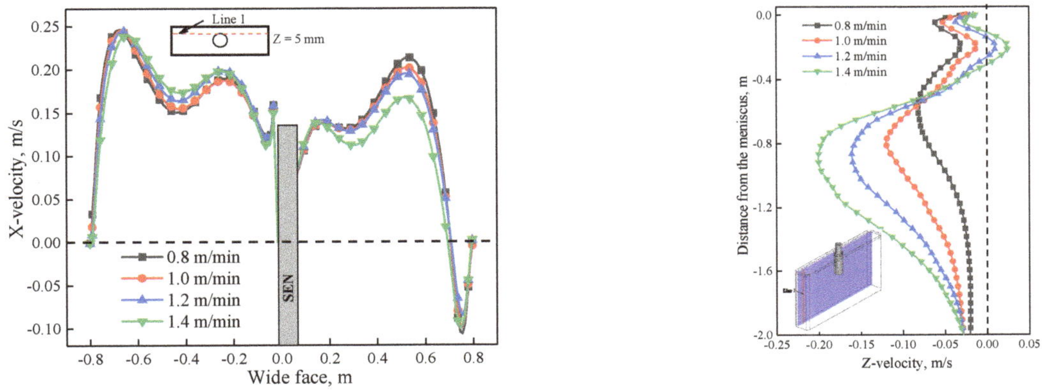

(a) **(b)**

Figure 11. Variation of velocity under different casting speeds at (600 A, 4 Hz): (**a**) X-velocity at line 1, (**b**) Z-velocity at line 2.

3.5. Effect of Casting Speed on Solidification

The variation of solidified shell thicknesses along the casting direction and center of the mold wide face and the narrow face under different casting speeds when the EMS current is 600 A are shown in Figure 12.

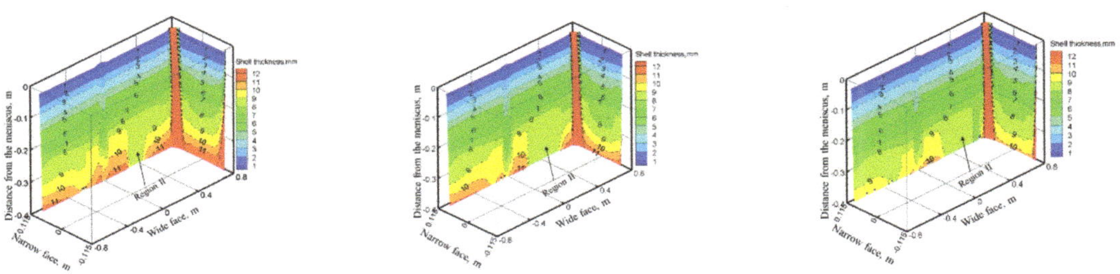

(a) (b) (c)

Figure 12. 3D variation of solidified shell thickness along the casting direction under different casting speeds at (600 A, 4 Hz): (**a**) 0.8 m/min, (**b**) 1.2 m/min, (**c**) 1.4 m/min.

As shown in Figures 12 and 13, the solidified shell thickness decreases with the increase of casting speed. For the mold wide face, a remelting concave has generated at region II because of the deviated jet flow induced by electromagnetic force gradient, and it is aggravated with the increase of casting speed. For the mold narrow face, it can be seen that with the increase of casting speed, the remelt phenomenon of solidified shell thickness aggravates due to the strong jet flow. When the casting speed increases from 0.8 to 1.4 m/min, the shell thickness of the wide face and narrow face at the mold exit changes from 20.72 and 20.4 mm to 15.95 and 17.31 mm, respectively.

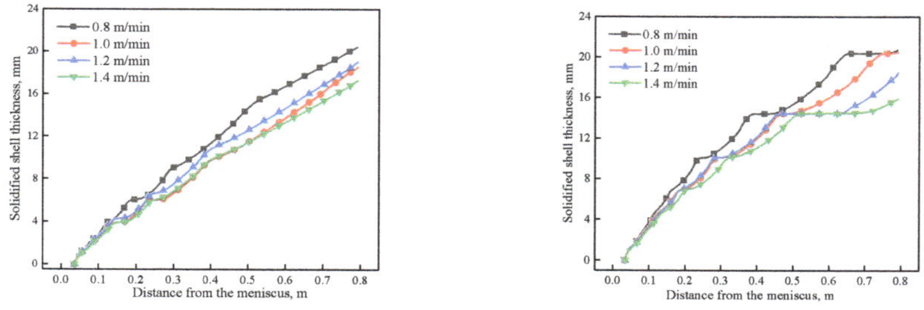

(a) (b)

Figure 13. Variation of the solidified shell at the center of, (**a**) wide face, (**b**) narrow face under different casting speeds at (600 A, 4 Hz).

The uniform index at z = −0.4 m under different casting speeds when the EMS current is 600 A is shown in Table 4. The uniform index at the mold narrow face and the wide face is shown in Figure 14. As shown in Figure 14, the uniform index at the mold narrow face is larger than that of the mold wide face, the main reason is that when EMS is applied, the jet flow which directly impacts the narrow face has been suppressed, and this may benefit uniform growth of the solidified shell, while for the mold wide face, the deviated jet flow can break the uniformity of the solidified shell. With the increase of casting speed, the uniform index at the mold wide face decreases firstly and then increases, while at the mold narrow face it increases but changes are not significant.

Table 4. Uniform index at z = −0.4 m under different casting speeds.

Casting Speed (m/min)	Uniform Index	
	Wide Face	Narrow Face
0.8	0.944	0.956
1.0	0.895	0.973
1.2	0.923	0.972
1.4	0.919	0.982

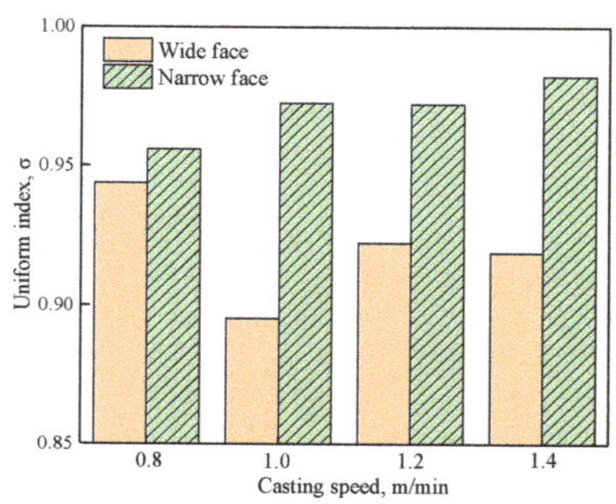

Figure 14. Uniform index at z = −0.4 m under different casting speeds.

4. Application Effects

According to the optimal parameters of the mathematical simulation, the industrial test including level fluctuation and number density of inclusion was carried out with and without EMS when the casting speed was 1.2 m/min. The relevant results are shown in Figure 15.

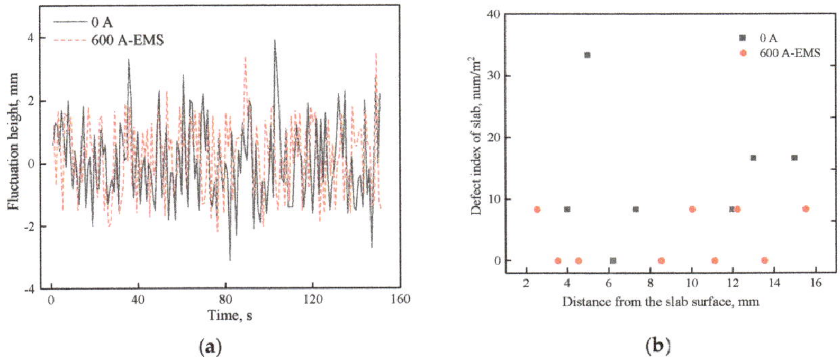

Figure 15. Industrial test: (**a**) transient level fluctuation at quarter of mold, (**b**) number density of inclusion.

As shown in Figure 15a, the largest level fluctuation is less than ±5 mm whether the EMS is on or off. However, when EMS is applied, the amplitude of fluctuation level is relatively small compared to that without EMS, it indicates that EMS can decrease the fluctuation level for a suitable EMS current. In addition, the inclusion has also been detected by magnifying glass and the size which is larger than 100 μm can be observed by this method. It can be seen that the number density within 15 mm from the slab surface have a decreasing trend when the EMS is applied, which demonstrates the benefits of the optimized flow control parameters.

5. Conclusions

A three-dimensional mathematical model is conducted to investigate the fluid flow, solidification and level fluctuations under different EMS currents and casting speeds. The uniform index of the solidified shell thickness has been introduced to judge the uniformity of the solidified shell. Conclusions of this study are summarized as follows:

(1) When EMS is applied, a horizonal recirculating flow has been generated, accelerated and decelerated regions exist in the mold. With the increase of EMS current, the difference of velocity near free surfaces decreases; large EMS current may generate plug flow near the mold wall.
(2) EMS can obviously improve the uniformity of the solidified shell, with the increase of EMS current, the uniform index at the mold narrow face increases while at the mold wide face it first increases and then decreases.
(3) Under the same EMS current, with the increase of casting speed, the difference of velocity near the free surface increases, the uniform index at the mold narrow face changes little, while at the mold wide face it first decreases and then increases.
(4) EMS can weaken the level fluctuation and reduce the number density of inclusion for a suitable EMS current through industry test.
(5) A rational EMS current range exists to obtain optimal steel quality. In the current study, when the casting speed is 1.2 m/min, the rational EMS current is 500–600 A.

Author Contributions: Methodology, Q.L. and J.Z.; formal analysis, Z.L. and H.L.; data curation, B.L. and C.J.; writing—original draft preparation, G.L. All authors have read and agreed to the published version of the manuscript.

Funding: This project was financially supported by the National Science Foundation of China (No. U1860107 and 52074181).

Institutional Review Board Statement: Not applicable.

Informed Consent Statement: Not applicable.

Data Availability Statement: Data sharing not applicable. No new data were created or analyzed in this study. Data sharing is not applicable to this article.

Acknowledgments: The authors express sincere thanks to Xiaoxuan Deng in Shougang Group Co., Ltd., Research Institute of Technology for his helpful assistance with structure and format of this paper.

Conflicts of Interest: The authors declare no conflict of interest.

References

1. Thomas, B.G. Review on Modeling and Simulation of Continuous Casting. *Steel Res. Int.* **2018**, *89*. [CrossRef]
2. Cho, S.-M.; Thomas, B.G. Electromagnetic Forces in Continuous Casting of Steel Slabs. *Metals* **2019**, *9*, 471. [CrossRef]
3. Jin, K.; Thomas, B.G.; Ruan, X. Modeling and Measurements of Multiphase Flow and Bubble Entrapment in Steel Continuous Casting. *Metall. Mater. Trans. B* **2016**, *47*, 548–565. [CrossRef]
4. Li, Z.; Zhang, L.; Ma, D.; Wang, E. Numerical Simulation on Flow Characteristic of Molten Steel in the Mold with Freestanding Adjustable Combination Electromagnetic Brake. *Met. Mater. Trans. A* **2020**, *51*, 2609–2627. [CrossRef]
5. Zhang, L.-S.; Zhang, X.-F.; Wang, B.; Liu, Q.; Hu, Z.-G. Numerical Analysis of the Influences of Operational Parameters on the Braking Effect of EMBr in a CSP Funnel-Type Mold. *Met. Mater. Trans. A* **2014**, *45*, 295–306. [CrossRef]
6. Dubke, M.; Tacke, K.-H.; Spitzer, K.-H.; Schwerdtfeger, K. Flow fields in electromagnetic stirring of rectangular strands with linear inductors: Part II. Computation of flow fields in billets, blooms, and slabs of steel. *Met. Mater. Trans. A* **1988**, *19*, 595–602. [CrossRef]
7. Yamada, T.; Satoh, S.; Fujisaki, K.; Shouji, S.; Takahiro, Y. Consideration of heat transfer and solidification in 3-D MHD calculation. *IEEE Trans. Magn.* **2000**, *36*, 1300–1304. [CrossRef]
8. Okazawa, K.; Toh, T.; Fukuda, J.; Kawase, T.; Toki, M. Fluid Flow in a Continuous Casting Mold Driven by Linear Induction Motors. *ISIJ Int.* **2001**, *41*, 851–858. [CrossRef]
9. Li, B.; Lu, H.; Zhong, Y.; Ren, Z.; Lei, A.Z. Influence of EMS on Asymmetric Flow with Different SEN Clogging Rates in a Slab Continuous Casting Mold. *Metals* **2019**, *9*, 1288. [CrossRef]
10. Wang, B.; Yang, Z.G.; Zhang, X.F.; Wang, Y.T.; Nie, C.P.; Liu, Q.; Dong, H.B. Analysis of the effects of electromagnetic stirring on solidification structure of bearing steel. *Metalurgija* **2015**, *54*, 327–330.
11. Yang, Z.-G.; Wang, B.; Zhang, X.-F.; Wang, Y.-T.; Dong, H.; Liu, Q. Effect of Electromagnetic Stirring on Molten Steel Flow and Solidification in Bloom Mold. *J. Iron Steel Res. Int.* **2014**, *21*, 1095–1103. [CrossRef]
12. Yang, Z.G.; Wang, B.; Zhang, X.F.; Wang, Y.T.; Dong, H.B.; Liu, Q. Effects of electromagnetic stirring on the flow field and level fluctuation in bloom molds. *J. Univ. Sci. Technol. Beijing* **2014**, *36*, 1354–1360.
13. Fang, Q.; Ni, H.; Wang, B.; Zhang, H.; Ye, F. Effects of EMS Induced Flow on Solidification and Solute Transport in Bloom Mold. *Metals* **2017**, *7*, 72. [CrossRef]
14. Ren, B.; Chen, D.-F.; Wang, H.-D.; Long, M.-J.; Han, Z.-W. Numerical simulation of fluid flow and solidification in bloom continuous casting mould with electromagnetic stirring. *Ironmak. Steelmak.* **2015**, *42*, 401–408. [CrossRef]
15. Wang, X.; Zheng, S.-G.; Liu, Z.-H.; Zhu, M.-Y. Numerical Simulation on Multiple Physical Fields Behaviors in Billet Continuous Casting with Different Stirrer Positions. *Steel Res. Int.* **2019**, *91*, 1900415. [CrossRef]
16. Maurya, A.; Jha, P.K. Influence of electromagnetic stirrer position on fluid flow and solidification in continuous casting mold. *Appl. Math. Model.* **2017**, *48*, 736–748. [CrossRef]
17. Maurya, A.; Jha, P.K. Numerical investigation of in-mold electromagnetic stirring process for fluid flow and solidification. *COMPEL-Int. J. Comput. Math. Electr. Electron. Eng.* **2017**, *36*, 1106–1119. [CrossRef]
18. Maurya, A.; Jha, P. Two-phase analysis of interface level fluctuation in continuous casting mold with electromagnetic stirring. *Int. J. Numer. Methods Heat Fluid Flow* **2018**, *28*, 2036–2051. [CrossRef]
19. FFujisaki, K.; Ueyama, T. Magnetohydrodynamic calculation for free surfaces. *J. Appl. Phys.* **1998**, *83*, 6356–6358. [CrossRef]
20. Fujisaki, K.; Ueyama, T.; Toh, T.; Uehara, M.; Kobayashi, S. Magnetohydrodynamic calculation for electromagnetic stirring of molten metal. *IEEE Trans. Magn.* **1998**, *34*, 2120–2122. [CrossRef]
21. Li, B.; Lu, H.; Zhong, Y.; Ren, Z.; Lei, Z. Numerical Simulation for the Influence of EMS Position on Fluid Flow and Inclusion Removal in a Slab Continuous Casting Mold. *ISIJ Int.* **2020**, *60*, 1204–1212. [CrossRef]
22. Moffatt, H.K. Electromagnetic stirring. *Phys. Fluids A Fluid Dyn.* **1991**, *3*, 1336–1343. [CrossRef]
23. Zhang, W.; Luo, S.; Chen, Y.; Wang, W.; Zhu, M. Numerical Simulation of Fluid Flow, Heat Transfer, Species Transfer, and Solidification in Billet Continuous Casting Mold with M-EMS. *Metals* **2019**, *9*, 66. [CrossRef]
24. Chen, W.; Ren, Y.; Zhang, L. Large Eddy Simulation on the Fluid Flow, Solidification and Entrapment of Inclusions in the Steel along the Full Continuous Casting Slab Strand. *JOM* **2018**, *70*, 2968–2979. [CrossRef]
25. Liu, G.; Liu, Q.; Ji, C.; Chen, B.; Li, H.; Liu, K. Application of a Novel Chamfered Mold to Suppress Corner Transverse Cracking of Micro-Alloyed Steel Slabs. *Metals* **2020**, *10*, 1289. [CrossRef]
26. Li, B.; Okane, T.; Umeda, T. Modeling of biased flow phenomena associated with the effects of static magnetic-field application and argon gas injection in slab continuous casting of steel. *Metall. Mater. Trans. B* **2001**, *32*, 1053–1066. [CrossRef]

27. Vives, C.; Ricou, R. Experimental study of continuous electromagnetic casting of aluminum alloys. *Metall. Mater. Trans. B* **1985**, *16*, 377–384. [CrossRef]
28. Qiu, S.; Liu, H.; Gan, Y. Numerical Simulation of Solidification Process of Slab Casting Based on Continuum Model. *J. Iron Steel Res.* **2003**, *15*, 16–20.
29. Lu, H.; Li, B.; Li, J.; Zhong, Y.; Ren, Z.; Lei, Z. Numerical Simulation of In-mold Electromagnetic Stirring on Slide Gate Caused Bias Flow and Solidification in Slab Continuous Casting. *ISIJ Int.* **2021**, *61*, 1860–1871. [CrossRef]
30. Li, B.; Lu, H.; Shen, Z.; Sun, X.; Zhong, Y.; Ren, Z.; Lei, Z. Physical Modeling of Asymmetrical Flow in Slab Continuous Casting Mold due to Submerged Entry Nozzle Clogging with the Effect of Electromagnetic Stirring. *ISIJ Int.* **2019**, *59*, 2264–2271. [CrossRef]
31. Yin, Y.; Zhang, J.; Wang, B.; Dong, Q. Effect of in-mould electromagnetic stirring on the flow, initial solidification and level fluctuation in a slab mould: A numerical simulation study. *Ironmak. Steelmak.* **2018**, 1–10. [CrossRef]

materials MDPI

Article

Multi-Parameter Characteristics of Electric Arc Furnace Melting

Michał Moskal [1,*], Piotr Migas [2] and Mirosław Karbowniczek [2]

1 COGNOR SA Branch HSJ in Stalowa Wola, Kwiatkowskiego 1, 37-450 Stalowa Wola, Poland
2 Department of Ferrous Metallurgy, Faculty of Metals Engineering and Industrial Computer Science, AGH-University of Science and Technology, 30 A. Mickiewicza A., 30-059 Krakow, Poland; pmigas@agh.edu.pl (P.M.); mkarbow@agh.edu.pl (M.K.)
* Correspondence: mmoskal@hsjsa.pl

Abstract: The article presents the results of analyses of numerical modelling of selected factors in electric arc furnace melts. The aim of the study was to optimise the melting process in an electric arc furnace using statistical-thermodynamic modelling based on, among other things, multiple linear regression (MLR). The article presents tools and methods which make it possible to identify the most significant indicators of the process carried out on the analysed unit from the point of view of improvement. The article presents the characteristics of the process and creation of the MLR model and, by applying its numerical analyses and results of calculations and simulations for selected variables and indicator, identifying the operation of a selected furnace. Developed model to demand of electric energy identification was used for calculations of energy balances, the distribution of the energy used in the furnace was presented.

Keywords: electric arc furnace; energy efficiency modelling; process optimization; mass and energy balances

Citation: Moskal, M.; Migas, P.; Karbowniczek, M. Multi-Parameter Characteristics of Electric Arc Furnace Melting. *Materials* **2022**, *15*, 1601. https://doi.org/10.3390/ma15041601

Academic Editors: Qing Liu and Jiangshan Zhang

Received: 31 January 2022
Accepted: 17 February 2022
Published: 21 February 2022

1. Introduction

Observed climatic changes in the environment are causing recycling technologies for steel production to increase their share in the total balance of crude steel production. This is particularly noticeable in the European Union. Therefore, the electric arc furnace is becoming one of the leading processing units for steel scrap used in crude steel production [1]. Electricity consumption in this process still accounts for about 50–80% of the total energy consumption during melting. Technological progress and the development of information technology create opportunities to reduce the unit demand for electrical or chemical energy.

Mathematical descriptions and characterisations of the arc furnace smelting process are complicated due to its nature resulting from, among others, high temperature and chemical composition gradients, turbulent flows, multiphase and heterogeneity of the system, significant mass flow of substrates, high speed of phase transformations and nonlinearity of phenomena. Therefore, the development of modelling of selected stages or the entire arc furnace smelting process remains justified [2].

The models used to describe the melting technology can be physical models—reproducing the processes on a smaller scale (e.g., laboratory and/or using other media than liquid metal and slag with the use of criterion numbers). An example of the application of a high-temperature model can be the studies on the performance of burners in an electric arc furnace carried out by Yonmo Sung [3]. They presented modification of oxy–gas burners, which was validated on a test bench.

Studies of scrap melting phenomena were also carried out based on water models [4] (using ice, three heating rods, and steam nozzles, which reflected scrap, electric arc, and oxy–gas burners, respectively).

Numerical methods can also be used to model the process. Mombeni [5] used a CFD model calculated in the commercial software ANSYS FLUENT to address the crack formation of water-cooled panels installed in the furnace vault. CFD modelling can also be used to determine the temperature distribution in a metal bath in the context of burner placement and media flow rates [6].

However, the mathematical models are the most universal method, which take into account physicochemical processes, thermodynamic and theoretical equations, mass and energy balances, or only technological data from the process [7]. They can be divided into two categories: linear and non-linear.

Linear models are created as a linear combination of predictor variables [8] based on average statistical input and output data from the process, or assumed theoretical values. They describe the relationship between individual parameters and their importance for the outcome of the calculation. Linear models are static and refer to only two points in time; they are the result of input data and, as mentioned above, they can only be used with reference to average values of a given process [9].

The most popular is the Kohle model, which describes the influence of process parameters on electricity demand.

Another model based on a multivariate linear regression equation is the one developed by Haupt [10]. He determined the influence of scrap quality on the melting process in an electric arc furnace.

However, non-linear models have the ability to describe the process more accurately. R. Morales created one of the first models, which included heat and mass transfer calculations and chemical reaction kinetics in terms of use the direct reduced iron [11]. The author created a dynamic model, which took into account the relationships between individual phenomena in terms of their physical course and not just the result, which was a statistical outcome of many variables. Thus, it makes it possible to analyse single melts and to follow the course of change of the result in time.

Attempts have been made to simplify the models and organise their construction. Clear boundaries of individual zones in the furnace were set. In terms of both: furnace construction, melting stages, and division into individual phases of materials/substances present in the furnace space. Bekker [12] specified two groups of solids and liquids in the constructed model. On the other hand, Logar [13,14] created a system of dependent modules representing particular groups of phenomena (including solid scrap zone, solid slag zone, liquid slag zone, liquid metal zone, and wall and roof zone). Models using neural networks are becoming increasingly popular. The first versions were developed in the 1990s. Ledoux and Bonnard, in their work, built a model based on a multi-layer perceptron. The model described the dynamics of an electric arc during the melting process.

Due to the process complexity, it is necessary to select models in such way that sufficient representation of the processes is maintained with the best possible ability to interpret and understand the simulation results [15]. The balance between these two characteristics depends largely on the type of models and the modelling methods [16].

2. Materials and Methods

2.1. Theoretical Mathematical Models

Mathematical models are the most commonly used in EAF simulation. They can be both linear and non-linear in nature. This classification results from taking into account the above-described variability of process efficiency as well as the variability of the parameters themselves during the melt [17]. Taking into account the kinetics of the reaction or the variation in the efficiency of media use or their control transforms the model into a non-linear form. Non-linearity is understood as effects ranging from media consumption dependent on melting programmes to the dependence of heat losses on the number and height of scrap heaps and thus arc exposure. When creating mathematical models, theoretical data or available technological and statistical data are used—which, depends on their availability and quality.

2.2. Linear Modelling of Steel Melting in Electric Arc Furnace

Regression linear models are created as a linear combination of predictor variables based on statistical mean input and output data from a process, or assumed theoretical values. This means that they refer to (only) two points in time, they are the result of input data and, e.g., output data, and as mentioned above, and they can only be used with respect to the average values of a given process. Static models often do not take into account some process disturbances and micro-scale phenomena (of low weights) and, therefore, their use for the analysis of single melts becomes difficult or, even impossible, due to the high variability of the real process conditions [18].

The most widely used statistical model in the literature is the Köhle model [18,19]. Köhle's Equation (1) describing the demand for electricity:

$$\frac{W_r}{kWh/t} = 300 + 900\left[\frac{G_E}{G_A} - 1\right] + 1600\frac{G_Z}{G_A} + 0.7\left[\frac{T_A}{^{\circ}C} - 1600\right] + 0.85\frac{t_S + t_N}{min} - 8\frac{M_G}{m^3/t} - 4.3\frac{M_L}{m^3/t} \quad (1)$$

where W_r is the electric energy demand, kWh/Mg; G_E is the weight of ferrous materials, kg; G_A is the tap weight (mass of the bath), kg; G_Z is the weight of slag components, kg; T_A is the temperature before taping, °C; t_S is the power on time, min; t_N is the power off time, min; M_G is the natural gas consumption, Nm^3; and M_L is the oxygen consumption, Nm^3.

This model characterises the electric energy demand, which in this case (of the units analysed by Köhle) is between 380 and 600 kWh/Mg with an accuracy equal to the standard deviation of 5 kWh/Mg.

The disadvantage of this solution is the low sensitivity of the model to changes in process conditions (occurrence of disturbances, change of the melting programme for the arc control system, or burners) which translates into a high error for individual melts. However, it is also an advantage in terms of long-term prediction. The Köhle model, due to its hybrid empirical–theoretical character, remains easy to interpret while maintaining sufficient accuracy.

In industrial practice, problems are encountered due to, e.g., insufficient process metering and difficult definition of boundary conditions. The amount of electrical energy supplied is easy to determine, whereas the efficiency of heat transfer through the arc depends on a large number of factors, e.g., the melting stage, the phenomena associated with arc burning (arc length, arc stability), or the plasma temperature associated with the gas atmosphere of the furnace, as well as the parameters characterising the steel scrap. Identifying the efficiency of the use of chemical energy—dependent on the reactions taking place in the reactor, which is the volume of the furnace bowl—is also a challenge. Its determination requires the identification of the reactions taking place and the distribution of the heat obtained between the phases, as well as the dynamics of the process [11–13,20].

When constructing models, it is necessary to perform mass and energy balances. Carrying out correctly defined balance calculations (of mass and energy) is essential from the point of view of the overall characteristics of the energy efficiency of the process and the determination of process boundary conditions [11].

In the course of the analyses, a database of indicators and technological parameters was created, which included 1953 records from one specific EAF. Due to a wide range of steel grades, the database was based on the most frequently manufactured group of steel grades and data was selected by rejecting melts:

- from furnace start-up periods;
- with furnace operation disturbances;
- switching from or to grades with large differences in chemical composition, when there was a justified suspicion of wet pouring;
- with process parameter values that are outliers.

The parameters included in the database are batch, material, and energy parameters.

2.3. Mass Balances

The mass balance is based on the assumption that the sum of the masses of the inputs m_{in} equals the sum of the outputs m_{out}.

$$m_{scr} + m_{sf} + m_c + m_{fa} + m_{ox} + m_{CH_4} + m_{la} = m_{hm} + m_{sl} + m_f + m_{wg} \quad (2)$$

where m_{scr} is the mass of the scrap, kg; m_{sf} is the mass of the slag formers, kg; m_{ox} is the mass of the oxygen, kg; m_{CH4} is the mass of the natural gas, kg; m_c is the mass of the carburizer, kg; m_{fa} is the mass of the foaming agent, kg; m_{la} is the mass of the leaked air, kg; m_{hm} is the mass of the hot metal, kg; m_{sl} is the mass of the slag, kg; m_{wg} is the mass of the waste gases, kg; and m_f is the mass of the fumes, kg.

For the purposes of consideration, the assumption has been made that the mass of the hot heel does not change between heats. In reality, the mass of hot heel changes and is difficult to estimate due to the relatively small volume it occupies and the constant irregular changes in the scale changing the volume of the furnace hearth.

2.3.1. Slag Weight Calculations

Slag mass was calculated on the basis of the chemical composition of the scraps used, assuming that the elements contained in them: Si, Mn, Cr, C, P, and Al are oxidised to the average of the level characterising the composition of the metal bath, and the amount of oxidised Fe constitutes 5.5% of the charge mass. These elements and the oxygen needed to burn them together with the addition of CaO and MgO make up the slag mass. The slag mass also includes ash from carbon carriers.

Chemical compositions of the slags were identified via energy dispersive X-ray fluorescence (XRF Oxford X-MET 3000TX +). In order to identify the chemical composition of the steel, a spectrophotometric method of analysis was used on a FOUNDRY-MASTER spark spectrometer.

The correctness of assumptions and calculations was checked by calculating the slag mass on the basis of CaO introduced into the process, which, as a component not dissolving in the metal bath, theoretically remains entirely in the slag.

$$m_{sl} = \frac{(m_{CaO\ met.} \cdot 0.958 + m_{CaO\ inj.} \cdot 0.864) \cdot 100\%}{\%CaO_{slag}} \quad (3)$$

where $m_{CaO\ met.}$ is the mass of the metallurgical calcium, kg; $m_{CaO\ inj.}$ is the mass of the injected pulverized calcium, kg; and $\%CaO_{slag}$ is the percentage of mass weight of CaO in slag.

A calculation was then made of the mass of each element that was oxidised during bath melting and slag formation, thus reducing the metal yield.

$$m_{ae} = \frac{M_{A_e} \cdot \%tl \cdot m_{sl}}{M_{A_o}}. \quad (4)$$

where m_{ae} is the mass of the alloying element melted in slag, kg; %tl is the percentage of mass weight of given element in slag; M_{A_e} is the atomic mass of element, u; and M_{A_o} is the atomic mass of oxide of given element, u.

2.3.2. Calculation of Waste Gas Masses

In order to calculate the mass of waste gases taking into account the course of the reaction of combustion of coal and methane, Reactions (5)–(7) were written:

$$C + \frac{1}{2}O_2 \rightarrow CO \quad (5)$$

$$C + O_2 \rightarrow CO_2 \quad (6)$$

$$CH_4 + 2O_2 \rightarrow CO_2 + 2H_2O \quad (7)$$

The mass of carbon monoxide and carbon dioxide for each reaction for one kilogram of product burnt was calculated. The values obtained were, respectively, 2.33 kg, 3.67 kg, and 2.75 kg (from Reactions (5)–(7)). Due to the lack of measurement of the flue gas composition, it was assumed that all the substrates burned completely, and left entirely with the flue gases. Therefore, in the mass of waste gasses m_{wg}, the components according to the Equation (8) are included:

$$m_{wg} = m_{N_2} + m_{CO_2} + m_{H_2O} \quad (8)$$

2.3.3. Calculation of Metal Bath Weight

The theoretical weight of the metal bath was calculated as the difference between the weight of the scrap, the alloying elements contained in the bath (including iron), and the estimated amounts of fumes.

$$m_{hm} = m_{scr} - \Sigma m_{ae} - m_f \quad (9)$$

where m_{hm} is the mass of hot metal, kg; m_{scr} is the mass of the scrap, kg; m_{ae} is the sum of masses of alloying elements that went into slag, kg; and m_f is the mass of the fumes, kg.

2.4. Energy Balances

The energy balance assumes that the sum of the energy input E_{in} into the process is equal to the sum of the energy output from the process in the form of product heat Q_{out} and thermal losses Q_{loss}. The energy input to the system consists of electrical energy and chemical energy. In the other direction, the energy is transferred out of the process in the form of metal baths, liquid slag, waste gases, dusts, and heat losses.

$$E_{el.} + E_{chem.} = Q_{hm} + Q_{sl} + Q_{wg} + Q_f \quad (10)$$

where $E_{el.}$ is the electrical energy, MJ; $E_{chem.}$ is the chemical energy, MJ; Q_{hm} is the heat of hot metal, MJ; Q_{sl} is the heat of the slag, MJ; Q_{wg} is the heat of waste gases, MJ; and Q_f is the heat of the fumes, MJ.

The chemical energy was calculated on the basis of enthalpies of exothermic reactions—oxidation of bath components and combustion of methane. The Perry Nist Janaf method was used to calculate the enthalpies of the reactions.

2.4.1. Calculation of the Physical Heat of Hot Metal

Calculation of the physical heat of the hot metal is based upon the values of the specific heat of the hot metal, its mass, and the temperature rise. It also takes into account the phase transition by using the values of latent heat, liquid temperature, and differentiation of specific heat into that occurring in the liquid and solid state.

For the calculation of the physical heat of the metal bath, the Formula (11) is used:

$$Q_{hm} = m_{hm} \cdot (c_{ss} \cdot (T_m - T_a) + Q_{lst} + c_{ls}(T - T_m))/1000 \quad (11)$$

where, Q_{hm} is the physical heat of the metal bath at temperature T, MJ; m_{hm} is the mass of hot metal, kg; c_{ss} is the average specific heat of steel in the solid state, kJ/kg K; c_{ls} is the average specific heat of steel in the liquid state, kJ/kg·K; Q_{lst} is the latent heat of fusion of steel, kJ/kg; T is the steel tap temperature, K; T_m is the steel melting temperature, and K; T_a is the ambient temperature, K.

The specific heat of the liquid state was taken as 0.836 MJ/Mg·K and that of the solid state as 0.698 MJ/Mg·K. The latent heat of fusion for the steel was taken as 271.7 kJ/kg.

The melting point of steel with a hot metal composition was calculated according to the Formula (12):

$$T_m = T_{Fe} - \sum_{i=1}^{n} p_i k_i \quad (12)$$

where T_m is the melting point, K; T_{Fe} is the melting point of pure iron (1812 K); p_i is the percentage content of the element in the metal bath, %; and k_i is the temperature reduction factor (according to the Table 1) [21].

Table 1. Temperature reduction factor for individual elements [21].

Element	C	Mg	Si	P	S	Cu	Ni	V	Mo	Cr
k_i	65	5	8	30	25	5	4	2	2	1.5

2.4.2. Calculation of the Physical Heat of Slag

The calculation of the physical heat of slag was based on the values of the slag's specific heat, mass, temperature rise, and latent heat. The following Formula (15) was used to calculate the physical heat of slag:

$$Q_{sl} = \frac{m_{sl}(c_{sl}(T - T_a) + Q_{lsl})}{1000} \quad (13)$$

where Q_{sl} is the physical heat of slag at temperature T, MJ; m_{sl} is the slag mass, kg; c_{sl} is the specific heat of slag at temperature T, kJ/kg·K; Q_{lsl} is the latent heat of melting of slag, kJ/kg; T is the slag tap temperature, K; and T_a is the ambient temperature, K.

The specific heat of slag was calculated from the Formula (14):

$$c_{sl} = 0.736 + 2.93 \cdot 10^{-4} T \quad (14)$$

The assumed latent heat of fusion is 209 kJ/kg.

2.4.3. Calculation of the Physical Heat of Waste Gases

The calculation of the physical heat of the waste gases is based on the specific heat values of the gases of their mass and temperature rise. The formula was used to calculate the physical heat of the waste gases:

$$Q_{wg} = m_{wg} c_{wg} (T - T_a)/1000 \quad (15)$$

where Q_{wg} is the physical heat of the waste gases at temperature T, MJ; m_{wg} is the mass of the waste gases, kg; c_{wg} is the specific heat of the waste gases at temperature T, kJ/kg·K; T is the temperature of the gases, K; and T_a is the ambient temperature, K.

The average specific heat of the waste gases is 1.33 kJ/Nm^3, whereas 1450 °C was taken as the value of gas temperature.

3. Results

In the field of research, considering modelling (MLR) as a first step, an attempt was made to apply and subsequently modify the Köhle equation [22]. Based on the analyses carried out on a furnace operating under the industrial conditions, one should state the lack of universality for the form of the model proposed by Köhle. The obtained values differ from the measured ones by an average of 50% for the 1992 version and by an average of 10% for the 2002 version of the equation [23]. As part of the work performed, the model was modified, and one of the changes made was the replacement of the M_L coefficient relating to oxygen consumption by an equation taking into account the heat gain from the oxidation of individual elements. This is due to the previously mentioned large number of scrap classes, as well as the large share of the iron oxidation reaction as a source of chemical energy. Therefore, the replacement of the coefficient that represents the average heat gain by a separate function is justified. This function is written in the form:

$$E_M = \sum_M \left(M_{sl} \cdot \frac{M_{AM} \cdot avg\%_{MO.sl} \cdot 1000\frac{kg}{Mg}}{100\% \cdot M_{AMO}} \cdot \Delta H_M \right) \tag{16}$$

where M_{sl} is the mass of slag, Mg; M_{AM} is the atomic weight of considered metal, u; $avg\%_{MO.sl}$ is the average considered metal oxide content of slag, %; M_{AMO} is the atomic weight of considered metal oxide, u; and ΔH_M is the the enthalpy of the oxidation reaction of a considered element, converted to kWh/kg; (where M = Cr, Mn, Fe, Al, Si).

Figure 1 shows the comparison of the real values of electricity consumption with the model values obtained based on two versions of the Köhle model: the 2002 version and its modification. It can be noted that for the modified version a much higher model fit was obtained (average difference 1%, maximum 20%).

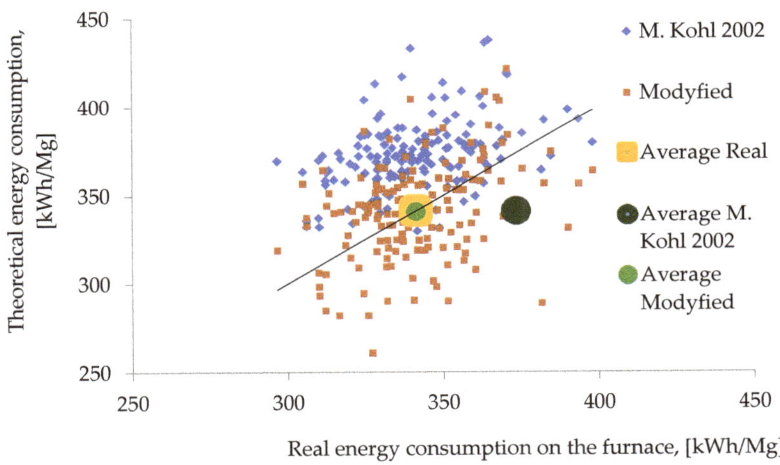

Figure 1. Comparison of Köhle model results with real electricity consumption [23].

In the realities of the furnace characterised, the model did not adequately provide quantitative answers as to what steps and extent of change should be taken to optimise the process.

Therefore, mass and energy balances were calculated and a linear multiple regression (MLR) model was undertaken.

For the calculation of mass and energy balances, it was necessary to identify the chemical composition of the liquid products (slag and steel). For this purpose, slag and metal samples were taken. A study of the chemical composition of slags was carried out for 200 melts in the same group of steel grades. The obtained average chemical compositions of slag and steel are presented in Tables 2 and 3.

Table 2. Average chemical composition of EAF slags before tap.

Oxide	CaO	MgO	MnO	Cr_2O_3	SiO_2	P_2O_5	Al_2O_3	TiO_2	FeO
% of weight	29.54	5.71	9.09	7.52	9.55	0.41	1.19	0.42	36.57

Table 3. Average chemical composition of liquid bath before tap.

Element	C	Mn	Cr	Ni	Si	P	S
% of mass	0.071	0.083	0.107	0.109	0.005	0.008	0.107

In the case of the unit under consideration, the insufficient metering, the high diversification of the steel grades, and the high variability of the charge (based on more than 40 scrap grades) led to some simplifications. For example, assuming a constant hot heel level; ignoring the change in furnace geometry resulting from erosion of the lining and, thus, assuming that the oxygen stream always reaches the bath and is fully consumed; and assuming that the chemical composition of the scrap corresponds to the average of the ranges described in PN 85/H 15000.

The results of the mass and energy balances obtained, and their analysis, clearly indicate the areas which should be investigated in order to improve the scrap melting process, e.g., appropriate use of natural gas, problems with furnace leak-tightness, and use of oxygen. However, the predominance of chemical energy in the melting process suggests that its optimisation will have a significant impact on the overall energy balance of the melt.

The parameters used to create the MLR were time, which is related to the amount of heat loss in the process; oxygen and gas consumption, as sources of chemical energy; the amount of slag-forming materials used, whose melting requires a significant amount of energy; and the scrap yield, which gives information on the amount of iron oxidised.

Calculations and multiple regression analyses were carried out to obtain the equation identifying the demand for electric energy. Table 4 presents the results for the selected backward MLR analysis, which gave the selected statistical model—the form of Equation (17). Statistical calculations were performed for the total sample size, which was N = 1959; the significance level of $\alpha = 0.05$ was assumed. Statistical correlation between the considered independent variables was checked; the relationships between the variables present in the equation were statistically insignificant.

Table 4. Statistical parameters of calculated MLR model.

Variable	b*	Std. Err. of b*	B	Std. Err. of b	t(1953)	*p*-Value
Intercept			169.942	8.689	19.558	0.000
Tap-to-tap time	0.248	0.020	0.263	0.021	12.324	0.000
Oxygen to charge	0.403	0.020	3271	0.166	19.670	0.000
Natural gas to charge	−0.055	0.020	−2633	0.952	−2767	0.006
Slag formers to charge	0.064	0.020	0.113	0.036	3166	0.002
Scrap yield	0.123	0.021	48.787	8.377	5.82	0.000

b* is a structural parameters of the regression function; B is an estimators of obtained regression equation; t is a t-Student test probability distribution; and *p*-value is a probability level of variable.

Statistical results were obtained: the regression coefficient at the level of R = 0.497; comparing for $R_{table} = 0.1946$ (for N > 1000), one can state the significance of the calculated linear correlation. Fisher–Snedecor's F-value, at the $\alpha = 0.05$ level, F(5.1953) = 128; for Ho's hypothesis, comparing it with the table critical value of $F_{table} = 2.108$, H_o's hypothesis of statistical significance R^2 of the equation—can be accepted at the significance level of $\alpha = 0.05$, which implies the possibility of rejecting the alternative hypothesis. It can also be seen that for all selected variables, the probability p is less than the significance level of 5% ($p < \alpha$), and the independent variables present in the equation are individually statistically significant.

Following the above, a linear regression model was created (3), resulting in:

$$E_{el} = 169.94 + 0.26 \cdot t_{TTT} + 3.27 \cdot \frac{m_{ox}}{m_{scr}} - 1.98 \cdot \frac{m_{CH_4}}{m_{scr}} + 0.11 \cdot \frac{m_{sf}}{m_{scr}} + 48.79 \cdot \frac{m_{hm\text{ł}}}{m_{scr}} \quad (17)$$

where t_{ttt} is the tap to tap time, min.; m_{scr} is the mass of scrap in charge, kg; m_{ox} is the mass of oxygen, kg; m_{CH4} is the mass of natural gas, kg; m_{sf} is the mass of slag formers, kg; and m_{hm} is the mass of hot metal, kg

In Figure 2, the normality analysis of the distribution of the residuals was shown. It seems, Shapiro–Wilk test: SW–W = 0.9971, $p = 0.0009$. Variance constancy: fulfilled, the uniformity of the scatter of the residuals is constant across the width of the interval.

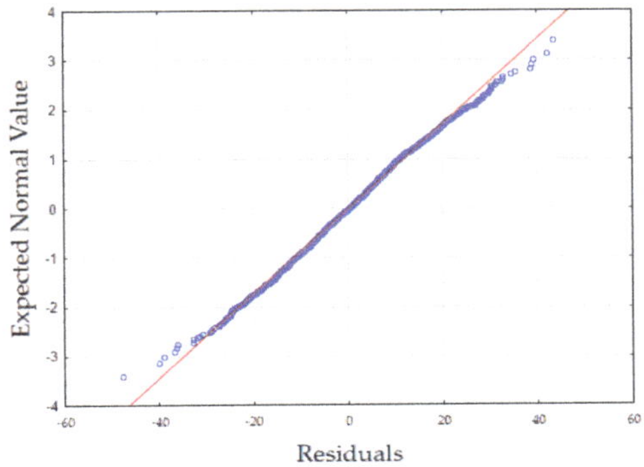

Figure 2. Normal distribution of residuals.

Simulations using the created MLR model confirmed the conclusions of the analysis of the balances performed, i.e., the predominance of chemical energy and the inverse proportionality of methane consumption to electrical energy consumption.

Taking into account the analysis results, which confirmed the conclusions of the mass and energy balance, the description of the process efficiency (defined as the ratio of the supplied energy to the amount of physical heat of the metal bath) as a function of the share of chemical energy was undertaken.

In Figure 3, the analysis of the dependence of process efficiency on the share of chemical energy was shown. A decrease in process efficiency (η_{EAF}) is visible with an increase in the share of chemical energy in the process. This trend in melt efficiency prompted an analysis of the efficiency of chemical energy use in the process under study.

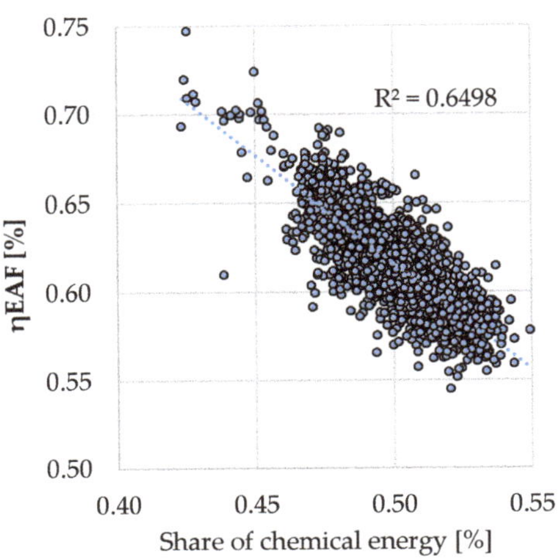

Figure 3. Dependence of process efficiency on the share of chemical energy.

As an example, the amount and timing of CH_4 addition was analysed as a function of the efficiency of the heat generated by its combustion.

It was assumed that the efficiency will change with the progress of scrap melting This is due to phenomena such as a change in the scrap surface with the progress of melting, but most importantly due to a decrease in the temperature difference between the flame and the charge, which heats up during the process. The change in burner efficiency was re-calculated on the basis of the approach presented in [21,24,25] and is presented in Figure 4.

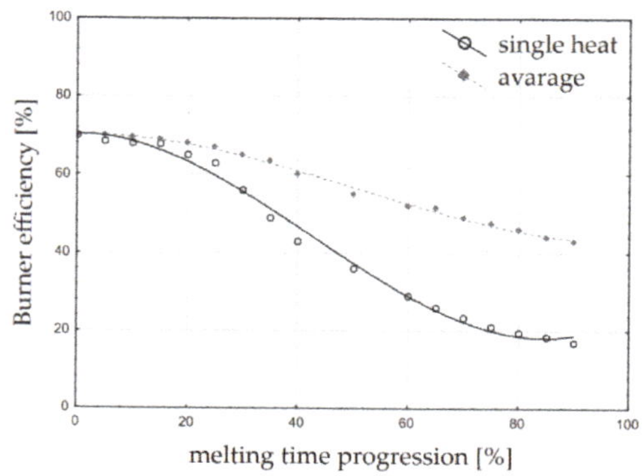

Figure 4. Burner efficiency as a function of melting time progression [%] (own study based on data from [21]).

The energy efficiency of the burners was considered first. This varies with the progress of melting. This efficiency is, on average, between 70 and 40%, and can drop as low as 20%. To identify the energy transfer efficiency, the Gottardi methodology was used [25], which made the energy efficiency of the burner dependent on, among other things, the temperature difference between the flame and the charge, which decreases with the progress of melting and, additionally, with a decrease in the scrap surface area. This dependence is described by the function (18):

$$\eta_{CH_4} = 70.272 - 8 \cdot 10^{-3} \cdot p_t^2 + 5 \cdot 10^{-5} \cdot p_t^3 \quad (18)$$

where η_{CH_4} is the energy efficiency of the burner [%]; and p_t is the melting progression [%].

Based on the mathematical description of the change in burner efficiency as a function of melting progress, three patterns of energy gain as a function of the average gas flow possible under the conditions of the unit analysed were simulated. Boundary conditions were established: minimum gas flow in burner mode, 90 Nm^3/h; the currently used average gas flow rate, 110 Nm^3/h; and near-maximum flow rate, 200 Nm^3/h. The maximum burner operation time—resulting from the melting stage characterised by the presence of scrap in the "cold regions" of the walls—is 30 min in total for all the baskets.

The change in energy contributed by the burners over time (Figure 4), the cumulative energy per 30 min (Figure 5a), and the cumulative energy as a function of the maximum possible amount of gas supplied (Figure 5b) were considered.

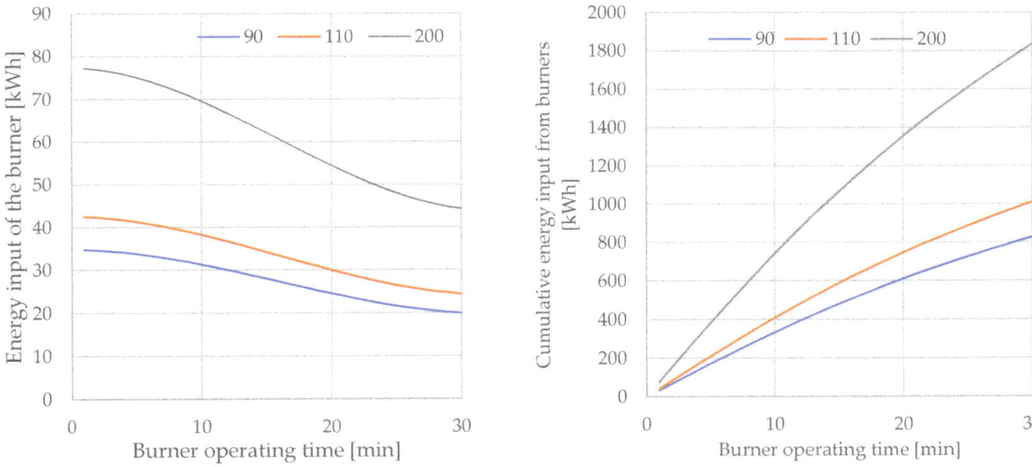

(a) (b)

Figure 5. (**a**) Variation of burner energy input as a function of burner operating time (**b**) cumulative energy as a function of burner operating time

From the simulations obtained, it can be concluded that it is beneficial to intensify the gas flow in the initial stage of the melt, as this allows the burner to use its moment of greatest efficiency. The calculated maximum achievable gas consumption is, respectively, 135, 165, and 300 Nm^3, and the energy contributed is 826 kWh, 1009 kWh, and 1835 kWh.

Finally, the equation was created to simulate production costs (in the sense of direct manufacturing costs), CO_2 emissions, and process efficiency (Mg of hot metal/h)—it is assumed that such a construction of the model will allow to control the technology in a manner depending also on economic objectives.

The attempt to consolidate the results in terms of obtaining the unit melt cost function contributed to the creation of Equation (19).

$$P_c = E_{el} \cdot X_1 + m_{CH_4} \cdot X_2 + (m_{scr} - m_{hm}) \cdot X_4 + m_{ox} \cdot X_3 + (m_C + m_{CH_4}) \cdot \frac{44}{16} \cdot X_5 + m_{sf} \cdot X_6 \quad (19)$$

where P_c is the production cost; X_1 is the cost per unit of electrical energy, PLN/kWh; X_2 is the cost per unit of natural gas, PLN/Nm^3; X_3 is the cos per unit of technical oxygen, PLN/Nm^3; X_4 is the average unit cost of scrap purchase, PLN/Mg; X_5 is the cost per emission of CO_2 unit PLN/Mg; and X_6 is the cost per unit of slag formers, PLN/kg.

Equation (5) was used to calculate production costs (in terms of direct production costs of liquid steel), CO_2 emissions, and process efficiency (Mg of hot metal/h)—assuming that such a model construction would allow to control the technology in a manner depending also on economic objectives. After calculating the production costs for the smelters under consideration (forming the database), a correlation analysis was carried out between the production cost (less the price of scrap metal forming the metal bath) and the amount of chemical energy input; the results are presented in Figure 6.

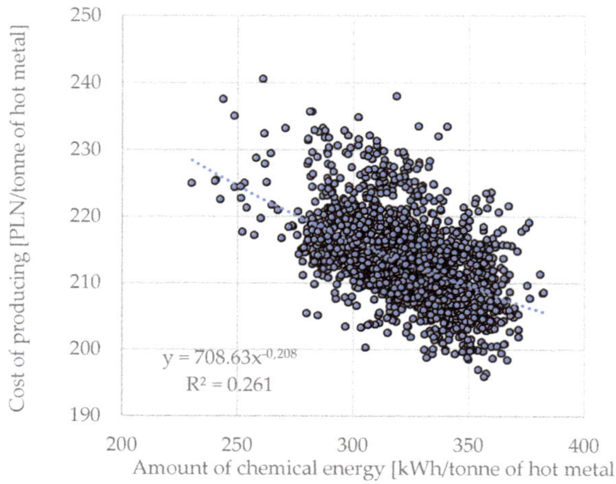

Figure 6. The amount of chemical energy in relation to the direct cost of producing a tonne of metal bath.

The correlation coefficient is R = 0.513 and the *t*-Student's *t*-value reaches $t = 41.51$ with $t_{tabular} = 1.960$. From the existence of this relationship (Figure 6) it should be concluded that it is advantageous to increase the share of chemical energy in the process, but taking into account the efficiency of its transfer by the oxygen treating bath stage.

4. Discussion

Modelling of the melting process in an electric arc furnace is possible under industrial conditions, even with incomplete metering. However, the individual characteristics of the unit under consideration have to be taken into account as existing models do not show universality. The use of models developed for other units requires their modification and taking into account the characteristics resulting from the proportion and share of chemical and electric energy as well as the input and technological conditions of the furnace.

Increasing the chemical energy input into the melt lowers the required total amount of electrical energy—methane consumption is inversely proportional to electrical energy consumption. However, due to the technological effect of the fact that oxygen is fed most intensively at the time of charge reheating—throughout this period—its correlation with electrical energy is positive.

Based on model calculations for the analysed furnace unit using a modified Köhle model from 2002, the model and actual data converged at the level of R = 0.36. The calculated electricity demand differed from the actual demand at the maximum level of 20%. However, it did not give sufficient guidance for process control.

Thanks to the mass and energy balance calculations performed, results were obtained from which it can be concluded that there is a predominance of chemical energy (52%) in the process compared to electrical energy.

It can be concluded that it is advantageous to intensify the gas flow at the initial stage of melting, as this makes it possible to use the moment of maximum efficiency of the burners (presented on Figure 5a,b). These conclusions require a complementary analysis; the real conditions of the process should be taken into account, which limit the maximum flow on the oxy–gas burners in the initial stages of the process (risk of burner damage by flame reflection from the scrap metal near the walls) and the fact that a higher energy input will lead to a faster melting of the deposited scrap metal

The statistical calculations carried out and the analyses of the statistical indices obtained made it possible to select the multiple linear regression model Equation (17)—as a method giving the possibility of an easy interpretation with an appropriate representation of the process in these conditions. The regression coefficient for the calculated model is R = 0.497, the difference between the calculated value and the real one is, at maximum, 17% (with mean 0). The obtained model was extended by the analysis of natural gas efficiency and consolidated with the cost calculation equation.

Author Contributions: Conceptualization, P.M. and M.K.; methodology, M.M.; resources, M.M.; data curation, M.M.; writing—original draft preparation, M.M.; writing—review and editing, P.M. and M.K. All authors have read and agreed to the published version of the manuscript.

Funding: This research received no external funding.

Institutional Review Board Statement: Not applicable.

Informed Consent Statement: Not applicable.

Data Availability Statement: Not applicable.

Acknowledgments: The present work was write as part of the second edition of "Implementation Doctorate" programme. The authors gratefully acknowledge their support.

Conflicts of Interest: The authors declare no conflict of interest.

References

1. World Steel Association. *Steel's Contribution to a Low Carbon Future and Climate Resilient Societies—World Steel Position Paper*; World Steel Association: Brussels, Belgium, 2020; ISBN 978-2-930069-83-8. Available online: https://www.worldsteel.org/en/dam/jcr:7ec64bc1-c51c-439b-84b8-94496686b8c6/Position_paper_climate_2020_vfinal.pdf (accessed on 10 November 2020).
2. Carlsson, L.S.; Samuelsson, P.B.; Jönsson, P.G. Using Statistical Modeling to Predict the Electrical Energy Consumption of an Electric Arc Furnace Producing Stainless Steel. *Metals* **2020**, *10*, 36. [CrossRef]
3. Sung, Y.; Lee, S.; Han, K.; Koo, J.; Lee, S.; Jang, D.; Oh, C.; Jang, B. Improvement of Energy Efficiency and Productivity in an Electric Arc Furnace through the Modification of Side-Wall Injector Systems. *Processes* **2020**, *8*, 1202. [CrossRef]
4. Xiaojun, X.; Shufeng, Y.; Jingshe, L.; Jinqiang, W.; Mengjing, Z.; Maolin, Y. Physical model experiment and theoretical analysis of scrap melting process in electric arc furnace combined blowing system. *Ironmak. Steelmak.* **2020**, *47*, 748–756.
5. Gharib Mombeni, A.; Hajidavalloo, E.; Behbahani-Nejad, M. Transient simulation of conjugate heat transfer in the roof cooling panel of an electric arc furnace. *Appl. Therm. Eng.* **2016**, *98*, 80–87. [CrossRef]
6. Coskun, G.; Yiğit, C.; Buyukkaya, E. CFD modelling of a complete electric arc furnace energy sources. *Innovations* **2016**, *4*, 22–24.
7. He, K.; Wang, L.; Li, X. Review of the Energy Consumption and Production Structure of China's Steel Industry: Current Situation and Future Development. *Metals* **2020**, *10*, 302. [CrossRef]
8. Faraway, J.J. Generalized linear models. In *International Encyclopedia of Education*, 3rd ed.; Peterson, P., Baker, E., McGaw, B., Eds.; Elsevier: Amsterdam, The Netherlands, 2010; pp. 178–183. [CrossRef]
9. Cameron, L.T.; Hangos, K.M. *Process Modelling and Model Analysis*, 1st ed.; Academic Press: San Diego, CA, USA, 2001; Volume 4, pp. 10–17; 253–275.
10. Haupt, M.; Vadenbo, C.; Zeltner, C.; Hellweg, S. Influence of Input-Scrap Quality on the Environmental Impact of Secondary Steel Production. *J. Ind. Ecol.* **2017**, *21*, 391–401. [CrossRef]
11. Morales, R.D.; Rodriguez, H.; Conejo, A.N. A Mathematical Simulator for the EAF Steelmaking Process Using Direct Reduced Iron. *ISIJ Int.* **2001**, *41*, 426–436. [CrossRef]
12. Bekker, J.G.; Craig, I.K.; Pistorius, P.C. Modeling and Simulation of an Electric Arc Furnace Process. *ISIJ Int.* **1999**, *39*, 23–32. [CrossRef]
13. Logar, V.; Dovžan, D.; Škrjanc, I. Modeling and Validation of an Electric Arc Furnace Part 1, Heat and Mass Transfer. *ISIJ Int.* **2012**, *52*, 402–412. [CrossRef]
14. Logar, V.; Dovžan, D.; Škrjanc, I. Modeling and Validation of an Electric Arc Furnace Part 2, Thermo-chemistry. *ISIJ Int.* **2021**, *52*, 413–423. [CrossRef]
15. Štrumbelj, E.; Kononenko, I. A General Method for Visualizing and Explaining Black-Box Regression Models. In *Adaptive and Natural Computing Algorithms*; ICANNGA 2011; Lecture Notes in Computer Science; Dobnikar, A., Lotrič, U., Šter, B., Eds.; Springer: Berlin, Germany, 2011; Volume 6594, pp. 21–30.
16. Banks, H.T.; Tran, H.T. *Mathematical and Experimental Modeling of Physical and Biological Processes*, 1st ed.; Chapman and Hall/CRC: New York, NY, USA, 2009; pp. 3–5.
17. Carlsson, L.S.; Samuelsson, P.B.; Jönsson, P.G. Predicting the Electrical Energy Consumption of Electric Arc Furnaces Using Statistical Modeling. *Metals* **2019**, *9*, 959. [CrossRef]
18. Köhle, S. Effects on the electric energy and electrode consumption of arc furnaces. *Stahl Eisen* **1992**, *112*, 59–67.
19. Köhle, S.; Lichterbeck, R.; Paura, G. *Verbesserung der Energetischen Betriebsführung von Drehstrom-Lichtbogenöfen*; European Commission: Brussels, Belgium, 1996.

20. Meier, T.; Gandt, K.; Echterhof, T.; Pfeifer, H. Modeling and Simulation of the Off-gas in an Electric Arc Furnace. *Metall. Mater Trans. B* **2017**, *48*, 3329–3344. [CrossRef]
21. Karbowniczek, M. *Electric Arc Furnace Steelmaking*, 1st ed.; CRC Press Taylor & Francis Group: Boca Raton, FL, USA, 2021; p. 239
22. Köhle, S. Recent improvements in modelling energy consumption of electric arc furnaces. In Proceedings of the 7th European Electric Steelmaking Conference, Venice, Italy, 26–29 May 2002.
23. Migas, P.; Moskal, M.; Karbowniczek, M.; Czyż, J. Analysis of the energy consumption during the electric arc process. In Proceedings of the Iron and Steelmaking 2019, Modern Metallurgy the XXVIII International Scientific Conference, Liptovský Mikuláš, Slovakia, 23–25 October 2019.
24. Pretorius, E.; Oltmann, H.; Jones, J. *EAF Fundamentals: Charging, Melting and Refining*; LWB Refractories: New York, NY, USA, 2005; pp. 16–21.
25. Bergman, K.; Gottardi, R. Design criteria for the modern UHP electric arc furnace with auxiliaries. *Ironmak. Steelmak.* **1990**, *17*, 282–287.

materials

MDPI

Article

Numerical Simulation Study of Gas-Solid Heat Transfer and Decomposition Processes of Limestone Calcined with Blast Furnace Gas in a Parallel Flow Regenerative Lime Kiln

Shaopei Duan [1,2], Baokuan Li [1,2,*] and Wenjie Rong [1,2]

1 School of Metallurgy, Northeastern University, No. 3-11, Wenhua Road, Heping District, Shenyang 110819, China; duanshaopei_neu@163.com (S.D.); rongwenjie@smm.neu.edu.cn (W.R.)
2 Key Laboratory of Data Analytics and Optimization for Smart Industry, Ministry of Education, Northeastern University, Shenyang 110819, China
* Correspondence: libk@smm.neu.edu.cn

Abstract: Quicklime is an essential reducing agent in the steel smelting process and its calcination from limestone is accompanied by considerable energy consumption. As a relatively economical lime kiln, the Parallel Flow Regenerative (PFR) lime kiln is used as the main equipment for the production of quicklime by various steel industries. PFR lime kilns generally use natural gas as the fuel gas. Although natural gas has a high calorific value and is effective in calcination, with the increasing price of natural gas and the pressure saves energy and protect the environment, it makes sense of exploring the use of cleaner energy sources or other sub-products as fuel gas. The use of blast furnace gas (BFG) as a low calorific value fuel gas produced in the steel smelting process has been of interest. This paper therefore develops a set of mathematical models for gas-solid heat transfer and limestone decomposition based on a Porous Medium Model (PMM) and a Shrinking Core Model (SCM) to numerically simulate a PFR lime kiln using BFG in order to investigate the feasibility of calcining limestone with low calorific fuel gas and to provide a valuable reference for future development of such processes and the kiln structure improvement.

Keywords: PFR lime kiln; lime calcination; BFG; numerical simulation; PMM; SCM

Citation: Duan, S.; Li, B.; Rong, W. Numerical Simulation Study of Gas-Solid Heat Transfer and Decomposition Processes of Limestone Calcined with Blast Furnace Gas in a Parallel Flow Regenerative Lime Kiln. *Materials* **2022**, *15*, 4024. https://doi.org/10.3390/ma15114024

Academic Editor: Irina Hussainova

Received: 7 May 2022
Accepted: 4 June 2022
Published: 6 June 2022

1. Introduction

The main component of quicklime is calcium oxide (CaO), which is widely used in the metallurgical, construction, agricultural and chemical industries because of its ease of preparation, low price and good reducing properties. Nearly half of the world's quicklime is used in metal smelting, so the metallurgical industry has the greatest demand for quicklime. Quicklime is obtained by calcining limestone at high temperatures, the main component of which is calcium carbonate ($CaCO_3$) with small amounts of silicon dioxide (SiO_2) and magnesium oxide (MgO).

Among all types of quicklime products, the quantity of quicklime used in metal smelting is not only huge, but there are also strict requirements for its high reducibility. At temperatures less than 1173 K the calcination does not occur and is called "under-burning"; at temperatures above 1423 K the surface porosity becomes smaller and thus less chemically active and is called "hard-burnt" or "over-burnt" At temperatures between 1173 K and 1423 K, calcination results in large surface porosity and strong chemical activity, making it suitable as an excellent reducing agent for metal smelting, which is known as "soft-burnt".

Many researches have been carried out on the association between calcination and the chemical activity of quicklime.Suju Hao finds the lime activity degree was determined by acid-base titration, and the lime pore distribution was measured by mercury intrusion porosimetry [1]. Vola deal with thermal analyses, burning trials and reactivity tests on 15 carbonate rocks, i.e., pure and impure carbonates, mud-supported and grain-supported limestones, crystalline marbles, and dolomites, used for the production of different lime products in industrial vertical shaft kilns worldwide [2]. Wang does many studies on the relationship between microstructure and physicochemical properties of limestone including, porosity, bulk density, pore size distribution, specific surface area and activity were studied under the condition of rapid heating from room temperature to 1350–1550 °C, the results showed that complete decomposition of limestone (diameter 12.5–15 mm) at 1350 °C, 1450 °C and 1550 °C took 11.7 min, 9.2 min and 6.9 min, respectively [3]. The subject of was to evaluate the influence of lime particle size and lime calcination level as well as slag properties (basicity, viscosity, CaO activity and density) on lime particle dissolution at temperature in the early stage of process blowing time of the basic oxygen furnace (1400 °C) [4].

The Parallel Flow Regenerative (PFR) lime kiln is widely used in the metallurgical industry due to its high daily production capacity (600 t/d–800 t/d), good chemical activity of the product and high thermal efficiency. At the same time, PFR lime kilns are constantly developing and evolving into a variety of improved models.

A three-dimensional steady-state model to predict the flow and heat transfer in a rotary lime kiln is presented [5]. Senegačnik et al. present a case of a remodelled annular shaft kiln, in which the air excess ratio can be reduced to its optimal level by recirculation of recuperator waste gas and its injection into the combustion chambers [6]. The objective of Gutiérrez is to analyze the energy and energy consumption of the calcination process in vertical shaft kilns, in order to identify the factors affecting fuel consumption [7]. The PFR lime kiln has established itself worldwide for the type of product. Piringer explains the functional principle of such lime kilns [8]. Krause present DEM-CFD simulations of the transient processes occurring in an industrial scale PFR-kiln [9].

There are also many CFD numerical simulations for lime kilns. The proposed predictive control method can make the output value of the calcination zone temperature of the lime rotary kiln fast and stable to track the change of the reference value [10]. Suárez builds a regression model for each one of the three prime variables (gas consumption, SO_2, and NO_x emissions) of a lime kiln employed in the paper manufacturing process using the multivariate adaptive regression splines (MARS) method in combination with the artificial bee colony (ABC) technique [11]. Metz evaluates the influence of electric arc furnace dust (EAFD) and lime kiln waste (LKW) [12]. Juneja covers the modeling and control aspects of Lime Kiln process [13]. CFD simulations of an industrial scale rotary kiln for cement clinker production are also conducted [14]. The impact of these coating layers on the clinker production process within a rotary kiln is investigated with CFD simulations [15]. Sagastume and Wenjie analyzed the thermal performance of the lime kiln using the worm-efficiency, pointing out that the performance of a new direction for the analysis of the performance of lime kilns [16,17].

Cheng, et al. [18,19] used the shrinking core model as the basis to find the gas-solid temperature profile inside the ASK by numerical simulation and analyzed the limestone decomposition rate related to the limestone type. Bluhm-Drenhaus [20] and Krause [21] combined the finite volume method (FVM) with the finite element method(FEM) to simulate the temperature field of the gas-solid flow inside the lime kiln and calculated the trajectory of the particles in the kiln with the flow process of the fuel gas. Ning Jingtao [22] studied the calcination of fine limestone using a shrinking core model and confirmed that 1200–1473 K is a more desirable calcination temperature for limestone. The coupled simulation of DEM (discrete element method) and CFD (finite volume method) is a method to solve the stacked bed problem in lime kilns, where the DEM model solves the limestone decomposition behavior while the CFD model simulates the gas-solid two-phase heat exchange and the pressure drop in the stacked bed [23]. Mikulčić, et al. [24] used this coupled approach to study the cement industry for CO_2 emissions. Meanwhile, a two-energy equation model based on a porous media model is developed by Long Huilong, et al. [25] to simulate the gas-solid chemical reaction in a non-stationary manner, which is in agreement with the experimental results. Shanshan Bu, et al. [26] studied the effect of three particle contact models on the heat transfer in a stacked bed and found that the three particle contact models have a large influence on the calculation results in turbulent flow. Liu, et al. [27] used a gas-burnt lime shaft kiln as a research object and solved the temperature field inside the lime kiln using a local non-thermal equilibrium model.

Previously, in order to ensure the calcination effect, natural gas was generally used as the fuel gas, mixed with air and then burned to generate heat and heat as the heat source for the calcination of limestone. While this traditional process has ensured a good quality of limestone calcination, it has also created significant problems. Firstly, the main component of natural gas is methane, and with the progressively stricter restrictions on greenhouse gas emissions in countries around the world, strict regulations have been made on the relevant equipment for metallurgical enterprises, and PFR lime kilns, as the main equipment consuming natural gas, have to consider energy saving improvements; secondly, the price of natural gas continues to rise due to fluctuations in global political and economic factors, bringing huge cost pressures on the relevant enterprises; finally, in some metallurgical enterprises, blast furnace gas is difficult to be widely used in various production processes because the combustion produces a temperature of only 1673 K to 1773 K, and the utilisation rate is low. In addition, the PFR lime kiln supports 40 mm to 120 mm particle size for calcination when using natural gas as the fuel gas, and there is still a lack of research into the range of particle sizes supported by the change of fuel. In summary, the consideration of low calorific value sub-products, represented by blast furnace gas, as an alternative to natural gas for limestone calcination is gaining interest in the industry. Therefore, this paper makes an attempt in this direction by replacing natural gas with blast furnace gas and establishing a reliable mathematical model of gas-solid heat exchange and calcium carbonate decomposition reaction for the study of its effect, providing a reference basis for future equipment design and the process improvement.

2. Mathematical models

The simulation of the PFR lime kiln is based on the Finite Volume Method (FVM), and for any scalar quantity such as temperature, the solution equation as follows:

$$\frac{\partial \rho \phi}{\partial t} + \frac{\partial}{\partial x_i}(\rho \vec{u_i} \phi - \Gamma \frac{\partial \phi}{\partial x_i}) = S_\phi \quad (1)$$

The mass conservation equation in fluid mechanics theory, also known as the continuity equation, is a fundamental equation of computational fluid dynamics based on a continuous model. It assumes that density, coordinates, time and other variables are continuous in space. Its expression is in the form of Equation (2).

$$\frac{\partial \rho}{\partial t} + \nabla \cdot (\rho \vec{v}) = S_m \quad (2)$$

Equation (2) is the basic form of the mass conservation equation. The model is used in PFR lime kilns to calculate the mass transfer between gas and solid, for example, the mixing of heating and cooling gases, and the emission of carbon dioxide during decomposition.

The Navier-Stokes equations are equations describing the conservation of momentum in the motion of incompressible fluid, i.e., the sum of the external forces acting on the control body is equal to the rate of change of momentum on the control body with time, and have the following basic form.

$$\frac{\partial (\rho \vec{v})}{\partial t} + \nabla \cdot (\rho \vec{v} \vec{v}) = -\nabla p + \nabla \cdot (\overline{\overline{\tau}}) + \rho \vec{g} + \vec{F} \quad (3)$$

In PFR lime kilns, the momentum equation is used to simulate the gas flow and wind resistance variations.

The space inside the lime kiln is filled with limestone particles and the particles move slowly downward, so it can be treated as a moving bed problem inside the lime kiln. The porous media model allows accurate calculation of the pressure drop problem for the moving bed/packed bed, which assumes that the space is occupied by a mixture of solids and fluids.

The introduction of the porous media model in the PFR lime kiln simulation is essentially the introduction of a source term for the momentum loss in the momentum equation to describe the pressure drop (flow wind resistance) generated by the fluid flowing through fluid zone. And the losses originate from the flow viscous resistance and inertial resistance, which is calculated as below.

Ergun equation

$$\frac{|\nabla p|}{L} = \frac{150\mu(1-\gamma)^2}{D_{\mathrm{P}}^2\gamma^3}v_\infty + \frac{1.75\rho(1-\gamma)}{D_{\mathrm{P}}\gamma^3}v_\infty^2 \tag{4}$$

$$\frac{1}{\alpha} = \frac{150(1-\gamma)^2}{D_{\mathrm{P}}^2\gamma^3} \tag{5}$$

$$C_2 = \frac{3.5(1-\gamma)}{D_{\mathrm{P}}\gamma^3} \tag{6}$$

The shrinking core model assumes that the reactants are spheres, that heat is supplied uniformly from all directions, and that the chemical composition of the raw materials is unifo

$$\frac{\partial r_{\mathrm{CaCO_3}}}{\partial t} = -k \cdot \frac{M_{\mathrm{CaCO_3}}}{\rho_{\mathrm{CaCO_3}}} \cdot R_{\mathrm{D}} \tag{7}$$

$$R_{\mathrm{D}} = k_{\mathrm{D}}(p_{\mathrm{eq}} - p_{\mathrm{co_2}}) \tag{8}$$

$$k_{\mathrm{D}} = 0.0001T_{\mathrm{p}}\exp(-4026/T_{\mathrm{p}}) \cdot Y_{\mathrm{T.C}} \tag{9}$$

$$p_{\mathrm{eq}} = 101325\exp[17.74 - 0.00108T_{\mathrm{i}} + 0.332\log(T_{\mathrm{i}}) - 22020/T_{\mathrm{i}}] \tag{10}$$

$$Y_{\mathrm{T.C}} = \begin{cases} \frac{480}{T_{\mathrm{P}}-958} & T_{\mathrm{P}} > 1150K \\ 2.5 & T_{\mathrm{P}} \le 1150K \end{cases} \tag{11}$$

$$\lambda = \frac{4\pi\lambda_1\lambda_2}{\lambda_1\left(\frac{1}{r_{\mathrm{c1}}} - \frac{1}{r_{\mathrm{c2.m}}}\right) + \lambda_2\left(\frac{1}{r_{\mathrm{c1.m}}} - \frac{1}{r_{\mathrm{c1}}}\right)} \tag{12}$$

Limestone decomposition rate is the volume fraction of decomposed calcium carbonate:

$$X_{\mathrm{S}} = 1 - \frac{r_{\mathrm{c2}}^3}{r_{\mathrm{c1}}^3} \tag{13}$$

The temperature equations for gas-solid convective heat transfer, internal and external particle heat conduction, and between solid particles

$$\frac{\partial}{\partial t}[\varphi_{\mathrm{CaCO_3}}\rho_{\mathrm{i}}c_{p\mathrm{i}}T_{\mathrm{i}}] + \nabla\left(\vec{v}_{\mathrm{down}}\varphi_{\mathrm{CaCO_3}}\rho_{\mathrm{i}}c_{p\mathrm{i}}T_{\mathrm{i}}\right) = \lambda(1-\gamma)(T_{\mathrm{o}} - T_{\mathrm{i}})/V_{\mathrm{s}} - k \cdot Q_{\mathrm{D}}\Delta H_{\mathrm{R}} \tag{14}$$

$$Q_{\mathrm{D}} = (1-\gamma)\frac{4\pi r_{\mathrm{c1}}^2}{V_p} \times R_{\mathrm{D}} \tag{15}$$

Energy equation of the external particle (calcium oxide)

$$\begin{gathered}\frac{\partial}{\partial t}(\varphi_{\mathrm{CaO}}\rho_{\mathrm{o}}c_{p\mathrm{o}}T_{\mathrm{o}}) + \nabla \cdot (\vec{v}_{\mathrm{down}}\varphi_{\mathrm{CaO}}\rho_{\mathrm{o}}c_{p\mathrm{o}}T_{\mathrm{o}}) = \\ \nabla \cdot ((k_{\mathrm{CaO}} + e_{\mathrm{b}})\nabla T_{\mathrm{o}}) + a_{\mathrm{v}}h_{\mathrm{v}}(T_{\mathrm{g}} - T_{\mathrm{o}}) - \lambda(1-\gamma)(T_{\mathrm{o}} - T_{\mathrm{i}})/V_{\mathrm{s}}\end{gathered} \tag{16}$$

The radiation equivalent thermal conductivity treatment of porous media used for radiation between solids [28]

$$e_b = 16\sigma T_o^3/(3\beta) \tag{17}$$

The convective heat transfer coefficient h_v is approximately calculated by flow-across tube bundle model in the heat exchanger model [29], which is calculated as following equations

$$h_v = \frac{Nu \cdot \lambda_g}{l_z} \tag{18}$$

$$Nu = Pr^{1/3} \cdot \frac{1.6274Re^{-0.575}}{\gamma} \tag{19}$$

$$l_z = 0.0178\rho^{0.596} \tag{20}$$

$$Re = \frac{\rho_g \cdot D_p \cdot u}{\mu} \tag{21}$$

$$Pr = \nu/a \tag{22}$$

$$a = \frac{\lambda_g}{\rho_g c_{pg}} \tag{23}$$

$$a_v = (1-\gamma) \times \frac{S_p}{V_p} \tag{24}$$

The gas energy equation

$$\frac{\partial}{\partial t}(\gamma \rho_g c_{pg} T_g) + \nabla\left(\gamma \vec{v} \rho_g c_{pg} T_g\right) = \nabla \cdot (k_g \nabla T_g) + a_v h_v (T_o - T_g) \tag{25}$$

$$\varepsilon = \left(1 - \frac{M_{CaO}/\rho_{CaO}}{M_{CaCO_3}/\rho_{CaCO_3}}\right) \times X_S \tag{26}$$

$$\varphi_g = (1-\gamma) \cdot \varepsilon + \gamma \tag{27}$$

$$\varphi_{CaCO_3} = (1-\gamma) \times (1 - X_S) \tag{28}$$

$$\varphi_{CaO} = 1 - \varphi_g - \varphi_{CaCO_3} \tag{29}$$

3. Physical Models

3.1. Geometric Model

The PFR lime kiln has a total height of 20 m and is divided into two chambers, which is connected by a cross channel. Each chamber is divided into a preheating zone, a calcination zone and a cooling zone from top to bottom. The preheating zone is 6 m high, The upper part is the limestone inlet and also the exhaust gas outlet, and the lower part has 8 fuel gas jets. The height of the calcination zone is 9 m. Below the calcination zone is the cooling zone, which is 5 m high and is surrounded by a channel in the upper outer ring, which gradually contracts downwards in the middle to the plane of the cooling air inlet and then gradually expands and extends to the CaO outlet (Figure 1). Chamber 2 has the same components as chamber 1.

A structural grid is used to discretize the entire computational domain. In the region of structurally complex calcination and preheating zones, the grid is sufficiently refined. The grid resolution affects not only the computational time and resources, but also the reliability of the simulation results. The sensitivity for grid numbers of 4.6 million, 5.2 million and 5.5 million was examined in this PFR lime kiln model to obtain grid independent results. The comparison results show that a denser grid has a minor effect on the computational results.

3.2. Operating Procedures

The PFR lime kiln consists of two chambers, with the fuel gas in Chamber 1 injected into the calcination zone to calcine the limestone while the fuel gas jet in Chamber 2 does not operate. The main component of limestone is the calcium carbonate, which is heated at high temperatures to decompose into calcium oxide and carbon dioxide, a process known as limestone calcination. When calcination begins, the high temperature gas heats the calcium carbonate, which starts to decompose into calcium oxide and carbon dioxide from the surface to the inside, and as the decomposition process proceeds, the material gradually moves down and from the calcination zone into the cooling zone. Cooling air is blown into the cooling zone from the bottom up by the cooling air inlet to quickly cool the material that has just entered the cooling zone so that its chemical activity is not reduced by the high temperature. The material is cooled as it gradually moves down and is eventually discharged through the CaO outlet. The high temperature gas is involved in the calcination and the temperature is reduced, while the cooling air is involved in the cooling and the temperature is increased. The two are mixed in the cross channel to form the exhaust gas. The heat carried by the exhaust gas has a preheating effect on the unreacted material in Chamber 2, while the cooling air is continuously blown into the Chamber 2 cooling gas inlet to enhance the exhaust gas convection and accelerate the efficiency of the exhaust gas upwards. Finally, after passing through the material in the calcining and preheating zones in Chamber 2, the exhaust gas is discharged by the upper exhaust gas outlet, while the limestone inlet in Chamber 1 is only feeding the material at this time and is not operating as an exhaust gas outlet. After 15 min of operation of Chamber 1, fuel gas is instead injected by the fuel gas jet in Chamber 2, while Chamber 1 functions as an exhaust gas outlet for 15 min. The PFR lime kiln completes one cycle every 30 min and continues to operate (Figure 2).

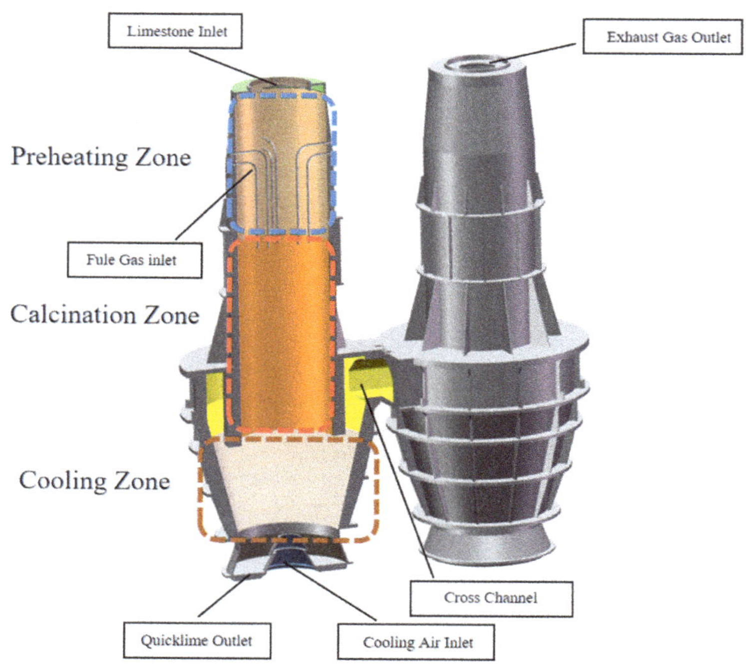

Figure 1. Schematic diagram of the PFR lime kiln components.

3.3. Boundary Conditions and Calculation Methods

As the outer walls of the PFR lime kiln are well insulated, the inner surface of the kiln is used as an adiabatic boundary. The interior of the kiln is divided into two fluid domains, the porous media area filled with the calcium carbonate and the gas area where only the exhaust gas is present. The boundary below the fuel gas nozzle is used as the fuel gas inlet boundary, while the rest of the nozzle is used as the exterior. Also, the temperature of the blast furnace gas after combustion is used as the fuel gas inlet temperature and other calculation conditions are shown in Table 1.

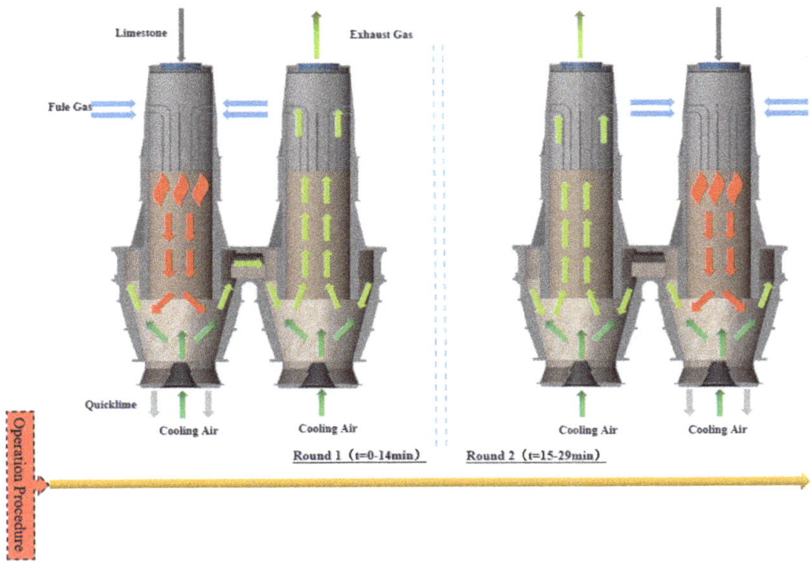

Figure 2. Schematic diagram of the operating procedure of the PFR lime kiln.

Table 1. Calculation conditions.

Calculation Conditions	Value	Unit
Fuel gas inlet velocity	20	m/s
Fuel gas inlet temperature	1673	K
Fuel gas nozzle diameter (8)	70	mm
Cooling air inlet velocity	10	m/s
Cooling air inlet temperature	300	K
Cooling air inlet diameter	1000	mm
Material movement speed	1.54	m/h
Initial temperature of material (limestone)	300	K
Average diameter of material	40, 80, 120	mm
Calcium carbonate density	3310	kg/m^3
Calcium oxide density	2810	kg/m^3
Calcium carbonate thermal conductivity	2.26	W/m·k
Calcium oxide thermal conductivity	0.07	W/m·k
Calcium carbonate decomposition temperature	1073	K
Void fraction	0.36, 0.41, 0.46	-

Density, thermal conductivity and void fraction is determined in the laboratory.

The semi-implicit method of coupled pressure equations (referred to as the SIMPLE algorithm) was used to solve the coupled velocity and the pressure problem in the process of solving the equations; two equations in the turbulence equations were solved in the first-order windward difference format, and the remaining equations were treated by the second-order windward difference; the convergence factor of each equation was set to 10^{-6}.

The following hypotheses will also be established in this study in order to exclude any interference to the results.

i. The limestone is composed exclusively of calcium carbonate, disregarding the influence of impurities that are in fact present in the limestone, such as MgO, on the calcination results.
ii. The limestone raw material is spherical and the particle size refers to the diameter.
iii. Ignoring the effect of airflow on the movement of the material.

4. Results and Analysis

A number of locations within the PFR lime kiln were taken for analysis, as shown in Figure 3, where the two intermediate lines are located in the geometric centre of the chambers called mid lines and the near wall lines are located 40 mm from the chamber side wall, with at least one material particle accommodated between the line and the side wall, which is not only farther away from the fuel gas inlet and poorly heated, but is also affected by the side wall effect, with differences in particle movement and decomposition downwards from mid lines. The exhaust gas temperature was analysed by taking the exhaust gas outlet above chamber 2. The central profile Plane 1 is taken to show the overall distribution of all fields. Plane 2 is taken in the centre of the intermediate channel to analyse the mixture of exhaust gas and cooling air after calcination. Plane 3 is taken at the intersection of the calcining and cooling zones and analyses the surface temperature of the solids after calcination, the core temperature and the decomposition of the limestone.

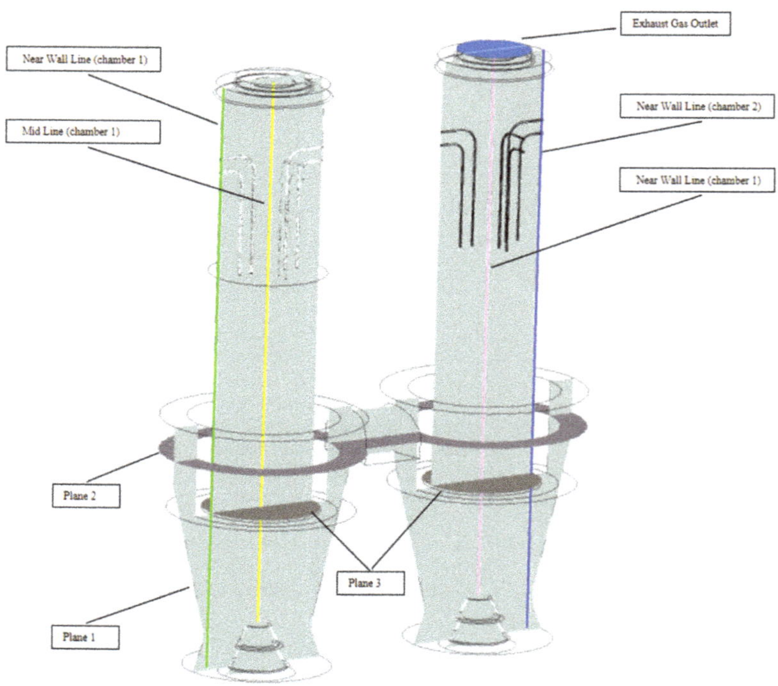

Figure 3. Schematic diagram of the analysis position.

4.1. Gas Velocity Field

The gas velocity distributions on the mid lines and near wall lines at 40 mm, 80 mm and 120 mm particle sizes are shown in Figure 4. At a distance of 14 m from the CaO outlet, fuel gas enters the calcination zone from the boundary to heat the limestone, and as the distance gradually decreases, the mid lines gas velocity increases, but is blocked by the solid material and the increase is small. Until about 5 m from the CaO outlet, the gas velocity decreases and then sharply increase to about 7 m/s at about 2 m from the CaO outlet due to the effect of cooling air blowing in. Due to the high temperature and low density of the fuel gas, some of it rises into the preheating zone after being blown into the calcination zone of the lime kiln. In addition, after the cooling air has cooled the high temperature material, some of the exhaust gas does not flow through the intermediate passage towards chamber 2, but enters the preheating zone via the calcination zone of chamber 1.

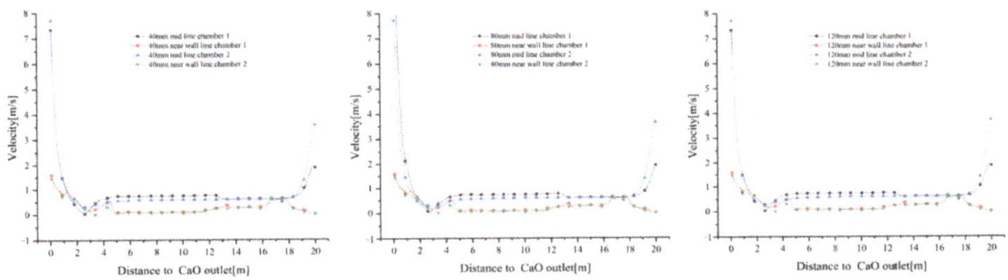

(**a**) (**b**) (**c**)

Figure 4. Gas velocity variation on the mid lines and near wall lines for 40 mm, 80 mm and 120 mm particle sizes. (**a**) Gas velocity variation at 40 mm; (**b**) Gas velocity variation at 80 mm; (**c**) Gas velocity variation at 120 mm.

The gas velocity distribution of the near wall line in the calcination zone shows an opposite trend to that of the mid line. As one approaches the CaO outlet, the near wall line gas velocity decreases and the gas reaches the cooling zone with a velocity close to that of mid lines gas velocity. However, at closer distances from the CaO outlet, the near wall line gas velocity does not show a sharp increase in velocity similar to mid line gas velocity, but increases slowly to below 2 m/s. Similarly, in the preheating zone, the near wall line gas velocity is less affected by the rising high temperature fuel gas and cooling air convergence, and the gas velocity is smaller, approaching 0 near the top limestone inlet. It should be noted that the velocity at 14 m from the CaO outlet is not the fuel gas inlet velocity, as neither mid line nor the near wall line passes through the fuel gas inlet boundary. (Figure 4a–c)

Similarly, the variation of gas velocity tends to be the same for the three particle sizes of 40 mm, 80 mm and 120 mm (Figure 5a–d), indicating that the effect of different void fractions on gas velocity is small.

Combined Figures 4 and 5 show that the gas velocity varies less with particle size. Whereas the velocity variations in the centre of the chamber (mid lines) and the kiln side wall (near wall lines) converge in the calcination zone, they show different variations in the preheating and cooling zones. The centre of the cooling air inlet is on the same axis as the centre of the chamber, which inevitably leads to higher velocities in the centre of the chamber. The side walls of the chamber, on the other hand, are less affected by the fuel gas and cooling air, and their velocity variation over the sections is also less.

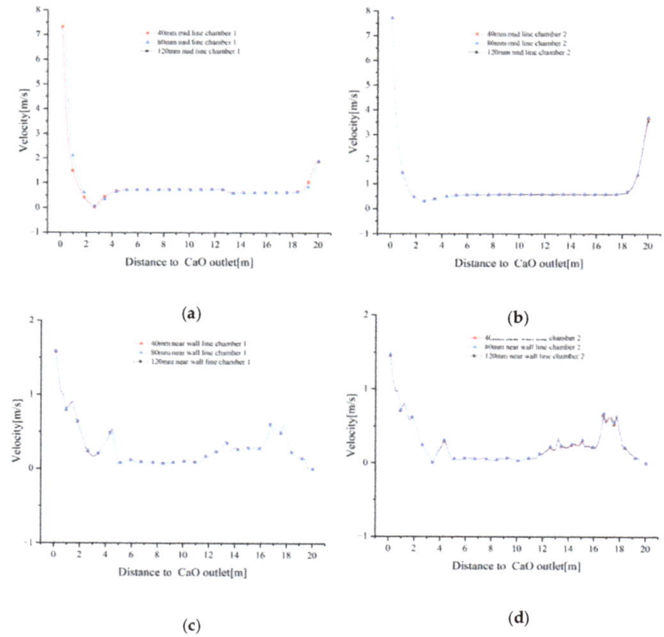

Figure 5. Comparison of gas velocities at mid lines and near wall lines in chamber 1 and 2. (**a**) Change in gas velocity at the mid line of chamber 1; (**b**) Change in gas velocity at the mid line of chamber 2; (**c**) Change in gas velocity at the near wall line of chamber 1; (**d**) Change in gas velocity at the near wall line of chamber 2.

4.2. Gas Temperature Field

As the gas temperature distribution is directly influenced by the gas velocity distribution, there is little difference in the gas temperature distribution at 40 mm, 80 mm and 120 mm particle sizes (Figure 6a–c). At 40 mm particle size, the temperature distribution in the calcination zone of chamber 2 (where calcination does not occur) is uniform (Figure 6a), while at 80 mm particle size, the temperature in the centre of the calcination zone of chamber 2 is significantly higher than that in the side walls (Figure 6b), and as the particle size increases to 120 mm, the temperature in the centre of both the calcination and preheating zones of chamber 2 is higher, with less difference between the temperature in the near wall zone and the center (Figure 6c).

As shown in Figure 7a, the temperature of the near wall line of chamber 1 at 40 mm particle size varies little from the preheating zone to the lower part of the calcination zone, until the junction of the cooling zone, where the gas temperature is influenced by the cooling air and decreases to around 350 K. The temperature of the mid line of chamber 1 remains similar to the near wall line in the preheating zone, but increases rapidly up to 1200 K upon entering the calcination zone. The mid line of chamber 1 remains at a similar temperature to the near wall line in the preheating zone, however, the temperature increases rapidly in the calcination zone due to the high temperature of the fuel gas up to 1200 K. After entering the cooling zone, the mid line gas temperature is also affected by the cooling air and decreases rapidly to around 350 K. The gas temperature at the mid line of chamber 2 varies more gently with the height of the chamber and only changes significantly between the cooling zone and the cross channel. The change in gas temperature in chamber 1 at 80 mm and 120 mm particle size is more or less the same as the change in gas temperature at 40 mm (Figure 7b,c).

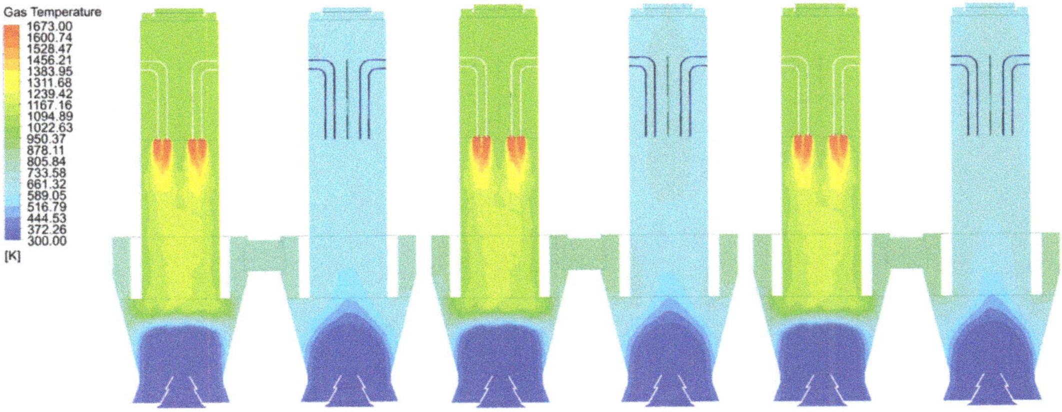

(a) (b) (c)

Figure 6. Gas temperature distribution at 40 mm, 80 mm and 120 mm particle sizes. (**a**) Gas temperature distribution at 40 mm (**b**) Gas temperature distribution at 80 mm (**c**) Gas temperature distribution at 120 mm.

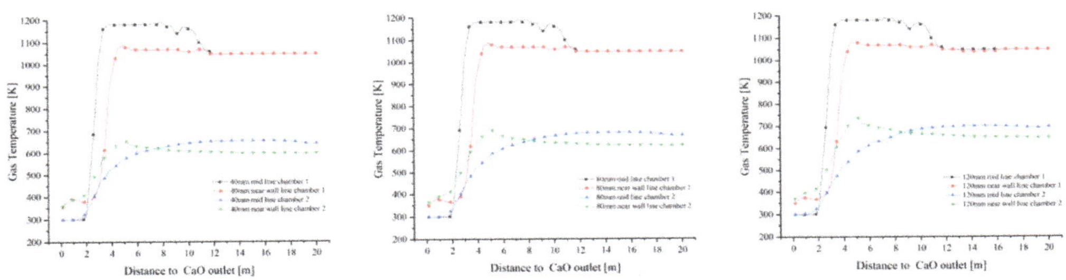

(a) (b) (c)

Figure 7. Gas temperature variation on the mid and near wall lines for 40 mm, 80 mm and 120 mm particle sizes. (**a**) Gas temperature variation at 40 mm (**b**) Gas temperature variation at 80 mm (**c**) Gas temperature variation at 120 mm.

The variation in mid line and near wall line temperatures in chamber 1 is essentially the same for all three particle sizes (Figure 8a,c). In chamber 2, the variation in mid line and near wall line gas temperatures is more pronounced with lime kiln height for all three particle sizes. 120 mm particle size chamber 2 mid line gas temperatures is higher than 80 mm and 40 mm chamber 2 mid line gas temperatures (Figure 8b). Similarly, the near wall gas temperature of chamber 2 for 120 mm particle size is higher than the near wall gas temperature of chambers 2 for 80 mm and 40 mm (Figure 8d).

The cross channel has only high temperature exhaust gas and no solid material, so the temperature variation is more pronounced than in the chambers, and is characterised by uneven gas mixing and temperature distribution in the outer ring of chamber 1, while entering the centre of the channel, the mixing is uniform and the average temperature is highest at this point, and gradually decreases as the gas enters the outer ring of chamber 2 (Figure 9).

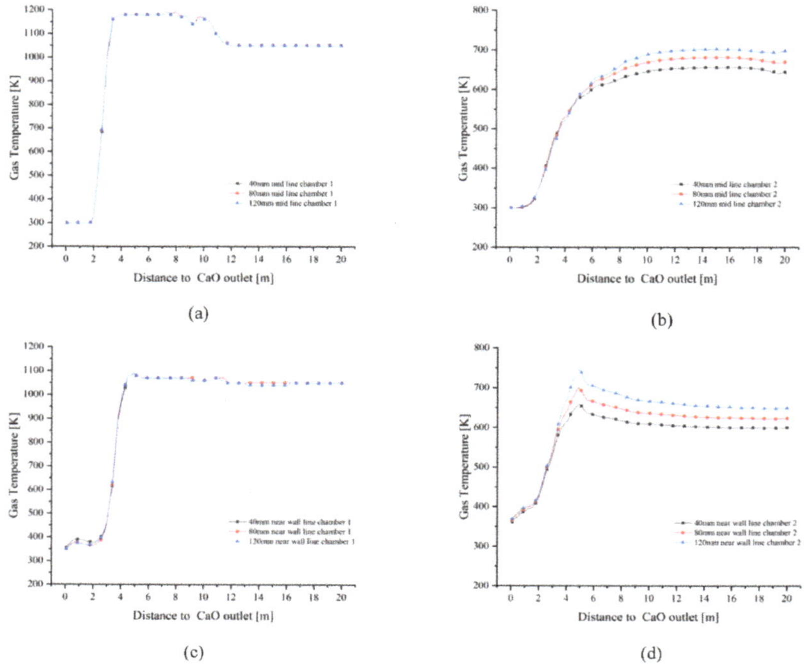

Figure 8. Comparison of gas temperature variations in the mid line and near wall line of chamber 1 and 2. (**a**) Change in gas temperature at the mid line of chamber 1; (**b**) Change in gas temperature at the mid line of chamber 2; (**c**) Change in gas temperature at the near wall line of chamber 1; (**d**) Change in gas temperature at the near wall line of chamber 2.

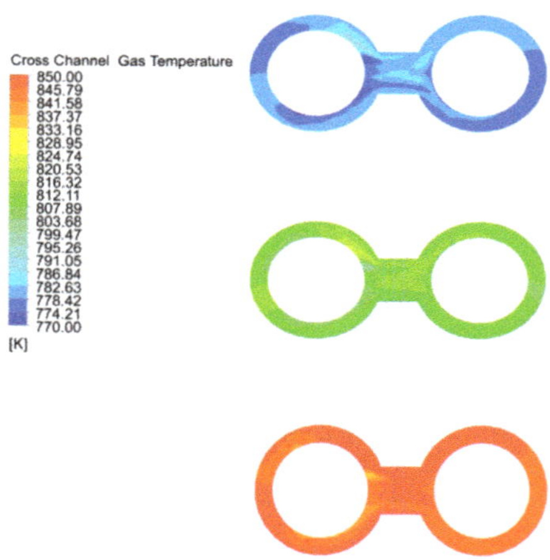

Figure 9. Gas temperature distribution and variation in the cross channel at 40 mm, 80 mm and 120 mm particle sizes.

The gas temperature distribution at the exhaust gas outlet in Figure 10 matches the gas temperature in Figure 6. 120 mm particle size has the highest gas temperature at the exhaust gas outlet, followed by 80 mm, while 40 mm particle size has the lowest gas temperature.

The gas temperature at the centre of the exhaust gas outlet is also higher than near wall, which is consistent with the results in Figure 7.

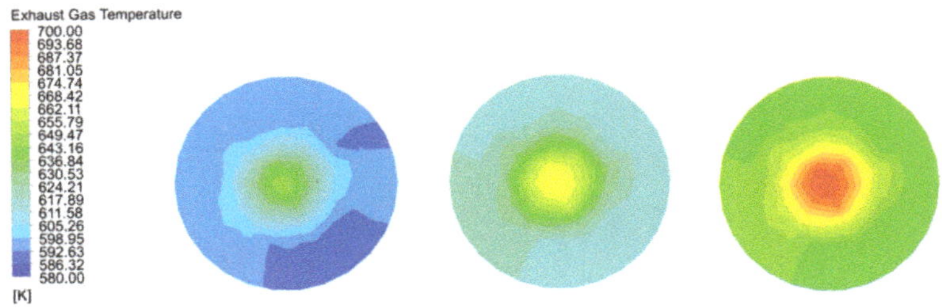

(**a**) (**b**) (**c**)

Figure 10. Gas temperature distribution and variation in exhaust gas outlet at 40 mm, 80 mm and 120 mm particle sizes. (**a**) Gas temperature in exhaust gas outlet of 40 mm; (**b**) Gas temperature in exhaust gas outlet of 80 mm; (**c**) Gas temperature in exhaust gas outlet of 120 mm.

Combined with the gas temperature variations and distributions in Figures 6–10, the gas temperature distributions in chamber 1 are similar for 40 mm, 80 mm and 120 mm particle sizes, while the gas temperature in chamber 2 increases as the particle size gradually increases, as reflected in the cross channel and the exhaust gas outlet cross section. This indicates that the gas temperature distribution varies with the particle size (void fraction) for a given gas velocity. The effect of particle size (void fraction) on gas temperature decreases as the gas velocity increases.

4.3. Solid Surface Temperature

For CaO used in reduction reactions in metallurgical processes, the chemical activity is the most important indicator, and maintaining a solid surface temperature of no more than 1423 K during calcination is a decisive parameter in maintaining chemical activity. If the solid surface pore size is reduced over a long period of time above 1423 K, the chemical activity decreases and is referred to as "hard-burnt" or "over-burnt", so it is important to maintain a reasonable solid surface temperature.

The 40 mm solid material briefly reaches about 1600 K near the fuel gas inlet in the calcination zone of chamber 1, then the temperature drops rapidly as the material moves down into the cooling zone, where the solid surface temperature is rapidly cooled to room temperature by the cooling air. In contrast, the solid surface temperature in the preheating zone does not exceed 1200 K (Figure 11a). The solid material surface temperatures at 80 mm and 120 mm particle size also show similar temperature distributions, but in chamber 2 the solid material surface temperature distribution at 80 mm and 120 mm particle size differs from that at 40 mm, which is more in line with the gas temperature distribution in Figure 6.

As shown in Figure 12a–c, the solid surface temperatures at the mid and near wall lines of chamber 1 are similar in the preheating zone and gradually increase as they move from the calcination zone towards the CaO outlet, with a greater temperature gradient at the mid line than at the near wall line. The mid line and near wall line solid surface temperatures are rapidly cooled by the cooling air until they reach the cooling zone, where the temperature rapidly decreases to around 300 K to 350 K. Chamber 1 shows similar variations for all three particle sizes: 40 mm, 80 mm and 120 mm. In chamber 2, however, the temperature of the near wall line is characterised by a larger particle size and a higher surface temperature due to the influence of the high temperature exhaust gas brought in by the cross channel at different temperatures.

(**a**) (**b**) (**c**)

Figure 11. Solid surface temperature distribution at 40 mm, 80 mm and 120 mm particle sizes. (**a**) 40 mm solid surface temperature distribution (**b**) 80 mm solid surface temperature distribution (**c**) 120 mm solid surface temperature distribution.

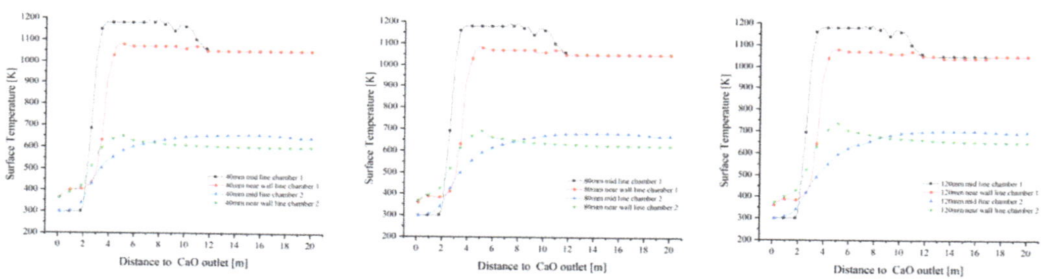

(**a**) (**b**) (**c**)

Figure 12. Variation of solid surface temperature on the mid lines and near wall lines for 40 mm, 80 mm and 120 mm particle sizes. (**a**) 40 mm solid surface temperature variation (**b**) 80 mm solid surface temperature variation (**c**) 120 mm solid surface temperature variation.

The mid line and near wall line solid surface temperatures of chamber 1 converge for each particle size (Figure 13a,c). In the preheating zone, the solid surface temperature remains around 1050 K. After calcinate in the calcining zone, the mid line solid surface temperature rises close to 1200 K and the near wall line solid surface temperature reaches 1100 K. Neither exceeds 1423 K, which does not lead to hard-burnt and reduced chemical activity. The trend in the mid line and near wall line solid temperatures in chamber 2 correlates with the change in gas temperature, with the largest solid surface temperature gradient at 120 mm particle size reaching 700 K in the mid line and 750 K in the near wall line, and up to 650 K in the mid line and near wall line line at 40 mm.

Plane 3 is the interface between the calcination and cooling zones, as shown in Figure 14 which reflects the temperature field of the solid material for 40 mm, 80 mm and 120 mm particle sizes before entering the cooling zone. At particle sizes 40 mm and 80 mm, the solid surface temperature distribution in chamber 1 is basically the same, with a maximum value of 1170 K, while the 80 mm particle size has a higher solid surface temperature gradient in chamber 2 than at 40 mm particle size, with a maximum temperature of 690 K, while the 40 mm particle size is only about 650 K. The maximum temperature in chamber 1 is still around 1170 K when the particle size is 120 mm, but the distribution is wider than at 40 mm and 80 mm, and the maximum solid surface temperature in chamber 2 reaches 740 K. In addition, the lowest surface temperature for all three particle sizes in chamber 2 is around 510 K.

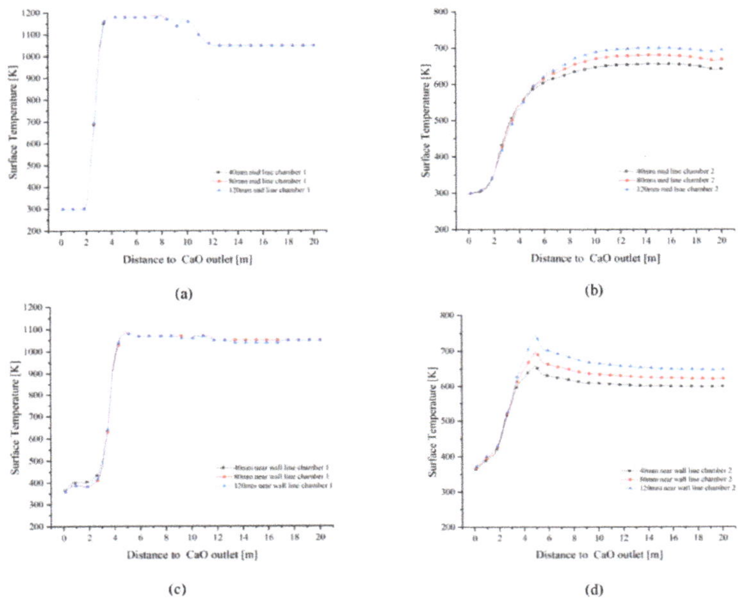

Figure 13. Comparison of solid surface temperature variations in the mid line and near wall line of chamber 1 and 2. (**a**) Change in solid surface temperature at the mid line of chamber 1; (**b**) Change in solid surface temperature at the mid line of chamber 2; (**c**) Change in solid surface temperature at near wall line of chamber 1; (**d**) Change in solid surface temperature at near wall line of chamber 2.

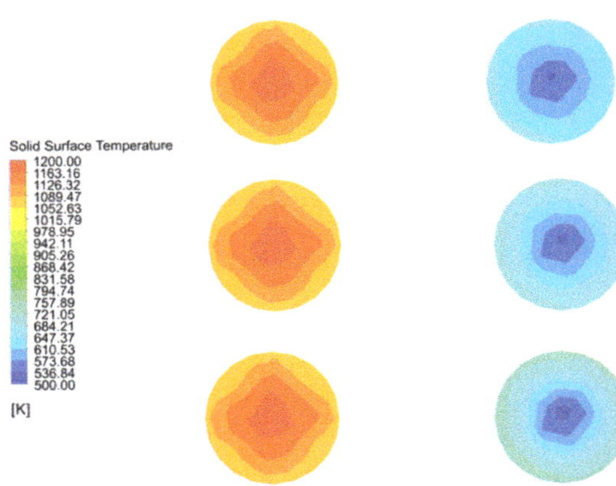

Figure 14. Distribution and variation of solid surface temperatures in double chambers on Plane 3.

The solid surface temperature did not reach the upper limit of 1423 K hard-burnt temperature, which is also related to the fact that the heat of combustion of BFG is lower than the heat of combustion of natural gas. In addition, the near wall line has a significant difference between the downward velocity and the temperature gradient of the material and the mid line material due to the side wall effect. Although the gas temperature fields converge for all three particle sizes, the solid surface temperature still shows a greater temperature gradient at 120 mm in chamber 2 as it gradually approaches the cooling zone In contrast, the solid surface temperature in chamber 1 does not change significantly with particle size.

4.4. Solid Core Temperature

During the calcination of limestone, the temperature must be greater than 1173 K before decomposition reactions begin. As the decomposition reaction progresses from the outside to the inside, the core temperature rises at a slower rate and eventually there may be a part of the core with a temperature less than 1173 K where no decomposition reaction takes place, a situation generally referred to as "under burning". The "under burning" means that the limestone has not decomposed completely.

As shown in Figure 15a, the core temperatures in both the calcination and preheating zones at 40 mm particle size reach above 1000 K, while the core temperature in the central part of the cooling zone reaches above 1000 K. The core temperature near wall drops rapidly. On the one hand, this is due to the fact that the combustion heat of the fuel gas cannot reach a long distance, and on the other hand, it is also due to the fact that the cooling air cools the solid in the cooling zone so quickly that not only the surface temperature of the material drops, but also the core temperature of part of the material. The core temperature in the cooling zone at 120 mm particle size is significantly smaller than that at 1000 K (Figure 15c). The core temperature distribution in chamber 2 for the three particle sizes increases with increasing particle size, resulting in a more pronounced increase in the area with higher core temperatures.

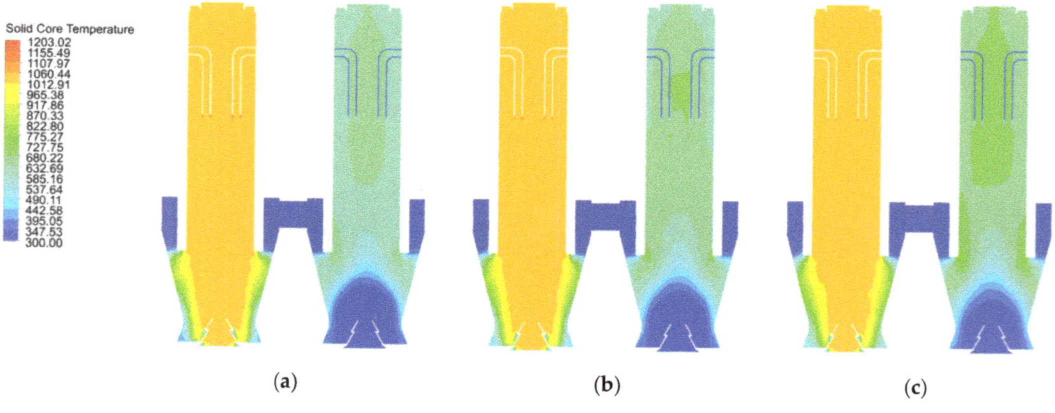

Figure 15. Solid surface temperature distribution at 40 mm, 80 mm and 120 mm particle sizes. (**a**) 40 mm solid core temperature distribution (**b**) 80 mm solid core temperature distribution (**c**) 120 mm solid core temperature distribution.

In combination with Figures 15a,b and 16a,b, the variation of the mid line and near wall line temperatures in chamber 1 are essentially the same for 40 mm and 80 mm grain sizes, while for 120 mm grain size, the mid line solid core temperature remains the same as the former, while the near wall line shows a greater drop as it enters the cooling zone (as in Figure 16c). For the solid core temperature in chamber 2, the changes in the mid line and near wall line are essentially the same, remaining at around 650 K and 550 K respectively.

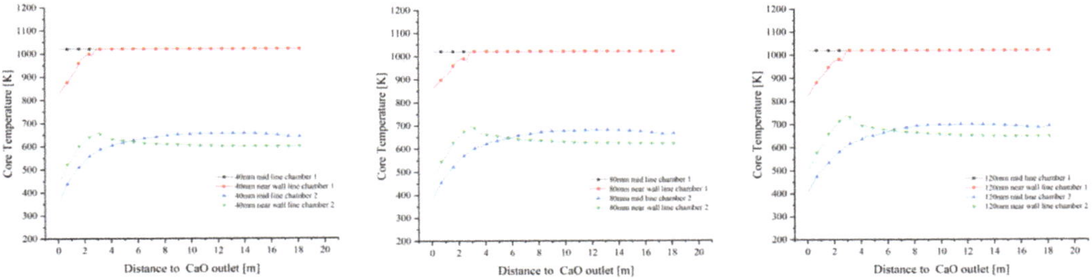

(**a**) (**b**) (**c**)

Figure 16. Solid core temperature variation on the mid line and near wall line for 40 mm, 80 mm and 120 mm particle sizes. (**a**) Variation in solid core temperature at 40 mm (**b**) Variation in solid core temperature at 80 mm (**c**) Variation in solid core temperature at 120 mm.

As shown in Figure 17a,b, the core temperatures in the mid line of chamber 1 are essentially the same, with minor differences in the near wall line only in the cooling zone. In contrast, the core temperatures for the 40 mm, 80 mm and 120 mm particle sizes at the mid line and near wall lines of chamber 2 show significant differences. During the upward movement away from the cooling zone of chamber 2, the solid core temperature reaches 700 K at the mid line of the 120 mm grain size and even 750 K at the near wall line at one point and stabilises around 650 K during the upward movement, while the temperature gradient decreases at the mid line and near wall line of the 80 mm and 40 mm grain sizes, with the maximum temperature gradient reaching over 50 K.

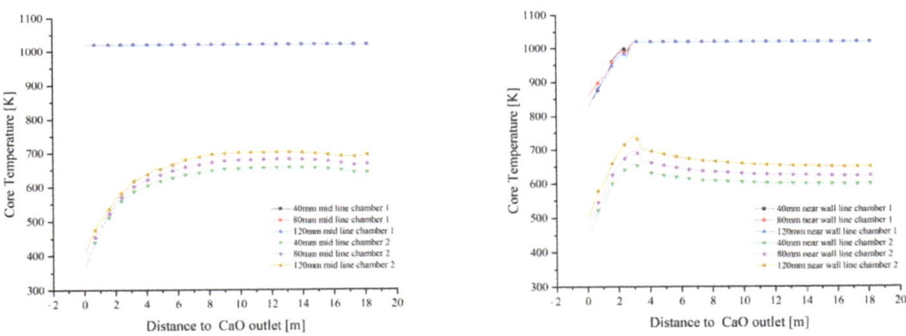

(**a**) (**b**)

Figure 17. Comparison of mid and near wall line solid core temperature variations. (**a**) Variation in solid core temperature at the mid line (**b**) Variation in solid core temperature at the near wall line.

Combining the temperature variations of chamber 1 in Figure 17a,b, it can be seen that the core temperature variations of Plane 3 at the interface between the calcination and cooling zones are mainly manifested in chamber 2, therefore the core temperature analysis of Plane 3 is also for chamber 2, as shown in Figure 18.

The maximum temperature of Plane 3 at the interface of the calcination and cooling zones in chamber 2 at 40 mm particle size is 650 K and the minimum temperature is 550 K. As the particle size increases to 80 mm, the core temperature reaches 740 K and 560 K respectively, and as the particle size continues to increase to 120 mm, the maximum core temperature basically remains unchanged at 740 K, but the minimum temperature rises to around 575 K. There is a significant reduction in the temperature difference.

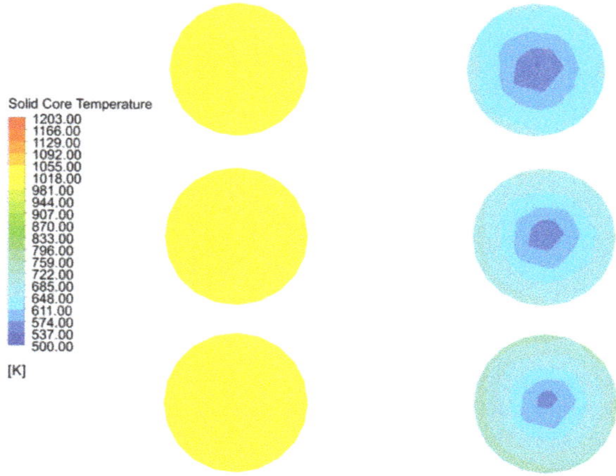

Figure 18. Solid core temperature distribution on double chambers and solid core temperature variation in chamber 2 on Plane 3.

The core temperature distribution and changes in the solids at 40 mm, 80 mm and 120 mm particle sizes shows that the core temperature in chamber 1 varies little with particle size, especially at 40 mm and 80 mm, and does not change much until 120 mm. The main reason for this is that as the decomposition reaction continues, the core temperature rises more slowly in the larger particle sizes, and by the time it reaches the cooling zone, the core of some of larger solids has not yet risen to a sufficient temperature and is affected by the cooling air and no longer continues to rise. The core temperature contrast in chamber 2 is even more pronounced, i.e., the temperature gradient is smaller for the smaller particle sizes and larger for the larger particle sizes, mainly because the smaller particle sizes warm up faster and the heat is more evenly distributed, while the larger particle sizes show an external hot and internal cold distribution of core temperature at the same time.

4.5. Limestone Decomposition Field

The calcination of limestone is completely dependent on the remaining calcium carbonate thickness and therefore the limestone decomposition is represented by the distribution of undecomposed thickness (e.g., Figure 19a–c). As the particle size gradually increases, the decomposition reaction becomes more difficult to carry out and the remaining limestone thickness no more decreases, indicating that the limestone reaction gradually increases in unreacted thickness as the particle size increases.

At 40 mm particle size, the undecomposed thickness of the chamber 1 calcination zone gradually decreases as the mid line moves down towards the CaO outlet, and the undecomposed thickness remains the same in the upper part of the preheating and calcination zones, remaining at around 2.5 mm after reaching the cooling zone. The undecomposed thickness near the wall line is above 3.5 mm in the preheating zone until it rapidly decreases and remains around 3 mm after entering the calcining zone. A similar trend was observed for the near wall line at 80 mm particle size, where the undecomposed thickness was close to 12 mm, while the undecomposed thickness of the mid line remained around 7 mm and did not change significantly with height (Figure 20b). At 120 mm grain size, the mid line undecomposed thickness is the same as at 80 mm, while the near wall line shows a large difference. The undecomposed thickness of the near wall line was as high as 30 mm in the preheating zone, and only after entering the calcination zone did the undecomposed thickness decrease significantly and was similar to the decomposition of the 40 mm and 80 mm grain sizes (Figure 20c).

(a) (b) (c)

Figure 19. Distribution of undecomposed thickness of $CaCO_3$ at 40 mm, 80 mm and 120 mm particle sizes. (**a**) undecomposed thickness of $CaCO_3$ at 40 mm (**b**) undecomposed thickness of $CaCO_3$ at 80 mm (**c**) undecomposed thickness of $CaCO_3$ at 120 mm.

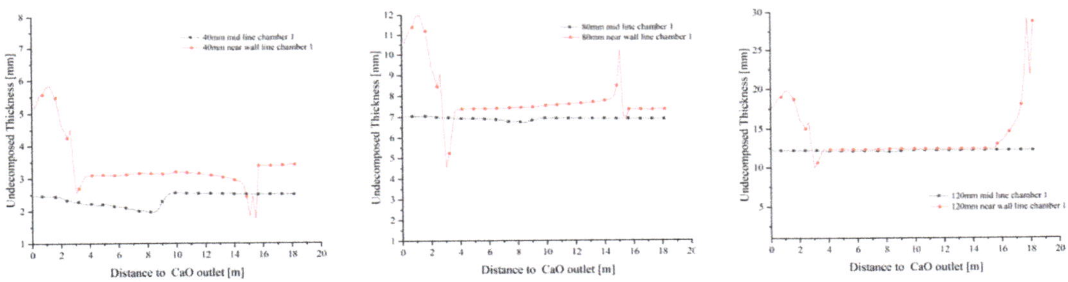

(a) (b) (c)

Figure 20. Variation of undecomposed thickness of $CaCO_3$ at 40 mm, 80 mm and 120 mm particle sizes. (**a**) Variation of undecomposed thickness of $CaCO_3$ at 40 mm (**b**) Variation of undecomposed thickness of $CaCO_3$ at 80 mm (**c**) Variation of undecomposed thickness of $CaCO_3$ at 120 mm.

At the CaO outlet, the undecomposed thickness distribution is shown in Figure 21. At a particle size of 40 mm, the undecomposed thickness near the inner side is small, with an average undecomposed thickness of about 3 mm, and increases to 7 mm as it moves towards the outer side, while at a particle size of 80 mm, the undecomposed thickness reaches 7 mm on the inner side and more than 13 mm on the outer side. The decomposition distribution at all grain sizes indicates that the side wall effect has a significant influence on the final decomposition, especially at larger grain sizes where the side wall effect has a greater influence on the decomposition process.

The variation of limestone decomposition rates for the three sizes is shown in Figure 22. At 40 mm particle size, the mid line limestone decomposition rate is close to 90%, while the near wall decomposition rate is between 75% and 85%. The mid line decomposition rate reaches about 83% with 80 mm particle size, while the near wall decomposition rate is the same as at 40 mm particle size, between 75% and 85%. When the particle size increases to 120 mm, the mid line decomposition rate remains around 20% and the decomposition effect is poor, while the near wall line can only reach 50% decomposition rate in the position between the preheating and calcination zones, and in other positions, the decomposition rate is generally below 30%. The main reason of its close its close to 50% decomposition rate is that the limestone inlet is closed when calcination takes place in chamber 1, i.e., a large amount of high temperature exhaust gas accumulates at the top, which promotes limestone decomposition to a certain extent, but this promotion is not stable and the final decomposition effect is far from the decomposition demand.

Figure 21. Variation of 40 mm, 80 mm and 120 mm undecomposed thickness of CaO outlet.

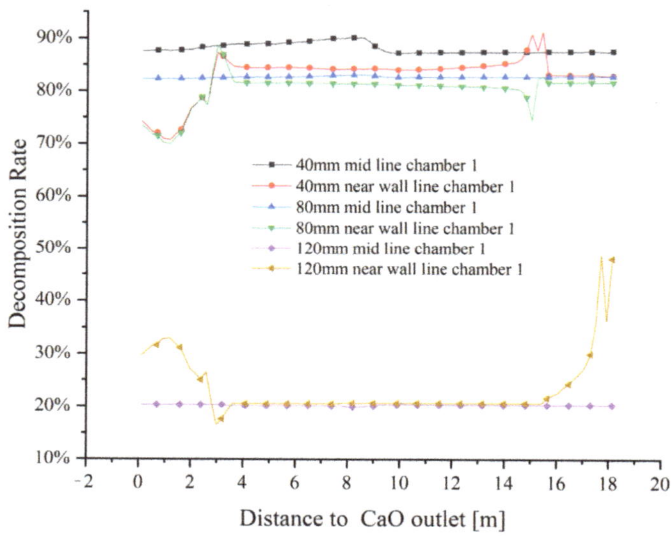

Figure 22. Variation in mid lines and near wall lines decomposition rates at 40 mm, 80 mm and 120 mm particle sizes.

The undecomposed thickness varies significantly from particle size. The larger the grain size, the more difficult the decomposition and the greater the effect on the final decomposition rate. The limestone decomposition rate is also strongly influenced by the side wall effect, with limestone decomposition in the mid line outperforming that in the near wall line. At the same time, the side wall effect has a greater influence on the small particle size than on the large particle size. At 120 mm particle size, the effect of not reaching the decomposition temperature at the core is much greater than the effect of the side wall effect on decomposition.

5. Conclusions

This paper presents a comparative analysis of the gas velocity field, gas temperature field, solid surface and core temperature field, and limestone decomposition field for limestone decomposition processes at 40 mm, 80 mm and 120 mm particle sizes. The results show that.

(i) Changing the fuel gas from natural gas to BFG while keeping the PFR lime kiln structure unchanged will reduce the range of supported calcined grain sizes from

40 mm to 120 mm to below 80 mm, and that limestone of 120 mm grain size is difficult to calcine and decompose under current conditions.

(ii) Even at 40 mm grain size, the decomposition rate is over 90% after calcination by natural gas, whereas with BFG calcination, the decomposition rate is 75% to 85% and the under burning rate is even higher.

(iii) By switching to calcination by BFG, the problem of hard burnt is largely avoided as the solid surface temperature does not exceed 1200 K.

(iv) Analysis of the exhaust gas outlet temperature shows that the exhaust gas is discharged at a temperature of no more than 700 K and even less than 500 K at 40 mm particle size, which is a further improvement in the utilisation of energy.

(v) The side wall effect has a greater impact on the calcination of small grain size limestone and is the main reason for the higher under burning rate.

(vi) There is room for further research and exploration in this study as follows. Firstly, it is clear that the original kiln structure is no longer suitable for BFG calcination of limestone and needs to be optimised for low calorific value fuel calcination of limestone; secondly, the particle size range for BFG calcination of limestone must be less than the range of 40 mm to 120 mm, which is suitable for natural gas calcination, but the exact particle size support range still needs to be further investigated in conjunction with experiments after optimising the PFR lime kiln structure. Finally, experimental validation of low calorific value gas-fired limestone calcination needs to be established and this part of the work will be necessary.

Author Contributions: S.D., writing—original draft, mathematical model establishing, programming; B.L., conceptualization, methodology, writing—review and editing; W.R., data curation, field measurements. All authors have read and agreed to the published version of the manuscript.

Funding: This work was supported by the National Natural Science Foundation of China (Grant No. 51934002).

Acknowledgments: Firstly: I would like to thank Baokuan Li for his excellent contribution to this paper. He was keenly aware of the importance of PFR lime kilns for energy efficiency and environmental protection, and has done a great deal of preliminary research in this area, as well as providing highly creative guidance on mathematical model building and development. Secondly, thanks Wenjie Rong's preliminary research on rotary lime kilns, which has been very informative for this paper, her research on sidewall effects and discrete elements has been very thorough and pioneering and has led to significant progress. Finally, thanks Cai Zhang for her significant contribution to the geometric modelling, her skilful BIM skills helped to convert the 2D drawings into a 3D model, which is the best PFR model by far. I have also benefited from her constant companionship in writing, which has been a constant source of inspiration and a relationship that I will continue to cherish for the rest of my life.

Conflicts of Interest: The authors declare that they have no conflict of interest.

Nomenclature

ρ_g Fluid density; D_p Diameter of the particles; α Permeability in porous media; C_2 Inertial drag coefficient; M_{CaCO_3} Relative molecular mass of calcium carbonate; d_{CaCO_3} Diameter of the internal core calcium carbonate; ρ_{CaCO_3} density of the calcium carbonate; R_D Limestone decomposition rate; k_D Reaction constants; p_{eq} Equilibrium partial pressure of carbon dioxide at the front of the reaction zone; p_{co_2} CO_2 partial pressure in the environment; T_p Average particle temperature; T_i Temperature of internal core calcium carbonate; $Y_{T.C}$ Reaction rate correction factor; λ_1 Thermal conductivity of internal core calcium carbonate, λ_2 Thermal conductivity of external core calcium oxide, r_{c1} Calcium carbonate radius $r_{c1.m}$ 1/2 calcium carbonate radius $r_{c2.m}$ calcium carbonate radius plus 1/2 calcium oxide layer radius; φ_{CaCO_3} Volume fraction of calcium carbonate, ρ_i density of calcium carbonate, c_i specific heat capacity of calcium carbonate; Q_D Particle reaction rate; ΔH_R Heat of decomposition of calcium carbonate(183,000 J/mol); V_p Volume of individual particle; φ_{CaO} Volume fraction of calcium oxide, ρ_o density of calcium oxide, c_o specific heat capacity of calcium oxide; a_v Specific surface area of limestone; h_v heat transfer coefficient of gas solid two phase flow; σ Stephen Boltzmanns constant;

5.6697×10^{-8} W/(K^4·m^2); β Radiation attenuation coefficient of porous media; Nu Nussel number, l_z Stacked bed feature length, Pr Prandtl number, Re Reynolds number, a Thermal diffusion coefficient; ν Kinematic viscosity of gas, μ dynamic viscosity of gas, u Velocity of gas movement.

References

1. Hao, S.; Hwang, J.Y.; Jiang, W.; Zhang, Y.Z. Preparation for High Activity Degree Lime and Effect on Desulfurization of Hot Metal Pretreatment. In *9th International Symposium on High-Temperature Metallurgical Processing*; The Minerals, Metals & Materials Series; Springer: Cham, Switzerland, 2018.
2. Vola, G.; Bresciani, P.; Rodeghero, E.; Sarandrea, L.; Cruciani, G. Impact of Rock Fabric, Thermal Behavior, and Carbonate Decomposition Kinetics on Quicklime Industrial Production and Slaking Reactivity. *J. Therm. Anal. Calorim.* **2018**, *136*, 967–993. [CrossRef]
3. Wang, G.; Ge, K.; Sun, T.; Shui, Z.; Hu, T.; Jiang, H.; Wang, Z. Comparison of Hydration of Thermally Activated Water-Washed Kaolins in Cement Mortar. *J. Test. Eval.* **2019**, *48*, 1837–1849. [CrossRef]
4. Vieira, L.M.; de Oliveira, H.C.C.; Telles, V.B.; Junca, E.; Vieira, E.A.; de Oliveira, J.R. Influence of Lime Particle and Slag Properties on Lime Dissolution in BOF Converter. *J. Mater. Res. Technol.* **2020**, *9*, 14878–14886. [CrossRef]
5. Georgallis, M.; Nowak, P.; Salcudean, M.; Gartshore, I.S. Modelling the Rotary Lime Kiln. *Can. J. Chem. Eng.* **2008**, *83*, 212–223. [CrossRef]
6. Senegačnik, A.; Oman, J.; Širok, B. Annular Shaft Kiln for Lime Burning with Kiln Gas Recirculation. *Appl. Therm. Eng.* **2008**, *28*, 785–792. [CrossRef]
7. Gutiérrez, A.S.; Martínez, J.B.C.; Vandecasteele, C. Energy and Exergy Assessments of A Lime Shaft Kiln. *Appl. Therm. Eng.* **2013**, *51*, 273–280. [CrossRef]
8. Piringer, H. Lime Shaft Kilns. *Energy Procedia* **2017**, *120*, 75–95. [CrossRef]
9. Krause, B.; Liedmann, B.; Wiese, J.; Bucher, P.; Wirtz, S.; Piringer, H.; Scherer, V. 3D-DEM-CFD Simulation of Heat and Mass Transfer, Gas Combustion and Calcination in An Intermittent Operating Lime Shaft Kiln. *Int. J. Therm. Sci.* **2017**, *117*, 121–135. [CrossRef]
10. Zhongda, T.; Shujiang, L.; Yanhong, W.; Xiangdong, W. SVM Predictive Control for Calcination Zone Temperature in Lime Rotary Kiln with Improved PSO Algorithm. *Trans. Inst. Meas. Control.* **2018**, *40*, 3134–3146. [CrossRef]
11. Suárez, V.M.G.; García-Gonzalo, E.; Bayón, R.M.; Nieto, P.J.G.; Antón, J.C.Á. Predictive Model of Gas Consumption and Air Emissions of A Lime Kiln in A Kraft Process Using The ABC/MARS-based Technique. *Int. J. Adv. Manuf. Technol.* **2018**, *100*, 1549–1562. [CrossRef]
12. Metz, J.C.; Maciel, E.F.; Garbin, M.; Modolo, R.C.E.; Moraes, C.A.M.; Gomes, L.B.; Brehm, F.A. Influence of Electric Arc Furnace Dust and Lime Kiln Waste in Portland Cement Hydration. *Ambiente Construído* **2020**, *20*, 225–241. [CrossRef]
13. Juneja, P.J.; Sunori, S.; Sharma, A.; Sharma, A.; Joshi, V. Modeling, Control and Instrumentation of Lime Kiln Process: A Review. In Proceedings of the 2020 International Conference on Advances In Computing, Communication & Materials (ICACCM), Dehradun, India, 21–22 August 2020.
14. Wirtz, S.; Pieper, C.; Buss, F.; Schiemann, M.; Schaefer, S.; Scherer, V. Impact of Coating Layers in Rotary Cement Kilns: Numerical Investigation with A Blocked-off Region Approach for Radiation and Momentum. *Therm. Sci. Eng.* **2020**, *15*, 100429. [CrossRef]
15. Pieper, C.; Wirtz, S.; Schaefer, S.; Scherer, V. Numerical Investigation of the Impact of Coating Layers on RDF Combustion and Clinker Properties in Rotary Cement Kilns. *FUEL* **2021**, *283*, 118951. [CrossRef]
16. Sagastume, A.; Vandecasteele, C. Exergy-based indicators to evaluate the possibilities to reduce fuel consumption in lime production. *Energy* **2011**, *36*, 2820–2827.
17. Rong, W.; Li, B. Energy and exergy analysis of an annular shaft kiln with opposite burners. *Appl. Therm. Eng.* **2017**, *119*, 629–638. [CrossRef]
18. Cheng, C.; Specht, E. Reaction rate coefficients in decomposition of lumpy limestone of different origin. *Thermochim. Acta* **2006**, *449*, 8–15. [CrossRef]
19. Duc, H.D.; Specht, E. Determination of reaction coefficient, thermal conductivity and pore diffusivity in decomposition of limestone of different origin. In Proceedings of the World Congress on Engineering and Computer Science, San Francisco, CA, USA, 19–21 October 2011; Volume 11, pp. 978–984.
20. Bluhm-Drenhaus, T.; Simsek, E.; Wirtz, S. A couple fluid dynamic-discrete element simulation of heat and mass transfer in a lime shaft kiln. *Chem. Eng. Sci.* **2010**, *65*, 2821–2834. [CrossRef]
21. Krause, B.; Liedmann, B.; Wiese, J. Coupled three dimensional DEM-CFD simulation of a lime shaft kiln Calcination, particle movement and gas phase flow field. *Chem. Eng. Sci.* **2015**, *134*, 834–849. [CrossRef]
22. Ning, J.-T.; Zhong, B.-J.; Fu, W.-B. Study on the Calcination of Fine Limestone Powder at High Temperature. *J. Combust. Sci. Technol.* **2003**, *9*, 205–208.
23. Wu, H.; Gui, N.; Yang, X.; Tu, J.; Jiang, S. Parameter analysis and wall effect of radiative heat transfer for CFD-DEM simulation in nuclear packed pebble bed. *Exp. Comput. Multiph. Flow* **2021**, *3*, 250–257. [CrossRef]
24. Mikulčić, H.; Vujanović, M.; Fidaros, D.K. The application of CFD modelling to support the reduction of CO_2 emissions in cement industry. *Energy* **2012**, *45*, 464–473. [CrossRef]

25. Long, H.; Xu, M.; Yu, D.; Ding, Y.; Wang, K. Two-temperature model of water gas shift reaction in porous media based on fluent. *Comput. Appl. Chem.* **2012**, *29*, 981–985.
26. LI, B.S.S.Y.J.; Qiu-Wang, S.Y.W. Effects of Point Contact Treatment Methods on Flow and Heat Transfer of Structured Packed Beds. *J. Eng. Thermophys.* **2013**, *34*, 534–537.
27. Liu, G.; Cui, G.; Ma, S.; Han, N. Simulation of the Temperature Field in a Lime Kiln by Using a Local Non-heat-equilibrium Model and Its Optimized Analysis. *J. Eng. Therm. Energy Power* **2018**, *33*, 96–105.
28. Lu, Z. Calculation of Effective Thermal Conductivity of Foam Porous Media. *J. Nanjing Univ. Sci. Technol.* **2001**, *25*, 257–261.
29. Zhu, L.; Bai, J.; Qiao, B. Calculation of Convective Heat Transfer for Active Limes Pre-heater. *Metall. Equip.* **2012**, 6–8.

materials

MDPI

Article

Effect of Preheating Temperature on Thermal–Mechanical Properties of Dry Vibrating MgO-Based Material Lining in the Tundish

Xiaodong Deng [1], Jianli Li [1,2] and Xiao Xie [3,*]

[1] The State Key Laboratory of Refractories and Metallurgy, Wuhan University of Science and Technology, Wuhan 430081, China
[2] Hubei Provincial Key Laboratory for New Processes of Ironmaking and Steelmaking, Wuhan University of Science and Technology, Wuhan 430081, China
[3] School of Computer Science and Technology, Wuhan University of Science and Technology, Wuhan 430081, China
* Correspondence: xiexiao@wust.edu.cn

Abstract: For the collapse of the working layer of dry vibrating material during preheating, the four-strand tundish of a steel plant was taken as a prototype for numerical simulation. The software ANSYS was used to calculate the temperature field and stress and strain field on the working layer under three preheating stages through the indirect coupling method. The results show that during the preheating process, the temperature field distribution on the hot surface of the working layer gradually develops toward uniformity with the increase in preheating temperature. However, the temperature gradient between the cold and hot surfaces increases subsequently, and the highest temperature between the cold and hot surfaces reaches 145.31 °C in the big fire stage. The stress on the top of the working layer is much larger than in other areas, and the maximum tensile stress on the top reaches 39.06 MPa in the third stage of preheating. Therefore, the damage to the working layer starts from the top of the tundish. In addition, the strain of the area near the sidewall burner nozzle in the casting area is much larger than that in the middle burner area with the increase in preheating temperature. Thus, the working layer near the sidewall burner nozzle is more prone to damage and collapse compared with the middle burner nozzle.

Keywords: dry vibrating material; collapsed ladle; mathematical modeling; temperature field; stress field; strain field

Citation: Deng, X.; Li, J.; Xie, X. Effect of Preheating Temperature on Thermal–Mechanical Properties of Dry Vibrating MgO-Based Material Lining in the Tundish. *Materials* **2022**, *15*, 7699. https://doi.org/10.3390/ma15217699

Academic Editor: Scott M. Thompson

Received: 26 September 2022
Accepted: 31 October 2022
Published: 2 November 2022

1. Introduction

Continuous casting tundish, working as a buffer and distributor of liquid steel between the ladle and continuous casting molds, plays a key role in affecting the performance of casting and solidification, as well as the quality of final products [1–3]. Therefore, the quality of the refractory material of the tundish is important for the full development and improvement of the metallurgical performance of the tundish. The development of the tundish working layer can be divided into four stages: (i) no working layer stage, (ii) insulation board stage, (iii) gunning material stage, and (iv) dry vibrating material stage [4]. Generally, tundish refractory lining mainly includes an insulation layer, a permanent layer, and a working layer. The working layer is in contact with molten steel. Thus, the working lining material should have good slag resistance, thermal shock resistance, volume stability, excellent thermal insulation performance, and high-temperature strength [5–7]. The dry vibrating material as the fourth generation of tundish working layer material, not only has the advantages of both insulation board and gunning material but also has the advantages of easy construction, high thermal efficiency, fast tundish turnover, long service life, and low energy consumption, which is widely used in the metallurgical industry of continuous casting tundish [8,9].

With the development of continuous casting technology, it is gradually realized that the thermal state of the tundish has an important role in controlling the temperature of the molten steel in the tundish, maintaining the stability of the superheat of the molten steel, improving the quality of the continuous casting billet, saving energy, and prolonging the service life of the tundish [10–12]. During the continuous casting process, the molten steel will be poured from the ladle into the tundish. There is conductive heat loss through the wall of the tundish and radiative heat loss through the bath surface [13,14]. In order to reduce the heat loss of the molten steel, the lining can be preheated to absorb a large amount of heat before pouring to reduce the temperature difference between the lining and the molten steel.

However, during the use of the dry vibrating material working layer, it was found that the dry vibrating material with phenolic resin as the low-temperature binding agent will undergo an oxidation reaction to form a decarbonized layer during the preheating process, and the original physical properties of the working layer of the tundish will be changed by the formation of the decarbonized layer [15]. When the preheating regime of the tundish is imperfect, the working layer is prone to large area collapse and local spalling, which leads to the production of instable running. In order to solve this problem, Shi et al. [16] analyzed the current status of refractory temperature and thermal stress research, and established and analyzed the temperature and thermal stress distribution of two types of refractories for ladles using APDL language. Li et al. [17] carried out a large number of follow-up analyses on the collapsed tundish of magnesium dry vibrating material tundish and optimized and improved the preheating process by testing and analyzing the strength of the dry vibrating material preheating process and the change law of the preheating temperature rise.

In the present study, a 1:1 model with a four-strand tundish of a steel mill as the prototype was established. Finite element simulation of the temperature field and the stress field during preheating in a tundish body was carried out by ANSYS software. According to the distribution and changes in temperature, stress, and strain fields on the working layer during the preheating process, the most vulnerable areas of the working layer were analyzed. This research aims to identify the stress distribution at the critical point and analyze the cause of the damage. The obtained results are of great significance to the normal operation of the continuous casting process.

2. Model Description

2.1. Fluid Flow and Heat Transfer

In the preheating process of tundish, the mixture of gas and air enters the tundish after ignition. Then, the combusted gas transfers heat to the tundish lining through convection and radiation in the flow process. During the preheating process fluid, flow, and heat transfer are modeled using conservation equations of mass Equation (1), momentum Equation (2), and energy Equation (3):

$$\frac{\partial \rho}{\partial t} + \frac{\partial(\rho u_j)}{\partial x_j} = 0 \tag{1}$$

$$\rho\frac{\partial u_i}{\partial t} + \rho u_j\frac{\partial u_i}{\partial x_i} = -\frac{\partial P}{\partial x_i} + \frac{\partial}{\partial x_j}\left[(\mu+\mu_t)\left\{\frac{\partial u_i}{\partial x_j} + \frac{\partial u_j}{\partial x_i}\right\}\right] + g_i(\rho-\rho_0) \tag{2}$$

$$\rho C_p\frac{\partial T}{\partial t} + \rho C_p\frac{\partial(u_jT)}{\partial x_j} = \frac{\partial}{\partial x_j}\left[\left(k_0 + \frac{C_p\mu_t}{Pr_t}\right)\frac{\partial T}{\partial x_j}\right] + S_T \tag{3}$$

where ρ is the density; C_P is the heat capacity; μ_t is the turbulent viscosity; Pr_t is the turbulent Prandtl number; S_T is the source term of the energy equation; u is the velocity; x is the Cartesian space coordinates, and subscripts i, j are for the coordinate directions.

The standard k-ε models [18] are shown in Equations (4) and (5):

$$\frac{\partial}{\partial t}(\rho k) + \frac{\partial}{\partial x_i}(\rho k u_i) = \frac{\partial}{\partial x_j}\left[\left(\mu + \frac{\mu_t}{\sigma_k}\right)\frac{\partial k}{\partial x_j}\right] + G_k + G_b - \rho\varepsilon - \gamma_M + S_K \tag{4}$$

$$\frac{\partial}{\partial t}(\rho \varepsilon) + \frac{\partial}{\partial x_i}(\rho \varepsilon u_i) = \frac{\partial}{\partial x_j}\left[\left(\mu + \frac{\mu_t}{\sigma_\varepsilon}\right)\frac{\partial k}{\partial x_j}\right] + C_1\frac{\varepsilon}{k}(G_k + C_3 G_b) - C_2\rho\frac{\varepsilon^2}{k} + S_\varepsilon \tag{5}$$

where k is the kinetic energy of turbulence per unit mass; ε_F is the turbulent energy dissipation rate; μ is the molecular viscosity; μ_t is the turbulent viscosity; G_k is the turbulent kinetic energy generated by the laminar velocity gradient; G_b is the turbulent kinetic energy generated by buoyancy [19]; γ_M is the fluctuation generated by the transition diffusion in compressible turbulence; σ_k and σ_ε are the turbulent Prandtl numbers of k and ε, respectively; C_1, C_2, and C_3 are the constants; and S_K and S_ε are the source term of the turbulent kinetic energy(k) and its dissipation rate (ε_F).

As mentioned by Launder and Spalding [18], the other values for model constants in this study were C_1 = 1.44, C_2 = 1.92, σ_k = 1.0, and σ_ε = 1.3. For buoyant shear layers for which the main flow direction is aligned with the direction of gravity, C_3 becomes 1. For buoyant shear layers that are perpendicular to the gravitational vector, C_3 becomes 0.

2.2. Solid Heat Transfer and Thermal Stress

During the preheating process, the heat transfer of the tundish package can be considered as a three-dimensional steady-state heat transfer, and the heat-transfer behavior is expressed by the following Equation (6):

$$\frac{\partial^2 T}{\partial x} + \frac{\partial^2 T}{\partial y} + \frac{\partial^2 T}{\partial z} = 0 \tag{6}$$

where T is the temperature; x, y, and z are the x, y, and z coordinate directions.

According to the temperature distribution and thermal expansive coefficient of each part of the tundish lining, the deformation is calculated in the special constraints. Then, the strain of each point of the tundish is calculated using the deformation of tundish displacement with the geometric equation. Finally, the stress in the tundish of each point is calculated through the strain according to the physical equation of tundish material [20,21].

The thermal stress field geometry equation is used to calculate the relationship between strain and displacement. It can be expressed by the following Equation (7).

$$\varepsilon = \begin{bmatrix} \frac{\partial}{\partial x} & 0 & 0 \\ 0 & \frac{\partial}{\partial y} & 0 \\ 0 & 0 & \frac{\partial}{\partial z} \\ \frac{\partial}{\partial y} & \frac{\partial}{\partial x} & 0 \\ 0 & \frac{\partial}{\partial z} & \frac{\partial}{\partial y} \\ \frac{\partial}{\partial z} & 0 & \frac{\partial}{\partial x} \end{bmatrix} \lambda \tag{7}$$

where $\varepsilon = \begin{bmatrix} \varepsilon_x & \varepsilon_y & \varepsilon_z & \gamma_{xy} & \gamma_{xz} & \gamma_{yz} \end{bmatrix}^T$ is the strain at any point in tundish; $\lambda = \begin{bmatrix} u & v & w \end{bmatrix}^T$ is the displacement along the directions of x, y, and z.

The physics equation of the stress field is used to calculate the relationship between strain and stress. It can be expressed by the following Equation (8).

$$\sigma = \frac{E(1-v)}{(1+v)(1+2v)}\begin{bmatrix} 1 & \frac{v}{1-v} & \frac{v}{1-v} & 0 & 0 & 0 \\ \frac{v}{1-v} & 1 & \frac{v}{1-v} & 0 & 0 & 0 \\ \frac{v}{1-v} & \frac{v}{1-v} & 1 & 0 & 0 & 0 \\ 0 & 0 & 0 & \frac{1-2v}{2(1-v)} & 0 & 0 \\ 0 & 0 & 0 & 0 & \frac{1-2v}{2(1-v)} & 0 \\ 0 & 0 & 0 & 0 & 0 & \frac{1-2v}{2(1-v)} \end{bmatrix} \quad (8)$$

where E is the elastic modulus; v is Poisson's radio; σ is the stress; and ε is the strain.

According to the above stress–strain relations, any stress of each point is calculated by each point of its inner strain obtained from the previous step, and the products are the object of the force, with the force meeting the balance equation.

2.3. Materials and Methods

In the present study, the characterization of the working and permanent layers required for the simulation calculations was made in the laboratory. The raw materials of the working and permanent layers were taken from the industrial raw materials of a steel mill. Test samples of the dry vibrating material were put into a mold with an inner size of 40 mm × 40 mm × 160 mm, followed by tamping until the density of raw materials was about 2.40 g/cm^3. These prepared samples were initially dried at 220 °C for 3 h to obtain handling strength, and then were fired in an electric furnace at 1000 °C for 2 h under air conditions (oxidizing atmosphere) and buried carbon conditions (reducing atmosphere), respectively, where the materials were heated at a rate of 5 °C/min, followed by cooling to room temperature in the furnace. Test samples of the permanent layer materials were put into a mold with the inner size of 40 mm × 40 mm × 160 mm, followed by vibrating. After heating for 3 h at 220 °C condition, all the specimens were fired at 1000 °C for 2 h in an electric furnace under an air atmosphere condition with the heated rate of 5 °C/min and then cooled down to room temperature. The bulk density of three test samples was evaluated using Archimedes' principle in kerosene medium according to GB/T 2997-2015 standard.

Thermal conductivity of the disc specimens with Φ180 mm × 20 mm at 350 °C, 600 °C, 800 °C, and 1000 °C was evaluated using the hot-wire method per the standard GB/T 5990-2006. The characterization of the dynamic elastic modulus at room temperature was evaluated according to GB/T 30758-2014. The cylindrical specimens of Φ8 mm × 50 mm were prepared, and the thermal expansion rate of the specimen was measured at 100–1400 °C using a high-temperature thermal (Precondar, Luoyang, China) expansion meter per standard GB/T 7320-2008.

2.4. Geometry, Mesh, and Boundary Conditions

In this study, a four-strand tundish of a steel mill is modeled as a prototype 1:1. Considering the symmetry of the tundish structure, half of the model is taken as the calculation area to reduce the calculation volume. The carbon element in the dry vibrating material of the working layer will react with the oxidizing substances in the preheating gas during the preheating process of the tundish, resulting in the delamination of the working layer and the generation of two layers of decarbonized layer and the original layer. Therefore, the working tundish lining with two layers was modeled, and the decarburization layer thickness was 25% of the whole working layer. The specific geometric model is shown in Figure 1a. The tundish body is composed of the original layer, decarbonized layer, permanent layer, insulation layer, steel shell, slag retaining wall, flow control device, and lid. The burner nozzle in the center of the pouring area on the lid is burner No. 1 and the burner nozzle in the pouring area near the side wall is burner No. 2. In addition, the burner nozzle in the impact area is burner No. 3. The physical properties of refractory linings used in this simulation are shown in Tables 1 and 2. The characterization of the insulation layer and the steel shell in Table 1 can be found by Zhang [22].

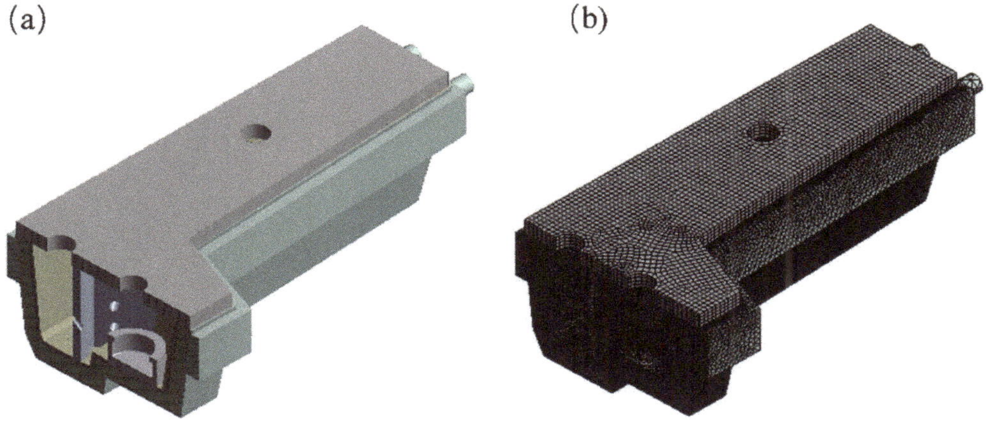

Figure 1. Geometry and mesh: (**a**) geometry; (**b**) grid of computational domain.

Table 1. Physical properties of refractory linings.

	P (Kg·m^{-3})	C_p (J·kg^{-1}k^{-1})	k (w·m^{-1}k^{-1})	α (°C^{-1})	E (MPa)
Original layer	3330	988.6	-	1.36×10^{-5}	7500
Decarbonized layer	2349	988.6	-	1.36×10^{-5}	880
Permanent layer	2600	800	-	6×10^{-7}	5700
Insulation layer	850	816.4	0.17	5×10^{-6}	1.2×10^5
Steel shell layer	7820	502	46.4	1.17×10^{-5}	2×10^5

Table 2. Thermal conductivity of materials at different temperatures.

Temperatures (°C)	Original Layer (W·m^{-1}k^{-1})	Decarbonized Layer (W·m^{-1}k^{-1})	Permanent Layer (W·m^{-1}k^{-1})
350	0.377	0.455	0.694
600	0.639	0.686	0.898
800	0.953	1.001	1.077
1000	1.015	1.13	1.27

The model is meshed using ANSYS MESH software. Based on the hybrid mesh division method, the areas of different complexity are divided into meshes of different sizes and densities to reduce the number of meshes as much as possible while ensuring calculation accuracy, so as to save the calculation resources and calculation time. The total number of meshes in the tundish model is about 8 million, in which the meshes of the working layer and burner of the tundish are encrypted. The mesh division results are shown in Figure 1b.

Non-slip conditions were applied on all wall boundaries for the fluid phase. A constant velocity flow was used at the inlet. At the outlet of the tundish, the outflow was applied. The heat loss is calculated based on the heat transfer coefficient at the side and bottom walls and the environment temperature. A summary of input parameters and boundary conditions used for computational fluid dynamics simulations is provided in Table 3. The heating process of the tundish is composed of three stages of preheating, namely, the first stage of preheating, the second stage, and the third stage of preheating. The preheating parameters for each stage are shown in Table 4. The data in Table 4 are from the steel mill site.

Table 3. Input parameters and boundary conditions used for CFD simulations.

Parameter	Value
Density	1.7878 $kg{\cdot}m^{-3}$
Viscosity	1.37×10^{-5} $kg{\cdot}m^{-1}s^{-1}$
Heat capacity	840.37 $J{\cdot}kg^{-1}K^{-1}$
Thermal conductivity	0.0145 $W{\cdot}m^{-1}k^{-1}$
Wall	no-slip
Side wall (heat loss coefficient)	15 $W{\cdot}m^{-2}$
environment temperature	25 °C

Table 4. The conditions of the burner nozzle.

	Flow Rate ($m^3{\cdot}h^{-1}$)	Velocity ($m{\cdot}s^{-1}$)	Temperature (°C)
The first stage	355	1.3951	400
The second stage	651	2.5583	800
The third stage	1071	4.2088	1150

3. Results

3.1. Temperature Field Distribution

Figure 2 shows the temperature field distribution of the hot surface of the wall in lengthwise direction of the working layer. The result shows that with the increase in preheating temperature, the temperature distribution on the hot surface of the working layer gradually tends to be uniform. As shown in Figure 2a,b, the temperature field distribution on the hot surface of the working layer is less uniform at the first and second stages of preheating. The red area is mainly concentrated in the area of the No. 1 burner nozzle and the bottom area of the No. 2 burner nozzle. The distribution area is small, and the temperature difference between zones is 16 °C and 20 °C. Figure 2c displays that the distribution of the temperature field of the hot surface tends to be uniform during the third stage of preheating, and the temperature difference between the red zones is only 11.6 °C.

Figure 3 shows the temperature field distribution of the cold surface. The result shows that the trend of temperature field change on the cold side of the working layer is slightly different from that on the hot side. The temperature field on the cold side does not develop toward uniformity with the increase in preheating temperature. The high-temperature area is mainly concentrated in the middle and lower regions of the wall, and in the region near the top of the tundish, the working layer temperature is lower. There is a larger temperature gradient compared with the middle and lower regions. In addition, the temperature on the cold side has a large temperature difference compared with the temperature on the hot side. The temperature gradient between the cold and hot sides further expands with the increase in preheating temperature. In the third stage of preheating, the difference between the highest temperature on the cold side in Figure 3c and the highest temperature on the hot side in Figure 2c is 145.31 °C, and the temperature gradient further increases in the orange temperature zone of both figures. The results of this calculation indicate that there is a large temperature gradient in the longitudinal direction in the working layer of the dry vibrating material during the preheating process, which increases the possibility of thermal shock damage to the material and is harmful to the pouring of the tundish [20].

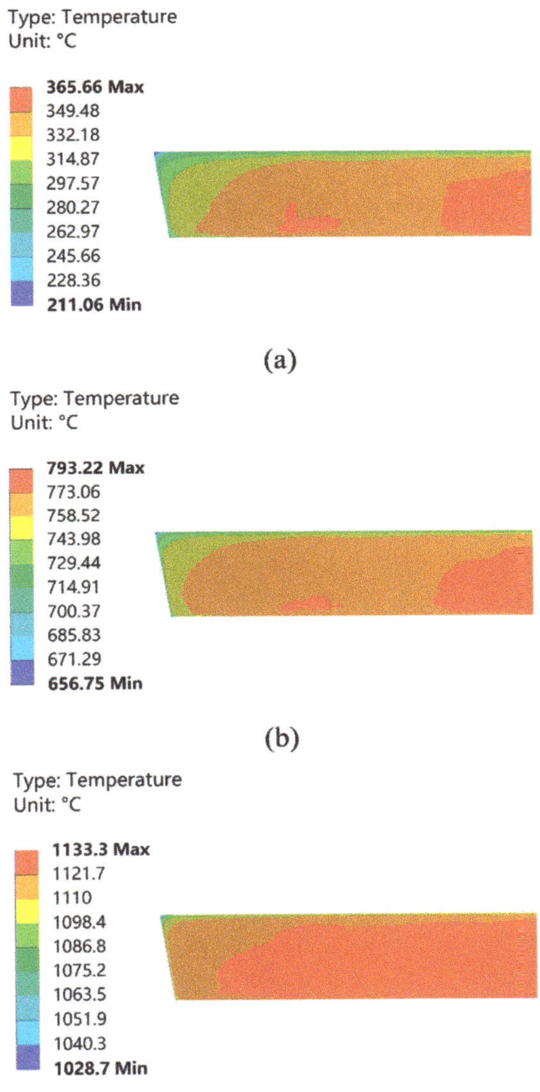

Figure 2. Temperature field distribution of the hot surface of the working layer. (**a**) The first stage of preheating; (**b**) the second stage of preheating; and (**c**) the third stage of preheating.

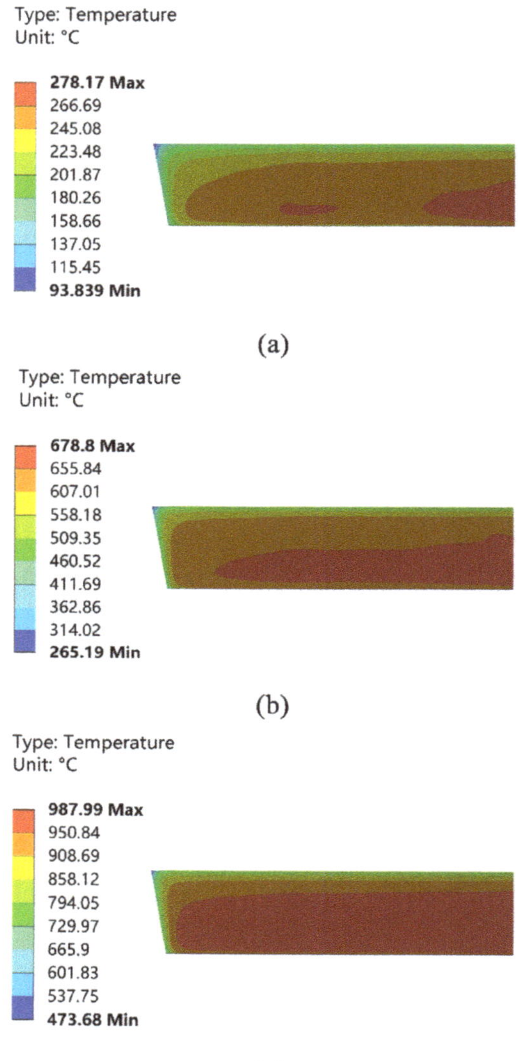

Figure 3. Temperature field distribution of the cold surface of the working layer. (**a**) The first stage of preheating; (**b**) the second stage of preheating; and (**c**) the third stage of preheating.

3.2. Stress and Strain Field Distribution

Figure 4 shows the distribution of the maximum principal stress applied to the hot surface of the wall in lengthwise direction at different preheating stages, which is oriented perpendicular to the wall of the working layer. When the maximum principal stress is positive, the stress is tensile stress and the direction is toward the fluid domain; when the maximum principal stress is negative, the stress is compressive stress and the direction is toward the permanent layer. As shown in Figure 4, the boundary region of the working layer belongs to the stress concentration region and the stress increases significantly with the increase in the preheating temperature. However, the maximum principal stress suffered by the working layer in the middle region has less change in stress value with the increase in the preheating temperature. The tensile stress at the top of the wall in the boundary region is significantly greater than that at the sides and bottom, which indicates that the possibility of cracks from the top of the working layer during preheating is much greater than that in other regions.

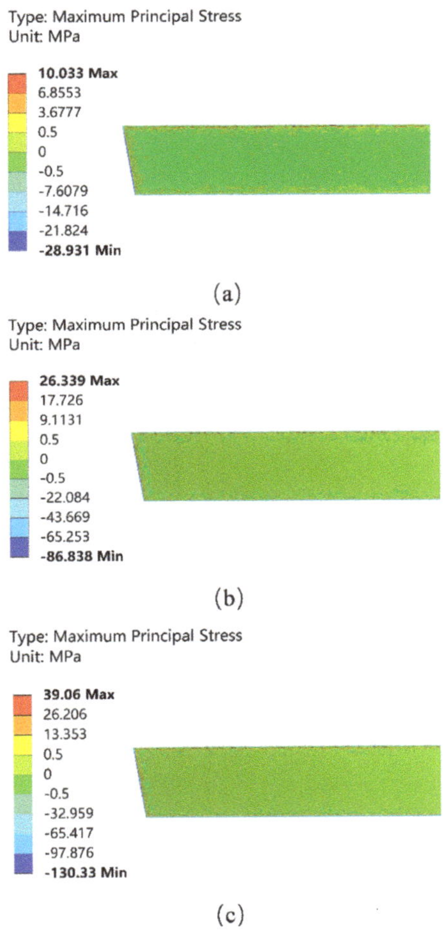

Figure 4. Distribution of the maximum principal equivalent stress field of the working layer. (**a**) The first stage of preheating; (**b**) the second stage of preheating; and (**c**) the third stage of preheating.

Figure 5 shows the maximum principal strain field on the cold surface of the working layer. The direction of the strain field is consistent with the direction of the maximum principal stress. As shown in Figure 5, the area where positive deformation occurred in the working layer was mainly concentrated in the middle of the wall, while the area where negative deformation occurred was distributed in the boundary area of the working layer. This indicates that during the preheating process, the working layer material deformed toward the fluid domain in the middle area, while the boundary area deformed toward the permanent layer, which exacerbated the stress concentration in the boundary area to a certain extent. The maximum positive deformation area on the cold surface of the working layer gradually shifts from the bottom to the side with the increase in preheating temperature. In the second and third stages of preheating, the maximum positive deformation area was concentrated in the area near the side wall. Moreover, the larger the value of positive deformation, the greater the possibility of detachment between the working layer and the permanent layer. When the deformation volume reaches a certain value, the working layer will be separated from the permanent layer, which is harmful to the pouring of the tundish.

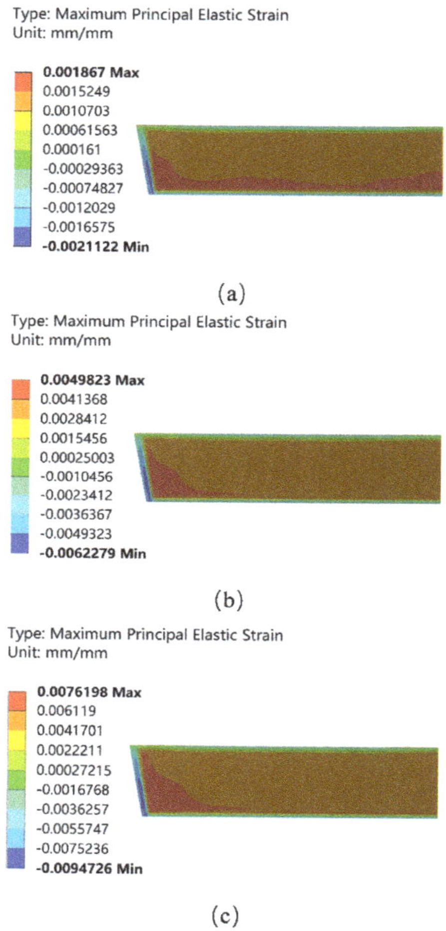

Figure 5. Distribution of the maximum principal elastic strain field of the working layer. (**a**) The first stage of preheating; (**b**) the second stage of preheating; and (**c**) the third stage of preheating.

4. Analysis and Discussion

In order to judge more intuitively the stress changes suffered by the working layer, combined with the simulation results of the temperature field and stress field of the dry vibrating material of the tundish in the third stage of preheating, a straight line called path 1 and path 2 was made in the area corresponding to the center of the burner nozzles 1 and 2 of the working layer, respectively, to calculate the temperature field and stress field distribution on the paths. The two paths are located on the hot surface of the working layer, as shown in Figure 6, while the distribution curves of temperature and maximum principal stress on the paths are shown in Figures 7 and 8. The horizontal coordinate 0 mm is the bottom of the middle pack, and L is the length on the path, and the larger its value, the closer the path is to the top of the pack. The red curve is the variation curve of the maximum principal stress with the path, and the black curve is the variation curve of temperature with the path.

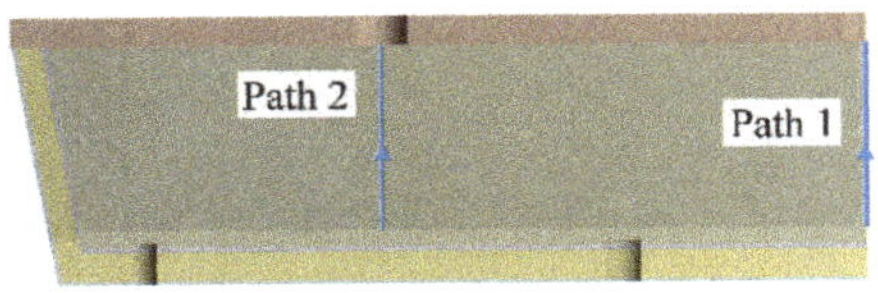

Figure 6. Path 1 and path 2.

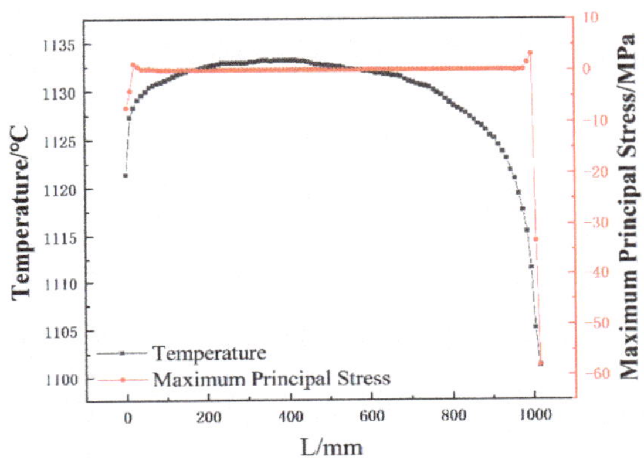

Figure 7. Temperature and maximum principal stress distribution on path 1.

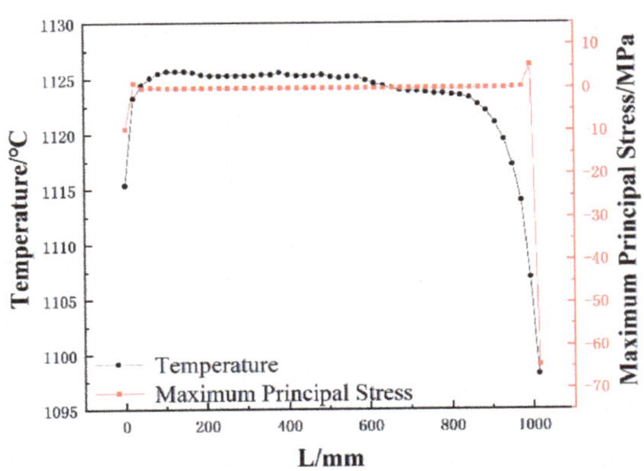

Figure 8. Temperature and maximum principal stress distribution on path 2.

As shown in Figure 7, the magnitude of the maximum principal stress in the region of 80–980 mm on path 1 varies extremely, and its stress value is close to 0 MPa, which cannot cause damage to the working layer. While in the bottom region of path 1, the maximum tensile stress is 1.336 MPa and the maximum compressive stress is 7.2 MPa. However, in the top region, the maximum tensile stress on path 1 is 3.03 MPa and the maximum compressive stress reaches 57.98 MPa. Figure 9 shows the strength characteristics of the dry vibrating material. As can be seen from the figure, the resin in the dry vibrating material is gradually cured with the increase in preheating temperature so that the working layer has a certain strength at low temperatures. However, the curing resin starts to oxidize and decompose at 200 °C, resulting in a loss in the strength of the working layer. After 1000 °C, the strength of the working layer gradually increases with the sintering of the dry vibrating material [23]. According to the temperature curve in Figure 7, it is known that the temperature at the top of the working layer is lower than in other areas. Thus, in the case where the dry vibrating material has lower strength and greater stress, the top of the working layer is more prone to be damaged, cracks are formed, and spalling occurs [17].

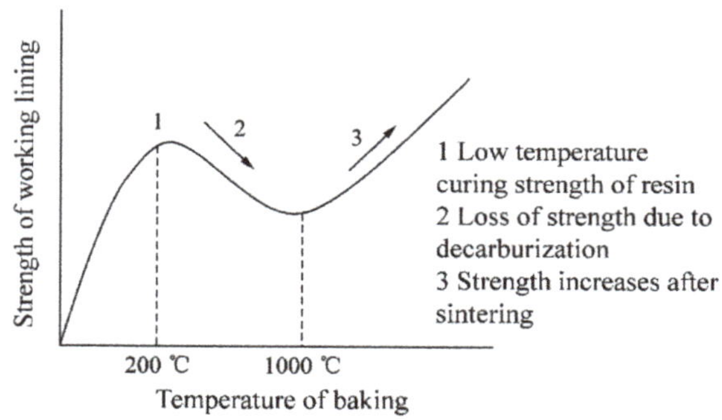

Figure 9. Strength characteristics of the dry vibrating material of the working layer [24].

Figure 8 shows that the maximum tensile stress is 1.14 MPa and the maximum compressive stress is 9.50 MPa in the bottom area of path 2, while the maximum tensile stress is 5.34 MPa and the maximum compressive stress is 64.6 MPa in the top area of path 2. Compared with path 1, the stress in the top area of path 2 has increased significantly, while the temperature has decreased slightly. Combined with the strength characteristics of the dry vibrating material in Figure 9, it is known that the working layer is more prone to be damaged and cracked at path 2. On the other hand, according to Figure 5c, there is a possibility of separation of the working layer from the permanent layer. When cracks are generated at the top of path 2 and the cracks extend to the bottom of the working layer, the steel will penetrate the working layer along the cracks during the pouring process, which will lead to large spalling and tundish collapse.

5. Conclusions

(1) The temperature field distribution on the hot surface of the working layer gradually develops toward uniformity with the increase in preheating temperature, while the temperature gradient between the cold and hot surfaces increases. The highest temperature difference between the cold and hot surfaces reaches 201.31 °C during the third stage of preheating, which increases the possibility of thermal shock damage to the refractory of the working layer.

(2) During the preheating process, the stress on the working layer is mainly concentrated in the boundary area. The maximum tensile stress at the top of the working layer reaches 39.06 MPa during the third stage of preheating, which is significantly greater than that in other areas. It can be seen that the possibility of damage from the top of the working layer to forming cracks is much greater than in other areas.

(3) In the high-temperature preheating stage, the maximum principal strain in the area of the No. 2 burner nozzle is 0.0076198 mm, which is much larger than in other areas. When cracks are generated at the top of the working layer and the cracks extend downward, the working layer will collapse and spall in a large area due to the penetration of steel.

Author Contributions: Methodology, X.D.; Supervision, J.L. and X.X.; Writing—original draft, X.D.; Writing—review & editing, X.X. and J.L. All authors have read and agreed to the published version of the manuscript.

Funding: This work was financially supported by the National Natural Science Foundation of China (51974210, 52274341), Hubei Provincial Natural Science Foundation (2019CFB697), and State Key Laboratory of Refractories and Metallurgy, Wuhan University of Science and Technology.

Conflicts of Interest: The authors declare that they have no conflict of interest.

Nomenclature

C_p	Heat capacity
E	Elastic modulus
G_b	Generation term of turbulent kinetic energy
G_k	Generation term of turbulent kinetic energy
k	Turbulent kinetic energy
L	Length
Pr_t	Turbulent Prandtl number
S_T	Source term of energy equation
$S\varepsilon$	Source term of turbulent energy dissipation rate
S_K	Source term of turbulent kinetic energy
T	Temperature
t	Time
u	Velocity
v	Poisson's radio
x	Cartesian coordinates
γ_M	Dilatation dissipation term
ε	Strain
ε_F	Turbulent energy dissipation rate
μ	Molecular viscosity
μ_t	Turbulent viscosity
ρ	Density
σ	Stress
σ_k	Turbulent Prandtl number for k
σ_ε	Turbulent Prandtl number foe ε

References

1. Sahai, Y. Tundish Technology for Casting Clean Steel: A Review. *Metall. Mater. Trans. B* **2016**, *47*, 2095–2106. [CrossRef]
2. Mazumdar, D. Review, Analysis, and Modeling of Continuous Casting Tundish Systems. *Steel Res. Int.* **2019**, *90*, 1800279. [CrossRef]
3. Sheng, D.Y. Design Optimization of a Single-Strand Tundish Based on CFD-Taguchi-Grey Relational Analysis Combined Method. *Metals* **2020**, *10*, 1539. [CrossRef]
4. Xu, Y. Research progress of dry vibrating mixes for continuous casting tundish. *Refractory* **2018**, *52*, 389–395.
5. Zhang, J.H.; Chen, Y.X.; Yang, D.Z. Process Design and Life Analysis of Tundish Lining. *Metalurgija* **2021**, *60*, 191–194.
6. Vdovin, K.N.; Tochilkin, V.V.; Yachikov, I.M. Designing Refractories for the Tundish of a Continuous Caster. *Refract. Ind. Ceram.* **2016**, *56*, 569–573. [CrossRef]
7. Biswas, S.; Sarkar, D. Modern refractory practice for clean steel. In *Introduction to Refractories for Iron- and Steelmaking*; Springer: Cham, Switzerland, 2020; pp. 409–425.
8. Musmeci, M.; Rendtorff, N.M.; Musante, L.; Martorello, L.; Galliano, P.; Aglietti, E.F. Characterization of MgO-Based Tundish Working Lining Materials, Microstructure and Properties. *Ceram. Int.* **2014**, *40*, 14091–14098. [CrossRef]
9. Wei, J.C.; Tu, J.B.; Wang, Y.L.; Zhang, L.J. Environment-Friendly Dry Vibration Refractories Used for Continuous Casting Tundish. *Adv. Mater. Res.* **2013**, *750–752*, 1335–1338.
10. Botnikov, S.A.; Khlybov, O.S.; Kostychev, A.N. Development of the Metal Temperature Prediction Model for Steel-Pouring and Tundish Ladles Used at the Casting and Rolling Complex. *Metallurgist* **2019**, *63*, 792–803. [CrossRef]
11. Sheng, D.Y.; Windisch, C. A Simulation-Based Digital Design Methodology for Studying Conjugate Heat Transfer in Tundish. *Metals* **2021**, *12*, 62. [CrossRef]
12. Yang, B.; Lei, H.; Bi, Q.; Jiang, J.; Zhang, H.; Zhao, Y.; Zhou, J.A. Fluid Flow and Heat Transfer in a Tundish with Channel Type Induction Heating. *Steel Res. Int.* **2018**, *89*, 1800173. [CrossRef]
13. Gastón, A.; Sarmiento, G.S.; Begnis, J.S.S. Thermal Analysis of a Continuous Casting Tundish by an Integrated Fem Code. *Lat. Am. Appl. Res.* **2008**, *38*, 259–266.
14. Sheng, D.Y.; Jönsson, P.G. Effect of Thermal Buoyancy on Fluid Flow and Residence-Time Distribution in a Single-Strand Tundish. *Materials* **2021**, *14*, 1906. [CrossRef] [PubMed]

15. Xu, L.; Gao, S.; Chen, M.; Wu, Y.; Shinozaki, K. Improvement in Microstructure and Thermo-Mechanical Properties of MgO-Based Dry Vibratable Material by Addition of Fe. *Mater. Chem. Phys.* **2020**, *253*, 123368. [CrossRef]
16. Shi, S.; Li, G.; Jiang, G.; Xie, L.; Liu, J. Temperature and Thermal Stress Analysis of Refractory Products. *Sens. Transducers J.* **2013**, *21*, 53.
17. Li, S.B.; Chen, Y.Q.; Huang, J.J.; Geng, H.L.; Li, Z.C.; Wang, D.W. Improvement of tundish baking process for magnesia dry material. *Steelmaking* **2021**, *37*, 46–49.
18. Jones, W.P.; Launder, B.E. The Prediction of Laminarization with a Two-Equation Model of Turbulence. *Int. J. Heat Mass Transf.* **1972**, *15*, 301–314. [CrossRef]
19. Chattopadhyay, K.; Isac, M.; Guthrie, R.I.L. Modelling of Non-Isothermal Melt Flows in a Four Strand Delta Shaped Billet Caster Tundish Validated by Water Model Experiments. *ISIJ Int.* **2012**, *52*, 2026–2035. [CrossRef]
20. Vasil'ev, D.V.; Grigor'ev, V.P. On the Problem of Thermomechanical Stresses in the Lining and Shell of Steel-Pouring Ladles. *Refract. Ind. Ceramc.* **2012**, *53*, 118–122. [CrossRef]
21. Li, G.; Liu, J.; Jiang, G.; Liu, H. Numerical Simulation of Temperature Field and Thermal Stress Field in the New Type of Ladle with the Nanometer Adiabatic Material. *Adv. Mech. Eng.* **2015**, *7*, 168781401557598. [CrossRef]
22. Zhang, L.L. Study of Flow and Heat Transfer Phenomena and its Influence on Deformation of Blank and Mold for Beam Blank Continuous. Ph.D. Thesis, Chongqing University, Chongqing, China, 2017.
23. Behera, S.; Sarkar, R. Sintering of Magnesia: Effect of Additives. *Bull. Mater. Sci.* **2015**, *38*, 1499–1505. [CrossRef]
24. Yan, P.Z.; Weng, X.Y.; Wu, K.D.; Tang, L.; Ye, X.X.; Jiang, X.X. Solution and control of collapse of dry vibration mix in tundish during heating. *World Iron Steel* **2011**, *11*, 60–62+67.

materials

Article

New Insights into the Mechanism of Nucleation of ZrO_2 Inclusions at High Temperature

Yutang Li [1,2], Linzhu Wang [1,2,*], Chaoyi Chen [1], Shufeng Yang [3] and Xiang Li [4]

1 School of Materials and Metallurgy, Guizhou University, Guiyang 550025, China
2 State Key Laboratory of Public Big Data, Guizhou University, Guiyang 550025, China
3 School of Metallurgical and Ecological Engineering, University of Science and Technology Beijing, Beijing 100083, China
4 Department of Materials and Metallurgy Engineering, Guizhou Institute of Technology, Guiyang 550003, China
* Correspondence: lzwang@gzu.edu.cn

Abstract: It is difficult to observe the nucleation mechanism of inclusions in real-time. In this study, the nucleation process of zirconium oxide inclusions was systematically studied by classical nucleation theory and first principles. Zr deoxidized steel with 100 ppm Zr addition was processed into metallographic samples for scanning electron microscopy energy-dispersive spectroscopy observation. The electrolytic sample was analyzed by micro X-ray diffraction and transmission electron microscopy, and the zirconium oxide in the sample was determined to be ZrO_2. The nucleation rate and radius of the ZrO_2 inclusions were calculated by classical nucleation theory, and they were compared with the experimental values. There was a considerable difference between the experimental and theoretical values of the nucleation rate. The effect of the nucleation size was analyzed by first-principles calculation, and the thermodynamic properties of ZrO_2 clusters and nanoparticles were analyzed by constructing $(ZrO_2)_n$ (n = 1–6) clusters. The thermodynamic properties of ZrO_2 calculated by first principles were consistent with the values in the literature. Based on two-step nucleation theory, the nucleation pathway of ZrO_2 is as follows: $Zr_{atom} + O_{atom} \rightarrow (ZrO_2)_n \rightarrow (ZrO_2)_2 \rightarrow$ core (ZrO_2 particle)–shell ($(ZrO_2)_2$ cluster) nanoparticle $\rightarrow (ZrO_2)_{bulk}$.

Keywords: inclusion; ZrO_2; nucleation mechanism; first principles

Citation: Li, Y.; Wang, L.; Chen, C.; Yang, S.; Li, X. New Insights into the Mechanism of Nucleation of ZrO_2 Inclusions at High Temperature. *Materials* **2022**, *15*, 7960. https://doi.org/10.3390/ma15227960

Academic Editor: Daniela Kovacheva

Received: 17 October 2022
Accepted: 3 November 2022
Published: 10 November 2022

1. Introduction

The mechanical properties and service life of steel are seriously affected by inclusions. After the proposal of the concept of oxide metallurgy [1], researchers have realized that it is important to control the size of inclusions [2–4] rather than increase the smelting cost [5–8]. Formation of inclusions begins with nucleation, and exploring the nucleation mechanism and properties of inclusions in steel is important to control the inclusion size. However, owing to the high speed of inclusion nucleation and high smelting temperature, the inclusion-nucleation process is difficult to detect and observe directly. Therefore, research on the inclusion of nucleation in steel is a challenge.

Researchers have found that Zr-containing inclusions can promote acicular ferrite transformation in Zr-deoxidized steel under certain conditions [9–13]. Zr-containing inclusions in Zr-deoxidized steel also play an important role in oxide metallurgy. On the one hand, they can induce nucleation of intragranular ferrite. There are several theories about the nucleation mechanism of ferrite. The change in the chemical composition of austenite around inclusions promotes nucleation. Inclusions and precipitates are coherent with ferrite to reduce the potential nucleation barrier and promote nucleation. In addition, the strain energy caused by the difference in thermal shrinkage between inclusions and austenite, as well as inclusions acting as an inert interface, promotes nucleation [14]. Among the proposed theories, the Mn-depleted-zone mechanism formed by the precipitation of MnS on ZrO_2 is considered to be one of the most effective mechanisms for intragranular ferrite nucleation [15,16]. On the other hand, Zr-containing inclusions play an important role in controlling the distribution of MnS inclusions in the steel. The density of ZrO_2 is close to that of liquid steel, and the volume is small. Therefore, it is not easy for ZrO_2 to float up in liquid steel, and it is easier for ZrO_2 to disperse and distribute in steel to improve the distribution of sulfide through heterogeneous nucleation. Theoretical calculation shows that the lattice mismatch degree between MnS and ZrO_2 is only 5.2%, and ZrO_2 is the most effective nucleation core to promote MnS nucleation [17]. The thermodynamics from first-principles calculation show that Mn will diffuse into ZrO_2 because ZrO_2 has cationic vacancies and can absorb Mn [18]. Whether it induces intragranular ferrite nucleation or acts as a heterogeneous nucleation core, the key is the size control of ZrO_2. However, there has been limited research on the nucleation mechanism and size control of ZrO_2. In addition, the inclusion-nucleation speed is fast and the smelting temperature is high, so the inclusion-nucleation process is difficult to detect and observe directly.

To study the evolution of alumina inclusions at the atom scale, Wang et al. [19–21] investigated the cluster structure from experimental and theoretical aspects, and they proposed a two-step nucleation mechanism. Using quenching and three-dimensional atomic-probe detection technology, Zhao et al. [22] captured the intermediate structure of titanium oxide, and they proposed the cluster-assisted nucleation mechanism. Yang et al. [23] simulated the growth process of clusters by molecular dynamics, and they found that clusters grow through collision. However, there has been limited research on the nucleation of zirconium oxide. Thus, it is necessary to study the nucleation process of zirconium oxide.

In this study, a high-temperature Zr deoxidation experiment was carried out, and the characteristics of the zirconium oxide inclusions in liquid steel, such as the composition, morphology, size, quantity, and area density, were statistically analyzed. According to classical thermodynamic nucleation theory, the relationships between the solute element activities and the nucleation radius and nucleation rate of inclusions were obtained, and the theoretical nucleation rate was compared with the experimental nucleation rate. Cluster models of zirconium oxide were constructed by Materials Studio software. The cluster structure and thermodynamic properties of nanoparticles after geometric optimization were calculated, and the accuracy of the first-principles calculation was verified. By combining the high-temperature experimental results, classical nucleation calculations, and first-principles analysis, the nucleation mechanism of zirconium oxide inclusions is proposed.

2. Materials and Methods

2.1. Sample Preparation

Pure iron was used as the raw material, and it was heated in a Si–Mo heating electric resistance furnace (Braveman Special Testing Furnace CO. LTD., Luoyang, Henan, China). The chemical composition of the pure iron sample is shown in Table 1. The pure iron sample was heated to 1873 K (1600 °C) in an alumina crucible in the Si–Mo heating electric resistance furnace. After the temperature was maintained at 1873 K for 30 min, the Zr–Fe alloy (60% Zr) wrapped in a high-purity iron belt was added to the melted pure iron, followed by stirring for 10 s to ensure uniform distribution of the Zr–Fe alloy. Finally, 120 s after adding the Zr–Fe alloy, samples were removed in quartz tubes, followed by quenching in water. The whole experimental process was protected by high-purity argon gas.

Table 1. Chemical contents of the impurities in the pure iron sample (wt%).

Composition	C	Si	Mn	P	S	Cr	Al	Cu	Ni	Ti	N	Fe
Amount	0.016	0.0033	0.01	0.0053	0.0017	0.0107	0.003	0.0037	0.0038	0.001	0.002	Bal.

The inclusions were extracted by electrolysis. A copper plate was used as the cathode, and the sample was the anode. After electrolysis, the anode was placed in anhydrous ethanol. By ultrasonic cleaning, the inclusions attached to the anode were dispersed in anhydrous ethanol. Finally, the inclusions extracted by electrolysis were analyzed by micro X-ray diffraction (μXRD, Bruker D8 Advance, Bruker, Berlin, Germany) and transmission electron microscopy (TEM, JEOL JEM-F200, JEOL, Tokyo, Japan).

2.2. Microstructure and Composition Characterization

To measure the composition and morphologies, the samples were processed into ∅5 mm × 15 mm metallographic samples. Scanning electron microscopy with energy-dispersive X-ray spectroscopy (SEM-EDS) (EM-30PLUS, COXEM, Daejeon, Korea) was then performed. In addition, the samples were processed into ∅5 mm × 10 mm bars for total oxygen and nitrogen content detection by the fusion-infrared absorption method. The oxygen and nitrogen contents were measured three times to investigate the uniformity of the total oxygen and nitrogen contents in the molten iron. In addition, the total Zr content was measured by inductively coupled plasma–atomic emission spectroscopy. The chemical contents of oxygen, nitrogen, and zirconium are given in Table 2.

Table 2. Chemical contents of O, N, and Zr in the steel sample after Zr addition.

Zr Addition	Holding Time/s	[O]/ppm	[N]/ppm	[Zr]/ppm
100 ppm	120	125	21	24
		89	28	31
		106	34	22

To measure the composition of the inclusions in the steel after zirconium addition, the inclusions extracted by electrolysis were observed by μXRD and TEM. The results provided an experimental reference for subsequent first-principles calculation of inclusion crystal-type selection.

2.3. Nucleation Calculation

According to classical nucleation theory, the critical nucleation size and nucleation rate I [$cm^{-3}\cdot s^{-1}$] can be calculated by [24]

$$\ln I = \frac{16\pi\gamma_{SL}^3 V_O^2}{3k_B R^2 T^3}\left(\frac{1}{(\ln S_O^*)^2} - \frac{1}{(\ln S_O)^2}\right) \tag{1}$$

$$r_C = -\frac{2\gamma_{SL}}{\Delta G_V} = \frac{2r_{SL} V_O}{RT\ln S_O} \tag{2}$$

where k_B is the Boltzmann constant (1.38×10^{-23} J/K), R is the gas constant (8.314 J/(mol K)), T [K] is the absolute temperature, and V_O [m^3/mol] is the molar volume of oxide. γ_{SL} [J/m^2] is the interfacial energy between the oxide and liquid steel, and it can be expressed by Young's equation:

$$\gamma_{SL} = \gamma_{SV} - \gamma_{LV}\cos\theta \tag{3}$$

$$\gamma_{LV} = 1.75 - 0.279\ln(1 + 140\cdot a_O) \tag{4}$$

where γ_{SV} is the surface energy of the solid inclusion, γ_{LV} is the surface energy of the liquid steel, which has been calculated in previous studies [25,26], and θ is the contact angle between liquid steel and the inclusion, as illustrated at Table 3.

Table 3. Data related to the calculation of the critical nucleation size and nucleation rate of oxide inclusions.

Inclusion	Θ (deg)	γ_{SV} (J/m^2)	V_O (m^3/mol)	$\log K_{eq}$
ZrO_2	111 [27]	1.395 [27]	10.5×10^{-6}	−57,000/T + 21.8 [28]

The experimental values of the nucleation rate I can be obtained by [24]

$$I = \frac{f_v}{\frac{4}{3}\pi r^3 \cdot t} \quad (5)$$

where t is the nucleation time (generally taken to be 0.2 s [29]) and r is the critical nucleation radius obtained by Equation (2). f_v is the volume fraction of oxide particles: [24]

$$f_v = \frac{\rho_{Fe}}{\rho_{ZrO_2}} \cdot \frac{M_{ZrO_2}}{x M_{Zr}} \cdot [\text{ppm insol. Zr}] \times 10^{-6} \quad (6)$$

where ρ_{Fe} and ρ_{ZrO2} are the densities of Fe and ZrO_2, respectively (ρ_{Fe} = 7.8 g/cm^3, ρ_{ZrO2} = 5.85 g/cm^3) [30], M_{ZrO2} is the molecular weight of ZrO_2, M_{zr} is the atomic weight of Zr, and [ppm insol. Zr] represents the Zr content.

2.4. First-Principles Calculation

The DMol [3] module based on density functional theory in the Materials Studio software package (Materials Studio8.0, Accelrys, California, America) was used for cluster optimization and thermodynamic property calculation. The Broyden–Fletcher–Goldfarb–Shanno mechanism was used for geometric optimization. The Perdew–Burke–Ernzerhof functional with the generalized gradient approximation was selected as the electron exchange–correlation potential function [31].

3. Results

3.1. Inclusion Characterization

To obtain the size distribution of the ZrO_2 inclusions, 100 SEM images were continuously taken at 5000× magnification, and the total observed area was 0.11 mm^2. The average size of the ZrO_2 inclusions in the sample was 0.56 μm, and the size distribution of ZrO_2 inclusions approximately approached a normal distribution (Figure 1).

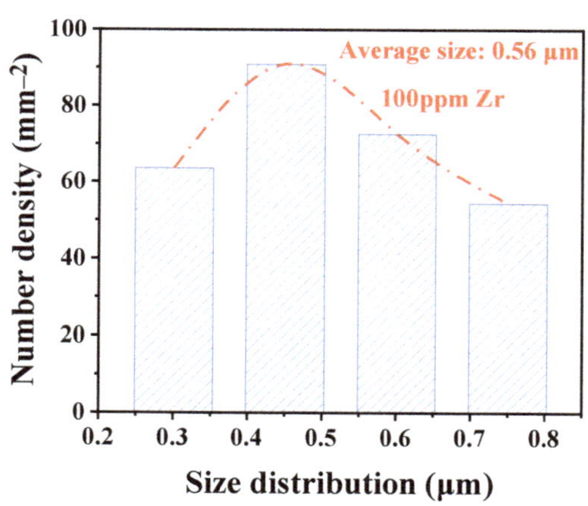

Figure 1. Size distribution of ZrO_2 in Zr deoxidized steel.

The inclusions extracted by electrolysis were characterized by μXRD, and the morphology and composition were analyzed by SEM-EDS (Figure 2). From the mapping image, the inclusions in the Zr deoxidized steel were zirconium oxide. A few Al inclusions were also detected because of the trace amount of Al in the raw materials. The results of μXRD suggested that the zirconium oxide was ZrO_2. In addition, both monoclinic and tetragonal ZrO_2 were detected. This may be because of the transformation of tetragonal ZrO_2 to monoclinic ZrO_2 during rapid cooling.

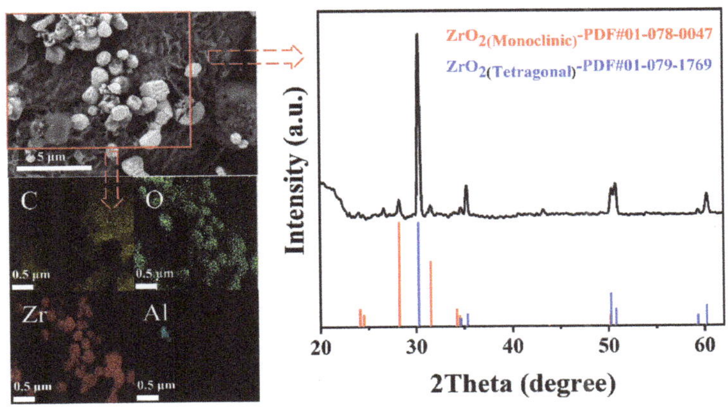

Figure 2. Morphology, composition, and X-ray diffraction pattern of the inclusions in Zr deoxidized steel.

3.2. Classical Nucleation Calculation

According to classical nucleation theory, the relationship between the critical nucleation radius of ZrO_2 and the activities of the solute elements at 1873 K is shown in Figure 3a. When the Zr activity is in the range 0.0001–1, and the oxygen activity is in the range 0.0001–0.1, the critical nucleation radius of ZrO_2 is 0.3–1.2 nm. The relationship between the nucleation rate and solute element activities is shown in Figure 3b. When the Zr activity is in the range 0.0001–1, and the oxygen activity is in the range 0.001–0.1, the critical nucleation rate of ZrO_2 is in the range 100–560 $cm^{-3}{\cdot}s^{-1}$. The points in Figure 3 are the experimentally measured Zr and oxygen activities.

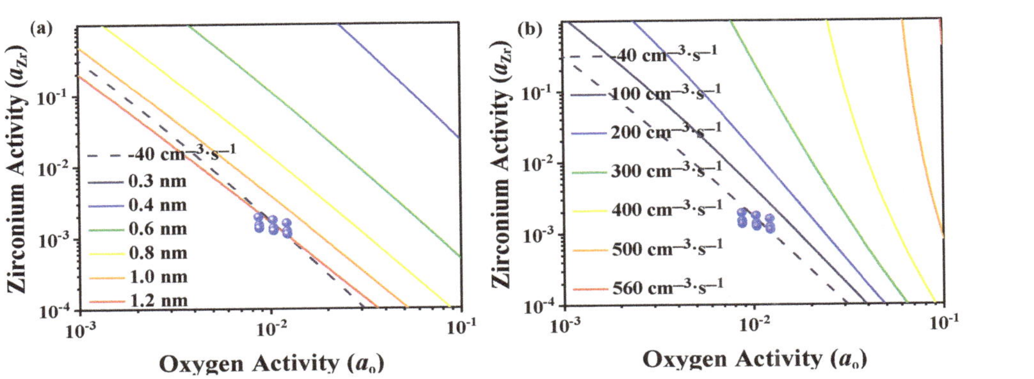

Figure 3. Relationships between the activities of the solute elements and the (**a**) critical nucleation radius and (**b**) nucleation rate of ZrO_2.

To obtain the activities of zirconium and oxygen, the composition of the zirconium deoxidized steel and corresponding thermodynamic data were substituted into

$$a_i = f_i[\text{mass\% } i] \quad (7)$$

$$\log f_i = \sum e_i^j[\text{mass\% } i] \quad (8)$$

where a_i, f_i, and [mass% i] are the activity, activity coefficient, and concentration of element i, respectively, and e_i^j is the first-order interaction coefficient (Table 4).

Table 4. Interaction coefficients of O and Zr at 1873 K [26].

e_i^j	Zr	O
O	−23	−0.17
Zr	0.032	−4

The experimentally estimated value of ln I can be calculated by Equations (5) and (6). The experimental value of ln I was 57 $cm^{-3}\cdot s^{-1}$. From Figure 3, the theoretical value of ln I is approximately −40 $cm^{-3}\cdot s^{-1}$. Therefore, the experimental value of the nucleation rate I was approximately 40 orders of magnitude larger than the theoretical value.

3.3. *First-Principles Calculations*

According to the two-step nucleation mechanism, the nucleation process of inclusions in liquid steel should include the multiphase deoxidation reaction, which can be expressed by the following two steps [32–35]. In the first step, the deoxidized elements in the molten steel melt and dissolve, and the deoxidized element atoms react with the dissolved oxygen in the molten steel to form oxide clusters. In the second step, the clusters combine to form cluster aggregates. The cluster aggregates then form critical crystal nuclei.

3.3.1. Structures of $(ZrO_2)_n$ Clusters

With increasing n value, the average bond length of the $(ZrO_2)_n$ cluster initially slightly increases, and it finally fluctuates at approximately 2.0 Å (Table 5). The nucleon binding ability in the nucleus is stronger and more stable for larger average binding energy (E_{bin}). The average binding energies of the $(ZrO_2)_n$ (n = 1–6) clusters are negative (Table 5), indicating that the binding between nuclei is relatively stable.

Table 5. Structures, average bond lengths, sizes, and average binding energies of $(ZrO_2)_n$ (n = 1–6) clusters.

Clusters	Bond Length(Zr-O) (nm)	Size (D_{max}) nm	E_{bin} eV
 $(ZrO_2)_1$	0.189	0.38	−5.55
 $(ZrO_2)_2$	0.192	0.48	−7.05
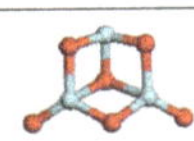 $(ZrO_2)_3$	0.204	0.60	−7.38
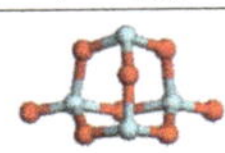 $(ZrO_2)_4$	0.20	0.66	−7.52
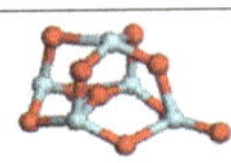 $(ZrO_2)_5$	0.206	0.70	−7.68
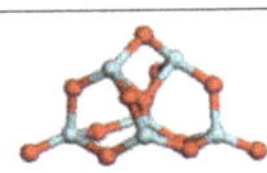 $(ZrO_2)_6$	0.203	0.93	−7.75

With an increasing number of atoms (n = 1–6), the average binding energy of the $(ZrO_2)_n$ cluster decreases, especially between the $(ZrO_2)_1$ and $(ZrO_2)_2$ clusters. This may be because of the energy error caused by the different numbers of atoms in different clusters. Therefore, the energy gap was used to compare the stabilities of the clusters. The energy gap is the difference between the lowest unoccupied molecular orbital (LUMO) and the highest occupied molecular orbital (HOMO). The energy gap reflects the ability of electrons to transition from an occupied orbital to an empty orbital, and it represents the ability of molecules to participate in chemical reactions. The system is more stable for a larger energy gap [36].

The energy gaps of the $(ZrO_2)_n$ clusters are given in Table 6. The $(ZrO_2)_2$ cluster has the largest energy gap, indicating that the $(ZrO_2)_2$ cluster is the most stable of the $(ZrO_2)_n$ clusters (n = 1–6). The LUMOs and HOMOs of the $(ZrO_2)_n$ clusters are shown in Figure 4. The blue and yellow area denote the orbitals of electron cloud, where the color is to distinguish the plus or minus of orbital wave function.

Table 6. LUMO–HUMO energy gaps of the $(ZrO_2)_n$ (n = 1–6) clusters.

Clusters	$(ZrO_2)_1$	$(ZrO_2)_2$	$(ZrO_2)_3$	$(ZrO_2)_4$	$(ZrO_2)_5$	$(ZrO_2)_6$
LUMO	−0.15927	−0.12413	−0.15856	−0.15835	−0.11532	−0.13762
HOMO	−0.18301	−0.2138	−0.21179	−0.19151	−0.20248	−0.22429
LUMO-HUMO	0.023742	0.089665	0.05323	0.033158	0.087153	0.086679

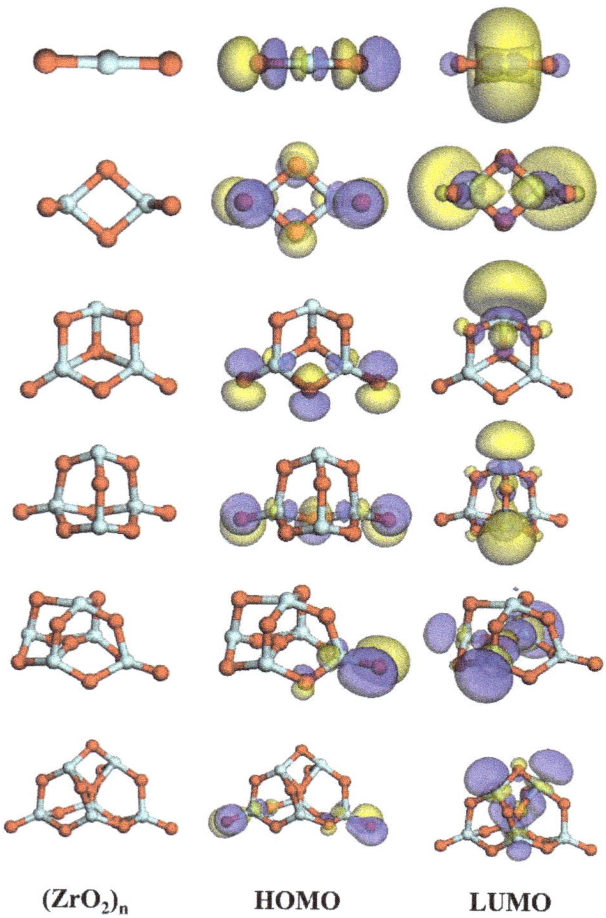

Figure 4. HOMOs and LUMOs of the $(ZrO_2)_n$ clusters.

3.3.2. Thermodynamic Properties of the $(ZrO_2)_n$ Clusters

The thermodynamic properties of ZrO_2 are shown in Figure 5. Where S is the entropy, and C_p is the heat capacity. In this study, the ZrO_2 crystal structure was tetragonal. The lines and points represent the calculated thermodynamic properties and values in the literature, respectively [37]. In the temperature range of 0–1000 K, there is a certain degree of deviation between the calculated and the values in the literature, but the variation trend of the thermodynamic properties with temperature is consistent. To the best of our knowledge, the reason that caused deviation is mainly the machine error derived from the thermochemical software package, and it was not possible to meet strict consistency criteria when combining data from various sources to form a data set for a substance. Another reason is that phase transformation occurred between monoclinic and tetragonal ZrO_2. In general, the calculated value is in good agreement with the value from the literature.

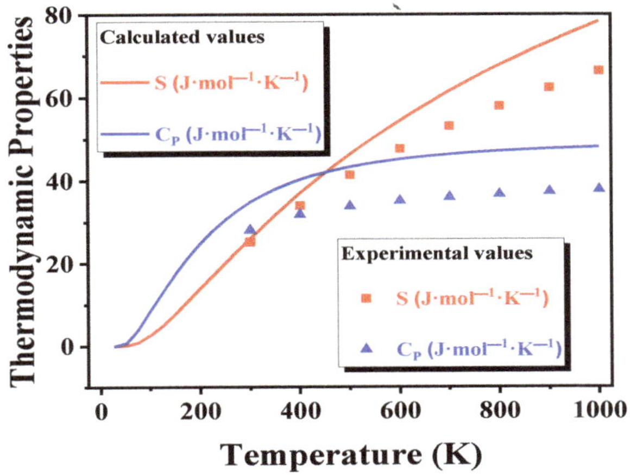

Figure 5. Thermodynamics properties of ZrO_2.

3.3.3. Gibbs Free Energy Changes of $(ZrO_2)_n$ Clusters and Nanoparticles

The formation Gibbs free energy (ΔG) curves of the $(ZrO_2)_n$ clusters and $ZrO_{2(s)}$ are shown in Figure 6a,b, respectively. The formation Gibbs free energies of $(ZrO_2)_n$ (n = 1–6) are negative, suggesting the $(ZrO_2)_n$ (n = 1–6) clusters form. However, the formation Gibbs free energy change from $(ZrO_2)_1$ to $ZrO_{2(s)}$ is positive, so the reaction from $(ZrO_2)_1$ to $ZrO_{2(s)}$ does not occur when the temperature is greater than 1000 K.

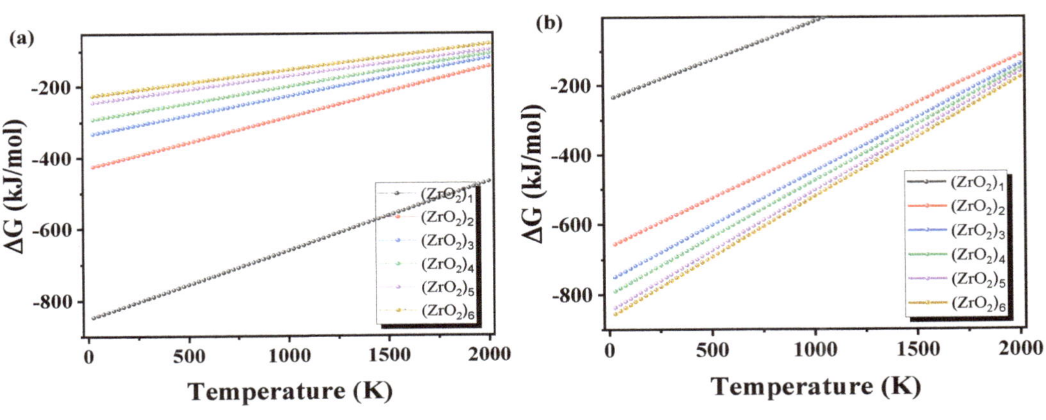

Figure 6. (**a**) Formation Gibbs free energy curves of $(ZrO_2)_n$. (**b**) Formation Gibbs free energy curve of $ZrO_{2(s)}$.

The Gibbs free energy changes from nanoscale ZrO_2 to $ZrO_{2(bulk)}$ and from the Zr and O atoms to nanoscale ZrO_2 are shown in Figure 7a,b, respectively. The Gibbs free energy of the macroscale ZrO_2 crystal is less than zero, indicating that nanoscale ZrO_2 will spontaneously transform to the macroscale ZrO_2 crystal.

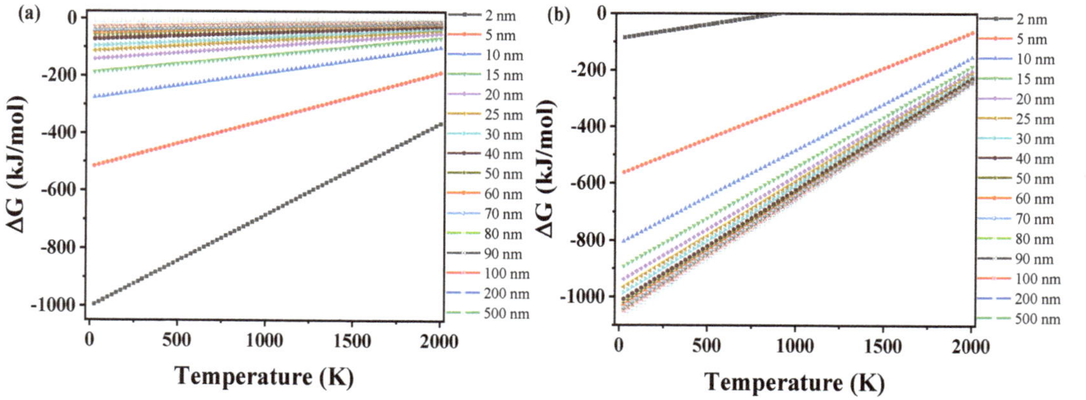

Figure 7. (**a**) Gibbs free energy change from nanoscale ZrO_2 to $ZrO_{2(bulk)}$. (**b**) Gibbs free energy changes from the Zr and O atoms to nanoscale ZrO_2.

4. Discussion

The formation of Gibbs free energy changes of the $(ZrO_2)_n$ clusters and various ZrO_2 nanoparticles are shown in Figure 8. Most of the Gibbs free energy change curves of the $(ZrO_2)_n$ clusters are higher than those of the ZrO_2 nanoparticles, which is in agreement with the law of the step-by-step change of the thermodynamic stability., from the perspective of the gradual decrease of the energy, this law indicates that the formation process of ZrO_2 occurs from the atoms to clusters to nanoscale crystal particles to the macroscale crystal.

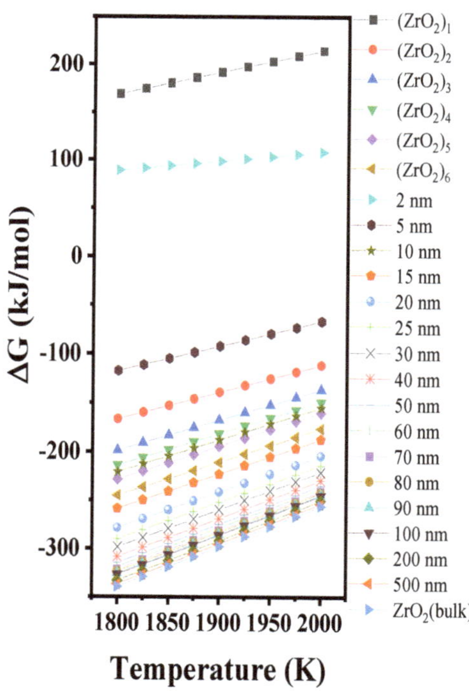

Figure 8. Formation Gibbs free energy changes of $(ZrO_2)_n$ clusters and various ZrO_2 nanoparticles.

The surfaces of nanoparticles usually contain two or three atomic layers [38,39], and the average length of the Zr–O bond is approximately 0.214 nm. Therefore, the surface atomic layer of zirconium oxide nanoparticles is approximately 0.428–0.642 nm thick. From Table 4, the sizes of the $(ZrO_2)_n$ clusters are also within this range. In addition, from Table 5, the energy gap of the $(ZrO_2)_2$ cluster is the largest, indicating that this cluster is the most stable. Therefore, the $(ZrO_2)_2$ cluster may be the atomic layer on the surface of the ZrO_2 nanoparticle. Based on the two-step nucleation theory, it is speculated that the nucleation pathway of ZrO_2 is $Zr_{atom} + O_{atom} \rightarrow (ZrO_2)_n \rightarrow (ZrO_2)_2 \rightarrow$ core (ZrO_2 particle)–shell ($(ZrO_2)_2$ cluster) nanoparticle $\rightarrow (ZrO_2)_{bulk}$. The nucleation process of ZrO_2 at 1873 K is shown in Figure 9.

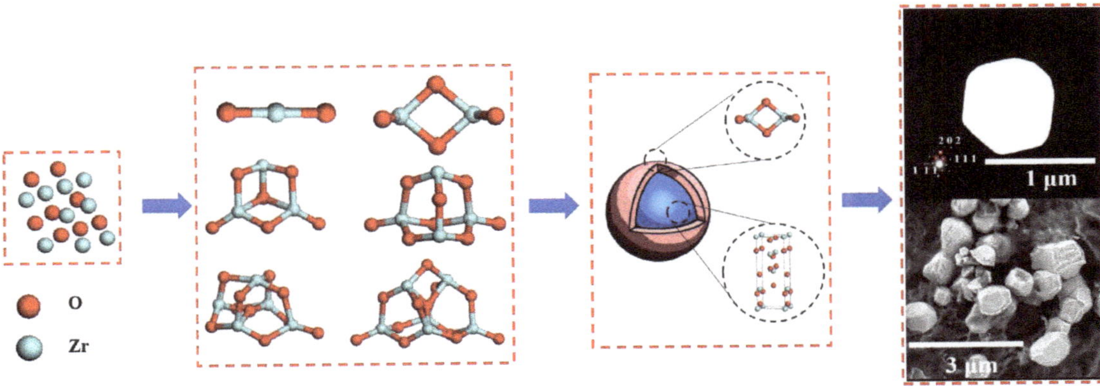

Figure 9. Nucleation process of zirconium oxide at 1873 K.

5. Conclusions

High-temperature deoxidation experiments and inclusion-extraction experiments have been performed. The nucleation process of ZrO_2 inclusions in Zr deoxidized steel was investigated by classical nucleation theory and first-principles calculation. The main conclusions are as follows.

When the Zr content was 100 ppm, SEM-EDS showed that the main inclusions in the steel were ZrO_2. µXRD analysis confirmed the existence of ZrO_2, and monoclinic and tetragonal ZrO_2 were simultaneously detected, which may be because tetragonal ZrO_2 transformed to monoclinic ZrO_2 during the rapid cooling process. The average size of the ZrO_2 inclusions was 0.56 µm. Through classical nucleation theory, the relationships between the solute element activities and the nucleation radius and nucleation rate of ZrO_2 were obtained. The theoretical value of the nucleation rate was compared with the values in the literature, and the experimental value of I was approximately 40 orders of magnitude larger than the theoretical value.

The thermodynamic properties of macroscale ZrO_2 were calculated by first principles, and the results were consistent with the experimental values. $(ZrO_2)_n$ (n = 1–6) cluster models were constructed, and the thermodynamic properties of the geometrically optimized cluster structures and nanoscale ZrO_2 particles were calculated, which verified the rationality of the existence of the pre-nucleation phase in terms of the thermodynamics. Based on two-step nucleation theory, the nucleation pathway of ZrO_2 is proposed to be $Zr_{atom} + O_{atom} \rightarrow (ZrO_2)_n \rightarrow (ZrO_2)_2 \rightarrow$ core (ZrO_2 particle)–shell ($(ZrO_2)_2$ cluster) nanoparticle $\rightarrow (ZrO_2)_{bulk}$.

Author Contributions: L.W., C.C. and S.Y. conceived the idea and designed the research. Y.L. performed the theoretical calculations. Y.L. and X.L. performed the μXRD analyses. X.L. performed the high-temperature experiments. Y.L. and L.W. wrote the original draft of the paper, and all authors contributed to the review and editing of the paper. L.W. acquired the funding and supervised the study. All authors have read and agreed to the published version of the manuscript.

Funding: This research was supported by the National Science Foundation of China (Grant Nos. 52064011 and 52274331) and the Science and Technology Planning Project of Guizhou (Grant Nos. Qian Ke He Ji Chu ZK [2021]258 and Qian Ke He Zhi Cheng [2021]342). This study was also supported by the Research Program for Talented Scholars of the Guizhou Institute of Technology (Grant No. XJGC20190962).

Data Availability Statement: Part of the data presented in this study are available on request from the corresponding author. The data are not publicly available due to intellectual property.

Acknowledgments: Thanks for the computing support of the State Key Laboratory of Public Big Data, Guizhou University.

Conflicts of Interest: The authors state that there are no conflict of interest to disclose.

References

1. Takamura, J.I.; Mizoguchi, S. Roles of oxides in steel performance. Proceedings of The Sixth International Iron and Steel Congress, ISIJ International, Nagoya, Japan, 21–26 October 1990; pp. 591–597.
2. Wang, L.; Yang, S.; Li, J.; Chen, C.; Li, C.; Li, X. Study on the Capillary Interaction Between Particles on the Surface of High-Temperature Melts. *Steel Res. Int.* **2021**, *92*, 2100013. [CrossRef]
3. Li, Y.; Wang, L.; Li, J.; Yang, S.; Chen, C.; Li, C.; Li, X. In-Situ Observation on the Agglomeration and Dispersion of Particles at the Interface of High-temperature Melts. *ISIJ Int.* **2021**, *61*, 753–762. [CrossRef]
4. Abraham, S.; Bodnar, R.; Raines, J.; Wang, Y. Inclusion engineering and metallurgy of calcium treatment. *J. Iron Steel Res. Int.* **2018**, *25*, 133–145. [CrossRef]
5. Liu, W.; Yang, S.F.; Li, J.S.; Wang, F.; Yang, H.B. Numerical model of inclusion separation from liquid metal with consideration of dissolution in slag. *J. Iron Steel Res. Int.* **2019**, *26*, 1147–1153. [CrossRef]
6. Peng, B.W.; Li, F.J.; Zheng, S.B.; Li, H.G.; Zhai, Q.J. Inclusion control study in high aluminum steel. *J. Iron Steel Res. Int.* **2018**, *25*, 796–802. [CrossRef]
7. Zhao, K.W.; Zeng, J.H.; Wang, X.H. Nonmetallic inclusion control of 350 km/h high speed rail steel. *J. Iron Steel Res. Int.* **2009**, *16*, 20–26. [CrossRef]
8. Wang, L.; Yang, S.; Li, J.; Wu, T.; Liu, W.; Xiong, J. Improving Cleanliness of 95CrMo Drill Rod Steel by Slag Refining. *Met. Mater. Trans. A* **2016**, *47*, 99–107. [CrossRef]
9. Yang, Y.; Zhan, D.; Lei, H.; Qiu, G.; Li, Y.; Jiang, Z.; Zhang, H. In situ observation of acicular ferrite nucleation and growth at different cooling rate in Ti-Zr deoxidized steel. *Metall. Mater. Trans. B* **2019**, *50*, 2536–2546. [CrossRef]
10. Chai, F.; Yang, C.F.; Su, H.; Zhang, Y.Q.; Xu, Z.; Yang, Y.H. Effect of Zr addition to Ti-killed steel on inclusion formation and microstructural evolution in welding induced coarse-grained heat affected zone. *Acta Metall. Sin. Engl. Lett.* **2008**, *21*, 220–226. [CrossRef]
11. Guo, A.M.; Li, S.R.; Guo, J.; Li, P.H.; Ding, Q.F.; Wu, K.M.; He, X.L. Effect of zirconium addition on the impact toughness of the heat affected zone in a high strength low alloy pipeline steel. *Mater. Charact.* **2008**, *59*, 134–139. [CrossRef]
12. Wu, H.B.; Hou, M.; Liang, G.L.; Tang, D. Effect of zirconium on the low-temperature toughness of CGHAZ in F40 ship plates containing titanium. *J. Univ. Sci. Technol. Beijing* **2012**, *34*, 137–142.
13. Min, J.; Hu, Z.; Wang, X.; Pak, J.J. Characterization of microstructure and non-metallic inclusions in Zr-Al deoxidized low carbon steel. *ISIJ Int.* **2013**, *53*, 1386–1391.
14. Bhadeshia, H.K.D.H.; Honeycombe, R.W.K. *Steels Micro-Structure and Properties*, 3rd ed.; Elsevier Ltd.: Oxford, UK, 2006; Volume 7, pp. 1–360.
15. He, K.; Baker, T.N. Zr-contaning precipitates in a Ti-Nb micro-alloyed HSLA steel containing 0.016% Zr addition. *Mater. Sci. Eng.* **1996**, *215*, 57–66.
16. Sarma, D.S.; Karasev, A.V.; Jonsson, P.G. On the role of nonmetallic inclusions in the nucleation of acicular ferrite in steels. *ISIJ Int.* **2009**, *49*, 1063–1074. [CrossRef]
17. Guo, Q.Y.; Song, B.; Song, M.M. Effects of Ti, Al and Zr deoxidation on morphology of sulfides in medium-sulfur non-quenched and tempered steel. *Trans. Mater. Heat Treat.* **2019**, *40*, 133–139.
18. Li, Y.; Wan, X.L.; Cheng, L.; Wu, K.M. First-principles calculation of the interaction of Mn with ZrO_2 and its effect on the formation of ferrite in high-strength low-alloy steels. *Scr. Mater.* **2014**, *75*, 78–81. [CrossRef]
19. Wang, G.; Zhang, L. Thermodynamics dependent on size in nucleation process of inclusions and calculation of critical nucleus radius in clean molten steel. *Iron Steel* **2012**, *47*, 22–26. [CrossRef]

20. Wang, G.C.; Wang, Q.; Li, S.L.; Ai, X.G.; Fan, C.G. Evidence of Multi-step Nucleation Leading to Various Crystallization Pathways from an Fe-O-Al Melt. *Sci. Rep.* **2014**, *4*, 5082. [CrossRef]
21. Lei, H.; Xiao, Y.; Wang, G.; Zhang, H.; Jin, W.; Zhang, L. Thermodynamic insight into the growth of nanoscale particle of Al-deoxidation in Fe–O–Al melt. *Sci. Rep.* **2020**, *10*, 16909. [CrossRef]
22. Zhao, D.; Bao, W.; Li, H.; Zheng, S.; Chou, K.-C. Cluster-assisted nucleation mechanism of titanium oxides in Fe-Ti supercooled alloys. *J. Alloys Compd.* **2018**, *744*, 797–800. [CrossRef]
23. Yang, L.; Zhang, W.; He, L.; Li, H.; Zheng, S. Study on the growth and morphology evolution of titanium oxide clusters in molten iron with molecular dynamics simulation. *RSC Adv.* **2019**, *9*, 32620–32627. [CrossRef] [PubMed]
24. Suito, H.; Ohta, H. Characteristics of Particle Size Distribution in Early Stage of Deoxidation. *ISIJ Int.* **2006**, *46*, 33–41. [CrossRef]
25. Wang, L.; Li, J.; Yang, S.; Chen, C.; Jin, H.; Li, X. Nucleation and Ostwald Growth of Particles in Fe-O-Al-Ca Melt. *Sci. Rep.* **2018**, *8*, 1135. [CrossRef] [PubMed]
26. Li, Y.; Wang, L.; Chen, C.; Li, J.; Li, X. Effect of Mg Treatment on the Nucleation and Ostwald Growth of Inclusions in Fe-O-Al-Mg Melt. *Materials* **2020**, *13*, 3355. [CrossRef] [PubMed]
27. Humenik, M.; Kingery, W.D. Metal-Ceramic Interactions: III, Surface Tension and Wettability of Metal-Ceramic Systems. *J. Am. Ceram. Soc.* **1954**, *37*, 18–23. [CrossRef]
28. Hino, M.; Ito, K. *Thermodynamic Data for Steelmaking*, 2nd ed.; Tohoku University Press: Tohoku, Japan, 2010; pp. 247–264.
29. Ohta, H.; Suito, H. Efects of Dissolved Oxygen and Size Distribution on Particle Coarsening of Deoxidation Product. *ISIJ Int.* **2006**, *46*, 42–49. [CrossRef]
30. Sakata, K.; Suito, H. Dispersion of fine primary inclusions of MgO and ZrO_2 in Fe-10 mass pct Ni alloy and the solidification structure. *Met. Mater. Trans. B* **1999**, *30*, 1053–1063. [CrossRef]
31. Perdew, J.P.; Burke, K.; Ernzerhof, M. Generalized Gradient Approximation Made Simple. *Phys. Rev. Lett.* **1997**, *77*, 3865. [CrossRef]
32. Wang, G.; Xiao, Y.; Zhao, C.; Jing, L.; Shang, D. Atomic Cluster Aggregates in Nucleation of Solid Alumina Inclusion in the Aluminum Deoxidation for Liquid Iron. *Metall. Mater. Trans. B* **2017**, *49*, 282–290. [CrossRef]
33. Wang, G.; Xiao, Y.; Song, Y.; Zhou, H.; Tian, Q.; Li, F. A density functional study on the aggregation of alumina clusters. *Res. Chem. Intermed.* **2017**, *43*, 1447–1463. [CrossRef]
34. Xiao, Y.; Lei, H.; Yang, B.; Wang, G.; Wang, Q.; Jin, W. Nucleation and growth for magnesia inclusion in Fe-O-Mg melt. *RSC Adv.* **2018**, *8*, 38336–38345. [CrossRef]
35. Xiao, Y.; Lei, H.; Yang, B.; Zhao, Y.; Wang, Q.; Wang, G. Thermodynamic insight into the growth of Calcia inclusions at the nanoscale: The case of Fe-O-Ca melt. *RSC Adv.* **2019**, *9*, 11135–11141. [CrossRef]
36. Aihara, J.-I. Reduced HOMOLUMO Gap as an Index of Kinetic Stability for Polycyclic Aromatic Hydrocarbons. *J. Phys. Chem. A* **1999**, *103*, 7487–7495. [CrossRef]
37. Chase, M.W. *NIST-JANAF Thermochemical Tables*; American Chemical Society: Washington, DC, USA, 1998; Volume 2, pp. 1880–1881.
38. Byun, J.S.; Shim, J.H.; Cho, Y.W.; Lee, D.N. Non-metallic particle and intragranular nucleation of ferrite in Ti-killed C-Mn steel. *Acta Mater.* **2003**, *51*, 1593–1606. [CrossRef]
39. Phillpot, S.R.; Wolf, D.; Gleiter, H. Molecular dynamics study of the synthesis and characterization of a fully dense, three dimensional nanocrystalline material. *J. Appl. Phys.* **1995**, *78*, 847–861. [CrossRef]

materials

MDPI

Article

Unveiling the Effect of CaF_2 on the Microstructure and Transport Properties of Phosphosilicate Systems

Yizhe Du [1,2], Zhidan Huang [1,2], Mujun Long [1,2], Huamei Duan [1,2] and Dengfu Chen [1,2,*]

1 Laboratory of Materials and Metallurgy, College of Materials Science and Engineering, Chongqing University, Chongqing 400044, China
2 Chongqing Key Laboratory of Vanadium-Titanium Metallurgy and New Materials, Chongqing University, Chongqing 400044, China
* Correspondence: chendfu@cqu.edu.cn

Abstract: As an effective flux, CaF_2 is beneficial in improving the fluidity of slag in the steel-making process, which is crucial for dephosphorization. To reveal the existence form and functional mechanism of CaF_2 in phosphosilicate systems, the microstructures and transport properties of CaO-SiO_2-CaF_2-P_2O_5 quaternary slag systems are investigated by molecular dynamics simulations (MD) combined with experiments. The results demonstrate that the Si-O coordination number does not vary significantly with the increasing CaF_2 content, but the P-O coordination number dramatically decreases. CaF_2 has a minor effect on the single [SiO_4] but makes the structure of the silicate system simple. On the contrary, F^- ions could reduce the stability of P-O bonds and promoted the transformation of [PO_4] to [PO_3F], which is beneficial for making the P element-enriched phosphate network structure more aggregated. However, the introduction of CaF_2 does not alter the tetrahedral character of the original fundamental structural unit. In addition, the results of the investigation of the transport properties show that the self-diffusion coefficients of each ion are positively correlated with CaF_2 content and arranged in the order of $F^- > Ca^{2+} > O^{2-} \approx P^{5+} > Si^{4+}$. Due to CaF_2 reducing the degree of polymerization of the whole melts, the viscosity decreases from 0.39 to 0.13 Pa·s as the CaF_2 content increases from 0% to 20%. Moreover, the viscosity of the melt shows an excellent linear dependence on the structural parameters.

Keywords: molecular dynamics simulation; CaF_2; melts' structure; transport properties; viscosity

Citation: Du, Y.; Huang, Z.; Long, M.; Duan, H.; Chen, D. Unveiling the Effect of CaF_2 on the Microstructure and Transport Properties of Phosphosilicate Systems. *Materials* **2022**, *15*, 7916. https://doi.org/10.3390/ma15227916

Academic Editor: Philippe Colomban

Received: 20 October 2022
Accepted: 8 November 2022
Published: 9 November 2022

1. Introduction

The physical and chemical properties of slag are crucial for mass transfer and chemical reactions between liquid steel and slag. It is well-known that the physical and chemical properties of slag are determined by its structural characteristics [1,2], and it is of extraordinary interest to study the structural information of slag to understand its performance.

There is a large quantity of experimental approaches that have been applied to study the structure information of slag, and they mainly include nuclear magnetic resonance, X-ray diffraction, neutron diffraction, Raman spectroscopy, etc. [3–5]. These methods help one to effectively understand the microstructure and unique properties of slag, which is a significant breakthrough in this research direction. In recent years, with the rapid development of computer technology, a large number of simulation techniques have gradually entered the field of vision of scholars. In particular, MD simulations are expected to provide an effective way to understand the slag structure from a microscopic point of view with its advantages. Specifically, it is not affected by experimental conditions, such as high temperature and pressure. At present, a large number of scholars have used the molecular dynamics (MD) simulation method to conduct studies on the microstructure and properties of metallurgical slag and achieved remarkable results [6–8].

At present, structural information of binary and ternary silicate and aluminate melts has been extensively and carefully studied. As for phosphate melt, it has not been systematically studied due to the complexity and diversity of its structure. However, phosphorus is one of the harmful elements in steel. The excessive amount of phosphorus in steel is detrimental to its quality and properties. Therefore, dephosphorization is one of the key tasks in steel-making. In addition, phosphorus removal relies on the reaction between steel and slag, so a comprehensive study of the microstructure and transport properties of phosphorus removal slag could help clarify the underlying causes of the alteration in its macroscopic properties. In previous studies, Diao et al. [9,10] studied the microstructure of the ternary slag system of CaO-SiO_2-P_2O_5 through MD simulation, and the results showed that silicon and phosphorus mainly formed a tetrahedral structure. Additionally, the concentration of free oxygen decreases significantly with increasing P_2O_5 content and the degree of polymerization of the melt increases. Fan et al. [11] reported the existence form of Si and P in CaO-SiO_2-P_2O_5 melt under high basicity, and the results showed that both Si and P tended to form complex anions, and P ions are more inclined to form a single tetrahedron structure. Moreover, Jiang et al. [12] used MD simulation to study the structure and properties of the molten CaO-SiO_2-P_2O_5-FeO slag system and concluded that the polymerization degree of the system decreased with the increase of basicity.

In our previous studies, a relatively deep understanding of the structural properties of binary phosphate systems has been obtained [1]. In actual production, additional components will be used to adjust the comprehensive physicochemical properties of the slag to meet the requirements of the metallurgical production process. CaF_2 is widely used as a conventional flux to reduce the viscosity of slag and improve its mobility. At present, some scholars have carried out studies on the structural properties of CaF_2-containing glasses. For example, Kansal et al. [13] studied the effect of the CaO/MgO ratio on the structure and thermal properties of CaO-MgO-SiO_2-P_2O_5-CaF_2, and found that CaF_2 always tends to combine with $[PO_4]$ to form composite structures. Pedone et al. [14] investigated the influence of halides on the structure of phosphosilica bioactive glasses by MD simulation. The results show that in the mixed-fluoride/chloride-containing glasses, fluorine tends to surround phosphate, whereas chloride moves toward the silicate network. Furthermore, interestingly, according to relevant reports, F^- can also directly participate in the dephosphorization reaction, thus directly affecting the structure of dephosphorization slag [15,16]. However, no reports have been found regarding the presence of CaF_2 in dephosphorized slag and its effects on structure and properties.

Therefore, to shed light on the morphological and functional mechanisms responsible for the presence of CaF_2 in dephosphorized slag, in this paper we focus on the quaternary slag system CaO-SiO_2-CaF_2-P_2O_5 and perform a comprehensive analysis of its microstructure using molecular dynamics simulations. Under the conditions of certain contents of $(xCaO)/(xSiO_2)$ and P_2O_5 in the slag, the influence of CaF_2 on the microstructure and transport performance of the slag system under high temperature is investigated combined with experiments. The results of this investigation can provide some valuable information to understand the microstructure of dephosphorized slag and clarify the intrinsic link between melt flow properties and the evolution of structural units.

2. Computational Methodology

2.1. Interatomic Potential

For molecular dynamics simulation, selecting appropriate potential function and corresponding parameters is the basis of accurate calculation. All molecular dynamics simulations were carried out using the Born-Mayer-Huggins (BMH) model [1,7,9–11] in this study, and the potential can be expressed as Equation (1):

$$U_{ij}(r) = \frac{Z_i Z_j e^2}{4\pi\varepsilon_0 r_{ij}} + A_{ij}\exp(-B_{ij}r) - \frac{C_{ij}}{r_{ij}^6} \quad (1)$$

where, Z_i and Z_j are the effect charges of ions i and j, respectively, e represents the charge of a single electron, ε_0 represents the vacuum permittivity, r_{ij} is the distance between atoms i and j, and A_{ij}, B_{ij}, and C_{ij} are the adjustable parameters for BMH potentials. The three items on the right side of the above formula represent the coulombic interaction, short-range repulsion interaction, and the van der Waals interactions. Various interaction potential parameters between the selected particles are listed in Table 1 [11,17,18].

Table 1. Born–Mayer–Higgins potential parameters for atomic pairs in CaO-SiO_2-CaF_2-P_2O_5 systems.

Atom 1	Atom 2	A_{ij} (eV)	B_{ij} (1/Å)	C_{ij} (eV·Å^6)
Ca	Ca	329,051.60	6.25	4.33
Ca	Si	26,674.68	6.25	0
Ca	P	164,585.80	12.5	0
Ca	O	717,827.00	6.06	8.67
Ca	F	496,191.60	6.06	8.67
Si	Si	2162.39	6.25	0
Si	P	1081.60	12.5	0
Si	F	43,406.00	6.06	0
Si	O	62,794.37	6.06	0
P	P	0	0.0	0
P	F	365,232.88	11.8	0
P	O	1847.70	3.45	0
F	F	730,722.80	5.88	17.34
F	O	1,046,135.40	5.88	17.34
O	O	1,497,049.00	5.88	17.34

2.2. Simulation Approach

The target sample composition was discussed in the full melting component range at the corresponding temperatures. The slag samples of CaO-SiO_2-CaF_2-P_2O_5 were divided into five groups, where the first group with 0% CaF_2 was used as a comparison with the other four groups. As Figure 1 shows, according to the liquid phase range of the CaO-SiO_2-CaF_2-P_2O_5 quaternary slag system at 1600 °C obtained through Factsage 8.0 thermodynamic calculation software, the chemical composition of each group in this study was determined. The varying number of atoms was then calculated based on the mole fraction of each group. Referring to the empirical formula in relevant literature and research results [19], the density of various groups at 1600 °C was calculated, respectively. The chemical composition, atomic number, and other information about each group of samples are listed in Table 2.

The computational methods used and the choice of parameters are critical factors in achieving efficient and accurate simulations. In this study, about 6000 atoms were randomly placed in a model box. Since the number of calculated atoms is always finite, periodic boundary conditions were performed on all faces of the model box to obtain an infinite system of atoms without boundaries. The results obtained with periodic boundary conditions are sufficient to reflect the actual situation. All MD simulations use a canonical ensemble (NVT), which means that the calculations were performed in a system with a constant atomic number (N), sample volume (V), and temperature (T). Additionally, the sum method of Ewald was used for the long-distance coulomb force, and the motion equation of atoms was explained by the jump integral method of a 1 fs timestep. The potential cutoff radius was set to 10 Å in the calculation of the repulsive force. Besides, the total time length of each group of simulations was determined to be 60 ps, equivalent to 60,000 steps. After the beginning of the simulation, the initial temperature was set at 5000 K (4727 °C) for 15,000 timesteps to agitate the atoms and eliminate the effect of the intentional distribution. Secondly, the temperature was cooled down to 1873 K (1600 °C) with 30,000 timesteps. Subsequently, the system was relaxed at 1600 °C for another 15,000 timesteps in the equilibrium calculation. The temperature, volume, and enthalpy remained nearly constant for 15,000 timesteps, demonstrating that the system has reached equilibrium.

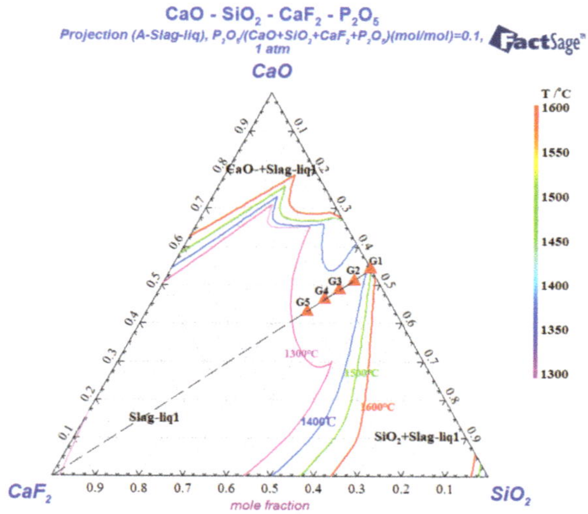

Figure 1. Liquid phase interval of the CaO-SiO_2-CaF_2-P_2O_5 system.

Table 2. Composition, number of atoms, and density of CaO-SiO_2-CaF_2-P_2O_5 melts at 1600 °C.

Groups	Mole Fraction (%)				Number of Atoms					Total	Density (g/cm^3)
	CaO	SiO_2	CaF_2	P_2O_5	Ca	Si	P	F	O		
G1	49	41	0	10	1010	845	412	0	3732	5999	2.5159
G2	46	39	5	10	1041	796	408	204	3551	6000	2.5109
G3	44	36	10	10	1095	730	405	405	3365	6000	2.5083
G4	41	34	15	10	1124	682	401	602	3191	6000	2.5018
G5	38	32	20	10	1152	636	397	795	3020	6000	2.4947

2.3. Statistics of Structural Information

The radial distribution function (RDF) is commonly used to investigate the character of the short-range order of melts. Equation (2) lists the mathematical expression of the RDF [20]:

$$g_{ij}(r) = \frac{V}{N_i N_j} \sum_j \frac{\langle n_{ij}(r - \Delta r/2, r + \Delta r/2) \rangle}{4\pi r^2 \Delta r} \quad (2)$$

where, V is the volume of MD-simulated cells, N is the number of particles, and n_{ij} is the average number of atom j surrounding the atom i within a distance $r \pm \Delta r/2$. The abscissa of the first peak and the first trough in the RDF curve represent the average bond length and cutoff radius of the corresponding atoms, respectively.

In addition, integration of the corresponding partial RDF of the particles generates the average CN function, which represents the number of atoms j around atoms i within the cutoff radius. The CN function is expressed as Equation (3):

$$N_{ij}(r) = \frac{4\pi N_j}{V} \int_0^r r^2 g_{ij}(r) dr \quad (3)$$

Finally, as far as the concentration of oxygen species and the distribution of structural units, Q^n, were concerned, this structural information will be counted by the Matlab program, based on the spatial atomic coordinates derived from the MD simulation.

2.4. Viscosity Calculation

Viscosity is one of the most significant physical parameters of slag. The viscosity of melts can reflect the degree of polymerization. Through statistical analysis of atomic coordination by MD simulation, a function of *MSD* would be generated, displayed as Equation (4):

$$MSD = \langle \Delta r(t)^2 \rangle = \frac{1}{N}\langle \sum_{i=1}^{N} |r_i(t) - r_i(0)|^2 \rangle \quad (4)$$

where, N is the number of particles, $r_i(t)$ represents the position coordinate of atom i at time t, and angular brackets denote a statistical average of many function values. The self-diffusion coefficient could be obtained from *MSD* as shown below [21]:

$$D = \lim_{t\to\infty} \frac{1}{6} \frac{d\left[\Delta \bar{r}(t)^2\right]}{dt} \quad (5)$$

Then, the shear viscosity information of the melts can be obtained by combining the self-diffusion coefficient, D, with the Stokes–Einstein equation [22,23]:

$$\eta = \frac{K_B T}{D\lambda} \quad (6)$$

where, K_B is the Boltzmann constant, which is 1.38×10^{-23} J/K, T is the system temperature, and λ is the particle transition step size, which is commonly considered $\lambda = 2R_O = 2.8$ Å [24–26]. Based on the above calculation method, the partial transport performance of the melts can be obtained, and the relationship between the structural information and performance can be established.

3. Experimental Method

Based on the mole fraction of each sample in Table 2, the composition of the experimental slag was obtained by mass conversion. The results are shown in Table 3. The reagents used in our experiments are all from a specialist chemical reagent company in Chongqing, China. The purity of the reagents (CaO, SiO_2, CaF_2, and P_2O_5) used was above 99.5 wt.%. The weighted sample powder was well-mixed and placed in a graphite crucible before the viscosity was measured.

Table 3. The chemical compositions of experimental slags.

Groups	Slag Composition (Mass Percent)			
	CaO	SiO_2	CaF_2	P_2O_5
G1	41.5	37.1	0	21.4
G2	38.3	34.8	5.8	21.1
G3	36.1	31.7	11.4	20.8
G4	33.1	29.5	16.9	20.5
G5	30.3	27.3	22.2	20.2

The viscosity was measured using the rotating cylinder method. The viscometer was calibrated at room temperature using an oil with known viscosity prior to the experiment. Approximately 250 g of each slag sample was placed into a graphite crucible to melt, and the average heating rate was 5 °C/min. Since P_2O_5 has a low boiling point, it is prone to volatilization and produces white smoke at high temperatures. Consequently, the other three components were firstly added into the crucible and heated to 1500 °C for 20 min. After the sample was fully melted, P_2O_5 was added, and the crucible was covered to prevent volatilization. Then, the crucible was opened after keeping it for 5 min. If there was no obvious white smoke, the crucible was reheated to 1600 °C and kept for 20 min to homogenize the chemical composition. Finally, the viscosity of each sample was determined from an average of 60 consecutive measurements.

4. Results and Discussion

4.1. Local Structural Characteristics

The local structure information of melts can be preliminarily obtained by RDF and CN. Taking G2 as an example, Figure 2a,b show the distribution of RDF and CN in the system of $CaO\text{-}SiO_2\text{-}CaF_2\text{-}P_2O_5$ at 1600 °C when CaF_2 content was 5%, respectively. According to the RDF curve, the average bond length of each atom pair in the melts can be concluded. As can be obtained from Figure 2a, the average bond lengths of Ca-O, Si-O, P-O, and Ca-F were 2.31, 1.62, 1.50, and 2.30 Å, respectively. The results are in good accordance with previous research obtained from MD simulations and experiments [9–11,27]. Table 4 shows the variation of the bond lengths for various pairs of atoms from G1 to G5.

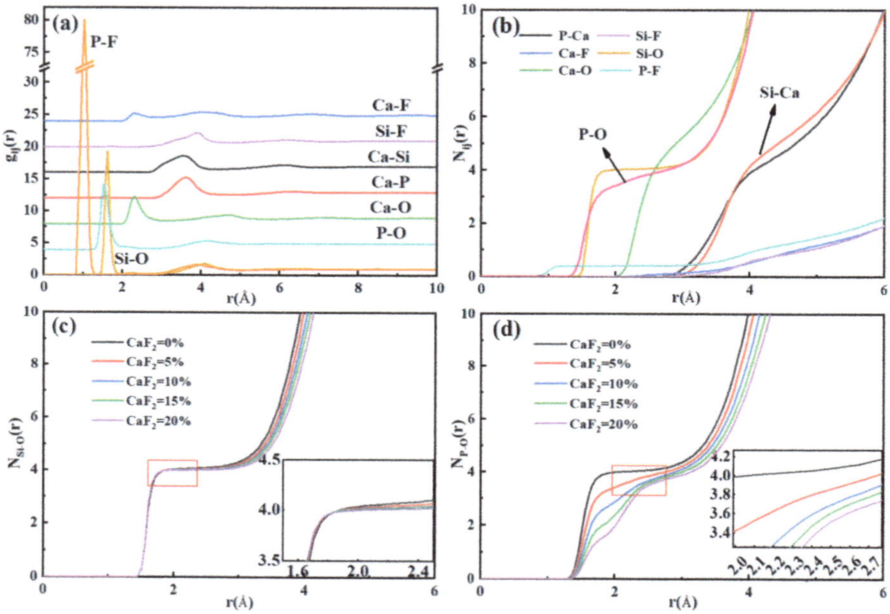

Figure 2. (**a**) RDF and (**b**) CN for different atom pairs in $CaO\text{-}SiO_2\text{-}CaF_2\text{-}P_2O_5$ systems and changes of P-O and Si-O coordination numbers: (**c**) Si-O and (**d**) P-O.

Table 4. Variation of average bond lengths of different atom pairs.

Pair	Ca-O	Si-O	P-O	P-F	Ca-F	P-Ca
R_{ij}/Å (G1)	2.31	1.61	1.50	—	—	3.60
R_{ij}/Å (G2)	2.31	1.61	1.52	1.01	2.31	3.60
R_{ij}/Å (G3)	2.31	1.61	1.54	1.01	2.31	3.60
R_{ij}/Å (G4)	2.32	1.61	1.55	1.01	2.32	3.61
R_{ij}/Å (G5)	2.32	1.61	1.56	1.01	2.32	3.61

In general, a strong and sharp peak in the RDF curve indicates a steep stabilization of the corresponding bond. Similarly, for CN curves, a broad flat plateau implies a large stability of the corresponding polyhedron. It can be observed in Figure 2a that both Si-O and P-O curves had a sharp peak, meaning Si and P tended to combine with O atoms and form stable structures. From Table 4, the average Si-O bond length remained constant with the increasing CaF_2 content in the melt, demonstrating that the Si-O bond was particularly stable and was not subject to CaF_2. However, the length of the P-O bond became longer, confirming the character of the P-O structure affected by CaF_2, while indicating a decrease in the strength of the P-O bond, which may lead to the evolution of the phosphate melt structure. Furthermore, the P-F bond appeared in the system due to the addition of CaF_2. Interestingly, the RDF curve for the P-F bond had an unusually sharp peak and the P-F bond length did not change significantly with CaF_2 content, indicating that the P-F bond was considerably more stable than the P-O bond, which is unprecedented. In addition, the Ca-O and Ca-F bonds had slightly increased lengths, indicating that they were more loosely bound with the addition of CaF_2.

As can be seen from Figure 2b, $CN_{Si\text{-}O}$, $CN_{P\text{-}O}$, and $CN_{Ca\text{-}O}$ were 4.04, 3.87, and 5.51, respectively. The plateau on the $CN_{Si\text{-}O}$ curve was smoother than that on the $CN_{P\text{-}O}$ curve, indicating that the stability of the Si-O structure was higher than that of the P-O in the $CaO\text{-}SiO_2\text{-}CaF_2\text{-}P_2O_5$ systems. Since Ca^{2+} is typically present as a network modifier, $CN_{Ca\text{-}O}$ exhibited a sloping plateau, meaning that no stable structure was formed between Ca-O, which is consistent with previous studies on slag or glassy structures containing CaO [10,28–30]. Additionally, it is worth noting that $CN_{P\text{-}F}$ had an extremely flat plateau between 0 and 1, which means that F^- and P^{5+} had a strong coordination tendency. Wang et al. [31] introduced CaF_2 into $CaO\text{-}SiO_2\text{-}Al_2O_3$ slag systems and found that F^- has a strong tendency to replace an O in the $[AlO_4]$ structure to form an Al-F bond. They attributed the phenomenon to the difference between the electronegativity of F^- and O^{2-}. Therefore, the addition of CaF_2 causes a shift in the original structure of the melts, especially for phosphate systems. The tendency of F^- to coordinate with P^{5+} is so strong that it may form a competitive relationship with O^{2-}, leading to a large-scale transformation of the P-O structure. This trend may be more significant in high-temperature conditions. Besides, the CN curves for all pairs of atoms except Si-O, P-O, and P-F did not have a clear plateau, suggesting that they do not typically form stable structures, and they are therefore not discussed in detail here.

Figure 2c,d show the alters of Si-O and P-O coordination numbers as CaF_2 content in slag from 0% to 20%, respectively. At 0% CaF_2 content, the coordination numbers of Si-O and P-O were close to 4.0, indicating that most of them exist as 4-coordinates and conform to the tetrahedral form. From Figure 2c, the coordination number changes of Si-O were not obvious in the range of the mole fraction of (CaF_2) = 0~20%, which were all around 4.0. Due to the high stability of $[SiO_4]$, it is difficult for F^- to break through the bond energy barrier between Si-O to coordinate with Si^{4+}, which was also discussed in previous studies [32,33]. However, when increasing the CaF_2 content from 0% to 20%, the coordination plateau of P-O became increasingly tortuous and the average coordination number decreased, with values of 4.05, 3.87, 3.74, 3.66, and 3.57. Moreover, it can be seen from the variation law of the bond length that P-O kept increasing, indicating that its stability decreased. In contrast, the P-F bond length was much smaller than the P-O bond length. All indications show that the affinity between P^{5+} and F^- is greater than that of O^{2-}, which confirms that CaF_2 will affect the coordination of P-O and alter the original $[PO_4]$ structure.

Figure 3 shows the coordination distributions for Si-O and P-O, with superscripts indicating coordination numbers. The content of SiI^V was always above 95%, which shows that $[SiO_4]$ is the main structural unit in silicate systems and the content of $[SiO_4]$ did not alter significantly with the increase of CaF_2 content, which is consistent with the findings of Fan et al. [34]. From Figure 3b, when the CaF_2 content was 0, virtually all P-O in G1 appeared in a 4-coordination structure, indicating that the majority of P exists in slag in the form of a $[PO_4]$ structure and serves as the basic structural unit of the phosphate systems. However, as the CaF_2 content increased, the PI^V content decreased and the P^V gradually disappeared, while the PIII content continued to increase. Therefore, in contrast to silicate systems, the structural units of phosphate systems absolutely change with increasing CaF_2 content, and new structures may emerge as the coordination number of P-O gradually evolves from high to low. With the gradual increase of tri-coordinated P content, combined with the coordination of P-F in Figure 2b, it indicated that the addition of CaF_2 prompted F^- to replace O^{2-}, and a $[PO_4]$ to $[PO_3F]$ structural transition occurred. A similar phenomenon also appeared in the study of the phosphate glass structure by

Rao et al. [35] and Touré et al. [36]. However, in Pedone et al.'s [14] work, no P-F/Cl bonds were found at room temperature. It may be that the particles become more active and their diffusion ability is enhanced at high temperatures compared with normal temperatures, which provides favorable thermodynamic and kinetic conditions for the bonding between P^{5+} and F^-.

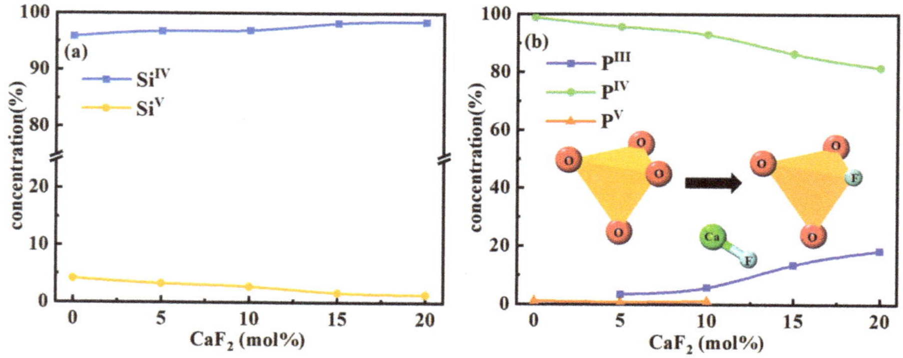

Figure 3. Coordination of Si-O and P-O under different CaF_2 contents: (**a**) Si-O and (**b**) P-O.

4.2. Distribution of Bond Angles

The distribution of bond angles is also a critical parameter to characterize the structure of the melt. Figure 4 shows the statistics of the O-Si-O and O-P-O bond angles' information with varying CaF_2 content. CaF_2 had a negligible effect on the distribution of O-Si-O and O-P-O bond angles in the $CaO\text{-}SiO_2\text{-}CaF_2\text{-}P_2O_5$ system. When CaF_2 content increased from 0% to 20%, the average bond angles of O-Si-O and O-P-O were 109.2° and 108.7°, respectively, which are very close to the theoretical value of 109.5° of standard tetrahedron. It indicates that although F^- tended to replace the position of an O^{2-} in $[PO_4]$ tetrahedron, it did not affect some structural characteristics of the original P-O bond, and the network structure with Si^{4+} and P^{5+} as the core still maintained the tetrahedral structure. Moreover, CaF_2 did not appear to cause large-scale rearrangements of the atoms in the whole systems, which consisted of a polymeric tetrahedral structure of Si^{4+}, P^{5+}, F^-, and O^{2-}, as well as network modifiers such as Ca^{2+} dispersed.

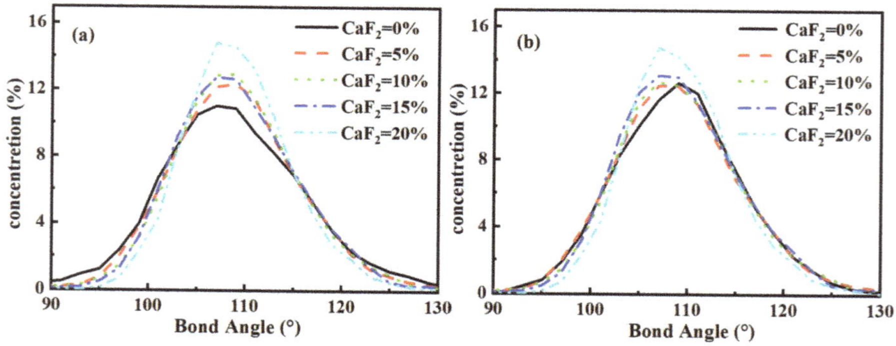

Figure 4. Distribution of bond angles in the system of varying CaF_2 content: (**a**) O-Si-O and (**b**) O-P-O.

4.3. Structural Unit Evolution

The silicate and phosphorene systems mainly consist of a network structure with O atoms connected to Si and P atoms. There are three types of distinct oxygen, which are divided into free oxygen (O_f), non-bridging oxygen (O_{nb}), and bridging oxygen (O_b).

Additionally, a unique tri-coordinated oxygen structure has been found in the aluminate system according to the literature [7]. Bridging oxygen with two tetrahedra, including Si-O-Si, Si-O-P, and P-O-P, improved the degree of polymerization of the system. Non-bridged oxygen was attached to only one tetrahedron, namely O-Si and O-P, while the other end was attached to a metallic cation. They function in the opposite way to bridging oxygen. The free oxygen is not connected to any tetrahedron. The cutoff radii of Si-O and P-O were selected to be 2.3 and 2.5 Å, respectively, and the distribution of various oxygen types in the melts was collected in Figure 5a. With the increase of CaF_2 content, the amount of free oxygen in the melts slightly increased, while the shift of the number of bridging oxygen and non-bridging oxygen had no obvious rule and was approximately in dynamic equilibrium.

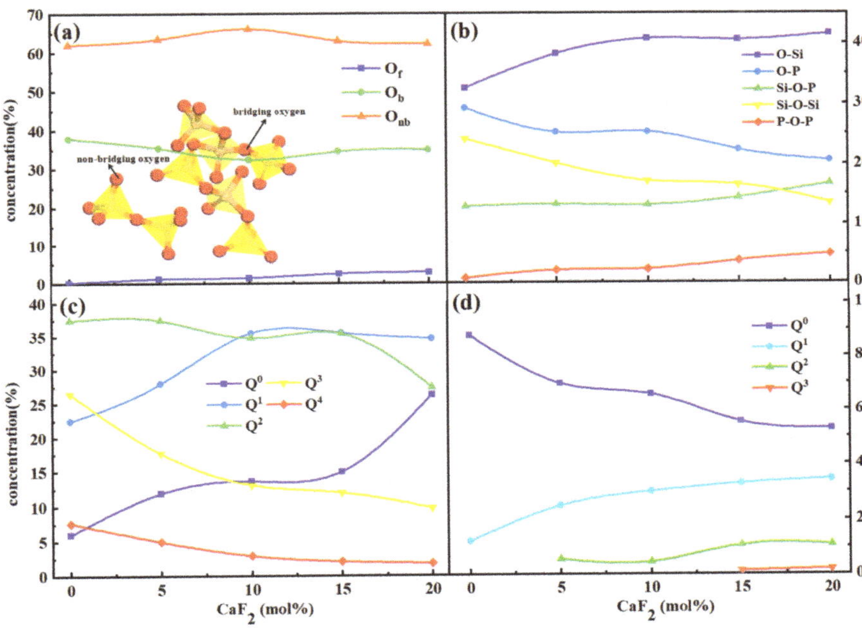

Figure 5. Structural unit evolution. (**a**) Types of oxygen atoms and (**b**) types of oxygen atoms after subdivision. (**c**,**d**) Distribution of Q^n with different CaF_2 contents: (**a**) Q^nSi and (**b**) Q^nP.

O_b and O_{nb} were further subdivided in the melts, as shown in Figure 5b. There are three types of O_b: Si-O-Si, Si-O-P, and P-O-P. As the CaF_2 content increased, Si-O-Si decreased from 24.1% to 20.0%, Si-O-P increased from 13.0% to 16.7%, and P-O-P increased from 0.7% to 4.8%. It has been shown that CaF_2 is beneficial in disrupting Si-O-Si and losing the initially polymeric silicate network structure. The Si-O-P structure in the system increased; that is, the addition of CaF_2 promoted the connection between $[SiO_4]$ and $[PO_4]$ or $[PO_3F]$, resulting in a silicophosphate composite structure that was more easily established in the systems. Moreover, the increase of P-O-P also indicates that the connectivity of the phosphate network structure became higher, which makes the phosphate melt structure more complex.

To further quantitatively analyze the influence of CaF_2 on the network structure of the systems, Q^n was introduced to characterize the polymerization degree of silicate and phosphate systems respectively, where n represents the number of bridging oxygen (O_b) in a single tetrahedral unit. The current results show that Q^n can be classified into five types: Q^0, Q^1, Q^2, Q^3, and Q^4, indicating that 0, 1, 2, 3, and 4 O_b are connected in a tetrahedral element. Figure 5c,d show the distribution of Q^n in silicate and phosphate systems, respectively. As CaF_2 content increased, the Q^0 and Q^1 in the silicate system increased, while the Q^2, Q^3, and Q^4 decreased, again confirming that CaF_2 breaks the high connectivity between [SiO_4] tetrahedral structures, simplifying the structure of the silicate systems. Besides, only Q^0 and Q^1 structures originally existed in the phosphate system, indicating that [PO_4] normally exists in the form of a single tetrahedron or pairings, which is consistent with the research results of Fan et al. [11]. However, as the CaF_2 content increased, the Q^0 rapidly decreased and the Q^1 increased. In addition, several Q^2 and Q^3 structures appeared and continued to increase. The results indicate that the original phosphate structure was not complicated, and the connectivity between the [PO_4] tetrahedra was low. However, the addition of CaF_2 reduced the number of single [PO_4] tetrahedral elements and the current [PO_3F] structure tended to form a chain or network composite structure, which increased the connectivity of the phosphate network. Macroscopically, higher connectivity is beneficial to the enrichment of P elements. In other words, CaF_2 can enrich the phosphate network, which is favorable for dephosphorization.

4.4. Transport Properties and Viscosity

The above results indicate that increasing the CaF_2 content simplified the structure of the silicate system in CaO-SiO_2-CaF_2-P_2O_5 melts but complicated the structure of the phosphate system. Therefore, to further understand the effect of CaF_2 on the degree of polymerization of the whole melt and to assess the changes in macroscopic properties, it is necessary to quantitatively analyze the transport properties of the system. Liquid molecules do not stay in a fixed position but are constantly moving [37]. The self-diffusion coefficient is a momentous parameter that reflects the diffusivity of the particles in the melt. As shown in Figure 6a, based on the *MSD* function and the Einstein relation, the self-diffusion coefficients of distinct ions can be obtained. It can be seen that the order of the self-diffusion coefficients of different ions was $F^- > Ca^{2+} > O^{2-} \approx P^{5+} > Si^{4-}$, and they were all in direct proportion to the content of CaF_2, indicating that the addition of CaF_2 can make each ion become more active. It mainly results from the depolymerization of the network structure in the melt by CaF_2, which lowers the energy barrier for the migration of ions in the melt and enhances the mobility of each particle. In addition, these phenomena can also lead to changes in the macroscopic properties of the melts. It is worth noting that the diffusion capacity of F^- in the melts was most prominent and much larger than that of O^{2-}, indicating that the substitution of F^- for O^{2-} improved the overall mobility of the phosphate structural units. Moreover, the diffusion coefficients of P^{5+} and O^{2-} were equivalent, which means that P and O always maintained the stable structure of [PO_4] or [PO_3F] and diffused cooperatively throughout the melts.

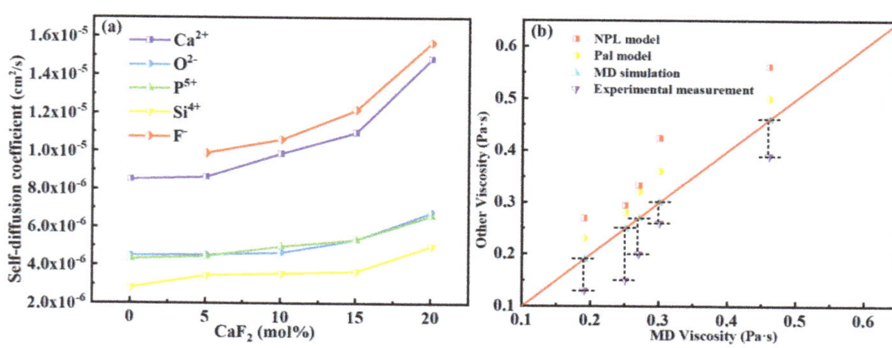

Figure 6. (**a**) The self-diffusion coefficient of each atom with different CaF_2 contents. (**b**) Viscosity comparison of MD simulation, models, and experimental measurement.

The melts' viscosity was calculated from the self-diffusion coefficient of each ion and compared with the experimental measurements. The results of the MD simulation and experiment in Figure 6b both show that with the increase of CaF_2 content, the viscosity of the $CaO-SiO_2-CaF_2-P_2O_5$ systems decreased and led to an improvement of melt liquidity. Clearly, the viscosity, which reflects the viscous resistance of the melt during the flow and depends prominently on the degree of polymerization of the melt, would be reduced in a melt with simple structural units. Besides, the NPL model [38] and Pal model [39] were also used to compare the calculation results. It can be observed that although there were some errors between the calculated viscosity consequences and the experimentally measured ones, the trends were in perfect agreement, which indicates that the MD simulations were able to predict the viscosity of the system accurately to some extent and reflects the reliability of the MD simulations. The predictions of both models differed significantly from the experimental data due to discrepancies in some of the components. Consequently, the MD viscosity calculations are in better agreement with the experimental results compared to both models.

In the process of steel-making dephosphorization, P is usually enriched in $2CaO{\cdot}SiO_2$-$3CaO{\cdot}P_2O_5$ (C_2S-C_3P) solid solution [40–42]. Dephosphorization depends on the concentration of phosphorus in the solid solution, and the flow properties of the dephosphorized product in the slag also determine whether phosphorus can be efficiently removed from the slag. As can be seen from the above analysis, the introduction of CaF_2 directly changed the basic structural units of the phosphate melt, making the phosphate network units more easily enriched. On the other hand, CaF_2 reduced the viscosity and improved the fluidity of the slag, so that the dephosphorized products enriched in P could be better transported to and removed from the slag layer. CaF_2 is thus favorable for dephosphorization both from the microscopic reaction point of view in slag and from the macroscopic flow properties. Our study links the microscopic to the macroscopic and essentially defines the critical role of CaF_2 in the dephosphorization of slag.

4.5. Correlation between Viscosity and Structural Properties

The viscosity of the slag depends on its degree of polymerization. Researchers have proposed two common approaches to describe the complexity of melts. The first one amounts to counting the number of non-bridged oxygen atoms, denoted as NBO, based on the results of molecular dynamics simulations. The parameter NBO/T, which reflects the degree of melts' polymerization, can be obtained by combining the number of network formers, T (Si or P), in the system [43]. The larger the NBO/T, the higher the ratio of non-bridging oxygen in the melts, that is, the simpler the structure of the melts is, then the viscosity and other parameters of the melts will also change accordingly. The second is to judge the melts' complexity according to the evolution of total Q^n. It is usually expressed by the ratio of high-complexity Q^n to low-complexity Q^n, such as, DOP = $(Q^3 + Q^4 + Q^5)/(Q^0 + Q^1 + Q^2)$. [44]. The higher the DOP, the more complex the systems. Comparing the calculated melts' viscosity with the above two parameters, we observed a correspondence between the viscosity and the two parameters, as shown in Figure 7.

It can be observed in Figure 7a that NBO/T increased with the increase of CaF_2 content, while DOP was the opposite. The results show that CaF_2 can effectively reduce the complexity of the system, and the variation of both quantities has a good correspondence with the trend of the viscosity value, suggesting that the melting viscosity is directly related to the complexity of the system. Specifically, the introduction of CaF_2 simplified some complex network units formed by interweaving $[SiO_4]$, $[PO_4]$, and $[PO_3F]$ structures in the whole melt, and formed simple structures such as single or chain, greatly reducing the connectivity of the whole melt. Furthermore, the complexity of the slag structure depends on the competing effects of silicates and phosphates on the polymerization of the molten slag. In the $CaO-SiO_2-CaF_2-P_2O_5$ systems, CaF_2 promoted the disaggregation of complex network units into small units, which made the diffusion of micro-particles easier. The macroscopic manifestation of this phenomenon is a reduction of the total viscosity.

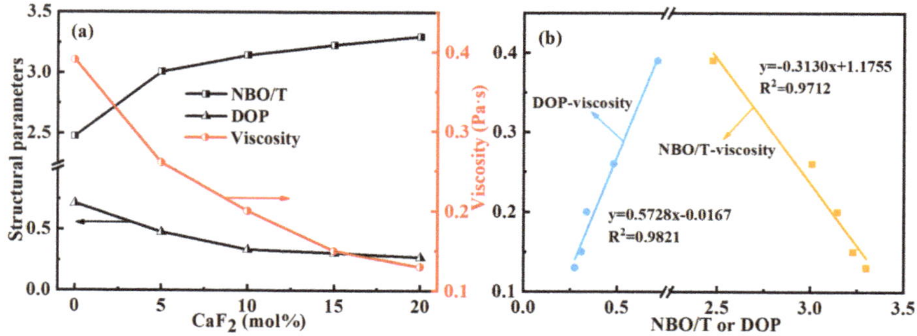

Figure 7. (**a**) Change of structural parameters and viscosity and (**b**) the relationship between viscosity and structural parameters.

In Figure 7b, the relationship between the viscosity and the above two parameters was obtained by linear fitting. For viscosity and DOP, y = 0.5728x − 0.0167, R^2 = 0.9821, and for viscosity and NBO/T, y = −0.3130x + 1.1755, R^2 = 0.9712. The correlation coefficients of the above two fitting results were high enough, so the relationship between viscosity and microstructure of CaO-SiO_2-CaF_2-P_2O_5 melts could be accurately described, and at the same time, the viscosity could also be predicted by the microstructure of the systems.

5. Conclusions

We have presented the microstructure information of the CaO-SiO_2-CaF_2-P_2O_5 melts at 1600 °C by MD simulation and explored the evolution of each structural unit with the increase of CaF_2 content. Combined with the analysis of microscopic particle transport and macroscopic flow properties, it is clear that the crucial role played by CaF_2 in phosphosilicate melts has been investigated.

By analyzing the distributional properties of the coordination and bond angles between different atoms, we found that both S^{4+} and P^{5+} were present in tetrahedral form in the molten CaO-SiO_2-CaF_2-P_2O_5 system. The coordination number of Si-O was maintained at around 4.0 when increasing the CaF_2 content from 0% to 20%, while the coordination number of P-O decreased from 4.05 to 3.57. Therefore, CaF_2 had little effect on the structure of $[SiO_4]$ but decreased the stability of the $[PO_4]$ structure. Specifically, F^- tended to replace O^{2-} and promote the transformation of $[PO_4]$ to a $[PO_3F]$ structure, and at the same time, it is beneficial to make the P element-enriched phosphate network structure more aggregated. However, the addition of CaF_2 did not lead to a large-scale rearrangement of the atoms in the whole system, and the network structure with Si^{4+} and P^{5+} as cores remained tetrahedral.

The results of the MD simulation and experiment showed that CaF_2 is beneficial for reducing the degree of polymerization of the melt and thereby reducing the melt viscosity, which decreased from 0.39 to 0.13 Pa·s as the CaF_2 content increased from 0% to 20%, and it had a good linear relationship with the structural parameters. In summary, CaF_2 is beneficial for dephosphorization both from the microscopic reaction point of view and from the macroscopic flow properties in slag.

Author Contributions: Formal analysis, methodology H.D.; funding acquisition, conceptualization D.C.; methodology, data curation Z.H.; resources, supervision M.L.; software, writing—original draft, Y.D. All authors have read and agreed to the published version of the manuscript.

Funding: The authors gratefully acknowledge the financial support provided by the National Natural Science Foundation of China, project No. 51874060 and No. 52274320.

Institutional Review Board Statement: The study did not require ethical approval.

Informed Consent Statement: Informed consent was obtained from all subjects involved in the study.

Data Availability Statement: Not applicable.

Conflicts of Interest: The authors declare no conflict of interest.

References

1. Du, Y.; Yuan, Y.; Li, L.; Long, M.; Duan, H.; Chen, D. Insights into structure and properties of P_2O_5-based binary systems through molecular dynamics simulations. *J. Mol. Liq.* **2021**, *339*, 116818. [CrossRef]
2. Liu, J.; Liu, Z.; Feng, J.; Li, B.; Chen, J.; Ren, B.; Jia, Y.; Yin, S. Reaction Mechanism of CA_6, Al_2O_3 and CA_6-Al_2O_3 Refractories with Refining Slag. *Materials* **2022**, *15*, 6779. [CrossRef] [PubMed]
3. Lee, S.; Min, D.J. Investigation of sulfide capacity of aluminosilicate slag based on ionic structure considerations. *J. Am. Ceram. Soc.* **2018**, *101*, 634–643. [CrossRef]
4. Yang, J.; Kim, Y.; Sohn, I. Gaining insights on high-temperature thermal conductivity and structure of oxide melts through experimental and molecular dynamics simulation study. *J. Mater. Res. Technol.* **2021**, *10*, 268–281. [CrossRef]
5. Yao, Y.; Wang, Y.; Wei, Q.; Cui, S.; Hao, L. Effect of the Formation of Amorphous Networks on the Structure and Hydration Characteristics of Granulated Blast Furnace Slag. *Materials* **2020**, *13*, 1462. [CrossRef]
6. Stoch, P.; Goj, P.; Wajda, A.; Stoch, A. Alternative insight into aluminium-phosphate glass network from ab initio molecular dynamics simulations. *Ceram. Int.* **2021**, *47*, 1891–1902. [CrossRef]
7. Bi, Z.; Li, K.; Jiang, C.; Zhang, J.; Ma, S.; Sun, M.; Wang, Z.; Li, H. Performance and transition mechanism from acidity to basicity of amphoteric oxides (Al_2O_3 and B_2O_3) in SiO_2–CaO–Al_2O_3–B_2O_3 system: A molecular dynamics study. *Ceram. Int.* **2021**, *47*, 12252–12260. [CrossRef]
8. Li, K.; Khanna, R.; Bouhadja, M.; Zhang, J.; Liu, Z.; Su, B.; Yang, T.; Sahajwalla, V.; Singh, C.V.; Barati, M. A molecular dynamic simulation on the factors influencing the fluidity of molten coke ash during alkalization with K_2O and Na_2O. *Chem. Eng. J.* **2017**, *313*, 1184–1193. [CrossRef]
9. Diao, J.; Fan, G.; Liu, X.; Xie, B. Computer simulation of anionic structures of molten CaO-SiO_2-P_2O_5 system. *Metall. Mater. Trans. B* **2014**, *45*, 1942–1947. [CrossRef]
10. Wang, Z.; Cai, S.; Zhang, M.; Guo, M.; Zhang, Z. Structural investigation of phosphorus in CaO-SiO_2-P_2O_5 ternary glass. *Metall. Mater. Trans. B* **2017**, *48*, 1139–1148. [CrossRef]
11. Fan, G.; Diao, J.; Jiang, L.; Zhang, Z.; Xie, B. Molecular dynamics analysis of the microstructure of the CaO-P_2O_5-SiO_2 slag system with varying P_2O_5/SiO_2 ratios. *Mater. Trans.* **2015**, *56*, 655–660. [CrossRef]
12. Diao, J.; Ke, Z.; Jiang, L.; Zhang, Z.; Zhang, T.; Xie, B. Structural Properties of Molten CaO–SiO_2–P_2O_5–FeO System. *High Temp. Mater. Process.* **2017**, *36*, 871–876. [CrossRef]
13. Kansal, I.; Goel, A.; Tulyaganov, D.U.; Rajagopal, R.R.; Ferreira, J.M. Structural and thermal characterization of CaO–MgO–SiO_2–P_2O_5–CaF_2 glasses. *J. Eur. Ceram. Soc.* **2012**, *32*, 2739–2746. [CrossRef]
14. Pedone, A.; Chen, X.; Hill, R.G.; Karpukhina, N. Molecular dynamics investigation of halide-containing phospho-silicate bioactive glasses. *J. Phys. Chem. B* **2018**, *122*, 2940–2948. [CrossRef]
15. Diao, J.; Xie, B.; Wang, Y.; Guo, X. Effect of fluorine on the minerals phase and citric acid solubility of CaO–SiO_2–FetO–P_2O_5–CaF_2 system. *ISIJ Int.* **2010**, *50*, 768–770. [CrossRef]
16. Ueda, S.; Watanabe, Y.; Ikeda, T.; Maeda, M. Phase equilibria in the CaF_2-CaO-P_2O_5 system at 1623 K. *Metall. Mater. Trans. B Process Metall. Mater. Process. Sci.* **2001**, *32*, 967–969. [CrossRef]
17. Belashchenko, D.; Ostrovskii, O. Computer Simulation of Noncrystalline Ionic–Covalent Oxides CaO–P_2O. *Inorg. Mater.* **2002**, *38*, 146–153. [CrossRef]
18. Hirao, K.; Kawamura, K. Material design using personal computer. *Shokabo Tokyo* **1994**, *52*, 52–54.
19. Burylev, B.P.; Laptev, D.M.; Moisov, L.P. Calculation of the viscosity of slag melts. *Svarochnoe Proizv.* **1999**, *10*, 8–11.
20. Ziman, J.M. *Principles of the Theory of Solids*; Cambridge University Press: Cambridge, UK, 1972.
21. Kubo, R. The fluctuation-dissipation theorem. *Rep. Prog. Phys.* **1966**, *29*, 255. [CrossRef]
22. Shimizu, N.; Kushiro, I. Diffusivity of oxygen in jadeite and diopside melts at high pressures. *Geochim. Cosmochim. Acta* **1984**, *48*, 1295–1303. [CrossRef]
23. Mungall, J.E. Empirical models relating viscosity and tracer diffusion in magmatic silicate melts. *Geochim. Cosmochim. Acta* **2002**, *66*, 125–143. [CrossRef]
24. Tinker, D.; Lesher, C.E.; Baxter, G.M.; Uchida, T.; Wang, Y. High-pressure viscometry of polymerized silicate melts and limitations of the Eyring equation. *Am. Mineral.* **2004**, *89*, 1701–1708. [CrossRef]
25. Reid, J.E.; Poe, B.T.; Rubie, D.C.; Zotov, N.; Wiedenbeck, M. The self-diffusion of silicon and oxygen in diopside ($CaMgSi_2O_6$) liquid up to 15 GPa. *Chem. Geol.* **2001**, *174*, 77–86. [CrossRef]
26. Lv, X.; Xu, Z.; Li, J.; Chen, J.; Liu, Q. Molecular dynamics investigation on structural and transport properties of Na_3AlF_6–Al_2O_3 molten salt. *J. Mol. Liq.* **2016**, *221*, 26–32. [CrossRef]
27. Goj, P.; Stoch, P. Influence of CaO on structural features of polyphosphate P_2O_5-Fe_2O_3-FeO glasses by molecular dynamics simulations. *J. Non-Cryst. Solids* **2020**, *537*, 120014. [CrossRef]

28. Sasaki, Y.; Iguchi, M.; Hino, M. The role of Ca and Na ions in the effect of F ion on silicate polymerization in molten silicate system. *ISIJ Int.* **2007**, *47*, 638–642. [CrossRef]
29. Christie, J.K.; Ainsworth, R.I.; de Leeuw, N.H. Ab initio molecular dynamics simulations of structural changes associated with the incorporation of fluorine in bioactive phosphate glasses. *Biomaterials* **2014**, *35*, 6164–6171. [CrossRef]
30. Park, J.H.; Min, D.J.; Song, H.S. The effect of CaF_2 on the viscosities and structures of CaO-SiO_2 (-MgO)-CaF_2 slags. *Metall. Mater. Trans. B* **2002**, *33*, 723–729. [CrossRef]
31. Wang, Z.; Tang, P.; Wen, G.; Liu, Q. Effect of F− Replacing O2− on Crystallization Behavior of CaO–SiO_2–Al_2O_3 Continuous Casting Mold Flux. *ISIJ Int.* **2019**, *59*, 367–374. [CrossRef]
32. Takeda, O.; Okawara, T.; Sato, Y. Development of a new viscometer based on rotating crucible method and viscosity measurement of SiO_2–CaO–CaF_2 system. *ISIJ Int.* **2012**, *52*, 1544–1549. [CrossRef]
33. Sasaki, Y.; Iguchi, M. The Coordination of F Ions around Mg and Ca Ions in Molten CaO–CaF_2–MgO–SiO_2 System at 1873K. *ISIJ Int.* **2009**, *49*, 602–604. [CrossRef]
34. He, S.; Wang, S.; Jia, B.; Li, M.; Wang, Q.; Wang, Q. Molecular dynamics simulation of the structure and properties of CaO-SiO_2-CaF_2 slag systems. *Metall. Mater. Trans. B* **2019**, *50*, 1503–1513. [CrossRef]
35. Venkateswara Rao, G.; Shashikala, H. Structural, optical and mechanical properties of ternary CaO-CaF_2-P_2O_5 glasses. *J. Adv. Ceram.* **2014**, *3*, 109–116. [CrossRef]
36. Touré, A.B.; Mele, E.; Christie, J.K. Atomic-scale clustering inhibits the bioactivity of fluoridated phosphate glasses. *Biomed. Glas.* **2019**, *5*, 76–84. [CrossRef]
37. Jiang, C.; Zhang, H.; Xiong, Z.; Chen, S.; Li, K.; Zhang, J.; Liang, W.; Sun, M.; Wang, Z.; Wang, L. Molecular dynamics investigations on the effect of Na_2O on the structure and properties of blast furnace slag under different basicity conditions. *J. Mol. Liq.* **2020**, *299*, 112195. [CrossRef]
38. Mills, K.; Sridhar, S. Viscosities of ironmaking and steelmaking slags. *Ironmak. Steelmak.* **1999**, *26*, 262–268. [CrossRef]
39. Pal, S.K.; Ray, H.S.; Choudhury, P.; Das, S.; Majumdar, M. An approach towards theoretical prediction of viscosity of glasses over a wide range of temperature. *Trans. Indian Ceram. Soc.* **2005**, *64*, 81–86. [CrossRef]
40. Song, Y.; Hu, X.; Chou, K. The Reaction Behavior of 2CaO· SiO_2 with CaO–SiO_2–FeO–P_2O_5 Slag. *Materials* **2022**, *15*, 6594. [CrossRef]
41. Xia, Y.; Li, J.; Fan, D.; Hou, G. Effects of interfacial oxygen potential and slag phase changing during slag formation process on dephosphorization behavior. *ISIJ Int.* **2019**, *59*, 1519–1526. [CrossRef]
42. Suito, H.; Inoue, R. Behavior of phosphorous transfer from CaO–FetO–P_2O_5 (–SiO_2) slag to CaO particles. *ISIJ Int.* **2006**, *46*, 180–187. [CrossRef]
43. Wu, T.; He, S.; Liang, Y.; Wang, Q. Molecular dynamics simulation of the structure and properties for the CaO–SiO_2 and CaO–Al_2O_3 systems. *J. Non-Cryst. Solids* **2015**, *411*, 145–151. [CrossRef]
44. Jiang, C.; Li, K.; Zhang, J.; Qin, Q.; Liu, Z.; Liang, W.; Sun, M.; Wang, Z. Molecular dynamics simulation on the effect of MgO/Al_2O_3 ratio on structure and properties of blast furnace slag under different basicity conditions. *Metall. Mater. Trans. B* **2019**, *50*, 367–375. [CrossRef]

materials

MDPI

Article

Characterization of Hot Deformation of near Alpha Titanium Alloy Prepared by TiH_2-Based Powder Metallurgy

Rongxun Piao [1], Wenjin Zhu [1], Lan Ma [2], Peng Zhao [3,*] and Biao Hu [4]

1 School of Mechanical Engineering, Anhui University of Science and Technology, Huainan 232001, China
2 Department of Vanadium and Titanium, Panzhihua University, Panzhihua 617000, China
3 School of Materials Science and Engineering, Shandong Jianzhu University, Jinan 250101, China
4 Anhui International Joint Research Center for Nano Carbon-Based Materials and Environmental Health, Huainan 232001, China
* Correspondence: zhaopeng3@hotmail.com

Abstract: TiH_2-basd powder metallurgy (PM) is one of the effective ways to prepared high temperature titanium alloy. To study the thermomechanical behavior of near-α titanium alloy and proper design of hot forming, isothermal compression test of TiH_2-based PM near-α type Ti-5.05Al-3.69Zr-1.96Sn-0.32Mo-0.29Si (Ti-1100) alloy was performed at temperatures of 1123–1323 K, strain rates of 0.01-1 s^{-1}, and maximum deformation degree of 60%. The hot deformation characteristics of alloy were analyzed by strain hardening exponent (n), strain rate sensitivity (m), and processing map, along with microstructure observation. The flow stress revealed that the difference in softening/hardening behavior at temperature of 1273–1323 K and the strain rate of 1 s^{-1} compared to the lower deformation temperature and strain rate. The strain hardening exponents at temperatures of 1123 K are all negative under all strain rates, and the most severe flow softening with minimum value of n was observed at 1123 K and 1 s^{-1}. The strain rate sensitives showed that the peak region with m value greater than 0.5 generally appeared in the high temperature range of 1273–1323 K, while strain rate sensitivity at low temperature behaved differently with strain rates. The processing map developed for strain of 0.6 exhibited high power dissipation efficiency at high temperatures of 1273–1323 K and a low strain rate of 0.01 s^{-1}, due to microstructure evolution of β phase. The decrease of strain rate at 1323 K resulted in the formation of globularization of α lamellae. The instability domain of flow behavior was identified in the temperature range of 1123–1173 K and at the strain rate higher than 0.01 s^{-1} reflecting the localized plastic flow and adiabatic shear banding, and inhomogenous microstructure. The variation of power dissipation energy (η) slope with strain demonstrated that the power dissipation mechanism during hot deformation has been changed from temperature-dependent to microstructure-dependent with the increase of temperature for the alloy deformed at 0.1 s^{-1}. Eventually, the optimum processing range to deform the material is at 1273–1323 K and a strain rate range of 0.01–0.165 s^{-1} ($\ln\dot{\varepsilon} = -4.6$––1.8).

Keywords: near-α titanium alloy; TiH_2-based powder metallurgy; hot compression; strain hardening exponents; strain rate sensitivity; processing map

Citation: Piao, R.; Zhu, W.; Ma, L.; Zhao, P.; Hu, B. Characterization of Hot Deformation of near Alpha Titanium Alloy Prepared by TiH_2-Based Powder Metallurgy. *Materials* **2022**, *15*, 5932. https://doi.org/10.3390/ma15175932

Academic Editor: Jae Wung Bae

Received: 23 July 2022
Accepted: 23 August 2022
Published: 27 August 2022

1. Introduction

Near-α titanium alloys are widely used in compressor discs and blades of jet engines because of their excellent high-temperature fatigue and creep properties, high strength-to-weight ratio, and good corrosion resistance [1]. Ti-1100 (Ti-6Al-2.75Sn-4Zr-0.4Mo-0.4Si, wt%) alloy has the highest capability to tolerate temperature up to 873 K without any degradation in mechanical properties among near-α titanium alloys [2]. Like most titanium alloys, near-α titanium alloy has poor formability and high cost, which has always been two key problems in a wide range of commercial applications [3]. Powder metallurgy (PM) is one of the most promising methods for production of titanium alloy components at low cost [4]. However, compared with ingot metallurgy (IM) titanium alloys, challenge of PM titanium alloy are high porosity and high oxygen contamination, which usually lead to detrimental effect on the final product by lowering ductility or other defects [5].

One of the effective ways to solve the two problems of insufficiently high density and oxygen contamination is to use TiH_2 powder as the main raw material. It has been demonstrated previously that PM parts developed using TiH_2 powder feedstock had better control over lower oxygen content, the microstructures, and good chemical homogeneity [6–8]. Using TiH_2 powder, density larger than 95% and good performance with fine grain microstructure can be easily obtained [6,8,9]. Previously, Hagiwara et al. [10] have prepared Ti-1100 alloy with ultra low chlorine hydride-dehydride (ELCL-HDH) titanium powder and master alloy powder as main raw materials through the means of blended elemental (BE) PM synthesis of cold pressing-vacuum sintering-heat treatment-hot isostatic pressing, and the prepared alloys have high density of 95% and better performance in tensile strength and high cycle fatigue strength compared to the conventional PM process. Zhang et al. [3] investigated the mechanical behavior of TiH_2-based near-α Ti–3Al–2Zr–2Mo titanium alloy prepared by a series process of cold pressing-sintering-hot extrusion-vacuum annealing-common annealing, and the prepared PM alloy samples showed a satisfying combination of superior tensile strength and excellent ductility. Using similar methods, recently, Zhang et al. [11] and Wu et al. [12] in the same research group studied the tensile properties of a hot-extruded TiH_2-based near-α Ti-3Al-2Zr-2Mo-0.36O and Ti-6Al-2Sn-titanium alloy, and the prepared material has shown high performance of tensile ductility.

For the proper design of thermal deformation of metal, it is necessary to study the thermomechanical behavior. The mechanism of flow stress softening as well as strain hardening phenomena determines the thermomechanical behavior of a material, depending on processing parameters such as strain, strain rate, and temperature. For the description of the hot deformation behavior of titanium alloys, the dynamic material model (DMM) [12,13] based processing map technology has been widely used to provide the optimization of processing parameters and processing windows with instability regions [14]. Likewise, the strain hardening, strain rate sensitivity, and temperature sensitivity of the alloys during deformation can provide the reference and guidance for better thermal deformation of the alloy. Previously, many studies have focused on the hot deformation of as-cast near-α titanium alloys. For example, Krishna et al. [15] investigated the hot deformation mechanism in near α titanium alloy 685 by constructing a processing map, obtaining the optimal condition of 1248 K and 0.001 s^{-1}, and proposing that dynamic recrystalization (DRX) occurred due to low oxygen level present in the alloy. Balasundar et al. [16] discussed geometric dynamic recrystallization behavior of a near-α titanium alloy TTTAN 29A (equivalent to IMI834) with acicular microstructure through hot compression tests at temperatures of 1123–1333 K and a strain rate of 3×10^{-4}–1 s^{-1}, and proposed the optimum condition is at 1193–1303 K along with strain rates of 3×10^{-4}–10^{-3} s^{-1} and dynamic recrystallization (DRX) or globularization of lamellar α phase is dominant mechanism of microstructure evolution. Zhou et al. [17] studied hot workability of near-α titanium alloy Ti–6Al–3Nb–2Zr–1Mo with an initial duplex microstructure by means of isothermal compressions. By integrating process maps and constitutive relationship, they proposed that the optimal hot working domain appeared at condition of 1198-1248 K/0.01–0.1 s^{-1}, and the main mechanism of microstructure evolution in $\alpha + \beta$ phase field is dynamic recrystallization of β phase and super-plasticity deformation (SPD). Su et al. [18] have study the hot deformation behavior of near-α DsTi700 titanium alloy and revealed that the deformation mechanisms are mainly kinking of α lamellas or flow localization at high strain rate of 0.5–1 s^{-1} in $\alpha + \beta$ two phase field, while DRV is the dominated deformation mechanism in β single phase. Morakabati and Hajari [19] studied the high temperature deformation behavior of the Ti−5.7Al−2.1Sn−3.9Zr−2Mo−0.1Si (Ti-6242S) alloy with an initial acicular microstructure by using processing map and revealed that the dynamic globularization occurred in lower temperature conditions of α/β two-phase region ($T \leq 1223$ K) and DRV was observed at high temperature (T = 1273 K). All above studies have been focused on as-cast alloy. However, the study of hot deformation behavior for PM alloy is rarely reported at present. Ding et al. [20] investigated the PM Ti600 alloy prepared by a HDH titanium powder-based PM-extrusion-annealing method and revealed that the microstructure was predominant by equiaxed grains after annealing below β transition temperature, while the bi-model structure was dominant after annealing at 1323 K in the β single phase. In our previous study [21], hot deformation of TiH_2-based PM Ti-1100 at a relatively low temperature range of 973–1173 K and strain rate of 0.01–1 s^{-1} was investigated. The results of strain rate sensitivity and temperature sensitivity analysis showed that the flow instability mainly occurred at strain rates of 0.01 and 1 s^{-1} under all temperature range and the safe working window was limited at a narrow zone of 1073–1173 K with strain rate of 0.1 s^{-1}.

This paper mainly focused on the hot deformation behavior of TiH_2-based PM near-α titanium alloy Ti-1100 in the upper α or α/β two phase region of 1123–1323 K and strain rates of 0.01–1.0 s^{-1}. For the characterization of hot deformation behavior of Ti-1100 alloy, the strain hardening exponents, strain rate sensitivity, microstructure, and processing map were investigated.

2. Materials and Methods

Titanium hydride powder (Ti-4.5H-0.12O-0.015C-0.023N-0.016Fe in mass%) with a particle size of 45 μm and high grade of Al, Sn, Zr, Mo, and Si powders with a particle size of 75 μm were well mixed by the ball-mill machine. The mixture was subjected to cold isostatic pressing (CIP) at molding pressure of 200 MPa and holding time of 180 s to obtain the cylindrical lump alloy, and then the vacuum sintering was conducted at 1423 K for 240 min. under Ar atmosphere of 80~120 Pa. Finally, the CIP sintered lump was subjected to hot isostatic pressing (HIP) at 1223 K and 200 MPa for 180 min.

In the hot compression test, a cylindrical sample with dimensions of φ 8 mm × 12 mm (in accordance to ASTM E209, preserving the aspect ratio) was machined along the axis of the HIP cylinder with a wire-electrode cutting machine. The hot compression tests of 15 samples were conducted by THERMECMASTER–Zthermos–simulation machine at deformation temperatures ranging from 1123 K to 1323 K with an interval of 50 K and strain rates of 0.01, 0.1, and 1 s^{-1}. The whole isothermal compression test process was carried out under a vacuum degree lower than 10 Pa to prevent the samples oxidation. The specimens were heated to the target temperature at a rate of 10 K/s and held for 180 s. The fixed height of each sample was reduced by 60%, and the samples were quickly cooled to room temperature in order to retain high temperature deformation microstructure. To minimize friction, 0.1 mm thick mica wafer was placed at both end faces between the specimens and die. The true stress–strain curves were acquired automatically by a computer-equipped monitor.

The isothermal compression specimen was cut along the radial direction, and the cutting section of specimen was grinded with SiC paper followed by mechanical polishing with a solution of water and diamond polishing paste. Then, the specimens were subjected to the etching in Kroll's reagent (1 mL HF, 5 mL HNO_3, and 100 mL water). Lastly, the microstructures of the central area of the cut specimens were characterized by optical microscopy (OM, DMM-490C, Caikang Optical Instrument Co., LTD, Shanghai, China) and scanning electron microscope (SEM, TESCAN-MIRA 4, Tescan Co., LTD, Shanghai, China branch).

The microstructure of the as-sintered alloys is shown in Figure 1. It shows a typical two-phase microstructure, mainly consisting of lath α phase with a width of about 20 μm, and β phase in the α phase boundary. From preliminary density measurements, the density of as-sintered alloy was determined to be 97.1% of the theoretical density, and the chemical composition of sintered alloy is shown in Table 1. The β-transus temperature of the alloy was preliminarily determined as 1293 K by differential scanning calorimeter (DSC), and the details of DSC analysis can be found in Appendix A.

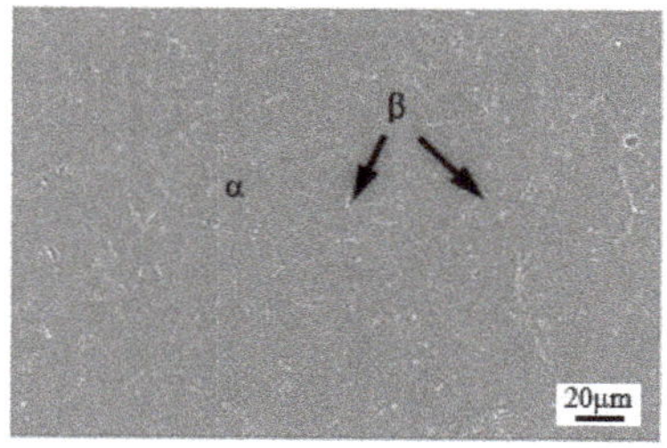

Figure 1. Microstructure of as-prepared Ti-1100 alloy.

Table 1. Chemical composition of sintered near-α Ti-1100 alloy (wt%).

Ti	Al	Zr	Sn	Mo	Si	C	N	H	O
88.3	5.05	3.689	1.96	0.32	0.29	0.0071	0.0013	0.0016	0.12

3. Results and Discussion

3.1. True Stress–Strain Curves

3.1.1. General Behavior

Shown in Figure 2 is the macro-appearance of deformed samples at different processing conditions. It can be seen from the top and side views in Figure 2 that severe cracks were observed in the specimens deformed at 1123 K/0.1 s^{-1} and 1173 K/1 s^{-1}, indicating that the sample under these conditions may not be suitable for thermal deformation. Likewise, slight cracks were observed on the side of the drum-shaped samples deformed at high temperature of 1273–1323 K and strain rate of 0.01 s^{-1}. It should be noted that slight cracks on the side of the specimen are usually caused by compressive stress and additional tensile stress, which is different from the deformation behavior of the central part of the specimen subjected to a single pressure, so the hot workability in this case cannot be judged by these macroscopic appearances alone [22]. Under other experimental conditions, there were no obvious macro-cracks and micro-cracks, and the deformation of the sample is uniform.

Figure 2. Macro-appearance (top and side views) of deformed alloys at different processing conditions.

Based on the hot compression experimental data of Ti-1100 alloy, the flow stress–strain curve was obtained, as shown in Figure 3. As seen at the beginning of the hot deformation process, the flow stress increases rapidly to the peak stress (σ_P). This hardening effect is mainly attributed to the dislocation proliferation and intersection [23].

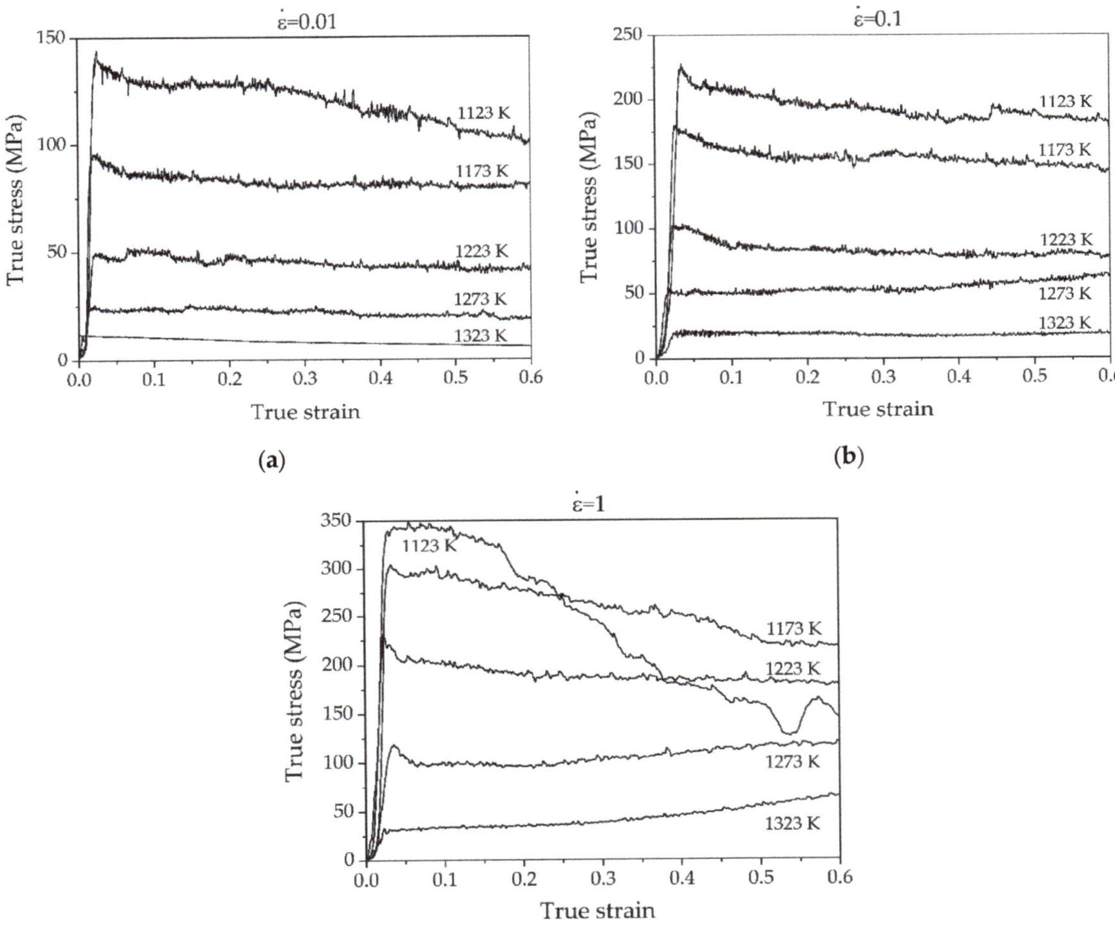

Figure 3. The true strain–stress curves of Ti-1100 alloy at different deformation conditions: (**a**) 0.01 s^{-1}, (**b**) 0.1 s^{-1}, and (**c**) 1 s^{-1}.

Beyond the peak stress, the flow stress exhibits the different behaviors. At a strain rate of 0.01 s^{-1}, the flow stress exhibits softening response under all deformation temperatures, and this softening effect is more obvious at low temperature range. At a strain rate of 0.1 s^{-1}, the flow stress exhibits the general behaviors of softening response in the low part of α/β region (T $\leq$ 1223 K). By comparison, the flow curves reached to steady-state condition at upper part of α/β region (T = 1323 K). This behavior can be explained by the fact that at a high temperature of 1323 K, the volume fraction of β-BCC phase with low mechanical property and high diffusivity is increased [24]. As the strain rate increases to 1 s^{-1}, the flow stress curves show increasing tendency in the high temperature range of 1273–1323 K, implying work hardening effect was dominant at this condition. Obviously, these hardening effects are more significant at high strain rate (1 s^{-1}) compared to that at 1273 K and 0.1 s^{-1}. These behaviors indicated that dynamic softening by temperature and the work-hardening by strain rate interact each other, resulting the different behaviors under the different temperature and different strain rate conditions. As consequence, it shows the softening effects at low strain rate of 0.01 s^{-1}, while both softening and hardening effects were observed at relatively high strain rate of 0.1–1 s^{-1} depending on temperature and strain rate.

3.1.2. Effect of Temperature on Flow Stress

The true stress of the alloy is sensitive to the deformation temperature and strain rate. Figure 4 shows a typical example of temperature dependence on flow stress at a strain of 0.1 under the different strain rates. As seen in Figure 4, at constant strain rate, the true flow stress decreases continuously with the increase of temperature. Obviously, the decreasing rate of the flow stress decreases with increasing temperature, and this behavior can be attributed to the increasing the volume fraction of β phase with higher diffusivity and high stacking fault energy against to α phase [20,24]. Another possible cause of temperature dependence of flow stress is the high kinetics of β to α transformation. Thus, as β phase has lower strength and higher active slip system than α phase [25], the true flow stress decreases by an increase in deformation temperature. Similar behavior can be found in the deformation of IMI834 alloy [26] and Ti-6242S alloy [19].

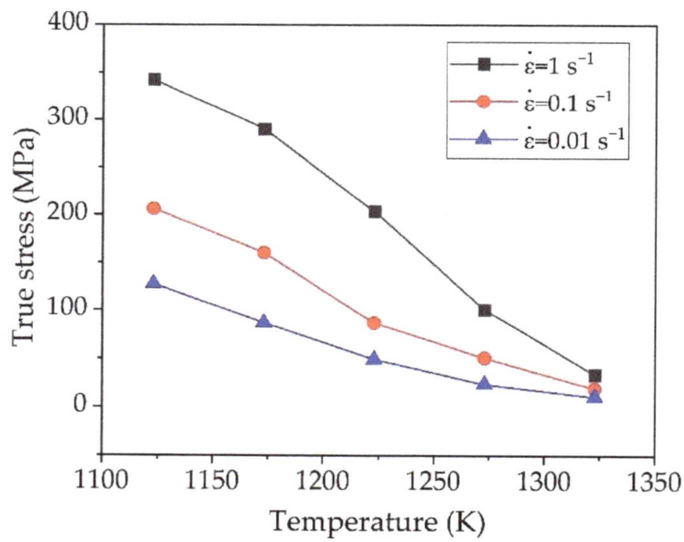

Figure 4. Temperature dependence of the flow stress at a strain of 0.1 under various strain rates.

3.2. Strain Hardening Exponents

To analyze the characteristics of hot work hardening of alloy, the concept of strain hardening was introduced based on the following functional model [27]:

$$\sigma = \sigma_0 + K\varepsilon^n \tag{1}$$

where σ_0 is the yield strength (MPa); K is the strength coefficient; ε is the strain; n is the strain hardening exponent, which can be express by Equation (1) [28]:

$$n = \left[\frac{\partial(lg\sigma)}{\partial(lg\varepsilon)}\right]_{\dot{\varepsilon},\, T} \tag{2}$$

The parameter n reflects the comprehensive effect of strain hardening and flow stress softening process [28,29]. It can be known from Equation (2) that the negative value of n represents the trend of stress softening, while the positive value of n represents the work hardening behavior. By the data processing and substitution into Equation (2), the strain hardening exponent n was obtained. Figure 5 shows the influence of temperature on the strain hardening exponent n under different strains. At a relatively low strain rate of 0.01 s^{-1}, the values of n under a strain of 0.1 shown in Figure 5a fluctuates near the balance state line of $n = 0$ with temperature. However, the value of n at a strain rate of 0.01 s^{-1} are mostly negative as the strain rate increases to 0.3 and 0.5, and it becomes more negative at higher temperature. When the strain rate is 0.1 s^{-1} and 1 s^{-1}, the values of n shown in Figure 5b,c are all negative in the temperature range of 1123–1223 K, and it becomes positive at temperature above 1223 K, meaning that the softening effect has been changed to the hardening effect near at 1223 K. Where the temperature is 1323 K and strain rate is 1 s^{-1}, the value of n increases from the minimum value of −0.8 to its maximum value of 0.8. Since the strain hardening exponent n is a parameter of comprehensive effects of strain hardening and flow stress caused by thermal effects, the increase of the n value is basically caused by the decreases of softening effect as the deformation temperature increases [28]. Similar behavior of the increasing tendency with the temperature was reported at SP700 superplastic titanium alloy under the high strain rate of 0.1 s^{-1} [30].

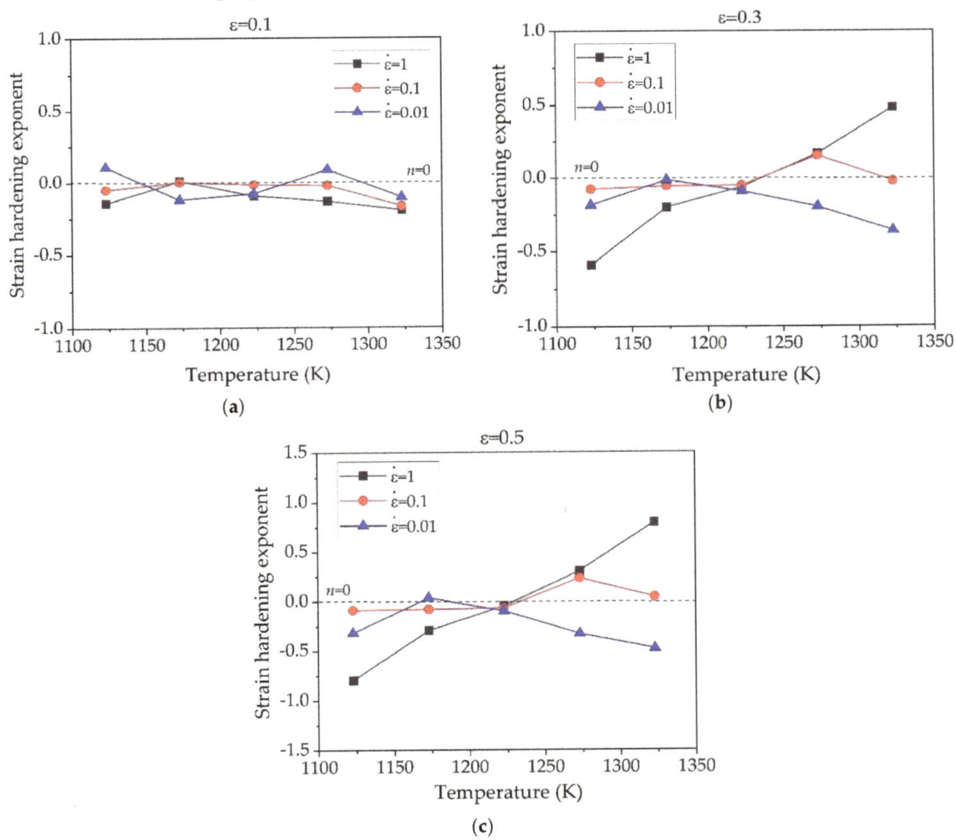

Figure 5. Influence of deformation temperature on the strain hardening exponent n of Ti-1100 titanium alloy under different strains: (**a**) $\varepsilon = 0.1$, (**b**) $\varepsilon = 0.3$, and (**c**) $\varepsilon = 0.5$.

Figure 6 shows the variation of the strain hardening exponent of the alloy with the change of strain under the different strain rates. It can be found that at temperatures of 1123 K under all strain rates, the curve of n value generally decreases during strain of 0.6 and approaches to stable state, indicating that the flow softening was gradually increased due to the temperature rise effect, and then decreased as temperature rise effect disappears at the high strain [30]. Moreover, under all strain conditions, the value of n decreases more negatively with the increase of strain rate. Similarly, at temperature of 1173 K, the n-value curve under all strain conditions moves to a more negative region with the increase of strain rate. As seen in Figure 6a at the temperature of 1273 and 1323 K, the thermal softening effect is more significant when the strain rate is 0.01 s^{-1}. However, when the strain rate increases to 0.1 s^{-1}, the strain hardening effect is dominant, and it becomes more significant as the strain rate further increases to 1 s^{-1} as shown in Figure 6c. It was revealed from Figure 6b,c that, the higher temperature the more prone to strain hardening and the lower temperature the more prone to thermal softening. At the temperature of 1223 K, the value of n is negative under all strain conditions and strain rates, and slightly increases with increase of strain rate. No significant change of n value with strain was observed at this temperature, showing better uniform plastic deformation ability.

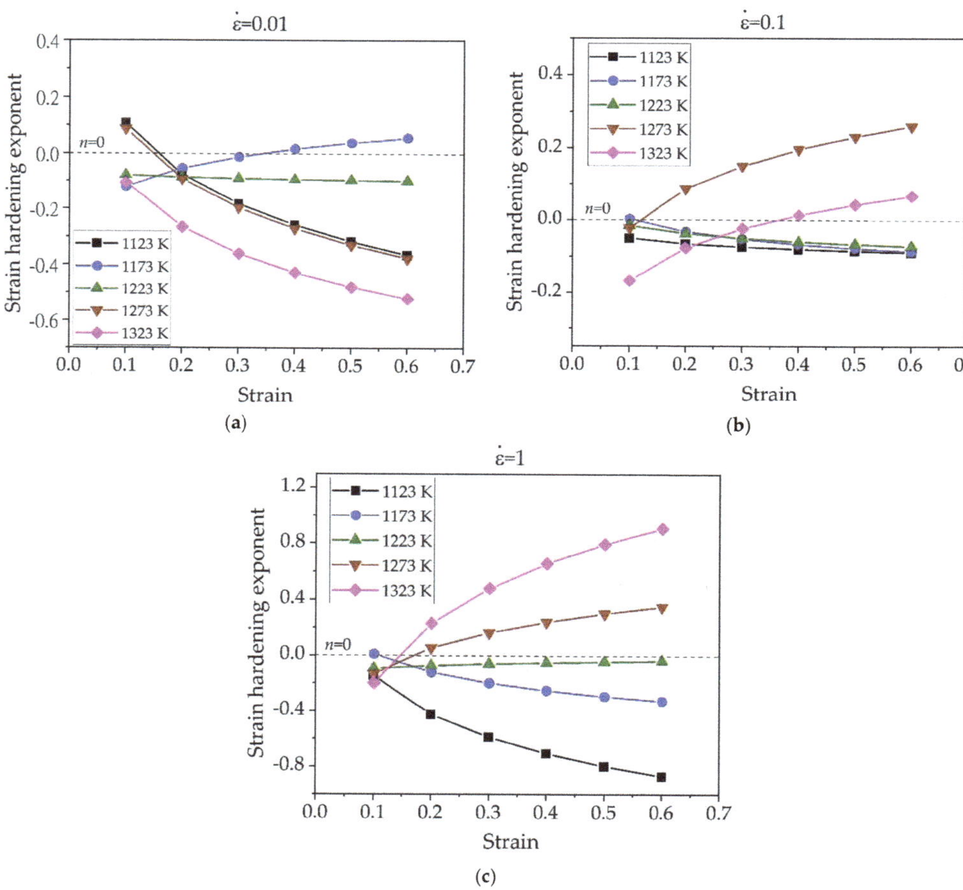

Figure 6. Variation of the strain hardening exponent as a function of strain under different strain rates: (**a**) $\dot{\varepsilon}$ = 0.01 s^{-1}, (**b**) $\dot{\varepsilon}$ = 0.1 s^{-1}, and (**c**) $\dot{\varepsilon}$ = 1 s^{-1}.

It is noted that the softening effect in term of negative n-value only considers the temperature rise effect, but not the microstructural evolution related softening effects such as dynamic recrystallization (DRX) or dynamic recovery (DRV). In general, the softening caused by DRX or DRV is favorable to hot deformation. However, the softening caused by temperature rise is not conducive to hot deformation. It is generally known that the temperature rise during alloy deformation will generate deformation heat, leading to instability phenomena such as local flow or adiabatic shear [31]. In this work, the most severe flow softening with minimum value of n occurred at low temperature of 1123 K with a high strain rate of 1 s^{-1}, so the deformation in this condition may be dangerous and should be avoided. In addition, obvious flow softening effect were also observed at high temperature of 1323 K with low strain rate of 0.01 s^{-1}. This is mainly due to the formation of the β phase as discussed in detail in the following section.

3.3. Microstructure Observation

As seen in Figure 3, at low temperature range of α/β region (T $\leq$ 1273K), the strain–stress curves show a peak stress at relatively low strain followed by flow softening, while flow curves show steady-state behavior at high temperature (T = 1323 K). In single-phase alloys, these two behaviors are often attributed to occurrence of DRV, i.e., the occurrence of dislocation climb and DRX, i.e., the nucleation and growth of new grains, respectively. However, different microstructural evolution is known to be responsible for the observed flow softening in two-phase Ti-alloys [32].

The microstructures of Ti-1100 alloy compressed at 1123 K, 1273 K, and 1323 K under the strain rates of 1~0.01 s^{-1} are presented in Figure 7(a1–a3), (b1–b3), and (c1–c3), respectively. According to Figure 7, the evolution of randomly distributed α lamellar phase with a length of 10~20 μm and a thickness less than 10 μm, during isothermal deformation, was a prevailing microstructure feature at deformation temperature below 1273 K, while at 1323 K, the various form of transformed β phase with non-lamellar α structure is its main feature, mainly due to an increase in volume fraction of β-BCC phase. During deformation at 1123–1273 K, the α colony, characterized by the fine parallel array structure, was observed, and tends to rotate perpendicular to the compressive axis. This, α colony formation is caused by local overheating. Due to the poor thermal conductivity of Ti-1100 alloy, the distortion energy caused by deformation cannot be released in the form of heat immediately, which may further lead to local overheating and the formation of fine parallel structure [33]. There is no specific grain size change with the change of strain rate. In addition, the bending/kinked lamella and sub-grain formation are the general characteristics of microstructural evolution. It is generally known that the thick initial lamellar microstructure undergoes buckling and elongates, while thin microstructure is more easily fragmented, so the bending/kinked lamella are possibly formed under the different deformation conditions. The sub-grain formation may be caused by boundary-splitting mechanism through intense localized shear during hot deformation [20]. Moreover, the globular shape or equiaxed grains with a size less than 5 μm are usually observed near the α lath boundary, indicating that dynamic recrystallization (DRX) of α phase occurs. This is because the fractured lamellar grains during hot deformation promote the formation of refined equiaxed microstructure through recrystallization of α grains. [4].

Differently, as observed in Figure 7(c1–c3), the microstructure of the alloy deformed at 1323 K has the characteristics of transformed β phase with different structures. As seen, the microstructure of the specimen deformed at the strain rate of 1 s^{-1} contained transformed β phase with basketweave (widmanstätten) structure, and discontinuous α layers at the transformed β grain boundary, and also within grain. This morphorlogy of the transformed microstructure is usually formed near β-transus temperature [34]. Clearly, compared to lower temperature, the volume of β phase at this condition was increased and the boundary of α phase becomes more irregular. As the strain rate decreases to 0.1 s^{-1}, the proportion of transformed β increased and the α/β grain boundary became blurred. When the strain rate is further decreased to 0.01 s^{-1}, the enrichment of transformed β near the α phase boundary was observed. In addition, the globularized α phase was also observed. In this case, the globularization of α phase is mainly controlled by termination migration mechanism. It is believed that the β phase diffuses into the boundary as an interlayer phase, so it can reduce the dihedral angle and further leads to grooves in the α-layers. In other words, the mass transfer from the curved surfaces of the lamellar grain terminations to the vicinal flat surface of the lamellae drives the termination migration [35]. More clear observations and phase identifications of this condition can be seen from the SEM micrographs in Figure 8. Figure 8a,c shows the secondary electron image (SE) of Ti-1100 alloy deformed at 1323 K/ 0.01 s^{-1} and 1323 K/1 s^{-1}, respectively, in which both α and β phases can be easily recognized. Figure 8b presents back-scattering electron (BSE) image of alloy deformed at 1323 K/1 s^{-1}. As shown by the white arrow in Figure 8b, due to the shear strain applied during deformation, there exists sub-boundary in "A" layer. As an unstable interface, this sub-boundary provides favorable conditions for β phase penetration. As a consequence, the α lamella is fractured and the α/β interface is formed. Similar observations have been reported previously for globularization of Ti-6242S alloy [20]. For phase identification, EDS (energy dispersive spectroscopy) analysis was conducted for matrix α phase (point 1 in Figure 8b) and transformed β phase (point 2 in Figure 8b), and the results are shown in Figure 8d. It can be seen from the EDS analysis that the α-stable element of Al is higher in the matrix, while the β-stable element of Mo or Si is rich in phase with fine parallel structure, thus confirming that the matrix is α phase and the other is the transformed β phase.

In general, DRX or DRV is the main mechanism of the flow stress softening in hot deformation. Usually, DRX occurs more obviously in the sample deformed at lower temperature and lower strain rate [22]. In present study, the flow stress softening is mainly controlled by DRX and microstructural evolution. As seen in Figure 7, dynamic recrystallization occurred in the alloys deformed in lower part of two-phase α/β region (T $\leq$ 1273 K), while microstructural evolution of transformed β grain and dynamic globalization of α by diffusional control are the dominant mechanism of flow softening at high temperature of 1323 K.

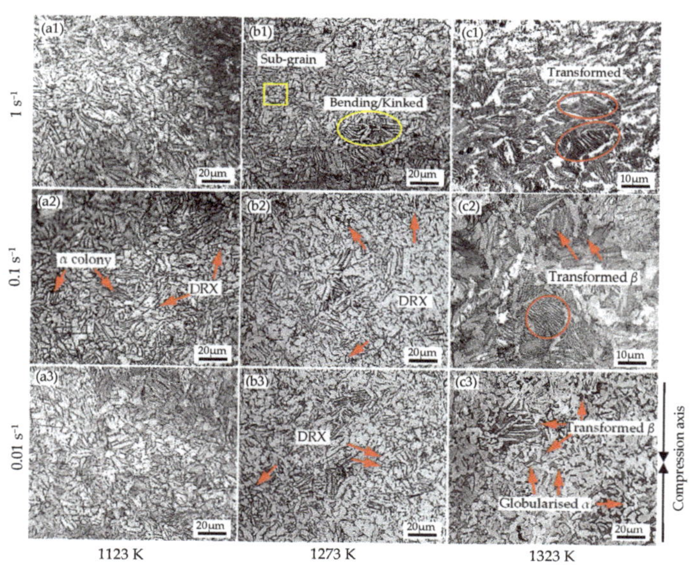

Figure 7. Microstructure of Ti-1100 alloy with a strain of 0.6 at (**a1**) 1123 K/1 s^{-1}, (**a2**) 1123 K/0.1 s^{-1}, (**a3**) 1123 K/0.01 s^{-1}; (**b1**) 1273 K/1 s^{-1}, (**b2**) 1273 K/0.1 s^{-1}, (**b3**) 1273 K/0.01 s^{-1}; and (**c1**) 1323 K/ 1 s^{-1}, (**c2**) 1323 K/0.1 s^{-1}, (**c3**) 1323 K/0.01 s^{-1}.

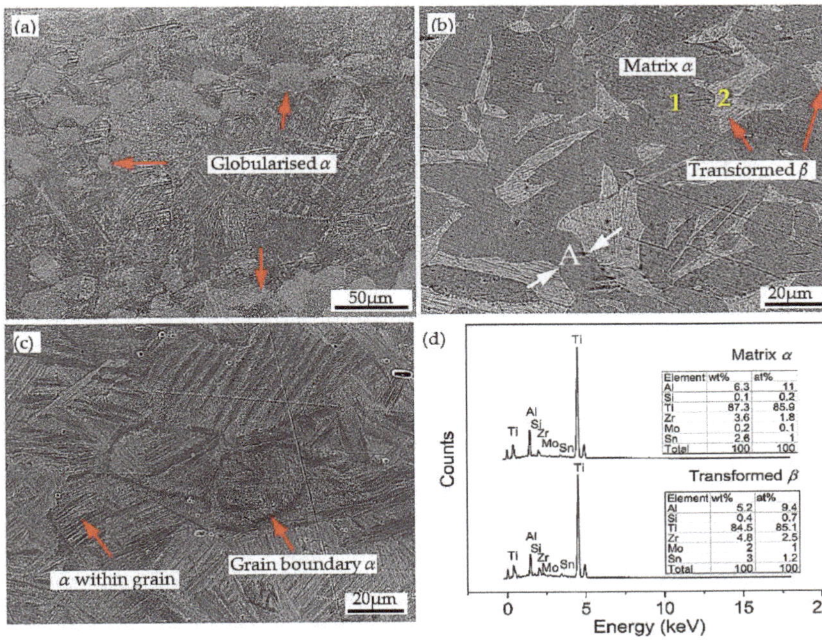

Figure 8. SEM micrographs of Ti-1100 alloy compressed at 1323 K: (**a**) SE image at 0.01 s^{-1}, (**b**) BSE image at 0.1 s^{-1}, (**c**) SE image at 1 s^{-1}, and (**d**) EDS analysis of point 1 and 2 in Figure 8b.

3.4. Strain Rate Sensitivity

It is widely known that the strain rate sensitivity (m) is a parameter of great importance to determine the deformation behavior of the material. The high strain rate sensitivity can either induce the occurrence work hardening or flow stress softening depending on the temperature and strain [21]. In general, the positive m value is favorable to the plasticity of the material, while the negative m value is adverse to hot deformation because the material is prone to the defects [30]. In addition, when the m value is greater than 0.5, the superplasticity occurs [36].

In terms of the high temperature deformation theory, the strain rate sensitivity is given in the following [37]:

$$m = \left[\frac{\partial ln\sigma}{\partial ln\dot{\varepsilon}}\right]_{T,\,\varepsilon} \approx \left[\frac{\Delta ln\sigma}{\Delta ln\dot{\varepsilon}}\right]_{T,\,\varepsilon} \tag{3}$$

After data processing, the values of m were obtained. Figure 9 shows the variation of strain rate sensitivity under different deformation conditions. As seen in Figure 9a at the strain rate of 0.01 s^{-1}, the m values are all positive and greater than 0.18. It generally shows increasing tendency with the increase of strain under most of temperature conditions, indicating the plasticity of material is enhanced as the strain increases. When the strain rate is increased to 0.1 s^{-1} and 1 s^{-1}, the strain rate sensitivity behaves differently depending on deformation temperature. As seen in Figure 9b,c under low temperature of 1123–1173 K, the m value decreases with the increase of strain, and this reduction of m is more significant at higher strain rates. These phenomena implied that the deformation mechanism has been changed as increasing the strain rate. When the high strain rate is 1s^{-1}, the value of m becomes negative as the strain approaches above 0.3, showing poor plasticity of material at this domain.

In all cases, the peak zone of m value mostly appeared in the high temperature range of 1273–1323 K and the strain of 0.6, indicating good plasticity of material. The maximum

m-value of 0.64 was obtained at 1273 K and strain rate of 0.6, and the maximum m-values of 0.52 and 0.56 under strain rate of 0.1 s^{-1} and 1 s^{-1} were obtained at 1273 K and strain rate of 0.6. These values are all greater than 0.5, indicating the material has superplasticity. This is mainly due to the increase in volume fraction of β phase transformed near α/β temperature region, resulting in higher m value and easy deformation. According to superplastic theory, the strain rate sensitivity value can be affected by the deformation parameters such as T, ε, and $\dot{\varepsilon}$, alloy composition, and phase fraction. Earlier studies on Ti60 [38], and Ti-6242S [20] alloys have shown that the maximum m-value was obtained at 0.93 T_β. Similarly, in the present study, the maximum m-value was obtained in the temperature range of 1273–1323 K, near the β-transus temperature.

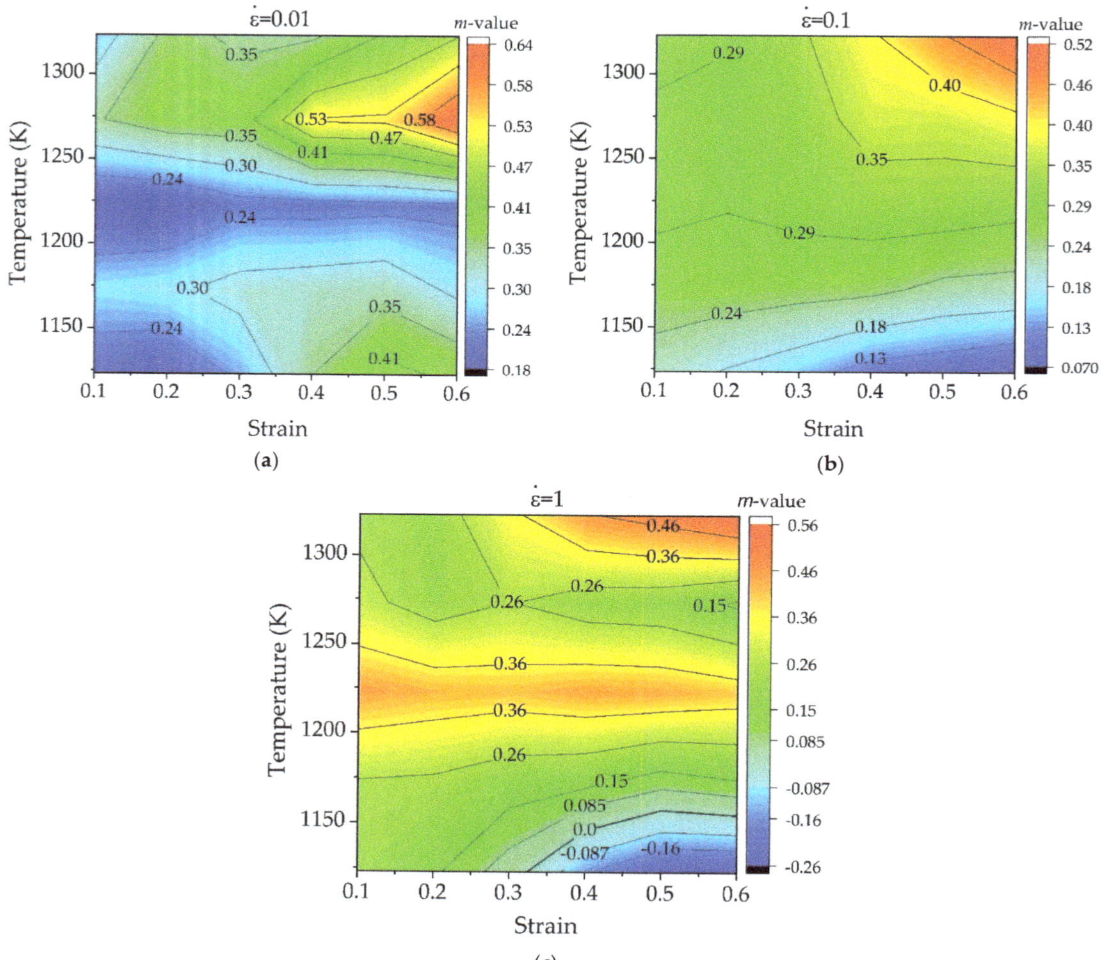

Figure 9. Contour maps of strain rate sensitivity under different strain rates: (**a**) $\dot{\varepsilon}$ = 0.01 s^{-1}, (**b**) $\dot{\varepsilon}$ = 0.1 s^{-1}, and (**c**) $\dot{\varepsilon}$ = 1 s^{-1}.

The strain rate sensitivity near 1223 K is relatively stable with strain under a fixed strain rate, but it increases with the increase of strain rate. When the strain rate is 0.01 s^{-1} and 0.1 s^{-1}, the m values are in the range of 0.18-0.30, which is a typical deformation controlled by climb-limited glide processes characteristic of power-low creep or dislocation glide [34]. When the strain rate is 1 s^{-1}, the m values are higher than 0.3, which is probably related to grain-boundary sliding (GBS) [28].

3.5. DMM Processing Map

In order to characterize the hot working behavior of materials under plastic deformation without losing stability at high temperature, processing map of a material based on the dynamic materials model (DMM) [39] was developed. This theory assumes that the workpiece is regarded as power dissipation in high temperature deformation and the power observed by workpiece essentially dissipates during the hot deformation process. In this map, the microstructure evolution such as DRV, DRX, and superplasticity are considered as "safe" deformation conditions, while the "unsafe" deformation conditions are various defects such as shear band and flow localization. To develop the processing map of the alloy, the efficiency of power dissipation (η) representing the power dissipation capacity is defined as Equation (4):

$$\eta\ (\mathrm{T},\ \dot{\varepsilon}\ ) = \frac{2m}{m+1} \tag{4}$$

where m is the strain rate sensitivity coefficient for a given flow stress and expressed by $m = \partial(\ln\sigma)/\partial(\ln\dot{\varepsilon})$. Variation of power dissipater efficiency with temperature and strain rate constitutes the power dissipater map, which presents various power domains correlated to specific microstructural mechanisms.

To identify the microstructural instability, Ziegler flow instability [40] was used as Equation (5):

$$\xi(\dot{\varepsilon}) = \frac{\partial \ln[m/(m+1)]}{\partial ln\dot{\varepsilon}} + m \tag{5}$$

where $\xi(\dot{\varepsilon})$ is a dimensionless instability parameter. This model assumes that when the instability parameter is negative, the flow instability of material occurs. The variation of instability criteria with strain rates and temperature constitutes the instability map. By superimposing power dissipater map and instability map, the processing map of a material is developed. Figure 10a,b show the processing maps of the alloy at the strain of 0.3 and 0.6, respectively. As seen, the value of η generally increased with increasing temperature, and there are multiple peak domains in high strain rate region and low strain region. As shown in Figure 10a, the instability mainly occurred in the area of temperature below 1173 K and higher strain rate, and this instability areas coincided with low efficiency regions. This instability region gradually disappears as the temperature increases and the strain rate decreases. This reduction of instability may be attributed to the reduction of deformation heating effects, which is considered as main factor of an adiabatic shear banding [31]. With the increase of strain, the instability regions expanded to the temperature of 1173 K and to the strain rate of 0.023 s^{-1} ($\ln\dot{\varepsilon} = -3.78$) as show in Figure 10b. In addition, another instability zone was appeared at a high strain rate of 1s^{-1} and a temperature of 1273 K. The optimum processing range of material deformation is 1273–1323 K along with strain rate range of 0.01–0.165 s^{-1} ($\ln\dot{\varepsilon} = -4.6 \sim -1.8$) since the power dissipation efficiency generally shows high values at this region and the instability criteria has a positive value.

Figure 11 shows the microstructures of specimens deformed under various deformation conditions. It can be known from microstructural observation that the shear deformation is the dominant instability mechanism. As seen in Figure 11a,b, the microstructure exhibited a severe flow localization at near 45° to the compression axis and large amount of shear band. In general, the localized plastic flow or shear deformation are easy to form at low temperatures and high strain rates [41], and similar results have been observed in Ti-6242S alloy [19]. As described in Section 3.2, the minimum value of strain hardening exponents n at 1123 K also demonstrated severe flow softening due to temperature rise. Another instability is inhomogenous microstructure as shown in Figure 11c at 1273 K/1 s^{-1}. This may be due to the insufficient phase transformation behavior occurring at high strain rate and high temperature, which leads to the inhomogeneity of microstructure.

These microstructures are undesirable in achieving appropriate mechanical properties, so it should be avoided during hot deformation.

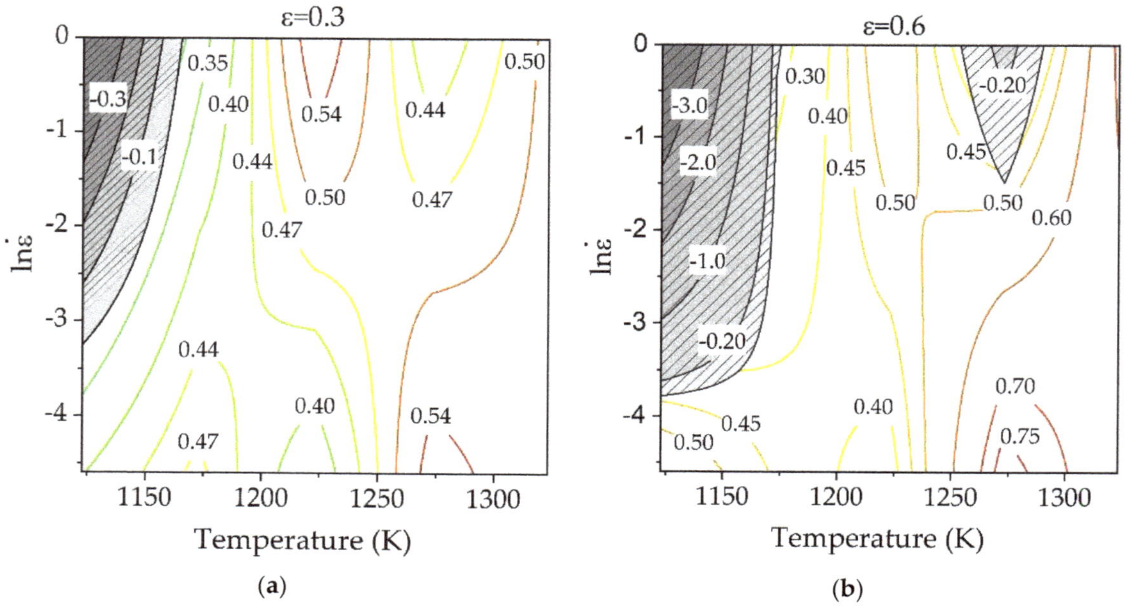

Figure 10. Processing maps of Ti-1100 alloy at strains of (**a**) 0.3 and (**b**) 0.6 (shaded zone represents instability domain; numbers represent power dissipation efficiency (%)).

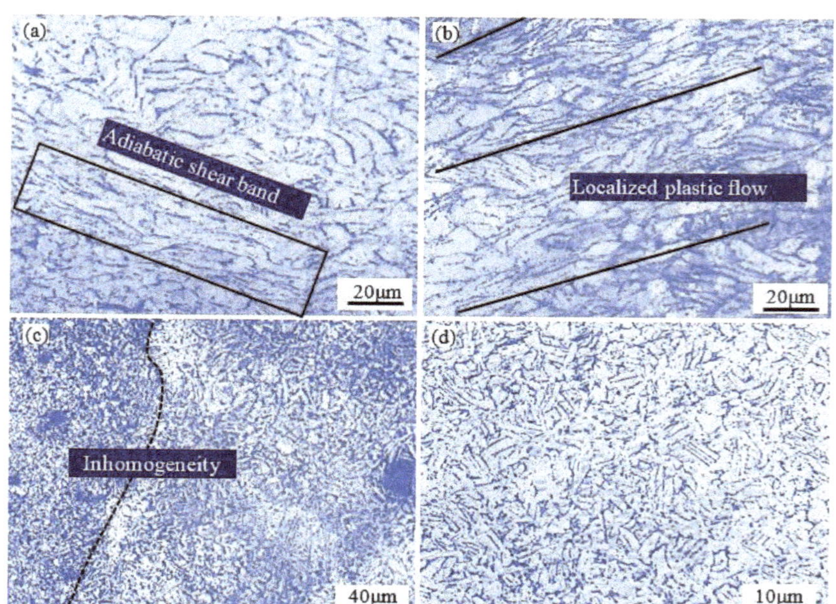

Figure 11. Optical micrographs of specimens deformed under different conditions: (**a**) 1123 K, 1 s^{-1}, (**b**) 1123 K, 0.1 s^{-1}, (**c**) 1273 K, 1 s^{-1}, and (**d**) 1273 K, 0.1 s^{-1}.

According to DMM, η is related to the microstructural mechanism which occurs dynamically to dissipate power, such as DRV, phase transformation, and grain refinement [42]. The higher η leads to a better dissipative behavior from the workpiece associated with microstructure evolution, resulting better deformation capacity. On the other hand, the lower η leads to worse dissipative behavior due to temperature rise effect. Therefore, it was worth to investigate the variations of η under different processing conditions. Shown in Figure 12a is the variation of η with strain at different temperatures. At 1123 K shown in Figure 12a, the values of η are all lower than 0.5, and it decreased more with the increase of strain. This behavior implies that the deformation power was mainly dissipated into viscoplastic heat due to temperature rise effect, indicating low formability at warm deformation region. So, this is the reason why the alloy is prone to defects when it is deformed at low temperature as discussed above. As shown in Figure 12a at moderate temperature of 1223 K, the value of η remains constant with strain, but it shows increasing tendency with strain at higher temperatures of 1273 K and 1323 K. Obviously, the slope of η with respect to strain gradually increases with increase of temperature, and this means that the power dissipation mechanism during hot deformation has been changed from temperature-dependent to microstructure-dependent with the increase of temperature. The variations of power dissipation efficiency with temperature under different strains and a fixed strain rate of 0.1 s^{-1} is shown in Figure 12b. Clearly, the power dissipation efficiencies increased with increasing temperature and the increase in η values is more significant with increase of strain. As seen in Figure 12a,b at fixed strain of 0.6, the value of η increases with increase of temperature, implying microstructure optimization were more likely to occur at higher temperature. These findings exactly explain the different temperature-dependent behavior of true stress–strain curves as discussed in preceding sections. Figure 12c shows the variations of power dissipation efficiency with strain under different strain rates and the fixed temperature of 1223 K. The η values remained constant with the strain at each strain rate. The η values at a strain rate of 0.01 s^{-1} remained about 0.6, implying microstructural changes such as DRX or DRV are more likely to occur in this case. Metallographic observation under this condition shown in Figure 7 also confirmed the DRXed microstructures. On the other hand, the η values at a strain rate of 1 s^{-1} are in the range of 0.3–0.4 and lower than 0.5, implying the microstructural optimization is difficult to occur in this case, but mostly consumed by viscoplastic heat or local overheating which may induce the instability such as shear deformation or localized plastic flow.

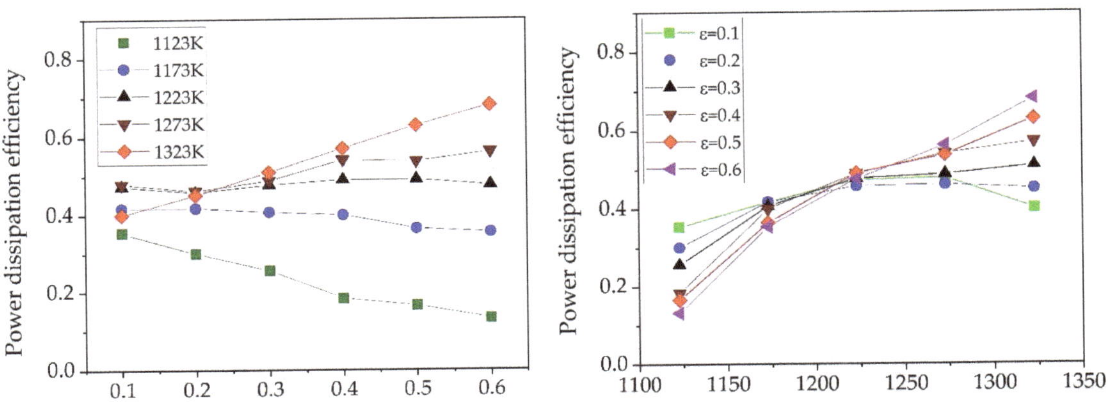

Figure 12. *Cont.*

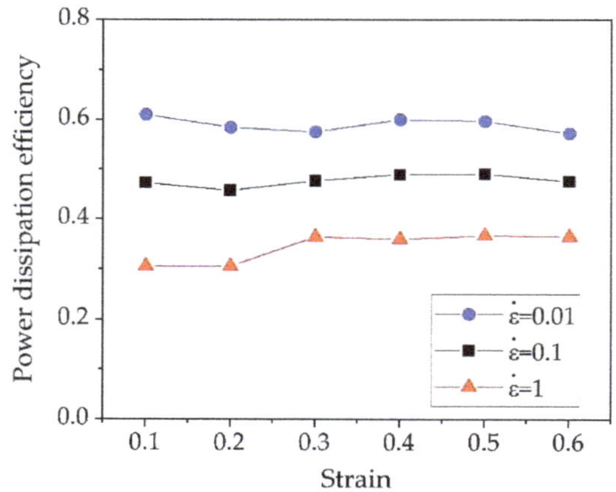

(c)

Figure 12. Variations of power dissipation efficiency with (**a**) strain at 0.1 s^{-1}, (**b**) temperature at strain rate of 0.1 s^{-1}, and (**c**) strain at 1223 K.

4. Conclusions

1. The true stress–strain curves showed that at a strain rate of 0.01 s^{-1}, the flow stress exhibited softening response under all deformation temperatures. At relatively high strain rate of 0.1 s^{-1} and 1 s^{-1}, the flow stress exhibits the general behaviors of softening response in the low part of α/β region (T $\leq$ 1223 K), while both softening and hardening effects were observed at upper part (T > 1273 K).
2. The strain hardening exponent (n) analysis revealed that the hardening/softening effects become more significant as the strain increases. In addition, the higher temperature the more prone to strain hardening and the lower temperature the more prone to thermal softening. When the strain rate is 0.1 s^{-1} and 1 s^{-1}, the values of n are all negative in the temperature range of 1123–1223 K and it becomes positive at temperature above 1223 K, implying that the softening effect has been changed to the hardening effect near at 1223 K. The most severe flow softening with minimum value of n occurred at low temperature of 1123 K with high strain rate of 1 s^{-1}, and high temperature of 1323 K with low strain rate of 0.01 s^{-1}.
3. Lamellar phase with a length of 10~20 μm and a thickness less than 10 μm was the main structure of alloy deformed at temperature below 1273 K, and the various form of transformed β phase with irregular shape of α structure is the main feature at temperature of 1323 K. The dynamic recrystallization of α *is* the main softening mechanism in lower part of temperature (T $\leq$ 1273 K), while microstructural evolution of transformed β grain and dynamic globalization of α by diffusional mainly control the softening effect of alloy.
4. The developed processing map demonstrated that the deformation in the temperature range of 1273–1323 K and strain rates of 0.01–0.165 s^{-1} was desirable and led to high efficiencies. At low temperature of 1123 K, increasing strain rate led to the increase of flow instability, which was primarily manifested as localized plastic flow, adiabatic shear bands and inhomogenous microstructure. The variation of power dissipation energy (η) with strain demonstrated that the power dissipation mechanism during hot deformation has been changed from temperature-dependent to microstructure-dependent with the increase of temperature for the alloy deformed at 0.1 s^{-1}.

Author Contributions: Conceptualization, R.P. and L.M.; methodology, L.M. and B.H.; software, R.P. and W.Z.; validation, R.P.; investigation, R.P. and W.Z.; writing—original draft preparation, R.P.; writing—review and editing, P.Z.; funding acquisition, P.Z. All authors have read and agreed to the published version of the manuscript.

Funding: This research was funded by the National Natural Science Foundation of China (NSFC) (No. 51804007), Natural Science Research Project of Anhui Universities (Grant No. KJ2019A0127), and Anhui University of Science and Technology introduction of talent research start-up fund project of (No. 13200456).

Institutional Review Board Statement: Not applicable.

Informed Consent Statement: Not applicable.

Data Availability Statement: Not applicable.

Acknowledgments: School of Materials Science and Technology, Anhui University of Science and Technology for the supports.

Conflicts of Interest: The authors declare no conflict of interest.

Appendix A

The β-transus temperature (T_β, the temperature at which the α + β-phases transform to the β-phase) of the as-sintered Ti-1100 titanium alloy was determined by differential scanning calorimeter (DSC, NETZSCH STA 449F5, Germany). The main parameters of DSC analysis are as follows:

(1) Sample mass (mg): 25.18
(2) Type of crucible: Al_2O_3 (85 μL) with lid
(3) Heating rate (K/min): 10

As seen in Figure A1, T_β was determined as 1293 K (or 1019.7 °C).

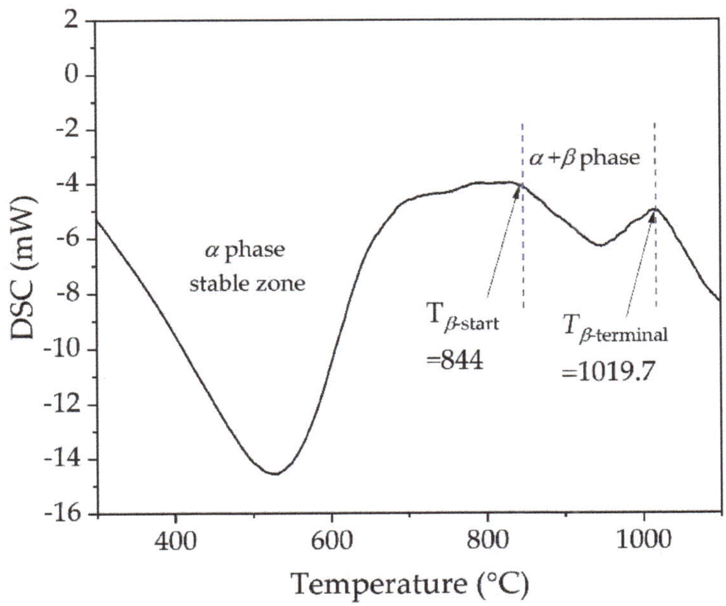

Figure A1. DSC curve of the as-sintered Ti-1100 titanium alloy.

References

1. Williams, J.C.; Starke, E.A., Jr. Progress in structural materials for aerospace systems. *Acta Mater.* **2003**, *51*, 5775–5799. [CrossRef]
2. Bania, P. *Metallurgy and Technology of Practical Titanium Alloys*; TMS: Warrendale, PA, USA, 1994; pp. 9–18.
3. Zhang, H.R.; Niu, H.Z.; Zang, M.C.; Yue, J.K.; Zhang, D.L. Microstructures and mechanical behavior of a near α titanium alloy prepared by TiH2-based powder metallurgy. *Mater. Sci. Eng. A* **2020**, *770*, 138570. [CrossRef]
4. Fang, Z.Z.; Paramore, J.D.; Sun, P.; Ravi Chandran, K.S.; Zhang, Y.; Xia, Y.; Cao, F.; Koopman, M.; Free, M. Powder metallurgy of titanium–past, present, and future. *Int. Mater. Rev.* **2018**, *63*, 407–459. [CrossRef]
5. Fang, Z.Z.; Sun, P.; Wang, H.T. Hydrogen sintering of titanium to produce high density fine grain titanium alloys. *Adv. Eng. Mater.* **2012**, *14*, 383–387. [CrossRef]
6. Joshi, V.V.; Lavender, C.; Moxon, V.; Duz, V.; Nyberg, E.; Weil, K.S. Development of Ti-6Al-4V and Ti-1Al-8V-5Fe alloys using low-cost TiH_2 powder feedstock. *J. Mater. Eng. Perform.* **2013**, *22*, 995–1003. [CrossRef]
7. Azevedo, C.R.F.; Rodrigues, D.; Neto, F.B. Ti–Al–V powder metallurgy (PM) via the hydrogenation–dehydrogenation (HDH) process. *J. Alloy. Compd.* **2003**, *353*, 217–227. [CrossRef]
8. Ivasishin, O.M.; Savvakin, D.G. The impact of diffusion on synthesis of high-strength titanium alloys from elemental powder blends. *Key Eng. Mater.* **2010**, *436*, 113–121. [CrossRef]
9. Zhang, S.Z.; Liu, J.W.; Zhao, Q.Y.; Zhang, C.J.; Bolzoni, L.; Yang, F. Microstructure characterization of a high strength Ti–6Al–4V alloy prepared from a powder mixture of TiH_2 and 60Al40V masteralloy powders. *J. Alloy. Compd.* **2020**, *818*, 152815. [CrossRef]
10. Hagiwara, M.; Emura, S. Blended elemental P/M synthesis and property evaluation of Ti-1100. *Mater. Sci. Eng. A* **2003**, *352*, 85. [CrossRef]
11. Zhang, H.R.; Niu, H.Z.; Liu, S.; Zang, M.C.; Zhang, D.L. Significantly enhanced tensile ductility and its origin of a<0001> micro-textured extrusion bar of a powder metallurgy near alpha titanium alloy. *Scr. Mater.* **2022**, *213*, 114633.
12. Prasad, Y.V.R.K.; Seshacharyulu, T. Processing maps for hot working of titanium alloys. *Mater. Sci. Eng. A* **1998**, *243*, 82–88. [CrossRef]
13. Seshacharyulu, T.; Medeiros, S.C.; Frazier, W.G.; Prasad, Y.V.R.K. Hot working of commercial Ti-6Al-4V with an equiaxed α-β microstructure: Materials modeling considerations. *Mater. Sci. Eng. A* **2000**, *284*, 184–194. [CrossRef]
14. Babu, K.A.; Mandal, S.; Kumar, A.; Athreya, C.N.; Boer, B.; Sarma, V.S. Characterization of hot deformation behavior of alloy 617 through kinetic analysis, dynamic material modeling and microstructural studies. *Mater. Sci. Eng. A* **2016**, *664*, 177–187. [CrossRef]
15. Krishna, V.G.; Prasad, Y.V.R.K.; Birla, N.C.; Rao, G.S. Hot-deformation mechanisms in near-alpha titanium alloy 685. *JOM* **1996**, *48*, 56–59. [CrossRef]
16. Balasundar, I.; Raghu, T.; Kashyap, B.P. Hot working and geometric dynamic recrystallisation behaviour of a near-α titanium alloy with acicular microstructure. *Mater. Sci. Eng. A* **2014**, *600*, 135–144. [CrossRef]
17. Zhou, D.; Zeng, W.; Xu, J.; Chen, W.; Wang, S. Characterization of Hot Workability for a Near Alpha Titanium Alloy by Integrating Processing Maps and Constitutive Relationship. *Adv. Eng. Mater.* **2019**, *21*, 1801232. [CrossRef]
18. Su, Y.; Kong, F.; You, F.H.; Wang, X.; Chen, Y. The high-temperature deformation behavior of a novel near-α titanium alloy and hot-forging based on the processing map. *Vacuum* **2020**, *173*, 109135. [CrossRef]
19. Morakabati, M.; Hajari, A. Hot working behavior of near alpha titanium alloy analyzed by mechanical testing and processing map. *Trans. Nonferrous Met. Soc. China* **2020**, *30*, 1560–1573. [CrossRef]
20. Ding, C.; Shi, Q.; Liu, X.; Zheng, L.; Li, R.; Hang, Z.; Yu, B.; Wu, W. Microstructure and Mechanical Properties of PM Ti600 Alloy after Hot Extrusion and subsequent Annealing Treatment. *Mater. Sci. Eng. A* **2019**, *748*, 434–440. [CrossRef]
21. Piao, R.; Yang, S.L.; Zhu, Y.L.; Ma, L. Hot Deformation Behavior of Near-α Titanium Alloy Ti-1100 Prepared by TiH_2-based Powder Metallurgy. *Rare Met. Mater. Eng.* **2020**, *49*, 3314–3324.
22. Yang, C.; Liu, X.; Si, J.Y.; Liu, F.; Jiang, L. Processing map and cracking behavior of powder metallurgy superalloy FGH96 during hot compression. *Chin. J. Nonferrous Met.* **2015**, *25*, 2707–2719.
23. Lei, J.; Zhu, W.; Chen, L.; Sun, Q.; Xiao, L.; Sun, J. Deformation behaviour and microstructural evolution during the hot compression of Ti-5Al4Zr8Mo7V alloy. *Mater. Today Commun.* **2020**, *23*, 100873. [CrossRef]
24. Li, A.; Huang, L.; Meng, Q.; Geng, L.; Cui, X. Hot working of Ti−6Al−3Mo−2Zr−0.3Si alloy with lamellar α+β starting structure using processing map. *Mater. Des.* **2009**, *30*, 1625–1631. [CrossRef]
25. Hossenin, R.; Morakabati, M.; Abbasi, S.M.; Hajari, A. Development of a trimodal microstructure with superior combined strength, ductility and creep-rupture properties in a near alpha titanium alloy. *Mater. Sci. Eng. A* **2017**, *696*, 155–165. [CrossRef]
26. Wanjar, P.; Jahazi, M.; Monajati, H.; Yue, S.; Immarigeon, J.P. Hot working behavior of near-α alloy IMI834. *Mater. Sci. Eng. A* **2005**, *396*, 50–60. [CrossRef]
27. Holloman, J.H. Tensile deformation. *Met. Technol.* **1945**, *12*, 268–290.
28. Luo, J.; Li, M.; Yu, W.; Li, H. The variation of strain rate sensitivity exponent and strain hardening exponent in isothermal compression of Ti-6Al-4V alloy. *Mater. Des.* **2010**, *31*, 741–748. [CrossRef]
29. Luo, J.; Li, M.Q. Strain rate sensitivity and strain hardening exponent during the isothermal compression of Ti60 alloy. *Mater. Sci. Eng. A* **2012**, *538*, 156–163. [CrossRef]
30. Qiu, Q.; Wang, K.; Li, X.; Wang, J.; Gao, X.; Zhang, K. Hot deformation behavior and processing parameters optimization of SP700 titanium alloy. *J. Mater. Res. Technol.* **2021**, *15*, 3078–3087. [CrossRef]

31. Wang, J.; Wang, K.L.; Lu, S.Q.; Li, X.; OuYang, D.L.; Qiu, Q. Softening mechanism and process parameters optimization of Ti-4.2Al-0.005B titanium alloy during hot deformation. *J. Mater. Res. Technol.* **2022**, *17*, 1842–1851. [CrossRef]
32. Ghavam, M.H.; Morakabati, M.; Abbasi, S.M.; Badri, H. Flow behavior modelling of IMI834 titanium alloy during hot tensile deformation. *Trans. Nonferrous Met. Soc. China* **2015**, *25*, 748–758. [CrossRef]
33. Cui, X. Study on Microstructure Evolution and Properties of TC11 Titanium Alloy during High Temperature Deformation Process. Master's Thesis, Harbin Institute of Technology, Harbin, China, 2008.
34. Semiatin, S.L.; Bieler, T.R. The effect of alpha platelet thickness on plastic flow during hot working of TI−6Al−4V with a transformed microstructure. *Acta Mater.* **2001**, *49*, 3565–3573. [CrossRef]
35. Stefansson, N.; Semiatin, S. Mechanisms of globularization of Ti−6Al−4V during static heat treatment. *Metall. Mater. Trans. A* **2003**, *34*, 691–698. [CrossRef]
36. Ma, L.; Wan, M.; Li, W.; Shao, J.; Bai, X.; Zhang, J. Superplastic deformation mechanical behavior and constitutive modelling of a near-a titanium alloy TNW700 sheet. *Mater. Sci. Eng. A* **2021**, *817*, 141419. [CrossRef]
37. Narayana Murty, S.V.S.; Nageswara Rao, B.; Kashyap, B.P. Instability criteria for hot deformation of materials. *Int. Mater. Rev.* **2014**, *45*, 15–26. [CrossRef]
38. Jia, W.J.; Zeng, W.D.; Zhou, Y.G.; Liu, J.R.; Wang, Q.J. High-temperature deformation behavior of Ti60 titanium alloy. *Mater. Sci. Eng. A* **2011**, *528*, 4068–4074. [CrossRef]
39. Prasad, Y.V.R.K.; Gegel, H.L.; Doraivelu, S.M.; Malas, J.C.; Morgan, J.T.; Lark, K.A.; Barker, D.R. Modeling of dynamic material behavior in hot deformation: Forging of Ti-6242. *Metall. Trans.* **1984**, *15A*, 1883–1892. [CrossRef]
40. Ziegler, H. *Progress in Solid Mechanics*; John Willey and Sons: New York, NY, USA, 1963; p. 93.
41. Yang, J.; Wang, G.; Jiao, X.; Li, Y.; Liu, Q. High-temperature deformation behavior of the extruded Ti-22Al-25Nb alloy fabricated by powder metallurgy. *Mater. Charact.* **2018**, *137*, 170–179. [CrossRef]
42. Qiang, L.; Song, H.; Ke, T.; Yu, Y.; Ye, W.; Song, S.Y. Investigation of high temperature behavior and processing map of Ti-6Al-4V-0.11Ru titanium alloy. *J. Alloy. Compd.* **2019**, *787*, 527–536.

materials

Article

In Situ Observation and Phase-Field Simulation Framework of Duplex Stainless-Steel Slab during Solidification

Tong Wang [1], David Wexler [1], Liangliang Guo [2], Yangfan Wang [1] and Huijun Li [1,*]

1 School of Mechanical, Materials, Mechatronics and Biomedical Engineering, University of Wollongong, Northfields Ave., Wollongong, NSW 2522, Australia
2 BaoWu Steel Group, Baoshan, Shanghai 201900, China
* Correspondence: huijun@uow.edu.au

Abstract: The melting and solidification process of S32101 duplex stainless steel (DSS) was investigated using high-temperature confocal microscopy (HTCM). The method of concentric HTCM was employed to study microstructure evolution during the solidification process of S32101 DSS. This method could artificially create a meniscus-shaped solid–liquid interface, which dramatically improved the quality of in situ observations. During the heating stage, γ-austenite transformed to δ-ferrite, and this transformation manifested itself in the form of grain boundaries (GBs) moving. The effects of cooling rate on the solidification pattern and microstructure were revealed in the present research. An enhanced cooling rate led to a finer microstructure, and the solidification pattern changed from cellular to dendritic growth. As the temperature decreased, the commencement and growth of precipitates were observed. In this paper, the experimental data, including parameters such as temperature, cooling rate, and growth mode, were used as the benchmark for the simulation. A simulation framework using Micress linked to a 1D heat transfer model enabling consistent analysis of solidification dynamics in DSS across the whole cast slab was established. Simulating the dendrite growth and elemental segregation of DSS at specific cooling rates shows that this framework can be a powerful tool for solving practical production problems.

Keywords: high-temperature confocal microscope; phase-field simulation; duplex stainless steel; cooling rate

Citation: Wang, T.; Wexler, D.; Guo, L.; Wang, Y.; Li, H. In Situ Observation and Phase-Field Simulation Framework of Duplex Stainless-Steel Slab during Solidification. *Materials* **2022**, *15*, 5517. https://doi.org/10.3390/ma15165517

Academic Editors: Qing Liu and Jiangshan Zhang

Received: 14 July 2022
Accepted: 9 August 2022
Published: 11 August 2022

1. Introduction

S32101 (US designation) is a duplex stainless steel (DSS) with low content of Ni and high content of N. High contents of Mn and N were added to replace expansive Ni to reduce the cost and improve mechanical properties and corrosion resistance. It is widely used in the chemical industry or offshore technologies, where a combination of high corrosion resistance and good tensile strength is required [1,2].

Many factors, such as cooling rate, can affect the microstructure of steel during the solidification process. Importantly, the microstructure can affect the mechanical properties of S32101 DSS and finally influence the quality of the slab. The dual-phase structure is the root of the excellent properties of DSS. Mazumdar and Kary [3], from various perspectives, such as heat transfer during crystallization and solid–liquid interface tribology, expounded on the control of the continuous casting process and pointed out that controlling the solidification conditions controlled the ratio of δ and γ to obtain products with excellent quality.

High-temperature confocal microscopy (HTCM) was used to observe the solidification process of S32101 DSS in the present research, and this technique can provide a so-called in situ observation. Zhao and Sun [4] applied laser scanning confocal microscopy (LSCM) in their research on DSS. They observed and discussed the effect of different N contents on the solid-state δ ↔ γ phase transformations and concluded that N could hinder the migration of δ/γ interface boundaries (IBs).

Computational simulation can also be used predict the solidification process from experimental observation. The Microstructure Evolution Simulation Software (Micress®, Version 7.1) is a software package that calculates the growth of newly formed phases during solidification or phase transformation of a metallurgical system on a temporal or spatial scale. The software is based on the multiphase-field concept, developed by ACCESS e.V. (affiliates of Aachen University, Aachen, Germany) scientists since 1995. It is widely used in solidification, solid-state transformations, grain growth, recrystallization, heat treatment, etc. [5,6]. The solidification process and subsequent solid-state phase transformation always start with nucleation, which is the first step of forming a new phase. This process is dominated by thermodynamic driving forces, diffusion, and interfacial curvature. The backbone of Micress is the multiphase field method for multicomponent alloys, which enables the treatment of multiphase, multigrain, and multicomponent problems. As a well-established commercial phase-field simulation software, Micress provides a simulation platform for materials researchers without a computer science background.

In continuous casting, liquid steel is poured into a water-cooled copper mold, with the high heat extraction rate chilling a solid steel shell that contains the yet-to-be solidified liquid in the center. This shell is continuously extracted from the copper mold, the rate of which is an important operating parameter. Secondary cooling through water mist is applied to the shell as it exits the mold, thus inducing sufficient cooling to ensure complete solidification. The solidification process occurs with a liquid-to-solid phase transformation, and segregation of alloying elements such as Cr and Ni, as well as N and Mo for DSS in particular, could lead to defects. For the continuous casting of S32101 DSS, a deeper understanding of the effect of thermodynamic and kinetic parameters on the rate and mechanism is the key to improving the final product quality.

Many studies have been published about DSS with respect to the influence of the different alloying element contents. Scarce research has been conducted to investigate the thermodynamic process from melting to solidification. This paper aimed to reveal the melting process of S32101 DSS and the solidification process regarding different cooling rates through 'in situ' observation; furthermore, on the basis of the experiments, a phase-field simulation framework was established to determine whether concentric solidification can be used to provide a basis for benchmarking the relationship between cooling rate and solidification structure and to what extent this can be used to provide data for developing the Micress phase-field simulation framework.

2. Experimental and Modeling Approaches

2.1. Experimental Approaches

As an important tool widely used to observe solidification and phase transformation, HTCM sometimes reduces the observation surface because the sample forms semicircular droplets in the crucible. If there is further obstruction by oxides or precipitates, the availability of observational data will be significantly reduced. The design of a concentric solidification method can tackle this problem.

2.1.1. Concentric HTCM

In 1961, Minski [7] established the principles of LSCM, which are widely applied in biological sciences. This equipment has been applied to investigate many fields of research, including the morphology of solidification and an analysis of the δ/γ interphase boundary dynamic behavior in low-carbon steels [8,9], inclusion agglomeration and engulfment in steel melt/solid interface [9,10], and crystallization of oxide melts [11]. Since the concentric HTCM was designed, significant developments in peritectic reaction and transformation have been achieved. Stephan [12,13] employed the concentric HTCM method to investigate the peritectic transformation of Fe–C steel and cast new light on the reason for massive-type phase transformation with thermodynamic and kinetic arguments. Lasertec Corporation manufactures the HTCM equipment in Yokohama, Japan. A 1.0 kW halogen lamp is located at a focal point in a gold-plated ellipsoid cavity in the lower half of the infrared furnace.

The sample was heated by radiation in this chamber. A gold-coated ellipsoidal chamber concentrated the heat onto the surface of the sample located at the other focal point of the furnace chamber. Figure 1 shows the components of the 1LM21H type HTCM with an oxygen detector, monitors, and proportional–integral–derivative (PID) controller.

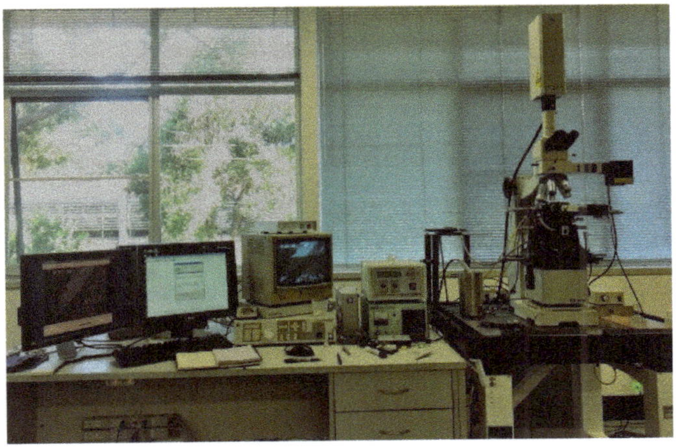

Figure 1. Picture of HTCM 1LM21H.

Reid et al. [14,15] developed and described the concentric solidification technique to improve the quality of in situ observation of HTCM. A thin sample is partially melted in the center, forming a pool of melt surrounded by a solid rim. A platinum sample holder holds the crucible in place, and a type B thermocouple is attached to the crucible. Since the in situ observation of HTCM is limited to the sample surface, the thickness becomes an important parameter. After several attempts with different sample thicknesses, ~250 μm was determined to be the minimum thickness for concentric solidification experiments on S32101 DSS. On the one hand, if the thickness is less than 250 μm, the liquid pool is easily ruptured. On the other hand, if the thickness exceeds 300 μm, the output power required to melt the sample will be too high to be achieved.

In DSS, initial melting is uneven due to segregation during the casting of the slab, which was not previously homogenized. Care needs to be taken for the rate of further heating to achieve a uniform liquid pool in the central region of the sample. With care, a pool of diameter around 5 mm can be achieved with a planar interface between liquid and solid. This then becomes the starting condition for all subsequent imposed cooling rates. The range of potential solidification microstructures possible includes planar at low cooling rates, and then cellular as the cooling rate increases. The subsequent transition is a dendritic microstructure which reverts to cellular and planar if a sufficiently high cooling rate is achieved. In the concentric solidification technique, the maximum controlled cooling rate that can be achieved is 500 to 1000 °C/min, far below the rate required for reverting to cellular/planar structures.

The sample and the holder were inserted into the top half of the furnace chamber, and high-purity argon gas (99.9999 vol.%) was flowed through the chamber. The gas was passed through a stainless-steel tube filled with titanium turnings to ensure a high-integrity inert gas, held at a temperature of 850 °C, before entering the furnace chamber. The oxygen level was reduced to 10–14 ppm before running the experiments. Moisture was also removed when the gas flowed through the Ti furnace.

The Omron ES100P digital PID controller panel was used to set the program profile before the experiments, as well as change the temperature during the experiments. The temperature program is shown in Figure 2a. All HTCM experiments were based on this procedure. The whole procedure can be divided into five stages. The first stage is to stabilize the equipment. The second stage is the heating stage, which involves heating to 1300 °C at a rate of 150 °C/min. The target temperature is below the melting point; one reason is to avoid overshooting which could leading to sample penetration, and another reason is to reserve enough operating range for the third step. Stage 3 is the most important because the liquid pool is formed at this stage, and it is necessary to artificially slowly adjust the temperature until a stable liquid pool of sufficient size is formed. The fourth stage is the heat preservation stage to form a stable solid–liquid interface. The last stage is the cooling stage with five different cooling rates, 5 °C/min, 10 °C/min, 50 °C/min, 80 °C/min, and 150 °C/min, to investigate the effect of cooling rate on solidification microstructures.

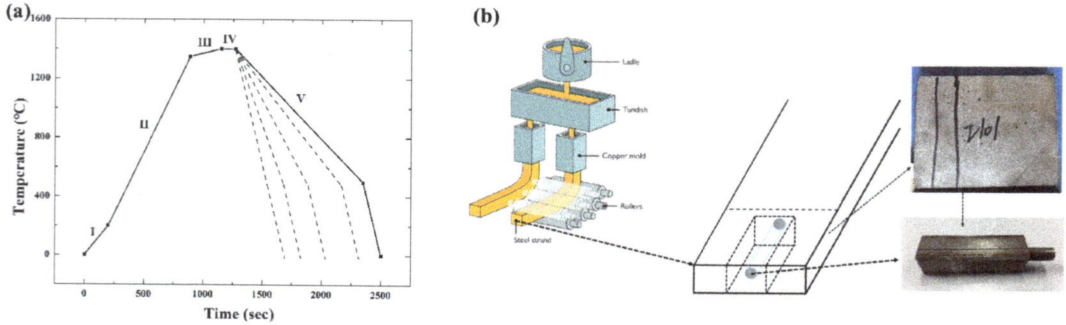

Figure 2. (**a**) Temperature profile of concentric HTCM experiments. I—Stabilizing stage, II—Fast heating stage, III—Slow heating stage, IV—Holding stage, and V—Cooling stage. Different dot lines represent different cooling rates; (**b**) schematic diagram and pictures of S32101 DSS samples.

2.1.2. Materials

The samples used in present research were provided by Baosteel (as shown in Figure 2b), named 1.4162 (X2CrMnNiN21-5-1) according to the EN standard. This grade of steel is widely used in extreme environments which require both excellent mechanical properties and high resistance to intergranular and pitting corrosion. The chemical composition of S32101 DSS is given in Table 1. All the samples were cut into slices. The diameter of a single slice was 9.7 mm, and the thickness was 0.25 mm. To prepare samples, a TechCut 5TM precision high-speed saw with water cooling was used to cut the sample into specific thicknesses before grinding on 800, 1200, 2400, and 4000 SiC papers. The samples were polished with standard metallographic techniques using a 6 μm diamond paste and 1 μm diamond paste. A fine sample surface can provide better images, especially in the heating stage or for solid-state studies.

Table 1. The chemical composition of S32101 (wt.%).

Grade	C	Mn	Cr	Ni	Mo	Cu	N
S32101	<0.04	5	21.5	1.5	0.5	0.5	0.2

2.2. *Modeling Approaches*

2.2.1. Thermodynamics: Thermo-Calc

Released over 30 years ago, Thermo-Calc has become one of the most popular thermodynamic calculation software packages in the world. Thermo-Calc simulations can improve design during manufacturing and aid in selecting heat treatment temperatures. Calculations can predict experimental outcomes and minimize the number of experiments. In this study, the 2021b version of Thermo-Calc with the TCFE9 database was used for the calculation of phasors and their compositions, transition temperatures, and phase diagrams.

Another use of Thermo-Calc was the generation of optimized thermodynamic and kinetic datasets encapsulated in a GES5 file. This was applied in the subsequent phase-field simulation as a basis for microstructure development.

2.2.2. Phase-Field Simulation: Micress

At the interfaces, the phase-field variables change continuously over an interface thickness η, which could be significantly large compared with the atomic interface thickness and tiny compared to the microstructure length scale. Their time evolution is calculated by a set of phase-field equations derived from the minimization of the free energy functional [16].

$$\dot{\varnothing}_\alpha = \sum_{\beta\neq\alpha}^{v} M_{\alpha\beta}\left[\frac{\sigma_{\alpha\beta}}{v}\left(\frac{1}{2}(\nabla^2\varnothing_\alpha - \nabla^2\varnothing_\beta) + \frac{\pi^2}{2\eta_{\alpha\beta}^2}(\varnothing_\alpha - \varnothing_\beta)\right) + \sum_{\gamma\neq\alpha\neq\beta}^{n} J_{\alpha\beta\gamma} + |\nabla\varnothing|\Delta_{\alpha\beta}\right], \quad (1)$$

$$J_{\alpha\beta\gamma} = \frac{1}{2}\left(\sigma_{\beta\gamma} - \sigma_{\alpha\gamma}\right)\left(\frac{\pi^2}{\eta^2}\varnothing_\gamma + \nabla^2\varnothing_\gamma\right), \quad (2)$$

$$M_{\alpha\beta} = \widetilde{M}_{\alpha\beta}\frac{8\eta}{\pi^2}, \quad (3)$$

where $J_{\alpha\beta\gamma}$ denotes the higher-order terms accounting for the interaction with additional phases in triple- or multiphase regions.

For the sake of simplicity, only the isotropic formulation is shown here. In Equation (3), $M_{\alpha\beta}$ is the mobility of the interface as a function of the interface orientation, $K_{\alpha\beta}$ is related to the local curvature of the interface, and $\sigma^*_{\alpha\beta}$ is the anisotropic surface stiffness. The thermodynamic driving force and the solute partitioning are calculated separately using the quasi-equilibrium approach [16] and are introduced into the equation for the multiphase fields (Equation (3)). If concentration coupling is activated, an explicit 1D temperature field in the z-direction can be optionally defined by the keyword '1d_temp'. With the 1D temperature field, heat flow and latent heat release are solved explicitly. For numerical reasons, the temperature in the 1D temperature field is calculated using a direct, explicit solver with a default kinetic coefficient [17]. The default value is preconfigured for typical casting processes and needs not to be changed in most cases.

The temperature field can be evaluated in a linear (cartesian), cylindrical (cylindrical), or polar (polar) coordinate system; however, in the present research, a linear setup was used. Furthermore, we ran simulations at 0 mm, 10 mm, and 50 mm in the cast slab.

As shown in Table 2, the thermodynamic data were provided by online coupling to the database TCFE9 using the elements given in Table 1 and the phases in Table 2. Similarly, diffusion data for γ and δ, including cross-terms, were taken from the MOBFE4 mobility database.

Table 2. Phases and phase-related parameters.

No.	Database ID	ID	Diffusion Data	$\mu_{1/x}$ (cm^4/Js)	$\mu_{2/x}$ (cm^4/Js)	$\sigma_{1/x}$ (J/cm^2)	$\sigma_{2/x}$ (J/cm^2)
1	FCC_A1	γ	MOBFE4		Diff Limit		0.8×10^{-4} to 1.2×10^{-4}
2	BCC_A2	δ	MOBFE4	Diff Limit		0.8×10^{-4}	
3	Liquid	L	Fixed	Diff Limit	1×10^{-5} to 1×10^{-10}	0.8×10^{-4} to 1.2×10^{-4}	0.3×10^{-4} to 1.2×10^{-4}

3. Results and Discussion

DSSs have a so-called dual-phase structure, which, for S32101, contains about 50% volume face-centered cubic (FCC) γ islands and 50% volume body-centered cubic (BCC) δ matrix. After the samples were ground and finally polished, the surface of S32101 DSSs was observed with an optical microscope using a 0.04 µm particle size silica gel cloth. For the S32101 DSS, as shown in Figure 3, even without etching, the dual-phase structure was prominent, with almost equal proportions for each phase.

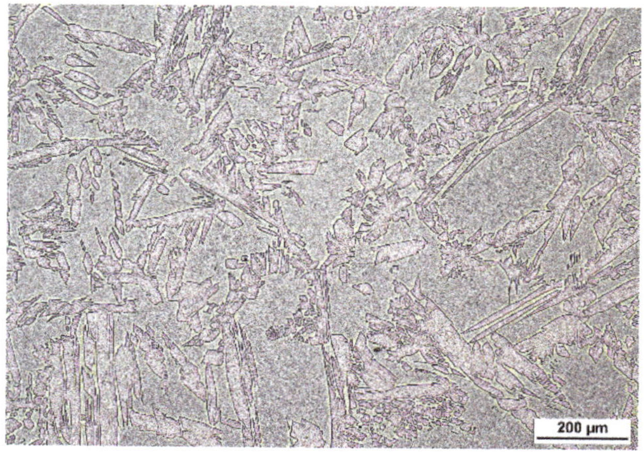

Figure 3. Optical microscopy of S32101 DSS cast with δ matrix and γ island structures.

A property diagram, which reveals the response of all phases to temperature change, was calculated using Thermo-Calc under equilibrium conditions in the present research. The phase equilibria were calculated with a temperature range from 500 °C to 1500 °C. As shown in Figure 4, as temperature decreased, the liquid phase first solidified to δ, and, after solidification was finished, δ remained for a while and then partially transformed to γ. As temperature decreased, under equilibrium solidification conditions, the precipitation of secondary phases, such as chromium nitrides, σ phase, and χ phase, proceeded.

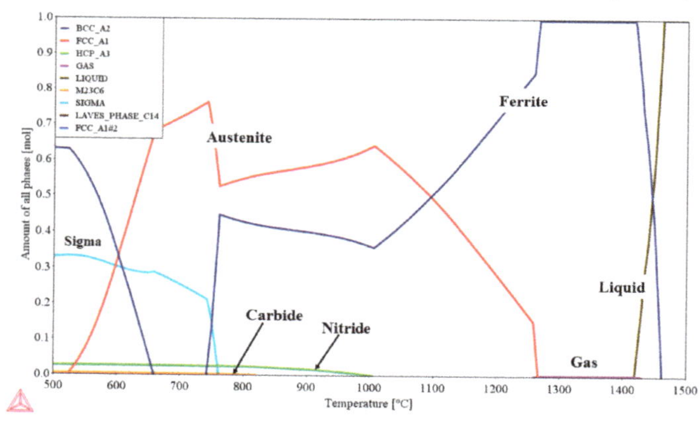

Figure 4. Property diagram of S32101 DSS calculated using Thermo-Calc showing the changes in all phases with temperature under equilibrium conditions.

According to the thermodynamic calculation of S32101 DSS, the δ-to-γ phase transformation occurred around 1270 °C. As shown in the figure, the gas phase formed after the steel fully solidified and before the commencement of δ-to-γ solid-state phase transformation, approximately in the range of 1280–1430 °C. For S32101 DSS, nitrogen is used as an alloying element, and its content is very high. The solubility of nitrogen was the most significant factor in the gas pore formation. The solubility of nitrogen in the solid phase is much smaller than in liquid. During the solidification process, the nitrogen was rejected from the liquid phase and formed nitrogen gas pores. The nitrogen gas pore formation was also influenced by nitrogen content, cooling rate, partial pressure, etc. Hence, the temperature range of gas-phase formation is not fixed.

3.1. Concentric HTCM Experimental Results

The HTCM results are discussed according to the timeframe from heating to cooling, and the time is counted from the start of the experiment. The temperature values shown in the concentric HTCM images are those of the sample edges. Since the diameter of the sample was 9.7 mm, the temperature of the center was higher than that of the edge.

3.1.1. Heating Stage

In Figure 5, a red circular area P is used as a reference marker. As shown in frames (a) and (b), the color of δ darkened, making it straightforward to distinguish γ and δ in this period. Frame (c) shows that the morphology of γ changed with the temperature rising. The γ and δ interface boundaries were apparent. The γ → δ phase transformation was not clear during the heating stage except for the morphology change. Frames (d) and (e) show the growth of δ/δ grain boundaries (GBs). The schematic diagram of frame (f) shows the morphology of γ. The γ accumulated along the grain boundaries of δ with a finger- or needle-like pattern. The intragranular γ-cells maintained an island pattern in the δ matrix. The grain boundary motion was in the form of a γ → δ phase transformation.

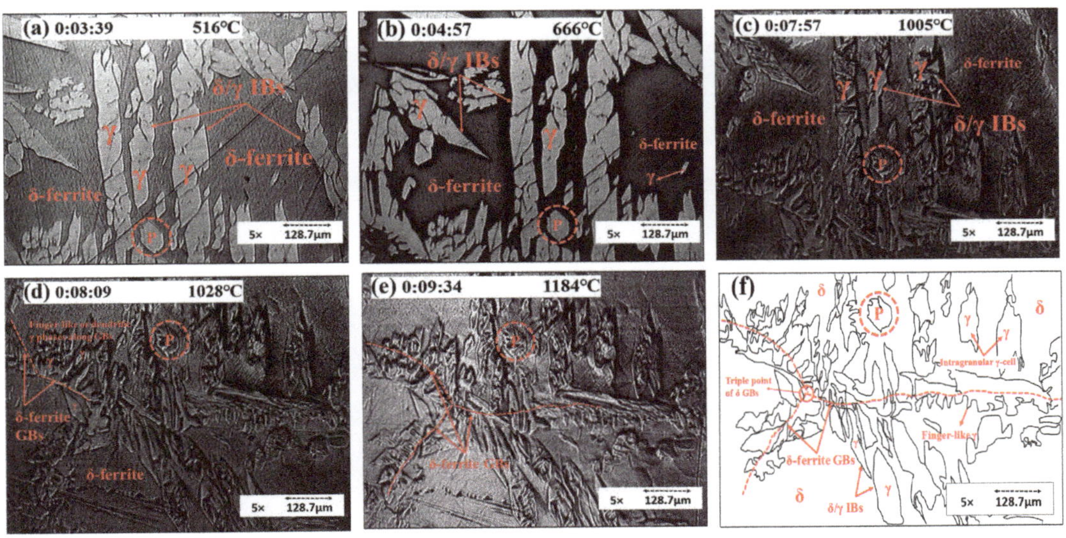

Figure 5. Concentric HTCM images revealing (**a–c**) the morphology changes of the γ phase and (**d–f**) the growth of δ/δ GBs; (**f**) a schematic diagram of the γ morphology of S32101 DSS.

As the temperature increased, γ fully transformed to δ, and δ GBs showed up. In Figure 6, the red dotted line represents the δ/δ GBs, while the white dotted line represents the solid/liquid IBs. In frame (b), a new δ GB showed up; after the growth of new δ GBs, the sample started melting, as shown in frame (c). The liquid phase in frame (d) was located along the δ GBs.

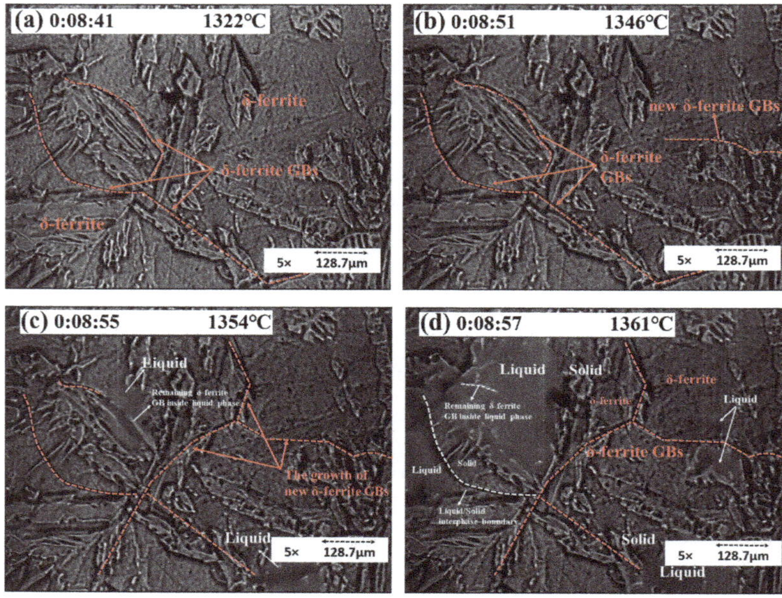

Figure 6. Concentric HTCM images revealing (**a–c**) the growth of δ GBs before melting and (**d**) the location of the liquid phase.

As the temperature increased, γ transformed to δ. The GBs enlarged and slipped before melting, and thermal etching appeared, as shown in Figure 7. These phenomena indicated that the steel would melt soon.

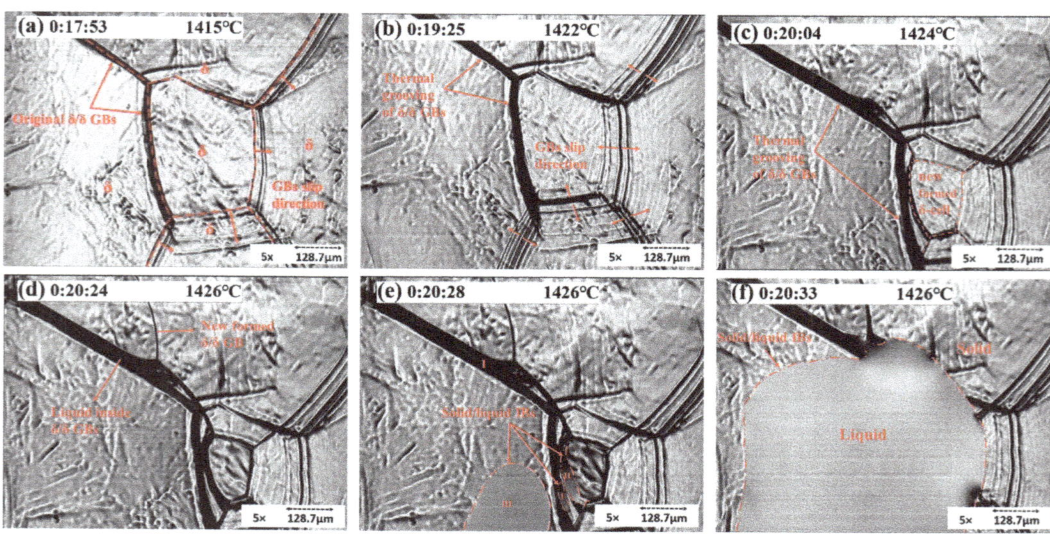

Figure 7. Concentric HTCM images revealing the δ/δ GBs enlarging and slipping (**a–c**) and S32101 DSS melting (**d–f**).

When polished samples are heated to high temperatures and exposed to an inert atmosphere (e.g., argon gas in present experiments), the GBs will slip. Figure 7 shows the thermal etching in the solid rim of concentric experiments. As indicated in Figure 7b, grooves formed at elevated temperatures and remained intact after cooling. Figure 7 also reveals mobile grain boundaries and stationary grain boundaries. In frame (a), the GBs slipped in one direction and later toward the opposite direction (frame (b)). Moreover, the stationary δ/δ GBs enlarged, which is called thermal grooving. The slipping of GBs is called thermal etching, leading to the formation of a new δ-cell, as shown in frame (c). After the temperature reached the melting point, the liquid phase firstly showed up inside the stationary δ GB. Frames (e) and (f) show the commencement and growth of the liquid phase. Areas I, II, and III represent the liquid phase in different places. I indicates the liquid phase inside GBs, II indicates the liquid within two δ GBs, and III is the inside of a δ-cell. Later, all the liquid phases merged into a liquid pool. The concentric solidification experiment improved the quality of in situ observations due to the minimization of the meniscus and the increase in the length of the visual interface between liquid and solid [15].

3.1.2. Solidification Process

In the continuous casting process, the quality of products is highly related to the grain structure during solidification, including grain size, morphology, and orientation [18]. The cooling rate can influence the solidification behavior and the solid shapes during the solidification process.

As shown in Figure 8, with a cooling rate of 5 °C/min, the sample solidified initially with a planar solidification front due to the required time for solute gradients to form as the solid grew and for instability to commence. After this initial planar solidification, the structure destabilized into a cellular solidification structure. As shown in frame (a), the sample started by solidifying from outside the liquid pool to inside gradually, and the solid/liquid interface boundary (S/L IB) moved slowly. The distance of S/L IB motion (D) under 5 °C/min was around 588 μm. Frame (b) shows that the δ solid grew with a columnar structure. The GBs were easy to observe, and the grain size was large. As shown in frame (c), the cellular structure could be observed as dendrites without secondary arms. Because of the prolonged cooling rate, the growth of primary arms was even toward the center of the liquid pool, with no space for secondary arms to grow. As shown in frame (d), the cellular structures were compact, and the content of alloy elements and gas precipitates was low; hence, gas pores were unlikely to form in this area. The temperature gradients in the liquid promoted the growth of cellular grains. The cooling rate also played an essential role in the growth of a columnar structure. As the cooling rate increased, the length of the columnar structure decreased.

Primary dendritic arms grew beneath the liquid surface, but secondary dendritic arms (SDAs) growing perpendicular to the growth direction appeared as a series of aligned islands at the liquid surface. When the cooling rate increased to 10 °C/min after the initial periods of planar and cellular solidification, there was a transition to dendritic solidification, as shown in frames (a–c). Solidification started in the S/L IB. As shown in frame (a), the D of S/L IB was 380.25 μm; right after the growth of the columnar δ, the secondary dendritic arms of δ commenced. In frames (b) and (c), the growth of primary and secondary dendritic arms can be observed. The appearance of SDAs followed the growth of a primary dendrite, and the growth of secondary arms was orientated. The arrows in frame (b) point to the preferred growth direction of primary and secondary arms. The microstructure under a 10 °C/min cooling rate showed large graininess and dendritic shapes.

For the sample cooled at 50 °C/min, as shown in Figure 9d, the D was 87.75 μm. As shown in frame (e), the dendrites grew in the center of the liquid pool and were surrounded by several δ GBs, which can be observed clearly in frame (f). Under the cooling rate of 50 °C/min, the grain size of dendritic and cellular δ was still significant.

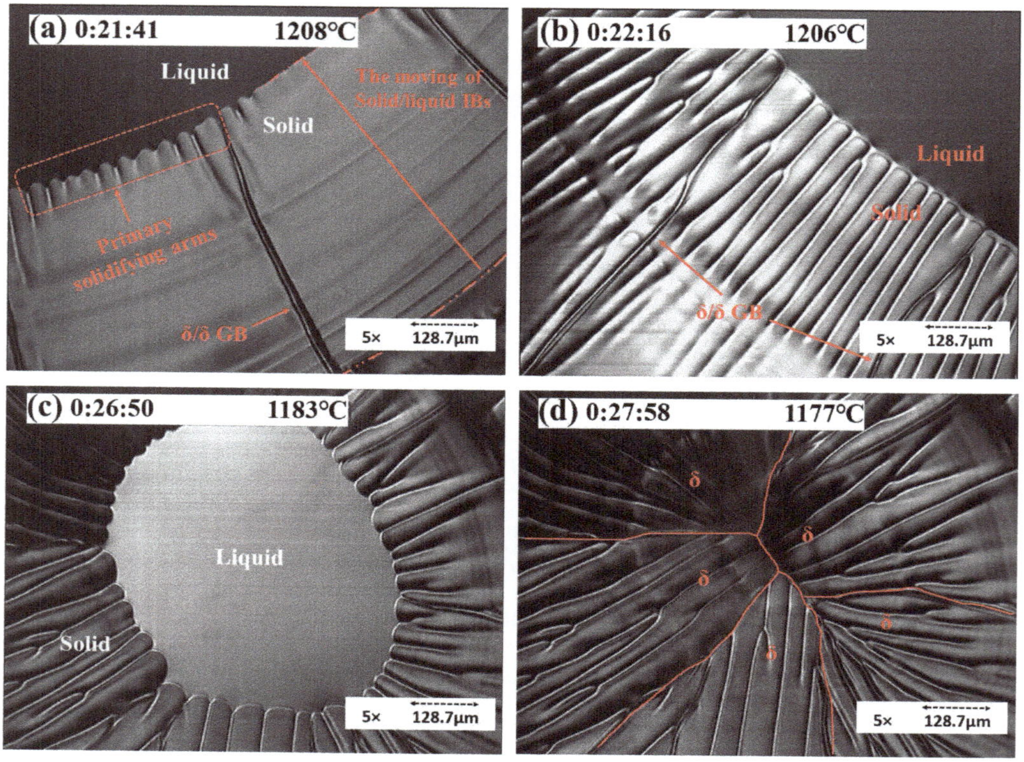

Figure 8. Concentric HTCM revealing the solidification process of S32101 under a 5 °C/min cooling rate: (**a**) the movement of S/L IB, (**b**) the growth of solidifying arms, (**c**) the residual liquid pool, and (**d**) the morphology of δ solids (cellular growth).

For the sample cooled at 80 °C/min, as shown in Figure 9g–i, solidification started in the S/L IB, and SDAs emerged from outside the liquid pool to the inside within 10 s. The microstructure comprised cellular structure domains with a small number of needle shape particles and no more dendritic microstructure. In frame (g), the D was 93.6 μm; right after the commencement of dendritic δ, the δ solids grew along the S/L IB. In frame (h), within the liquid pool of the sample, the solidification proceeded and then grew into a long and slender needle-like or dendritic pattern (frame (i)).

For the sample with a 150 °C/min cooling rate, as shown in Figure 9j–l, the region of cellular solidification was no longer observed. Instead, a small band of planar solidification destabilized directly to dendritic solidification with a smaller primary and secondary arm spacing, and solidification first started along the S/L IB. The solids simultaneously emerged both at the center of the liquid pool and at the edge of the S/L IB. The morphology comprised grains of finer cellular shape without evidence of a dendritic microstructure. The grain size became finer and smaller than outside of the liquid pool. In frame (a), the D was 76.05 μm, and the columnar δ stopped growing shortly after. The nucleation started inside the liquid pool and along the S/L IB shortly after. In frame (k), a number of secondary dendritic arms commenced inside the center liquid pool, whereas, in frame (l), all solid δ showed up with a cellular structure.

These results allowed studying the characteristics of the solidification microstructure response to thermal conditions. They demonstrated that concentric solidification offers a good platform for studying duplex stainless steels, forming a good testbed for subsequent simulations.

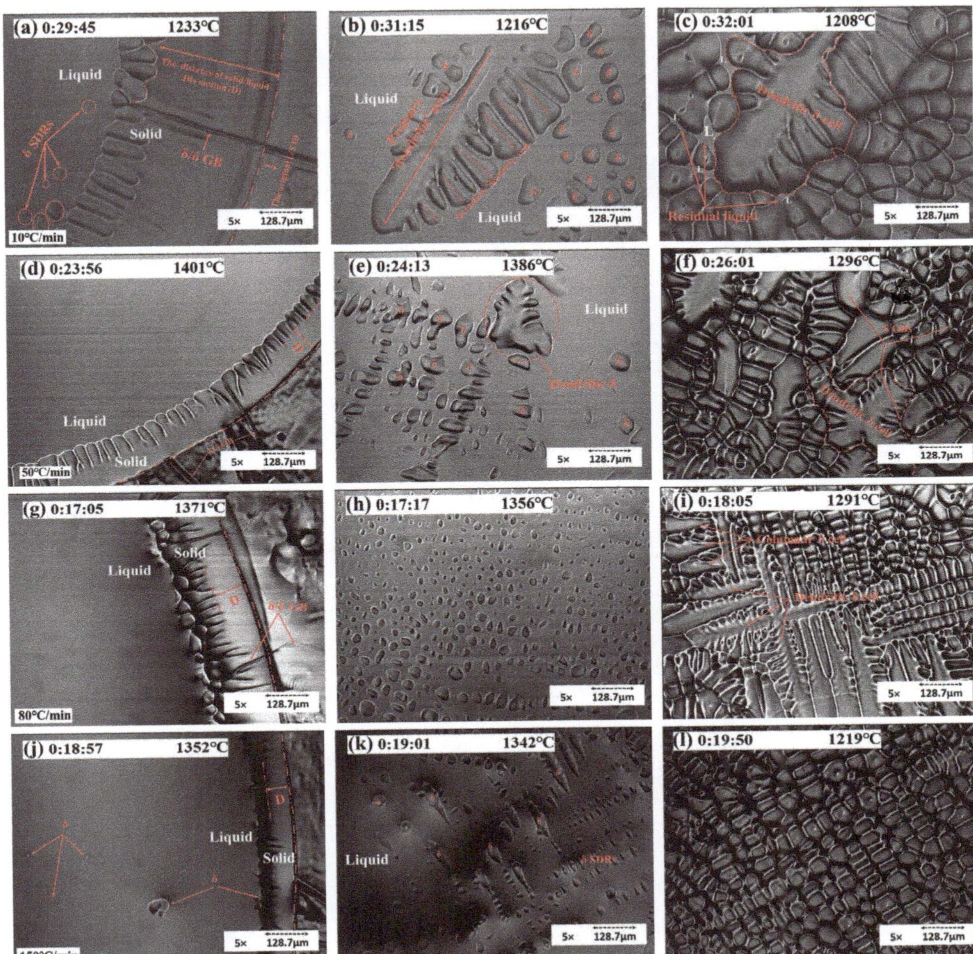

Figure 9. Concentric HTCM revealing the solidification process of S32101 under the cooling rates of (**a–c**) 10 °C/min (**d–f**) 50 °C/min (**g–i**) 80 °C/min, and (**j–l**) 150 °C/min.

3.2. Micress Simulation

The most significant development of the present research was establishing a framework to utilize the combination of experiments and simulation to study the solidification process of S32101 DSS. This simulation framework can reduce the experimental workload dramatically. Concentric HTCM was also used to authenticate the accuracy of the simulation.

Micress simulations followed the same conventions, with the bottom of the domain oriented in the direction of the slab surface and the top of the domain pointing to the center of the slab. Figure 10 shows the selected outputs from the region of the chill surface where (a) SDAs formed and started to grow, (b) and (c) dendrite competition and primary dendrite arm spacing selection occurred, and (d) the dendrite tips reached the end of the domain; from this point until the end of the simulation run, secondary arm coarsening and final solidification of the liquid were achieved.

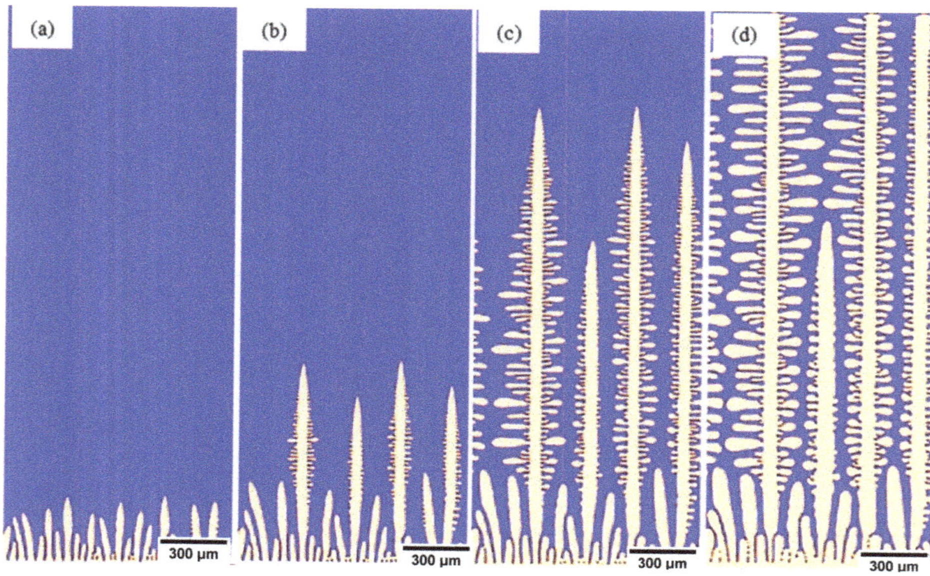

Figure 10. Micress simulation of S32101 chill surface with 100 °C superheat. (**a**) SDAs formation, (**b**) dendrite competition, (**c**) primary dendrite arm selection, and (**d**) dendrite tip across domain.

For each alloy, composition simulations were linked to a 1D heat transfer model allowing simulations to be run at a different position in the slab. In Figure 11, three such locations are used to illustrate this: (a) the chill surface, (b) 10 mm inside the slab, and (c) 50 mm inside the slab. The pertinent features are the coarsening of the microstructure; further into the slab, the simulation indicated that reducing heat transfer conditions led to a thicker solidified shell. Here, the primary difference was in the primary arm spacing increasing as the position moved into the cast slab. These simulations were further contrasted at a later point in the solidification progression in Figure 12 to highlight the impact on secondary arm spacing development in response to position in the slab ((a) chill surface, (b) 10 mm, and (c) 50 mm).

These results implied segregation into the liquid as solidification proceeds. As highlighted in the Thermo-Calc calculations, the segregation of elements is critical in issues such as nitrogen gas bubble formation. The results from a simulation of S32101 showing the segregation of each of the seven elements nominated in the project are shown in Figure 13. Micress provides the opportunity to probe how heat transfer conditions, microstructural development, and segregation interrelate. In the figure, the information is presented graphically with differences in concentration displayed by changing color levels. The underlying data at each of the 400,000 grid points is also accessible in Micress and can be manipulated using the Micress analysis software or extracted to conventional spreadsheets such as Excel for further analysis. In this way, the composition of the liquid phase during solidification can be plotted as a function of temperature (Figure 14).

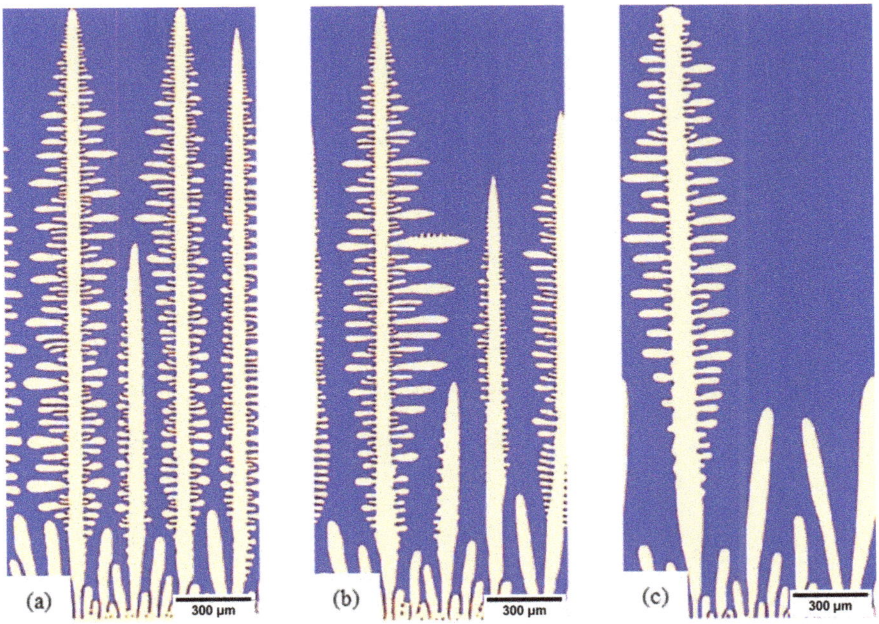

Figure 11. S32101, 100 °C superheated microstructure when growing dendrites reached the top of the simulation domain: (**a**) chill surface, (**b**) 10 mm in the slab, and (**c**) 50 mm in the slab.

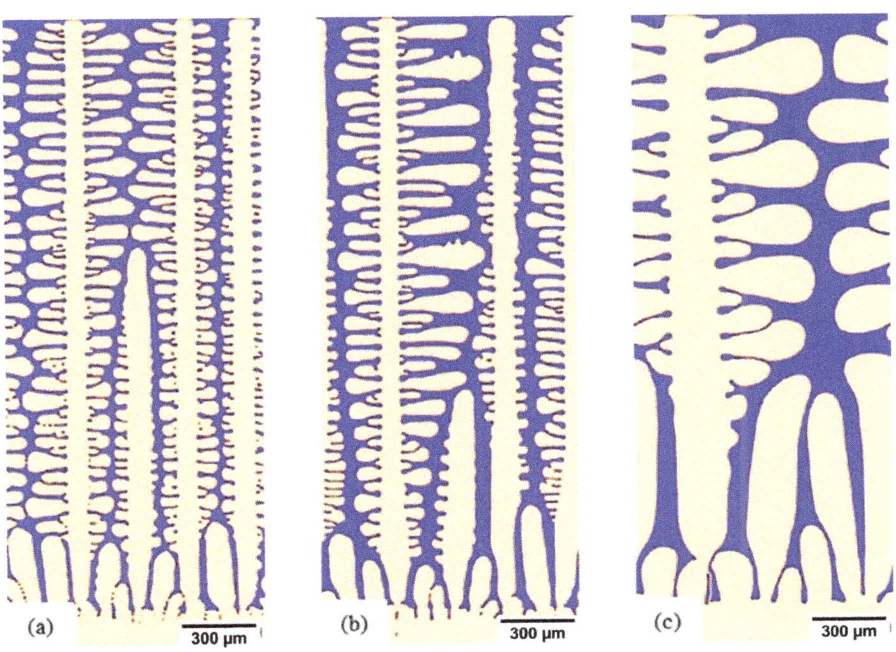

Figure 12. S32101, 100 °C superheated microstructure showing the development of SDAs: (**a**) chill surface, (**b**) 10 mm in the slab, and (**c**) 50 mm in the slab.

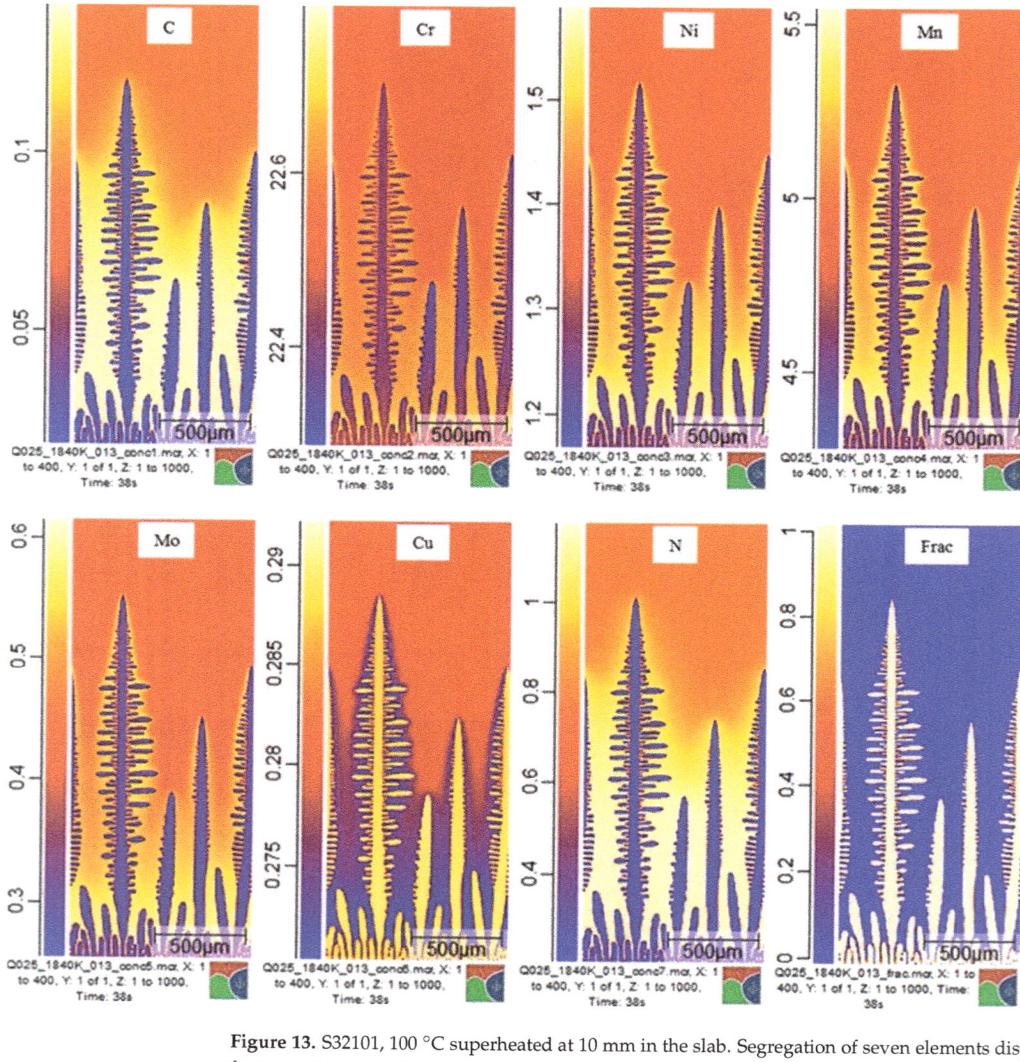

Figure 13. S32101, 100 °C superheated at 10 mm in the slab. Segregation of seven elements displayed from one timestep of a simulation run.

Running Micress simulations enables to determine the effects of changing process parameters on the subsequent solidification. For example, by testing two superheating temperatures on the casting of S32101, we could extract relationships between fraction liquid as a function of temperature and the nitrogen segregation in the liquid (Figure 15). In this case, increasing the superheat led to a lower nitrogen composition in the liquid at a given temperature compared to superheating at 50 °C. As the nucleation of nitrogen gas bubbles is related to the content of dissolved nitrogen in the liquid, such an analysis offers opportunities for probing defect formation minimization.

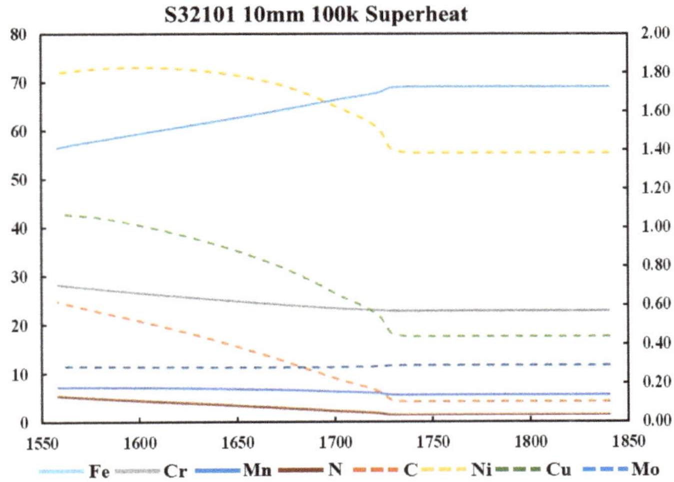

Figure 14. S32101, 100 °C superheated at 10 mm. The liquid phase's average composition as a temperature function. Solid lines, left-hand axis; dashed lines, secondary axis.

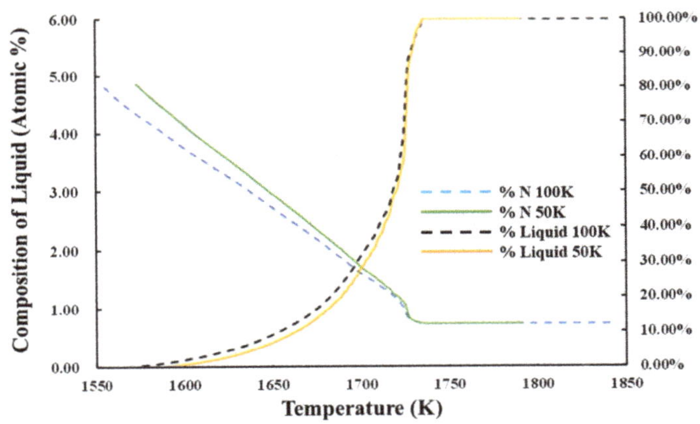

Figure 15. S32101 comparison of segregation of carbon and nitrogen in the liquid during solidification as a function of superheat (50 °C and 100 °C).

The simulation domains comprised 400,000 grid points, and Micress outputted data on the position of the interface (in this case, solid and liquid). The average occupancy of grid points is directly related to the fineness of a given microstructure. Therefore, we can compare the influence of melt superheat on the resulting microstructure. In Figure 16, it can be observed that a higher superheat led to a coarser microstructure. Alternatively, the simulations can address heat transfer conditions such as heat flux Q at the slab surface. The base simulations were conducted with a fixed heat transfer coefficient of 0.25 $W/cm^2{\cdot}K$, in line with conventional casting practices. A comparative simulation for S32101 with a superheating temperature of 50 °C but with the heat transfer coefficient increased to 0.35 $W/cm^2{\cdot}K$ was also run, and the simulation results at 10 mm in the slab are included in Figure 16. The critical point is that, through incorporating a 1D heat transfer model into the Micress simulations of solidification, cooling conditions in the caster can be adjusted consistently, allowing direct comparison to, for example, the measure of microstructure refinement that arises from increasing the heat transfer coefficient. However, all the other information entailed in the simulations, such as element segregation, secondary phase nucleation, and growth, can also be assessed simultaneously.

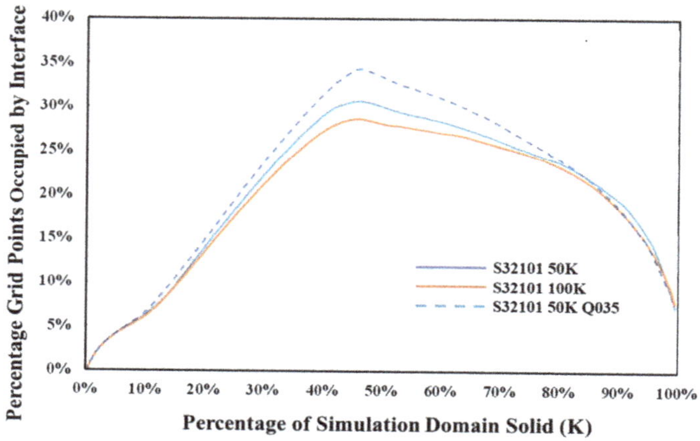

Figure 16. Influence of superheat on microstructural fineness of S32101, 50 °C and 100 °C superheat at 10 mm in the cast slab, when increasing heat flux at the boundary.

4. Conclusions

The concentric HTCM method was employed to study the heating and cooling stages of S32101 DSS. During the heating process, the GB presented $\gamma \rightarrow \delta$ phase transformation. As the temperature increased, γ fully transformed to δ. The GBs enlarged and slipped before melting, and thermal etching appeared. These phenomena indicate that the steel would melt soon. The liquid phase preferably formed inside of or close to the δ GBs. During the solidification process, the cooling rate affected the microstructure of S32101 DSS; as the cooling rate increased, the solid's size was finer, the morphology comprised smaller grains, and the solidification behavior in the body of the liquid pool commenced more quickly. The distance (D) of S/L IB motion decreased as the cooling rate increased. The size of the dendritic structure of δ decreased until it disappeared. The results demonstrate that concentric solidification offers a good platform for studying DSSs, forming a good testbed for subsequent simulations. The growth of precipitates was observed.

The experimental approach based on in situ HTCM was found to be helpful in studying the solidification process of S32101 DSS. Moreover, the data can be used as a benchmark for simulation. A simulation framework using Micress linked to a 1D heat transfer model that enables consistent analysis of solidification dynamics in duplex stainless steels across the whole cast slab was established. The analysis of phase transformations, solute segregation, and microstructural fineness, among other aspects, was demonstrated.

The establishment of the Micress framework to simulate the solidification process can dramatically address the limitations of experiments and reduce the experimental workload. Micress offers the opportunity to explore how heat transfer conditions, microstructural development, and separations relate. By revealing the elemental segregation of S32101 DSS under the cooling rates of 5 and 150 °C/min, Micress was proven to be a powerful tool to tackle the problems encountered in production.

Author Contributions: Conceptualization, T.W. and Y.W.; methodology, T.W.; software, T.W.; validation, D.W. and H.L.; formal analysis, T.W.; investigation, T.W.; resources, L.G.; writing—original draft preparation, T.W.; writing—review and editing, H.L. and D.W.; supervision, H.L. and D.W.; project administration, L.G.; funding acquisition, H.L. and L.G. All authors have read and agreed to the published version of the manuscript.

Funding: This research was funded by Baosteel-Australia Joint Research and Development Centre (BAJC), grant number BA17010.

Institutional Review Board Statement: Not applicable.

Informed Consent Statement: Not applicable.

Data Availability Statement: Data sharing is not applicable. No new data were created or analyzed in this study.

Conflicts of Interest: The authors declare no conflict of interest.

References

1. Sieurin, H.; Sandström, R.; Westin, E.M. Fracture toughness of the lean duplex stainless steel LDX 2101. *Metall. Mater. Trans. A* **2006**, *37*, 2975–2981. [CrossRef]
2. Escriba, D.; Materna-Morris, E.; Plaut, R.L.; Padilha, A.F. Intermetallic phase precipitation in duplex stainless steels during high temperature exposition. *Proc. Mater. Sci. Forum* **2010**, *636–637*, 478–484. [CrossRef]
3. Mazumdar, S.; Ray, S. Solidification control in continuous casting of steel. *Sadhana* **2001**, *26*, 179–198. [CrossRef]
4. Zhao, Y.; Sun, Y.; Li, X.; Song, F. In-situ Observation of $\delta \leftrightarrow \gamma$ Phase Transformations in Duplex Stainless Steel Containing Different Nitrogen Contents. *ISIJ Int.* **2017**, *57*, 1637–1644. [CrossRef]
5. Böttger, B.; Apel, M.; Eiken, J.; Schaffnit, P.; Steinbach, I. Phase-Field simulation of solidification and solid-state transformations in multicomponent steels. *Steel Res. Int.* **2008**, *79*, 608–616. [CrossRef]
6. Zhang, L.; Steinbach, I.; Du, Y. Phase-field simulation of diffusion couples in the Ni–Al system. *Int. J. Mater. Res.* **2011**, *102*, 371–380. [CrossRef]
7. Minsky, M. Microscopy Apparatus. U.S. Patent 3,013,467, 19 December 1961.
8. Yin, H.; Emi, T.; Shibata, H. Morphological instability ofδ-ferrite/γ-austenite interphase boundary in low carbon steels. *Acta Mater.* **1999**, *47*, 1523–1535. [CrossRef]
9. Chikama, H.; Shibata, H.; Emi, T.; Suzuki, M. "In-situ" real time observation of planar to cellular and cellular to dendritic transition of crystals growing in Fe–C alloy melts. *Mater. Trans. JIM* **1996**, *37*, 620–626. [CrossRef]
10. Shibata, H.; Yin, H.; Yoshinaga, S.; Emi, T.; Suzuki, M. In-situ observation of engulfment and pushing of nonmetallic inclusions in steel melt by advancing melt/solid interface. *ISIJ Int.* **1998**, *38*, 149–156. [CrossRef]
11. Wang, Y.; Sridhar, S. Liquid Al_2O_3-CaO-MgO inclusion entrapment at delta ferrite intercellular boundaries during solidification. *Ironmak. Steelmak.* **2003**, *30*, 223–228. [CrossRef]
12. Griesser, S.; Reid, M.; Bernhard, C.; Dippenaar, R. Diffusional constrained crystal nucleation during peritectic phase transitions. *Acta Mater.* **2014**, *67*, 335–341. [CrossRef]
13. Griesser, S.; Bernhard, C.; Dippenaar, R. Effect of nucleation undercooling on the kinetics and mechanism of the peritectic phase transition in steel. *Acta Mater.* **2014**, *81*, 111–120. [CrossRef]
14. Niknafs, S.; Phelan, D.; Dippenaar, R. High-temperature laser-scanning confocal microscopy as a tool to study the interface instability during unsteady-state solidification of low-carbon steel. *J. Microsc.* **2013**, *249*, 53–61. [CrossRef] [PubMed]
15. Reid, M.; Phelan, D.; Dippenaar, R. Concentric solidification for high temperature laser scanning confocal microscopy. *ISIJ Int.* **2004**, *44*, 565–572. [CrossRef]
16. Eiken, J.; Böttger, B.; Steinbach, I. Multiphase-field approach for multicomponent alloys with extrapolation scheme for numerical application. *Phys. Rev. E* **2006**, *73*, 066122. [CrossRef] [PubMed]
17. Böttger, B.; Eiken, J.; Apel, M. Phase-field simulation of microstructure formation in technical castings–A self-consistent homoenthalpic approach to the micro–macro problem. *J. Comput. Phys.* **2009**, *228*, 6784–6795. [CrossRef]
18. Phelan, D. In-Situ Studies of Phase Transformations in Iron Alloys. Ph.D. Thesis, University of Wollongong, Wollongong, Australia, 2002.

materials

MDPI

Article

Influence of Mineralogical Structure of Mold Flux Film on Heat Transfer in Mold during Continuous Casting of Peritectic Steel

Lei Liu, Xiuli Han *, Mingduo Li and Di Zhang

College of Mining Engineering, North China University of Science and Technology, Tangshan 063210, China; heutliulei@163.com (L.L.); limingduo@ncst.edu.cn (M.L.); ncstdearz@163.com (D.Z.)
* Correspondence: hanxl1965@126.com

Abstract: The mineralogical structure of flux films is a critical factor in controlling heat transfer in the mold and avoiding the longitudinal cracking of slabs during the continuous casting of peritectic steel. In this study, the layered structure, crystallization ratio, mineralogical species, and morphology features of flux films were characterized by optical microscopy, X-ray diffraction, and electron-probe microanalysis. Microstructural observation revealed that the normal flux films for peritectic steels present a multilayered structure and high crystallization ratio (60~90 vol%), mainly composed of well-developed crystalline akermanite and cuspidine. In contrast, the films with outstanding flux characteristics with abundant longitudinal cracks on the slab surface have a low crystallization ratio (<50 vol%) or vast crystallite content (>80 vol%). Furthermore, heat transfer analysis showed that the low crystallization ratio and the vast crystallite content of flux films worsen the heat transfer rate or uniformity in the mold, whereas the appropriate thickness and cuspidine content of flux films can improve the heat transfer performance. From the above results, it is concluded that using strong crystalline flux to obtain the ideal mineral phase structure of flux film is one of the important measures for reducing longitudinal cracks during continuous casting of peritectic steel slabs.

Keywords: crystallization ratio; microstructure; flux film; heat transfer; longitudinal crack; peritectic steels

Citation: Liu, L.; Han, X.; Li, M.; Zhang, D. Influence of Mineralogical Structure of Mold Flux Film on Heat Transfer in Mold during Continuous Casting of Peritectic Steel. *Materials* **2022**, *15*, 2980. https://doi.org/10.3390/ma15092980

Academic Editor: Adam Grajcar

Received: 17 March 2022
Accepted: 13 April 2022
Published: 20 April 2022

1. Introduction

Peritectic steel is widely used in manufacturing high-strength automobile plates and ship boards with excellent properties of thermoplasticity and machinability [1,2]. Because of the peritectic reaction during continuous casting of a peritectic steel slab, volume contraction and thermal stress congestion of the solidification shell always occur, more easily causing nonuniform heat transfer and longitudinal cracks [3–6]. Thus, avoiding longitudinal cracks during the continuous casting of peritectic steel slabs remains a challenge.

In order to improve the slab quality, a large amount of research has been carried out on steel solidification characteristics, continuous casting process conditions, and continuous casting flux [7–10]. Whereas it can be difficult to change the steel solidification characteristic and the casting process condition, the mold flux has become a central factor preventing slab quality defects [11–15]. In-depth knowledge regarding the heat transfer mechanism of mold flux films is becoming increasingly essential. Shibata et al. [16–18] suggested the average thickness of mold flux films for various steel grades can be 0.5–1.75 mm and the interfacial thermal resistance increases with increasing flux film thickness. Tsutsumi et al. [19–22] measured the surface roughness of solidified flux films using a stylus surface profiler and analyzed the surface roughness near the mold to correlate and evaluate the air gap formation. Choi et al. [23–28] considered the crystallization behavior of mold fluxes to be vital in controlling the overall heat transfer in the mold by means of the mold simulator, differential thermal analysis, hot thermocouple technique, and confocal laser scanning microscope. Furthermore, many researchers found that the crystallization ratio of mold flux is a key factor in controlling the heat transfer in the mold [29–32]. However, the effect of mineralogical structure on the heat transfer property of flux films has not been investigated through a systematic approach.

In this work, the microstructure characteristics of flux films matched for peritectic steels were investigated by optical microscopy, X-ray diffraction (XRD), and electron-probe microanalysis (EPMA). Then, the effect of mineralogical structure on the heat transfer property of flux films was analyzed. The aim of the present work is to ravel out the relationship between the mineralogical structure of flux films and the longitudinal cracks of slabs, and then to select the suitable mold flux for avoiding the longitudinal cracks during continuous casting of peritectic steel slabs. The results will provide a theoretical basis for optimizing mold flux and improving slab quality.

2. Experimental Section

2.1. Materials

Continuous casting experiments were carried out at Hebei Iron and Steel Group Co., Ltd. in Shijiazhuang, China. Samples of flux films (see Figure 1) were taken from the meniscus region (see Figure 2).

Figure 1. Samples of solidified flux films.

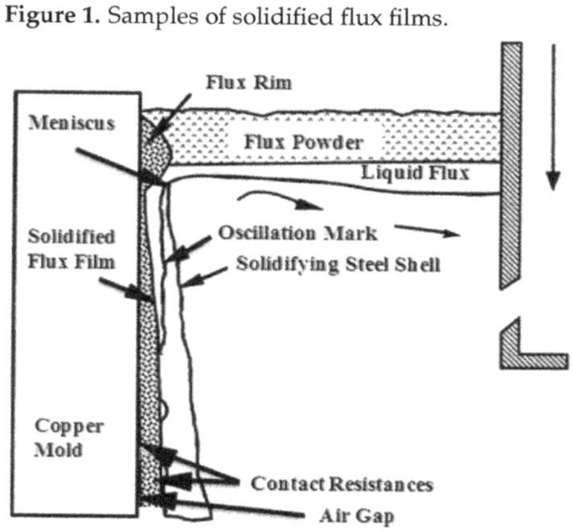

Figure 2. Position diagram of flux films in mold.

Combined with the surface quality of the slab, four typical samples of flux films were selected. Sample No. 1 represents flux films with normal slab quality for peritectic steel A, and sample No. 2 represents flux films with longitudinal cracks on the surface of the slab for peritectic steel A. It was the same with sample No. 3 and sample No. 4 for peritectic steel B. Some parameters of the typical flux films and the corresponding slab quality are listed in Table 1. Chemical compositions and physical properties of mold fluxes for peritectic steels A and B are listed in Tables 2 and 3.

Table 1. Parameters of mold flux films for peritectic steels.

Peritectic Steel Number	Flux Number	Film Number	Thickness of Liquid Flux (mm)	Consumption of Mold Flux ($kg \cdot t^{-1}$)	Slab Quality
A	F-A	No. 1	10.25	0.596	Normal
		No. 2	10.41	0.568	Longitudinal crack
B	F-B	No. 3	10.33	0.582	Normal
		No. 4	10.32	0.560	Longitudinal crack

Table 2. Chemical compositions of mold fluxes (wt%).

Peritectic Steel Number	Flux Number	SiO_2	Al_2O_3	MgO	CaO	Fe_2O_3	$K_2O + Na_2O$	MnO	F^-	C	H_2O
A	F-A	28.99	2.74	4.17	35.83	0.57	7.65	1.73	7.46	4.88	0.33
B	F-B	31.77	4.18	3.72	36.29	1.03	7.82	1.80	6.87	6.31	0.34

Table 3. Physical properties of mold fluxes.

Peritectic Steel Number	Flux Number	Alkalinity	Melting Point (°C)	Melting Speed (s)	Viscosity (Pa·s)	Molten Heavy (g/cm^3)
A	F-A	1.24	1050	21	0.108	0.80
B	F-B	1.14	1104	21	0.153	0.70

2.2. Methods

In this work, some samples of flux films from the same group were glued together by epoxy adhesive, and their section along the thickness direction was glued to the slide, ground to a thickness of 0.03 mm, and eventually made into a polished thin section. In addition, other samples of flux films and mold fluxes were completely ground to 0.074 mm before being used.

The thickness, mineralogical composition, crystallization ratio, and microstructure of flux films were analyzed using the polished thin section with a polarizing microscope (Axioskop A1 pol, Carl Zeiss Co., Ltd., Bangkok, Thailand). Then, using the polished thin section with an electron-probe microanalyzer (JXA-8230, JEOL Co., Ltd., Tokyo, Japan), the chemical compositions of different mineral crystals in flux films were confirmed.

The crystal phase composition of powder-like flux films was detected with an X-ray diffractometer (BDX-3200, Bruker Co., Ltd., Seongnam-si, Korea) using Cu-Ka radiation within the scanning range of 10° to 80°. The melting temperature and melting speed of powder-like mold fluxes were measured using a melting point and melting speed tester (RDS-04, Northeastern University in China).

The heat flow density (1673 K) and viscosity (1573 K) of mold fluxes were measured using a mold heat flux simulation and viscosity tester (HF-201, Chongqing University in China). The experimental procedure was as follows: First, 350 g prepared powder-like flux was put in a graphite crucible and heated to 1673 K with the $MoSi_2$ furnace of the tester. Then, the water-cooled sensor was immersed into the liquid slag and measured 10 data points of heat flow density within 45 s. Then, the sensor was taken out from the liquid slag, and the flux film adhered to the sensor was obtained (probe diagram of mold heat flow simulator can be seen in Figure 3). After a period of time, when the temperature dropped to 1573 K, a standard rotating spindle was immersed into liquid slag and measured 10 data points of viscosity.

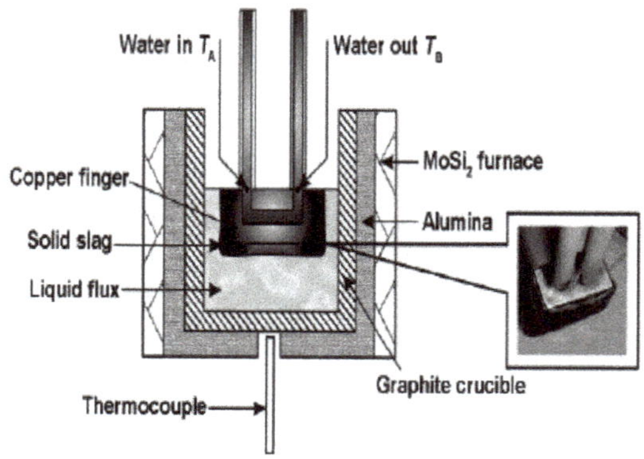

Figure 3. The probe diagram of the mold heat flow simulator.

To reduce the error, three or more samples were prepared for each mold flux, and the physical properties of each flux were determined from the average value of the measured data of samples; the difference among the measured data was no greater than 3%.

3. Results and Discussion

3.1. Mineralogical Composition and Crystallization Ratio of Flux Films

As the formation of flux films in a mold is affected by complex factors, the composition, content, morphology, and size of the crystalline minerals are obviously different, showing the variant heat transfer capacity. When the heat transfer capacity of flux films is poor, it will often lead to longitudinal cracks during casting crack-sensitive steel. In this work, mineralogical compositions, proportions, and crystallization ratio of flux films for the peritectic steels were quantitatively analyzed by using a polarizing microscope.

The polarizing microscope analysis results (Table 4 and Figure 4) show that the main crystalline minerals of flux films for steel A are akermanite, cuspidine, and wollastonite. The crystallization ratio of flux films for steel A varies widely. The crystallization ratio of flux films with good slab quality reaches 85~90%, and the content of akermanite is as high as 75~80%. The glass phase of flux films with longitudinal cracks on the slab surface increases obviously, and their crystallization ratio is only 45~50%.

Table 4. Mineralogical compositions, proportions, and crystallization ratio of flux films (vol%).

Peritectic Steel Number	Film Number	Slab Quality	Cuspidine	Akermanite	Wollastonite	Glass Phase	Crystallization Ratio
A	No. 1	Normal	10~15	75~80	1~5	5~10	85~90
	No. 2	Longitudinal crack	25~30	20~25	1~5	45~50	45~50
B	No. 3	Normal	40~45	20~25	<1	30~35	60~65
	No. 4	Longitudinal crack	12~17	80~85	<1	1~5	>95

In contrast, the main crystalline minerals of flux films for steel B are akermanite, cuspidine, and a small account of wollastonite. The crystallization ratio of flux films with good slab quality is as low as 60~65%, but the content of cuspidine is as high as 40~45%. Though the crystallization ratio of flux films with longitudinal cracks on the slab surface is as high as 95%, the content of well-shaped cuspidine is only 12~17%, and the remaining content is akermanite crystallites.

Figure 4. Mineralogical structures in flux films: (**a**) akermanite in film 1; (**b**) cuspidine in film 1; (**c**) akermanite in film 2; (**d**) akermanite in film 3; (**e**) cuspidine in film 3; (**f**) akermanite in film 4.

According to the backscattered electron (BSE) micrographs and energy-dispersive spectrometer (EDS) images from the electron-probe microanalyzer (Figures 5 and 6), it can be found that a large number of granular crystals and interlaced crystals are melilite, and a small number of spearhead-shaped crystals are cuspidine. Melilite has many isomorphisms such as akermanite ($2CaO{\bullet}MgO{\bullet}2SiO_2$) and gehlenite ($2CaO{\bullet}Al_2O_3{\bullet}SiO_2$). Their optical properties are similar, so it is difficult to distinguish them by microscope. However, it can be seen from the XRD analysis results (Figure 7) that the melilite in the flux film mainly is akermanite ($2CaO{\bullet}MgO{\bullet}2SiO_2$).

The crystallization ability of flux films directly affects the heat transfer in the mold. The flux film with low crystallization has low roughness, which makes thermal conductivity become high, and low precipitation of minerals with higher melting points decreases the interfacial thermal resistance, which may lead to surface longitudinal cracks in the peritectic steel. The mineralogical structure of the flux film determines the quality and production efficiency of the slab. The crystallization ratio of the flux film No. 2 for steel A with longitudinal cracks on the slab surface is obviously lower, worsening the heat transfer uniformity in mold, which is one of the causes of longitudinal cracks of slabs. However, the cuspidine content of the flux film No. 4 for steel B with longitudinal cracks on the slab surface is only 12~17%, which may bring about an excessive heat transfer rate causing longitudinal cracks to occur on the surface of the slab. Moreover, the vast akermanite crystallite content in the flux film No. 4 can cause heat transfer nonuniformity in the mold.

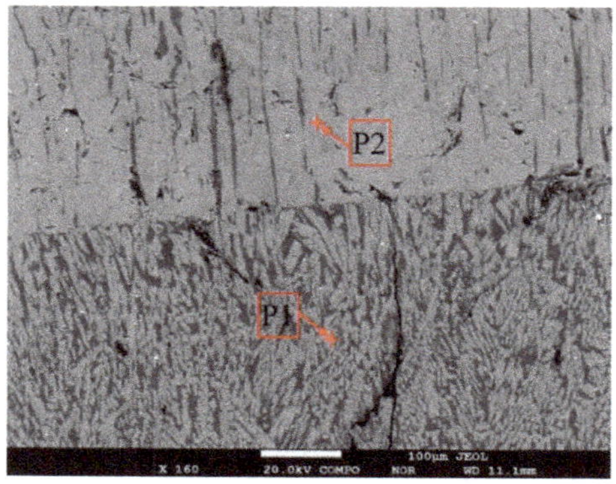

Figure 5. BSE micrographs of flux films by EPMA.

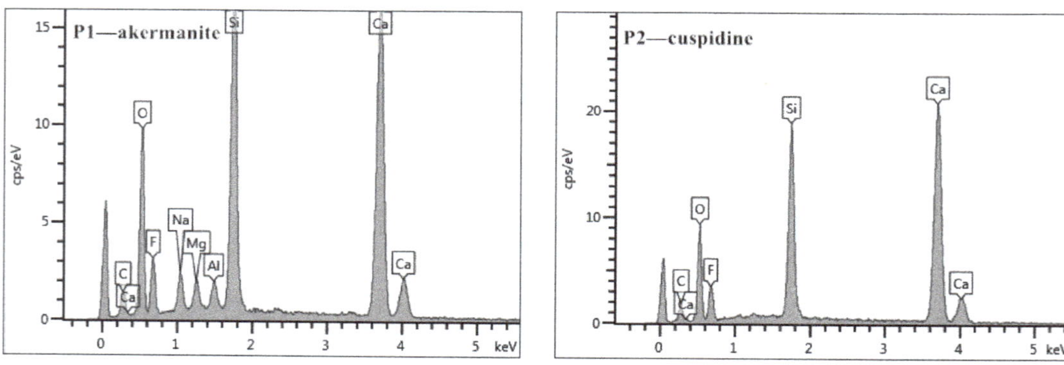

Figure 6. EDS spectra of different mineral crystals in flux films by EPMA.

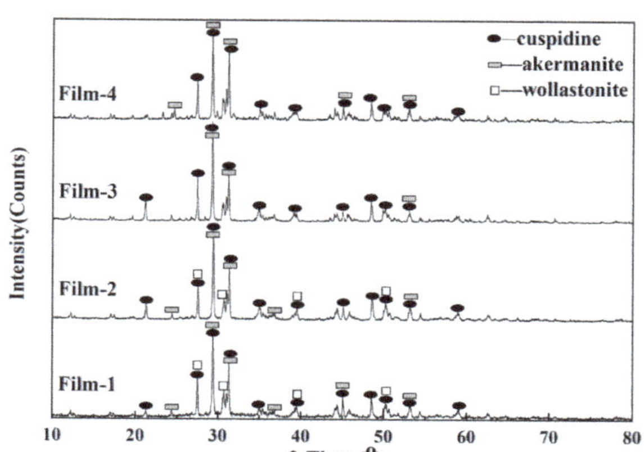

Figure 7. XRD results of flux films.

3.2. Microstructure of Flux Films

The layered structures, mineralogical species, and morphology features of flux films were analyzed and determined using polarizing microscopy. The layered structure of normal flux film for steel A is obvious, and the crystallizing layer accounts for a large area of the flux film (Figure 8a). The normal flux film of steel B exhibits a typical two-layered structure, and part of the flux film has a multilayered structure (Figure 8c). Compared with the glass layer near the mold side of the two normal flux films, the flux film of steel A has a higher degree of discontinuity than the flux film of steel B. The layered structure of flux film for steel A with longitudinal cracks is a typical three-layered structure with glass–crystalline–glass layers (Figure 8b). The crystallizing layer of flux film for steel B with longitudinal cracks accounts for all the flux film (Figure 8d).

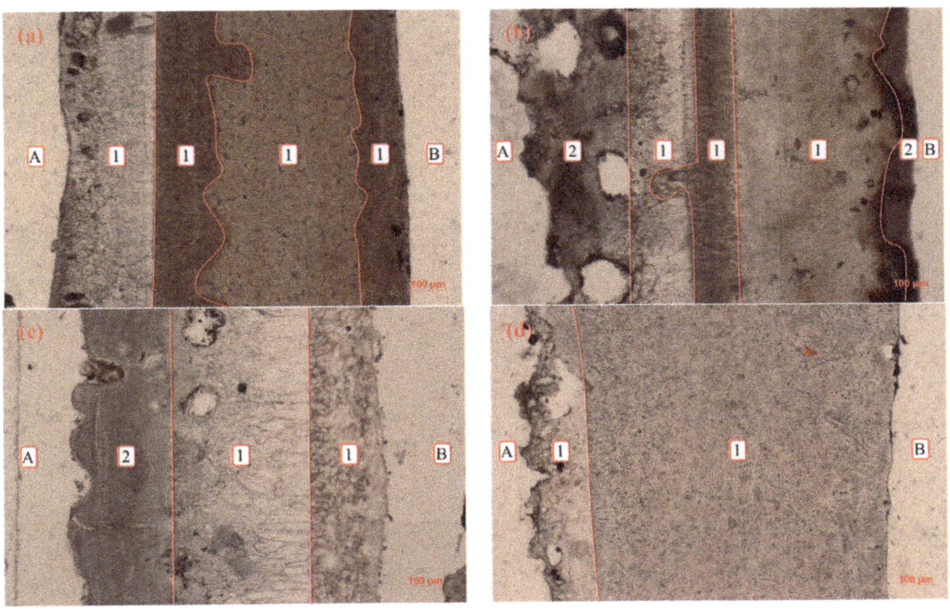

Figure 8. Layered structures of flux films: (**a**) film 1; (**b**) film 2; (**c**) film 3; (**d**) film 4; A—the mold side; B—the shell side; 1—crystallization layer; 2—glass layer.

The research results show that the flux films have differences in microstructure, especially in the mineral morphology and particle size. The akermanite and wollastonite in the crystal layer of the flux films are distributed alternately on the shell side, which has an obvious boundary with the cuspidine. The akermanite of the flux films for steel A is more prone to be intertexture-shaped near the shell side and to have a radial and chrysanthemum shape on the mold side. However, the akermanite of flux films for steel B is particle-shaped crystallites with small grain size, and the coarse granular and spearhead-shaped cuspidine of the flux films is concentrated on the mold side.

3.3. Heat Transfer Property of Flux Films

The heat flux density can be defined as the amount of heat that passes through the flux film per unit cross-sectional area and per unit time. In this study, the heat flow density of mold fluxes was measured by using the mold heat flux simulator when the temperature was maintained at 1673 K, and the measurements represent maximum and minimum heat transfer and heat transfer uniformity. The heat flow density of mold fluxes for peritectic steels was measured at different times by the simulator, as shown in Figure 9. The maximum heat flow density represents the heat transfer capacity of flux films at the meniscus, and the average heat flow density reflects the heat transfer capacity of flux films in the middle and upper part of the mold. The heat transfer test revealed that the heat flow density decreases gradually as the cooling water pass time is increased. The maximum and average heat flow densities of flux films for steel A are higher than those of flux films for steel B, and the same law applies to the heat transfer property for these films.

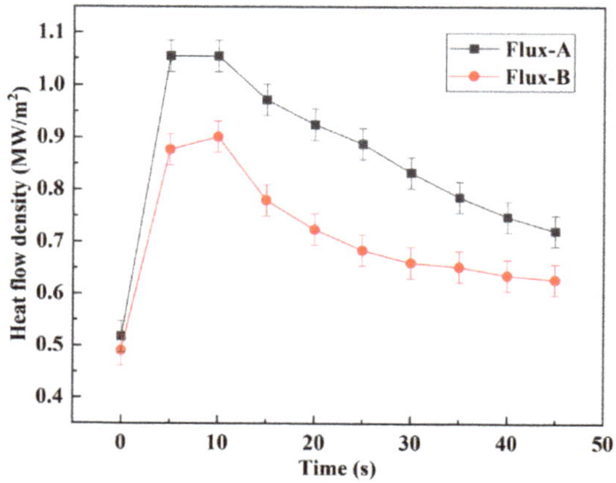

Figure 9. The heat flux density of mold flux films at different times.

3.4. Discussion

The relationship between the heat flow density and the mineralogical structure of flux films is presented in Figure 10. It can be seen that the greater the thickness of the slag film is, the lower the hindrance to the heat flux is. The greater the crystalline phase of flux film, the less the obstruction to the heat flow. Furthermore, the high crystallization of the flux film promotes an increase in its solidification thickness. However, early studies [33,34] on heat transfer ignored the influence of the mineralogical composition and proportion of flux films in the casting of peritectic steels. Instead, these studies considered that more resistance to heat transfer arises from the higher crystallization ratio or thickness of flux films. Thus, they were unable to explain the phenomenon of heat flow density dropping and becoming smaller than expected at the lower crystallization ratio and film thickness. As shown in Figure 10, the thickness and the heat flow density of flux films for steel A are higher than those of flux films for steel B. According to the thermal conductivity of minerals, cuspidine has the lowest thermal conductivity among all the mineral phases of flux films. With respect to the mineralogical structure of flux films for steel B, the content of cuspidine is as high as 40~45%, which can increase the thermal resistance of the flux film and maintain the suitable heat transfer rate and uniformity in mold all the time. In addition, there is an exception in Figure 10: the heat flow density drops and becomes smaller at the 90% crystallization ratio of the flux film; it seems perfectly natural to ascribe this to the bigger thickness and higher crystallization ratio of the flux film with the lower impact of cuspidine content. Therefore, considering the effect of the mineralogical structure of flux films on heat transfer in the mold and slab quality, this study carried out a comprehensive work and found that the mineralogical compositions and proportions can also be a key factor in controlling heat transfer suitable for the slab quality, so it is different from previous investigators' observations which incorrectly ascribed this effect to the crystallization ratio or thickness of flux films.

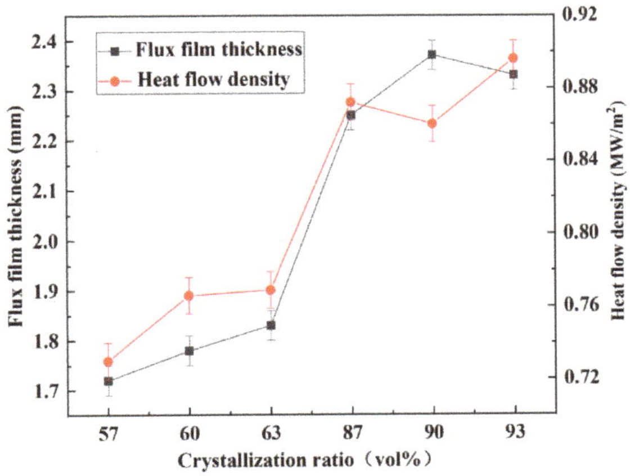

Figure 10. Relationship between the heat flow density and mineralogical structure of flux films.

Flux films of peritectic steel with longitudinal cracks generally have lower crystallization ratios and crystal growth levels than the normal flux films. The direct cause of vast longitudinal cracks on the peritectic steels is the abnormal microstructure of flux films, which may be caused by the inappropriate chemical components of flux and the cooling condition of the mold. The crystallization ratio of the unqualified films for steel A with longitudinal cracks may be only 45~50%. Compared with the reasonable flux films for steel B, the unqualified films have a mass of akermanite crystallite and the crystallization can abnormally reach 80% or more. All of these results suggest that the primary reasons for longitudinal cracking on the slab surface of the peritectic steels are the low crystallization ratio and the vast crystallite content, which can decrease the thermal resistance of flux films and worsen the heat transfer rate and uniformity.

In view of the strong crack sensitivity of peritectic steel, the longitudinal crack ratio of the slabs can be reduced by improving the mineralogical structure of the flux films. However, if the heat transfer is controlled only by enhancing the crystallization capacity, the balance between the heat transfer and lubrication properties of the flux films may be lost. So, during continuous casting of peritectic steels, the contradiction between heat transfer and lubrication should be considered comprehensively, and the optimum mineralogical structure of flux films should be obtained through reasonable composition regulation of the mold flux. In this work, it is recommended that the content of (CaO/SiO_2) or ($Na_2O + K_2O$) of the mold flux for steel A should be properly increased. While maintaining high alkalinity (CaO/SiO_2), the CaF_2 content of the mold flux for steel B should be increased appropriately.

4. Conclusions

The influence of the mineralogical structure of mold flux films on heat transfer in the mold during the continuous casting of peritectic steels for avoiding the longitudinal cracks of slabs has been investigated. The following conclusions can be drawn:

(1) The mineralogical structure of normal flux films during the continuous casting of peritectic steels presents a multilayered structure with a high crystallization ratio (60~90 vol%), mainly composed of well-developed crystalline akermanite and cuspidine.

(2) Flux films for peritectic steels with longitudinal cracks have a lower crystallization ratio and crystal growth level than the normal flux films have, showing the characteristic of low crystallization ratio (<50 vol%) or vast crystallite content (>80 vol%).

(3) The results from heat transfer analysis confirmed that the mineralogical composition and proportion of flux films is a key factor in controlling heat transfer during the casting of peritectic steels. The more glass phase and crystallite flux films have, the worse the heat transfer rate and uniformity in the mold are.

(4) Using strong crystalline flux to obtain an ideal mineral phase structure of flux films is one of the important measures for reducing the longitudinal cracking of peritectic steels, and the ideal mineral phase is characterized by a high crystallization ratio (>60 vol%), without the vast crystallite content, and as much cuspidine content as possible.

Author Contributions: Conceptualization, L.L. and X.H.; methodology, formal analysis, L.L. and M.L.; investigation, resources, L.L. and D.Z.; data curation, M.L. and D.Z.; writing—original draft preparation, L.L.; writing—review and editing, L.L. and X.H.; visualization, M.L. and D.Z.; supervision, X.H.; project administration, L.L. and X.H. All authors have read and agreed to the published version of the manuscript.

Funding: This research was funded by the National Natural Science Foundation of China (No. 51774140 and No. 51574105), the Operation Expenses for Universities' Basic Scientific Research of Hebei Province (No. JQN2020038), and the Postgraduate Innovation Funding Project of Hebei Province (No. CXZZBS2021101).

Institutional Review Board Statement: Not applicable.

Informed Consent Statement: Not applicable.

Data Availability Statement: Not applicable.

Conflicts of Interest: The authors declare no conflict of interest.

References

1. Hechu, K.; Slater, C.; Santillana, B.; Sridhar, S. The use of infrared thermography to detect thermal instability during solidification of peritectic steels. *Mater. Charact.* **2019**, *154*, 138–147. [CrossRef]
2. Long, X.; He, S.P.; Xu, J.F.; Huo, X.L.; Wang, Q. Properties of high basicity mold fluxes for peritectic steel slab casting. *ISIJ Int.* **2012**, *19*, 39–45. [CrossRef]
3. Suzuki, M.; Yamaoka, Y. Influence of carbon content on solidifying shell growth of carbon steels at the initial stage of solidification. *Mater. Charact.* **2003**, *44*, 836–844. [CrossRef]
4. Liu, T.; Long, M.J.; Chen, D.F.; Wu, S.X.; Tang, P.M.; Liu, S.; Duan, H.M.; Yang, J. Investigations of the peritectic reaction and transformation in a hypoperitectic steel: Using high-temperature confocal scanning laser microscopy and differential scanning calorimetry. *Mater. Charact.* **2019**, *156*, 109870. [CrossRef]
5. Jeong, H.; Hwang, J.Y.; Cho, J.W. In-depth study of mold heat transfer for the high speed continuous casting process. *Met. Mater. Int.* **2016**, *22*, 295–304. [CrossRef]
6. Hanao, M.; Kawamoto, M.; Yamanaka, A. Influence of mold flux on initial solidification of hypo-peritectic steel in a continuous casting mold. *ISIJ Int.* **2012**, *52*, 1310–1319. [CrossRef]
7. Zhao, H.; Wang, X.; Zhang, J. Research on mold flux for hypo-peritectic steel at high casting speed. *Int. J. Miner. Metall. Mater.* **2007**, *14*, 219–224. [CrossRef]
8. Wang, W.L.; Gu, K.; Zhou, L.J. Radiative heat transfer behavior of mold fluxes for casting low and medium carbon steels. *ISIJ Int.* **2011**, *51*, 1838–1845. [CrossRef]
9. Park, J.Y.; Ko, E.Y.; Choi, J.; Sohn, I. Characteristics of medium carbon steel solidification and mold flux, crystallization using the multi-mold simulator. *Met. Mater. Int.* **2014**, *20*, 1103–1114. [CrossRef]
10. Ko, E.Y.; Choi, J.; Park, J.Y.; Sohn, I. Simulation of low carbon steel solidification and mold flux crystallization in continuous casting using a multi-mold simulator. *Met. Mater. Int.* **2014**, *20*, 141–151. [CrossRef]
11. Zhang, Y.F.; Han, X.L.; Liu, L.; Liu, L.N. Effect of mineralogical structure of flux film on slab quality for medium carbon steel. *Trans. Indian Inst. Met.* **2018**, *71*, 1803–1807. [CrossRef]
12. Nakada, H.; Susa, M.; Seko, Y.; Hayashi, M.; Nagata, K. Mechanism of heat transfer reduction by crystallization of mold flux for continuous casting. *ISIJ Int.* **2008**, *48*, 446–453. [CrossRef]
13. Wen, G.H.; Tang, P.; Yang, B.; Zhu, X.B. Simulation and characterization on heat transfer through mould slag film. *ISIJ Int.* **2012**, *52*, 1179–1185. [CrossRef]
14. Liu, Y.Z.; Wang, W.L.; Ma, F.J.; Zhang, H.H. Study of solidification and heat transfer behavior of mold flux through mold flux heat transfer simulator technique: Part I. Development of the technique. *Metall. Mater. Trans. B* **2015**, *46*, 1419–1430. [CrossRef]
15. Mills, K.C. Structure and properties of slags used in the continuous casting of steel: Part 1. Conventional mould powders. *ISIJ Int.* **2016**, *56*, 1–13. [CrossRef]
16. Shibata, H.; Kondo, K.; Suzuki, M.; Emi, T. Thermal resistance between solidifying steel shell and continuous casting mold with intervening flux film. *ISIJ Int.* **1996**, *36*, 79–81. [CrossRef]
17. Mills, K.C.; Fox, A.B. The role of mould fluxes in continuous casting-so simple yet so complex. *ISIJ Int.* **2003**, *43*, 1479–1486. [CrossRef]

18. Ma, F.J.; Liu, Y.Z.; Wang, W.L.; Zhang, H.H. Study of solidification and heat transfer behavior of mold flux through mold flux heat transfer simulator technique: Part II. Effect of mold oscillation on heat transfer behaviors. *Metall. Mater. Trans. B* **2015**, *46*, 1902–1911. [CrossRef]
19. Tsutsumi, K.; Nagasaka, T.; Hino, M. Surface roughness of solidified mold flux in continuous casting process. *ISIJ Int.* **1999**, *39*, 1150–1159. [CrossRef]
20. Wen, G.H.; Zhu, X.B.; Tang, P.; Yang, B.; Yu, X. Influence of raw material type on heat transfer and structure of mould slag. *ISIJ Int.* **2011**, *51*, 1028–1032. [CrossRef]
21. Ozawa, S.; Susa, M.; Goto, T.; Endo, R.K.; Mills, K.C. Lattice and radiation conductivities for mould fluxes from the perspective of degree of crystallinity. *ISIJ Int.* **2006**, *46*, 413–419. [CrossRef]
22. Cho, J.W.; Blazek, K.; Frazee, M.; Yin, H.; Park, J.H.; Moon, S.W. Assessment of CaO-Al_2O_3 based mold flux system for high aluminum TRIP casting. *ISIJ Int.* **2003**, *53*, 62–70. [CrossRef]
23. Choi, S.Y.; Lee, D.H.; Shin, D.W.; Choi, S.Y.; Cho, J.W.; Park, J.M. Properties of F-free glass system as a mold flux: Viscosity, thermal conductivity and crystallization behavior. *J. Non-Cryst. Solids* **2004**, *345*, 157–160. [CrossRef]
24. Han, X.L.; Zhang, Y.F.; Liu, L.; Liu, L.N. Effects of soda ash on properties of mold flux and mineralogical structures of flux film. *J. Iron Steel Res. Int.* **2016**, *23*, 197–202. [CrossRef]
25. Lei, Y.; Xie, B.; Ma, W.-H. Analysis of crystallization behavior of mold fluxes containing TiO_2 using single hot thermocouple technique. *J. Iron Steel Res. Int.* **2016**, *23*, 322–328. [CrossRef]
26. Zhang, L.; Wang, W.L.; Sohn, I.L. Crystallization behavior and structure analysis for molten CaO-SiO_2-B_2O_3 based fluorine-free mold fluxes. *J. Non-Cryst. Solids* **2019**, *511*, 41–49. [CrossRef]
27. Zhang, S.D.; Li, M.; Zhu, L.L.; Wang, Q.Q.; He, S.P.; Wang, Q. Effect of substituting Na_2O for SiO_2 on the non-isothermal crystallization behavior of CaO-BaO-Al_2O_3 based mold fluxes for casting high Al steels. *Ceram. Int.* **2019**, *45*, 11296–11303. [CrossRef]
28. Kölbl, N. New mold slag compositions for the continuous casting of soft steels. *Steel Res. Int.* **2021**, *93*, 2100165. [CrossRef]
29. Wang, W.; Blazek, K.; Cramb, A. A Study of the crystallization behavior of a new mold flux used in the casting of transformation induced plasticity steels. *J. Metall. Mater. Trans.* **2008**, *39*, 66–74. [CrossRef]
30. Kang, Y.; Lee, J.; Morita, K. Thermal conductivity of molten slags: A review of measurement techniques and discussion based on microstructural analysis. *ISIJ Int.* **2014**, *46*, 2008–2016. [CrossRef]
31. Han, X.L.; Zhang, Y.F.; Liu, L.; Liu, L.N. Influence of chemical composition of mold flux on viscosity and texture of slag film. *Environ. Toxicol. Chem.* **2016**, *98*, 511–517. [CrossRef]
32. Costaa, I.T.; Carvalho, C.S.; Oliveira, J.R.; Vieirab, E.A. Physical properties characterization of a peritectic mold flux formed from the addition of calcitic marble residue in the commercial one. *J. Mater. Res. Technol.* **2019**, *8*, 3297–3305. [CrossRef]
33. Cho, J.W.; Emi, T.; Shibata, H.; Suzuki, M. Heat transfer across mold flux film in mold during initial solidification in continuous casting of steel. *ISIJ Int.* **1998**, *38*, 834–842. [CrossRef]
34. Cho, J.W.; Shibata, H.; Emi, T.; Suzuki, M. Thermal resistance at the interface between mold flux film and mold for continuous casting of steels. *ISIJ Int.* **1998**, *38*, 440–446. [CrossRef]

materials

MDPI

Article

The Effect of Heat Source Path on Thermal Evolution during Electro-Gas Welding of Thick Steel Plates

Jun Fu [1,2], Qing Tao [1,3], Xiaoan Yang [1,2], Bogdan Nenchev [1], Ming Li [2], Biao Tao [2] and Hongbiao Dong [1,*]

1 School of Engineering, University of Leicester, Leicester LE1 7RH, UK; jf320@leicester.ac.uk (J.F.); qt15@le.ac.uk (Q.T.); xy149@le.ac.uk (X.Y.); bn66@le.ac.uk (B.N.)
2 Nanjing Iron & Steel United Co., Ltd., Nanjing 210035, China; liming1@njsteel.com.cn (M.L.); taobiao@njsteel.com.cn (B.T.)
3 School of Materials and Physics, China University of Mining and Technology, Xuzhou 221116, China
* Correspondence: h.dong@le.ac.uk; Tel.: +44-(0)7718369708

Abstract: In recent years, the shipbuilding industry has experienced a growing demand for tighter control and higher strength requirements in thick steel plate welding. Electro-gas welding (EGW) is a high heat input welding method, widely used to improve the welding efficiency of thick plates. Modelling the EGW process of thick steel plates has been challenging due to difficulties in accurately depicting the heat source path movement. An EGW experiment on 30 mm thickness E36 steel plates was conducted in this study. A semi-ellipsoid heat source model was implemented, and its movement was mathematically expressed using linear, sinusoidal, or oscillate-stop paths. The geometry of welding joints, process variables, and steel composition are taken from industrial scale experiments. The resulting thermal evolutions across all heat source-path approaches were verified against experimental observations. Practical industrial recommendations are provided and discussed in terms of the fusion quality for E36 steel plates with a heat input of 157 kJ/cm. It was found that the oscillate-stop heat path predicts thermal profile more accurately than the sinusoidal function and linear heat path for EGW welding of 30 mm thickness and above. The linear heat path approach is recommended for E36 steel plate thickness up to 20 mm, whereas maximum thickness up to 30 mm is appropriate for sinusoidal path, and maximum thickness up to 35 mm is appropriate for oscillate-stop path in EGW welding, assuming constant heat input.

Keywords: electro-gas welding; high heat input; heat source movement path; finite element analysis; thermal evolution

Citation: Fu, J.; Tao, Q.; Yang, X.; Nenchev, B.; Li, M.; Tao, B.; Dong, H. The Effect of Heat Source Path on Thermal Evolution during Electro-Gas Welding of Thick Steel Plates. *Materials* **2022**, *15*, 2215. https://doi.org/10.3390/ma15062215

Academic Editor: Adam Grajcar

Received: 23 January 2022
Accepted: 15 March 2022
Published: 17 March 2022

1. Introduction

In the shipbuilding industry and offshore engineering, the size and structures of ships are increasing rapidly [1,2], thus giving rise to a growing demand for higher control in welding thick steel plates [3,4]. Electro-gas welding (EGW) has become an indispensable welding method for shipbuilding enterprises due to its outstanding advantages, such as large heat input and high welding efficiency [5]. As illustrated in Figure 1, EGW is an automatic welding process using a special flux-cored wire with CO_2 gas protection, and used for the welding of vertical position of steel plates [6]. During the welding process, the torch moves along the chosen weld path from bottom to top. A water-cooled copper slider is placed on the front of the weld, and a ceramic "backing" plate is positioned at the rear. The welding torch is also used to feed the welding metal into the groove. In thick plates welding, the welding pool is restrained by the weld pieces, backing plate and sliding copper shoe, so that single pass welding can be accomplished.

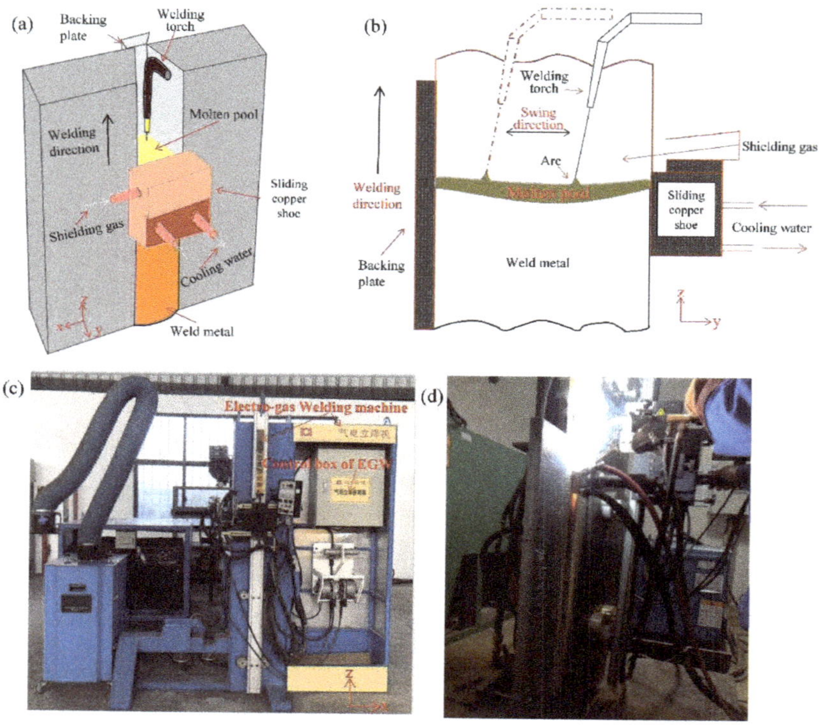

Figure 1. The electro-gas welding (EGW). (**a**) The 3D schematic diagram of EGW equipment, (**b**) The 2D schematic diagram of EGW equipment, (**c**) Photographs of EGW equipment, (**d**) Photographs of experimentation.

In EGW, heat transfer from the heat source to the plates occurs mainly through radiation and convection [7,8]. In thick steel plate welding, the efficiency of heat transfer in the "y" direction (with depth, Figure 1) of single pass EGW is a key factor affecting the quality of the joint [9,10]. In 2016, Hwang, Kim and Lee [11] introduced double ellipsoidal moving heat sources to model the temperature profile and residual stress distribution in EGW of marine steel. However, the ellipsoidal method oversimplifies the oscillate-stop heat source path of industrial welding process to 1 dimensional linear case, whereby the movement path in the "y" direction is not considered. Therefore, the heat source paths are inconsistent with the real EGW process, hence the predicted fusion lines do not agree with experimental results.

To include an oscillate-stop heat source path within weld modelling, in 2017, Xu, Pan and Wan [12] used a sinusoidal function to simulate metal active gas arc welding (GAW). In 2018, Yuan et al. [13] applied a piecewise function to model the real oscillate movement path of gas metal arc welding. In 2019, da Silva Pereira et al. [14] implemented weave patterns by path parameterization, which improved the accuracy and has been further extended to predict defects in welding [15]. However, the effect of different heat source paths on thermal evolution in modelling EGW of thick steel plates and further optimize heat source paths according to thickness of steel plate has not yet been investigated systematically.

This study investigates the effect of three different heat source paths (linear, sinusoidal and oscillate-stop) through FEA on the resulting thermal evolution in thick steel plate welding. The results for EGW welding of E36 marine steel plates with thicknesses ranging from 20–40 mm and a heat input of 157 kJ/cm are compared based on fusion efficiency. The FEA models are verified by comparing against experimental observations. Finally, practical industrial recommendations are provided for each heat source path's approach, according to variation in plate thickness.

2. Materials and Model Description

2.1. Electro-Gas Welding Experiments

This study investigates the effect of heat source path on thermal evolution during EGW of E36 marine steel plates. For the FEA modelling, experimental parameters, such as the geometry of the welding joints, thermo-mechanical and EGW process variables, and steel composition were considered as inputs, listed in Table 1. The temperature variation of material properties such as density and thermal conductivity of the E36 steel plate was also taken into account. JMatPro was implemented to calculate the material properties of the steel [16,17] for the E36 steel composition, shown in Table 2. In addition to the five main elements: carbon, manganese, phosphorus, and sulphur, microalloying elements such as niobium, titanium, and aluminium were added to improve the mechanical properties of steel. The dimensions of the steel plate model are shown in Figure 2. The thickness of the steel plate is 30 mm, while the length and width of the steel plate are 800 mm and 250 mm, respectively.

Table 1. Parameters of welding experiments.

Heat Input (kJ/cm)	**Welding Speed (cm/min)**	**Current (A)**	**Voltage (V)**	**Wire Diameter (mm)**	**Wire Feed Rate (m/min)**
157	6.9	420	43	1.6	13
Parameters of the torch movement of horizontal oscillations					
Internal stop (s)	External stop (s)	Amplitude (mm)	Frequency (s^{-1})		
1	1.5	10	2/11		

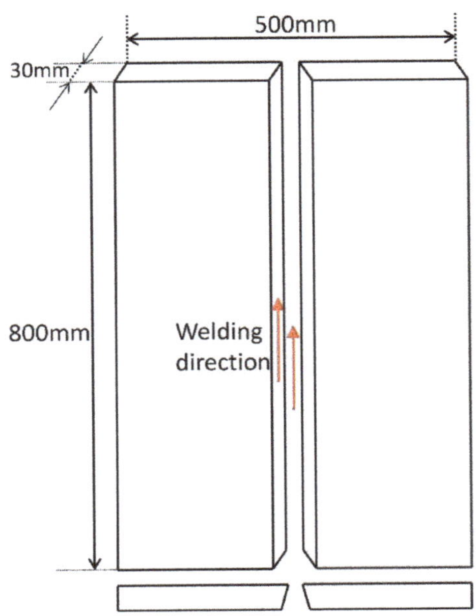

Figure 2. Geometry of welded steel plate.

Table 2. Chemical composition of E36 steel (wt %).

C	Si	Mn	S	P	Nb	Ti	Al
0.08	0.27	1.45	0.002	0.01	0.014	0.014	0.034

In the electro-gas welding process of 30 mm thick steel E36, flux-cored wire JIS Z3319 DW-S60G with a diameter of 1.6 mm was used. The chemical composition and the properties of the weld deposit are presented in Tables 3 and 4.

Table 3. Chemical composition of weld deposit.

Grade	C	Mn	Si	S	P	Cr	Ni	Mo	B	Ti
DW-S60G	0.07	1.68	0.33	0.006	0.011	0.02	0.77	0.26	/	0.02

Table 4. Mechanical properties of weld deposit.

Yield Strength Rp0.2 (MPa)	Tensile Strength Rm (MPa)	Elongation A (%)	Akv at −20 °C (J)
534	662	26	124, 139, 120

2.2. Three-Dimensional EGW Model for the Numerical Simulation

Heat transfer during welding is a complex process dependent on multiple thermo-physical and geometrical factors. In EGW, heat from the heat source is dissipated to the welding plates mainly by conduction, radiation, and convection. Thus, a balance must be found between the rate of heat generation (speed and energy of EGW) and the rate of dissipation, determined by the case specific geometry and chemistry.

The heat transfer equation is given as follows:

$$\rho C_p \frac{\partial T}{\partial t} + \nabla(-k\nabla T) = Q - Q_r - Q_c \tag{1}$$

where ρ—density of the material; C_p—specific heat capacity; T—absolute temperature; t—time; k—thermal conductivity; Q—heat source; Q_r—radiation heat loss; Q_c—convection heat loss. During the EGW process, the peak temperature of the heat-affected zone can reach more than 2000 °C, thus the temperature gradient between the steel plate and the environment is massive, causing significant radiation. The heat dissipation terms for radiation and convection are:

$$-Q_r = \varepsilon\sigma\left(T_{amb}^4 - T^4\right) \tag{2}$$

$$-Q_c = h_f(T_{amb} - T) \tag{3}$$

where ε—thermal emissivity; σ—Stephan Postman's constant: 5.67×10^{-8} [W/(m^2·K^4)]; T_{amb}—ambient temperature [K]; T—steel plate surface temperature [K]; q_r—radiant heat transfer flux [W/m^2]; q_c—convective heat transfer flux [W/m^2]; h_f—convection heat transfer coefficient between welding parts and the environment.

A three-dimensional EGW model was implemented, as shown in Figure 3a,b. This model uses a free tetrahedral mesh with the following principles designed for improving the computational efficiency:

- The weld region close to the weld heat source has a larger temperature gradient and, hence, was divided into a finer grid (minimum mesh tetrahedral edge length 7.67 mm).
- A coarser mesh (maximum mesh tetrahedral edge length 24.24 mm) was applied to the base steel plate areas far away from the heat source, where a smaller change in thermal gradients occurred.

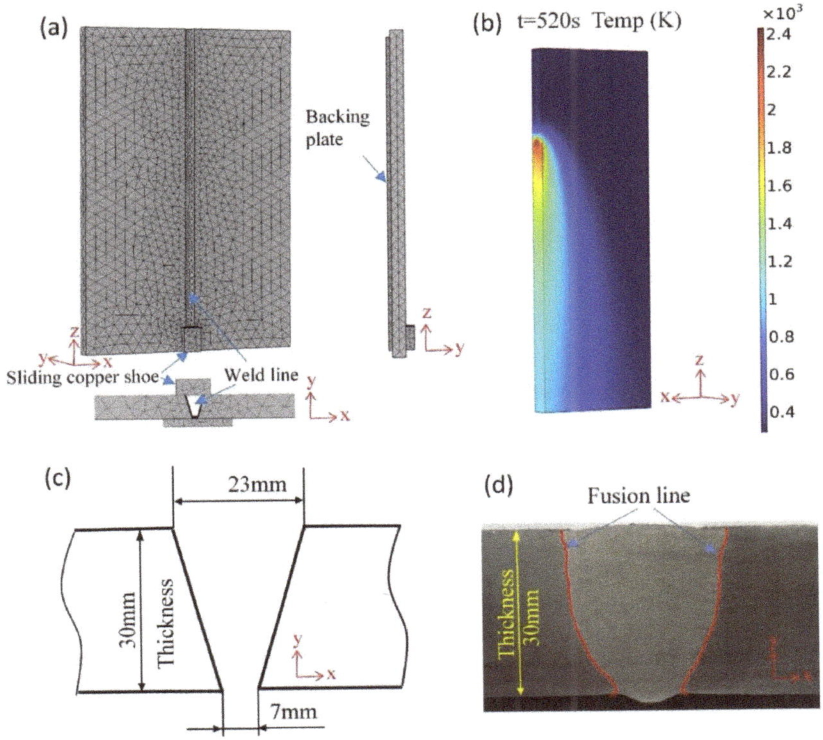

Figure 3. (**a**) Geometry and mesh of EGW model implemented in this study, (**b**) the transient temperature distribution simulated of steel plate at 520 s, (**c**) schematic diagram welding joint of pre-welding bevel, (**d**) welding joint of after-welding showing the fusion line.

Figure 3b illustrates the thermal profile and fusion line of a steel plate during the EGW process at 520 s at the midsection. The thermal distribution at surface and side of the steel plate is available from the diagram. Experimental fusion lines are used to verify the simulation results. A schematic diagram of the pre-welding bevel and micrograph of real joint after welding is shown in Figure 3c,d, respectively.

The filling of weld metal gradually with the movement of heat source is considered. The study simulates the 'activate' and 'inactive' of weld metal. Geometry of weld metal is pre-drawn in the work piece gap and properties of weld metal are activate point by point with the heat source movement.

For the welded area of the weld metal, the material properties of the geometry are activated.

For the unwelded area of the weld metal, the material properties of the geometry are inactivated, as shown in Figure 4.

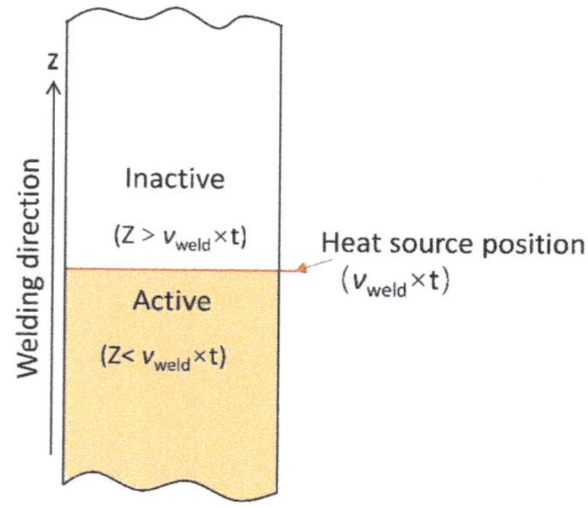

Figure 4. Schematic diagram of the activation process of the weld metal.

2.3. Heat Source Movement Path

2.3.1. Heat Source Model

The earliest and simplest heat source model is the point heat source model proposed by Rosenthal [18] and it has been widely used in welding simulation. Rosenthal's heat source model applies a quasi-steady state 3D semi-infinite geometry for point source. A velocity term was added to Rosenthal's model to simulate heat source movement by Lecoanet et al. [19–21]. To describe the heat source distribution, a double ellipsoidal heat source model [22,23] was proposed by Rouquette et al., combining two different ellipses, one in the front quadrant and the other in the rear quadrant [24,25]. Laser Welding Processes have been simulated by a double ellipsoidal heat source model in recent years [26–28]. The two heat source models, i.e., Gaussian, semi-ellipsoid and double ellipsoidal heat source model, predict similar temperature distribution and distortion [29,30]. A simplified Gaussian heat source model was to improve calculation efficiency by Cai and Norman [31,32].

In this study, the semi-ellipsoid heat source model was selected. The equation of semi-ellipsoid heat source model is shown below:

$$q(x,\ y,\ z) = \frac{6Q}{\pi r^3 \sqrt{\pi}} \exp\left(-\frac{3(x^2+y^2+z^2)}{r^2}\right) \tag{4}$$

where $q(x,y,z)$ represents the heat flow density distribution of (x,y,z), Q is the effective power of the arc, r is the radius of the semi-ellipsoid heating source. The schematic diagram of semi-ellipsoid heat source is shown in Figure 5.

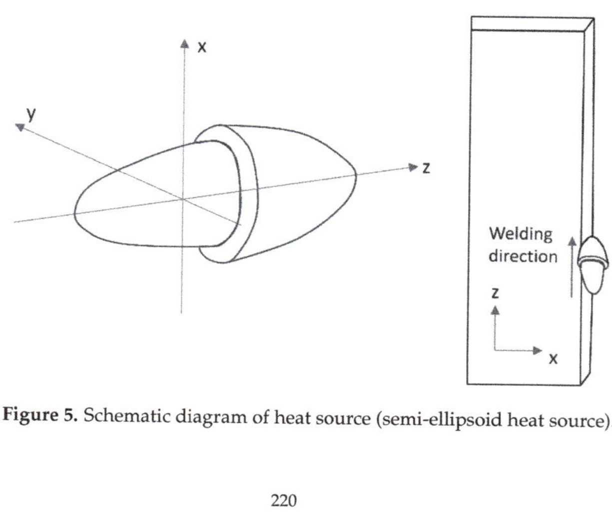

Figure 5. Schematic diagram of heat source (semi-ellipsoid heat source).

2.3.2. Heat Source Movement Path

The movement of the heat source is expressed using a linear, sinusoidal and oscillate-stop path. In this study, the movement speed in the z-axis is defined as v_1 and in the y-axis as v_2. As shown in Figure 6, three welding paths are modelled. Their exact coordinate heat source locations at each time step are expressed in terms of v_1, v_2 and then substituted in Equation (4).

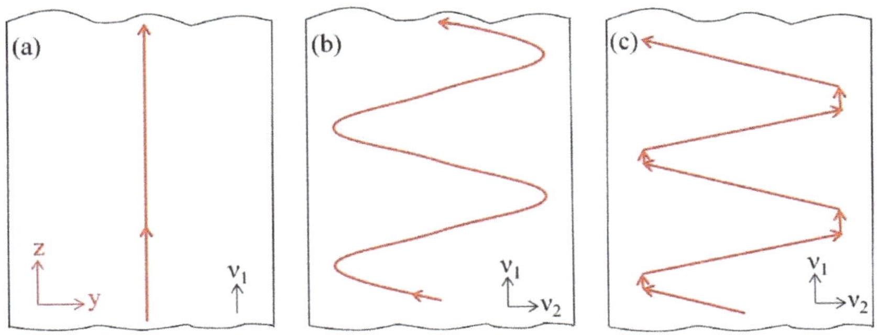

Figure 6. Schematic diagram of heat source path. (**a**) linear heat source path, (**b**) sinusoidal path heat source, (**c**) oscillate-stop heat source path.

1. Linear path heat source

The linear path heat source considers the welding heat source moving in the welding direction with a welding speed of v_1; the speed of the y-axis direction is zero. The expression of welding heat flux for its movement along the welding direction is shown in Equation (5):

$$q(x,y,z) = \frac{6Q}{\pi r^3 \sqrt{\pi}} \exp\left(-\frac{3\left(x^2 + y^2 + (z - v_1 t)^2\right)}{r^2} \right) \quad (5)$$

The simulation of the welding heat source moves from bottom to top along the welding direction with the welding speed during the electro gas welding process. The heat source moves along the centre of the steel plates for different thicknesses, with a welding speed of 6.9 cm/min, as given in Table 1.

2. Sinusoidal path heat source

The heat source, moving both in the welding direction with speed v_1 and in the thickness direction with speed v_2, is considered in the sinusoidal path heat source model. The schematic diagram of the path is shown in Figure 6b.

The defined oscillate range is d, the length of the weld beam is L, the welding speed is v_1, and the oscillate speed of the heat source in the thickness direction is v_2; the equation of the welding period is:

$$T = \frac{2d}{v_2} \quad (6)$$

Total welding time:

$$t = \frac{L}{v_1} \quad (7)$$

The heat source position of the y-axis direction is considered as:

$$y = \frac{d}{4} sin \frac{2\pi t}{T} \quad (8)$$

Steel plates with three different thicknesses of 30 mm, 35 mm and 40 mm were used in this study to investigate the applicable thickness for different heat source movement paths. The parameters of the sinusoidal for different thickness steel plates are shown in Table 5.

Table 5. Sinusoidal path parameters of welding heat source.

No.	Thickness of Plate (mm)	Amplitude (A) (mm)	Period (T) (s)	Oscillate Range (mm)	Offset (k) (mm)
1	30	10.0	5.5	20	20
2	35	12.5	5.5	25	20
3	40	15.0	5.5	30	20

According to the parameters of the sinusoidal heat source movement path, the position of the heat source changes with time during the welding process, as described in Equation (10). The amplitude of sinusoidal function is 10 mm, 12.5 mm and 15 mm for three thick plates, the period of the sinusoidal cycle is 5.5 s, respectively. The oscillate range is 20 mm, 25 mm and 30 mm. Due to the heat source needing to be close to the surface while leaving a certain distance for technological factors, the distance of 5 mm to the top surface and the bottom surface is appropriate.

$$P_1(t) = A\sin\left(\frac{2\pi t}{T}\right) + k \tag{9}$$

The welding heat flux of the sinusoidal path is described as:

$$q(x,y,z) = \frac{6Q}{\pi r^3\sqrt{\pi}}\exp\left(-\frac{3\left(x^2 + (y - P_1(t))^2 + (z - v_1 t)^2\right)}{r^2}\right) \tag{10}$$

According to Equation (10), the position of heat source change with time for three thickness plates are shows in Figure 7. The y-axis is the position along the thickness of steel plates and the z-axis is the position along the welding direction from the bottom to the top of steel plates. The movement path of the heat source under three thicknesses is drawn using solid lines with red, blue and green colour and the three thickness of 30 mm, 35 mm and 40 mm steel plates are indicated as dotted lines with red, blue and green colour, respectively. The positions of the heat sources for 1 s, 5.5 s, 6 s, 7 s and 11 s are marked in Figure 7.

3. Oscillate-stop heat source

According to the characteristics of the oscillate-stop path, the heat source not only moves from bottom to top during the welding process but also oscillates in the direction of the depth of the melt pool, which helps the welding bevels on both sides to obtain the same melting depth. According to the welding practice, the oscillate-stop parameters of the heat source for different thickness steel plates are shown in Table 6.

Table 6. Oscillate-stop parameters of welding heat source.

No.	Thickness (mm)	Internal Stop (s)	External Stop (s)	Oscillate Range (R) (mm)	Oscillate Center (C) (mm)	Period (T) (s)
1	30	1.0	1.5	20	20	5.5
2	35	1.0	1.5	25	20	5.5
3	40	1.0	1.5	30	20	5.5

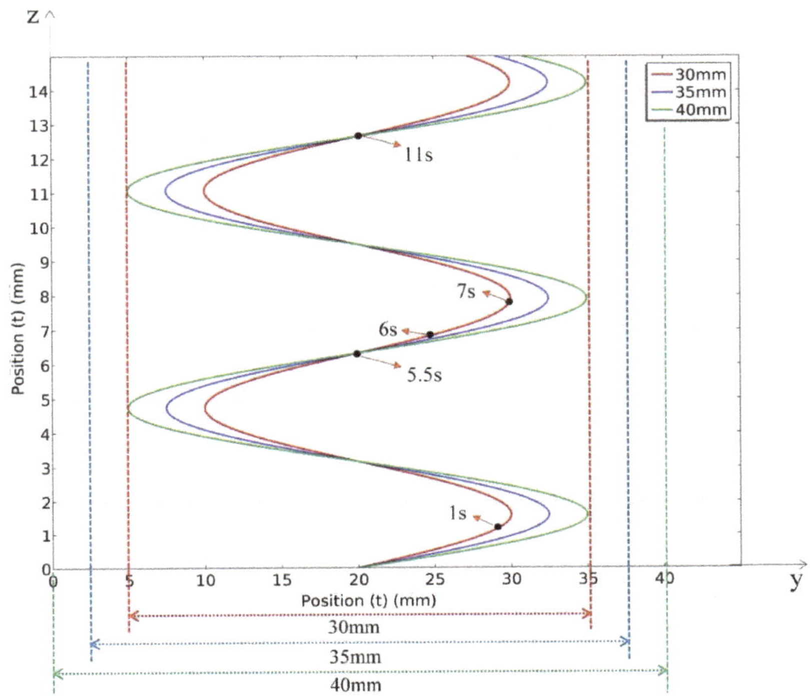

Figure 7. Sinusoidal heat source movement vs. time (t) for 30 mm, 35 mm and 40 mm thick steel plates.

Assume that the oscillate range is R, the distance of the oscillate centre to the origin of the coordinates is C, the period of the cycle is T, cycles is n. according to parameters of the oscillate-stop path; the position of heat source relative to the steel plate with different thickness in the welding period is described as Equation (12):

$$P_2(t) = \begin{cases} \frac{R}{1.5}t + 10 - 5.5(n-1)\frac{R}{1.5}, & T(n-1) < t \le T(n-1) + 1.5 \\ C + \frac{R}{2}, & T(n-1) + 1.5 < t \le T(n-1) + 2.5 \\ -\frac{R}{1.5}t + 30 - [5.5(n-1) + 2.5]\left(-\frac{R}{1.5}\right), & T(n-1) + 2.5 < t \le T(n-1) + 4 \\ C - \frac{R}{2}, & T(n-1) + 4 < t \le T(n-1) + 5.5 \end{cases} \quad (11)$$

Regarding their thickness, the oscillate centre C = 20 mm and period T = 5.5 s are constant, so the equation can be simplified to:

$$P_2(t) = \begin{cases} \frac{R}{1.5}(t - 5.5n + 5.5) + 10, & 5.5n - 5.5 < t \le 5.5n - 4 \\ 20 + \frac{R}{2}, & 5.5n - 4 < t \le 5.5n - 3 \\ -\frac{R}{1.5}(t - 5.5n + 30) + 30, & 5.5n - 3 < t \le 5.5n - 1.5 \\ 20 - \frac{R}{2}, & 5.5n - 1.5 < t \le 5.5n \end{cases} \quad (12)$$

Define a function for the cycle of the equation, $P(mod(t, 5.5))$ means the equation cycle one time every 5.5 s. The welding heat flux of oscillate stop path be described as:

$$q(x, y, z) = \frac{6Q}{\pi r^3 \sqrt{\pi}} \exp\left(-\frac{3\left(x^2 + (y - P_2(mod(t, 5.5))^2 + (z - v_1 t)^2\right)}{r^2}\right) \quad (13)$$

The position of the heat source changes with time for three thickness plates, as shown in Figure 8.

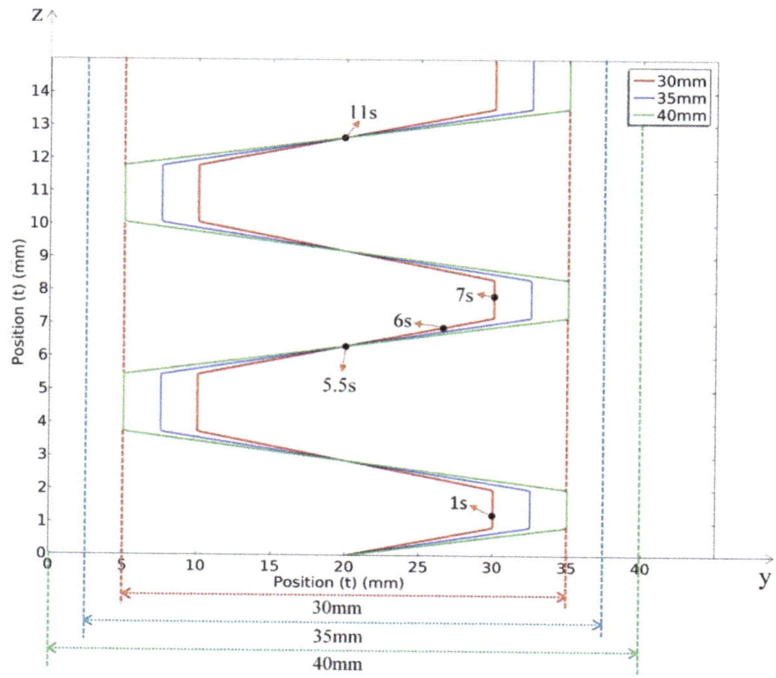

Figure 8. Movement path of oscillate-stop for thickness of 30 mm, 35 mm and 40 mm.

3. Model Verification via Experiment

To verify the accuracy of the model, a cross-section is taken along the thickness (y-axis) of the weld joint. As shown in Figure 9, the calculated melting pool and fusion line using different heat source path modes are compared against the experimental welding joints. The model fusion line is taken at the material melting point, 2055 K.

(1) As shown in Figure 9b, the application of the linear heat source path did not cause sufficiently high temperatures to melt the top and bottom of the joint; only the centre part of joint is melted, which differs to the experimental observation.

(2) The application of the sinusoidal path (see Figure 9c) leads to a fully melted weld joint in the thickness direction. The fusion line extended to the outside of the welding groove, but the area covered by the fusion line is smaller than that observed in the experiments (see Figure 9a).

(3) The application of the oscillate-stop path leads to a fully melted weld joint in the thickness direction. Additionally, the fusion line extended to the outside of the welding groove base metal near the fusion is melted to form a solid joint, and the area covered by the fusion line is similar to that observed in the experiments.

Quantitative analysis is carried out to measure the coordinates of the points of intersection between the fusion lines and top and bottom surface (marked as A, B, C and D in Figure 9). The coordinates for points A, B, C, and D are listed in Table 7. The difference between simulated and experimental coordinates is calculated, defined as "error" as shown in Table 7, where Error = (simulated results-experiment result)/experiment result. The "×" sign refers to the lack of fusion at the given location, i.e., no fusion line in the area.

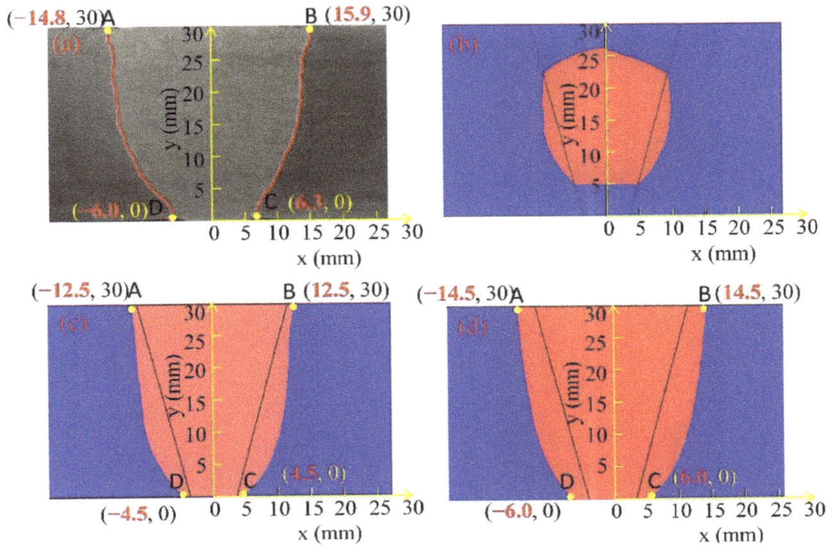

Figure 9. Melt pool and fusion line of experiments and simulated welded joints of 30 mm thickness steel plates, (**a**) experimental weld joint, (**b**) modelling using linear path heat source, (**c**) modelling using sinusoidal path heat source, (**d**) modelling using oscillate-stop path heat source.

Table 7. Comparison of coordinates of selected points between simulated and experimental welding melt pools.

Coordinate	Experiment (mm)	Linear (mm)	Error	Sinusoidal (mm)	Error	Oscillate-Stop (mm)	Error
A	(−15.9, 30)	×	×	(−12.5, 30)	21.4%	(−14.5, 30)	8.8%
B	(14.8, 30)	×	×	(12.5, 30)	15.5%	(14.5, 30)	2.0%
C	(6.3, 0)	×	×	(4.5, 0)	28.6%	(6.0, 0)	4.8%
D	(−6.0, 0)	×	×	(−4.5, 0)	25.0%	(−6.0, 0)	0

Note: Error = (simulated result—experiment result)/experiment result.

In the case of the linear heat source path, there are no fusion lines running through the top to the bottom of the welded joint. For the sinusoidal heat source path and the oscillate-stop heat source path, the welding arc not only moves from bottom to top during the welding process but also along with the weld oscillates in the direction of the depth of the melt pool, which helps the welding bevels on both sides to obtain the same melting depth. The "errors" of four selected points with sinusoidal heat source are: 21.4%, 15.5%, 28.6% and 25.0%, respectively, while the error of four selected points with oscillate-stop heat source is 8.8%, 2.0%, 4.8% and 0%, respectively. The "error" of the oscillate-stop heat source is greatly reduced. The heat source with oscillate-stop path holding on at near surface of steel plate for more times than sinusoidal path, which helps to transfer more heat to the surface of the joint during the welding process.

In summary, the path of the heat source can significantly affect the thermal profile of the weld joint, hence FEA models are essential in optimizing and predicting accurately the thermal profile, including the fusion line and shape of the melt zone in different thickness and path conditions.

4. Evolution of Thermal Profile in Heat Affected Zone

4.1. Simulated Heat Source Paths

Heat source paths calculated using the linear heat source path, sinusoidal path and oscillate-stop path are shown in Figure 10. In the figure, thermal profiles indicating heat sources at 340 s, 349 s, and 357 s are shown as examples. The linear heat source moves from the bottom to the top along the centre of the thickness of the steel plate, the sinusoidal heat source moves along with the weld direction and the thickness of the weld pool at the same time, and the oscillate-stop heat source path stays near the surface for a period of time when oscillating to the internal and external position. The numbers in brackets on the right side of every picture are the coordinates of the heat source centre of every picture. The x axis alone the welding line and the y axis alone is the plate thickness direction. When the welding time is 0 s, the heat source centre is in the starting point of the welding line (z) and the centre of plate thickness (y); the coordinates of the heat source centre are the origin coordinates (0, 0).

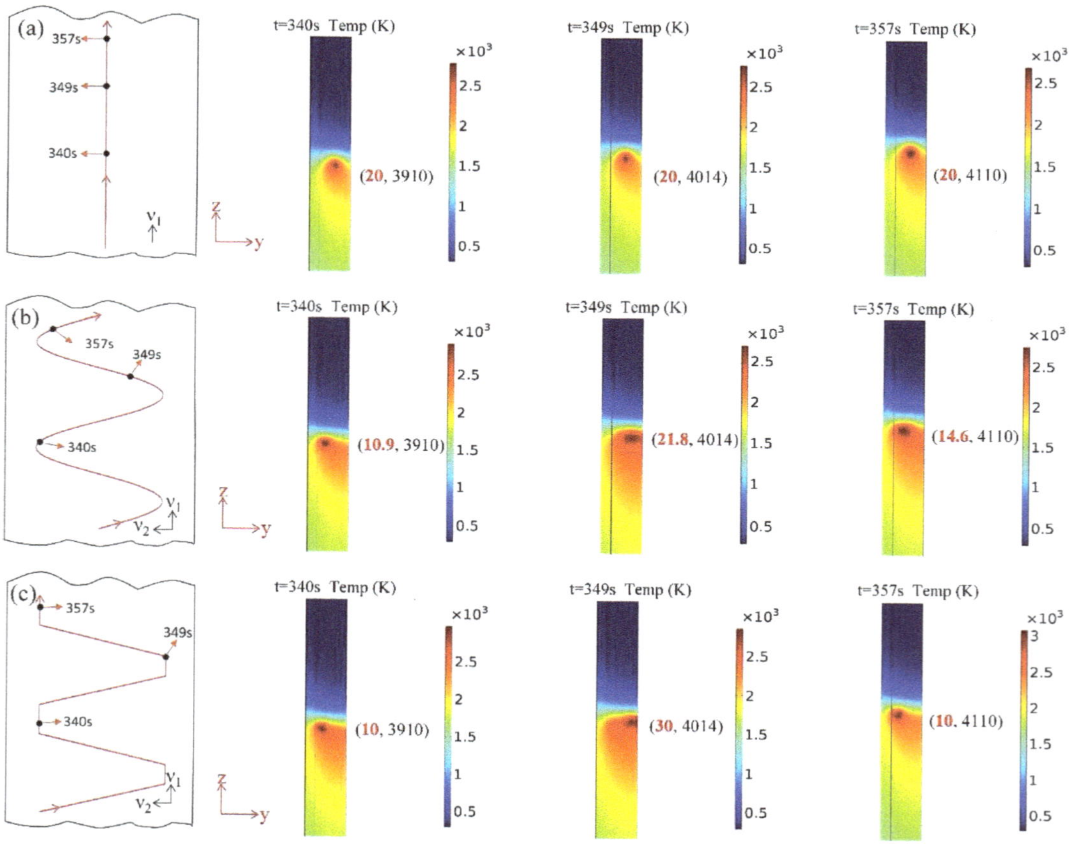

Figure 10. Calculated thermal profiles using three different heat source paths at different welding times of 340 s, 349 s and 357 s. (**a**) linear path (**b**) sinusoidal path, (**c**) oscillate-stop path.

4.2. Thermal Cycles in Heat Affected Zone

Figure 11a is a schematic diagram indicating the positions of the selected points in the heat affected zone (HAZ). Point 1, point 6 and point 8 are located in the groove which belongs to the molten pool, point 2, point 3, point 4, point 5 and point 7 are located in the base metal, the distance to the heat source of the five points is from 11 mm to 50 mm. Figure 11b–d show the simulated thermal cycle curve of the five selected points during the welding process with the linear path, sinusoidal heat source path and oscillate-stop heat source path, respectively.

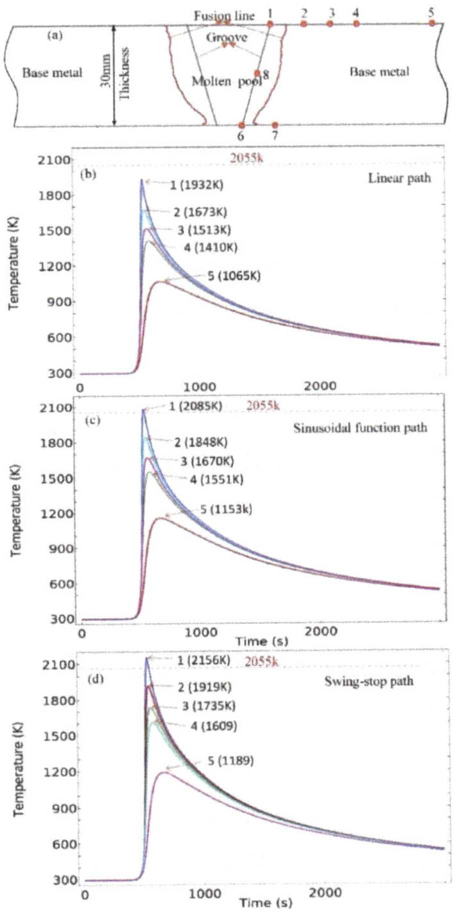

Figure 11. Calculated thermal cycles at selected points in heat affect zones (HAZ), (**a**) position of selected points in HAZ, (**b**) thermal cycle curves using linear path, (**c**) thermal cycle curves using sinusoidal path, (**d**) thermal cycle curves using oscillate-stop path.

As illustrated in Figure 11, the temperature quickly rises when the welding heat source approaches the selected point, reaching a peak value, and then it decreases gradually as the welding heat source moves away from the point. The peak temperature varies according to the distance from the welding central line to the points. As listed in Table 8, the peak temperature of point 1 with a linear heat source path is 1932 K, which does not reach the melting point of 2055 K. The peak temperature of point 1 with sinusoidal heat source path is 2085 K, which exceeds the melting point by 30 K, the point 1 will be melted. The peak temperature of point 1 with the oscillate-stop heat source path is 2156 K, which exceeds the melting point by 106 K, compared with the sinusoidal heat source path; its fusion line advances to a position further away from the groove to form a joint with better quality. The peak temperature values of the selected point are listed in Table 8.

Table 8. The peak temperature of selected point as shown in Figure 8.

Path \ Point	1 (K)	2 (K)	3 (K)	4 (K)	5 (K)	6 (K)	7 (K)	8 (K)
Linear path	1932	1673	1513	1410	1065	1810	1640	2420
Sinusoidal path	2085	1848	1670	1551	1153	2021	1811	2295
Oscillate-stop path	2156	1919	1735	1609	1189	2410	1960	2223

Using the sinusoidal path and the oscillate-stop path, the distance between the heat source and the selected points becomes closer when the heat source moves to the surface. While the oscillate-stop path stays near the internal and external position for a period of time in the cycle of movement. The welding arc can transfer more heat to the surface of steel plates and advance the fusion line in the position. So, the temperature in HAZ with the oscillate-stop path is higher than that with the sinusoidal path. EGW is a single pass method for welding thick steel plates, so the stop-over of the heat source at a near-surface position has great significance on the quality of the weld joint.

5. Recommended Heat Source Path Model for Simulating EGW Welding of Thick Steel Plates

Three different heat source paths have been used for simulating EGW welding of thick steel plates. In this section, the study will examine the effect of the heat source model on the shape of weld pool and the fusion line, so that a different applicable heat source model for simulating EGW thick plates can be defined.

The size of the cross section of the work piece gap is shown in Figure 12. The heat source movement path for 20 mm, 25, 30 mm, 35 mm and 40 mm thick steel plate were designed and listed in Tables 5 and 6 and Figures 7 and 8.

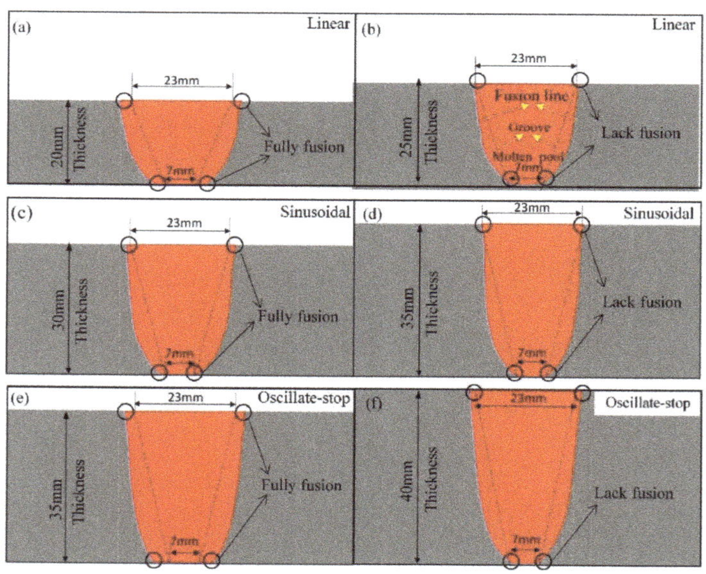

Figure 12. The simulated welding molten pool with three heat sauce paths. Linear heat source path model for 20 mm thickness (**a**), for 25 mm thickness (**b**); Sinusoidal path for 30 mm thickness (**c**), for 35 mm thickness (**d**); Oscillate-stop path for 35 mm thickness (**e**) and for 40 mm thickness (**f**).

Simulations were carried out on steel plates with different thicknesses and different heat source models, as listed below. Other modelling parameters are listed in Tables 3 and 4, and the heat input is 157 kJ/cm for all trials.

- 20 mm, 25 mm thickness steel plates—using the linear path (Figure 12a,b);
- 30 mm, 35 mm thickness steel plates—using sinusoidal path (Figure 12c,d);
- 35 mm and 40 mm thick steel plates—using oscillate-stop path (Figure 12e,f).

For the linear path model: Figure 12a shows a fully fusion welding joint for 20 mm thick steel plates, while Figure 12b shows the lack of fusion in the bottom of the joint for 25 mm thick steel plates. So, the maximum applicable thickness for the linear heat source path is estimated to be 20 mm.

For the sinusoidal heat source path model: Fully fusion of the weld pool was obtained for 30 mm thick steel plates, as shown in Figure 12c, while Figure 12d shows a lack of fusion in the bottom of the joint for 35 mm steel plate. So, the maximum applicable thickness for the sinusoidal function path is estimated to be under 30 mm.

For the oscillate-stop path: Figure 12e shows that a fully fusion of the weld pool is obtained for 35 mm thick steel plates, but the lack of fusion in the bottom of the joint for 40 mm thick steel plates. So, the maximum applicable thickness for the oscillate-stop heat source path is estimated to be around 35 mm.

6. Conclusions

1. Three different types of heat source path models (linear, sinusoidal function and oscillate-stop) were implemented to simulate the EGW process of marine steel with a heat input of 157 kJ/cm.
2. For EGW welding of 30 mm thickness steel plates, the model using the oscillate heat source path predicted a more accurate thermal profile (the shape of melt pool and the fusion line) than those using the sinusoidal and linear heat source paths.
3. The applicable heat source paths for modelling the EGW process of steel plates with different thicknesses were investigated. The linear path model can be used for simulating steel plate thickness up to 20 mm, a maximum thickness of 30 mm is appropriate for the sinusoidal path, and a maximum thickness of 35 mm is appropriate for the oscillate-stop path in EGW welding, with a heat input of 157 kJ/cm.

Author Contributions: Conceptualization, J.F. and H.D.; methodology, Q.T.; software, J.F.; validation, J.F., B.N. and M.L.; formal analysis, Q.T.; investigation, J.F.; resources, H.D.; data curation, X.Y.; writing—original draft preparation, J.F.; writing—review and editing, B.N., J.F.; supervision, H.D.; project administration, B.T.; funding acquisition, B.T. All authors have read and agreed to the published version of the manuscript.

Funding: This research received no external funding.

Institutional Review Board Statement: Not applicable.

Informed Consent Statement: Not applicable.

Data Availability Statement: For further data sets, please contact the corresponding author.

Acknowledgments: All authors gratefully acknowledge the support from Nanjing Iron & Steel United Co., Ltd. (NISCO), in particular for providing a PhD scholarship for this study.

Conflicts of Interest: The authors declare no conflict of interest.

References

1. Ichimiya, K.; Sumi, H.; Hirai, T. 460 MPa Yield Strength Class Steel Plate with JFE EWEL® Technology for Large Heat Input Welding. *JFE Tech. Rep.* **2008**, *5*, 7–12.
2. Kaplan, M.B.; Solomon, S. A coming boom in commercial shipping? The potential for rapid growth of noise from commercial ships by 2030. *Mar. Policy* **2016**, *73*, 119–121. [CrossRef]
3. Sasaki, K.; Suda, K.; Motomatsu, R.I.; Hashiba, Y.; Ohkita, S.; Imai, S. Development of Two-electrode Electro-gas Arc Welding Process. *Shinnittetsu Giho* **2004**, *90*, 57–64.
4. Fritzsche, A.; Avilov, V.; Gumenyuk, A.; Hilgenberg, K.; Rethmeier, M. High power laser beam welding of thick-walled ferromagnetic steels with electromagnetic weld pool support. *Phys. Procedia* **2016**, *83*, 362–372. [CrossRef]
5. Wang, G.; Li, J.B.; Sha, Y.Z.; Ma, J. Study on the penetration of electrogas welding. *Weld. Technol.* **2002**, *5*, 8–9.

6. Eakkachai, W.; Koei, H.; Motomichi, Y.; Kenji, S.; Kota, K.; Tadakazu, T.; Hiroshi, Y.; Tsutomu, F.; Shin, N.; Tetsuro, N.; et al. Welding phenomena during vertical welding on thick steel plate using hot-wire laser welding method. *Proc. Weld. Soc.* **2015**, *33*, 143s–147s.
7. Haelsig, A.; Mayr, P. Energy balance study of gas-shielded arc welding processes. *Weld. World* **2013**, *57*, 727–734. [CrossRef]
8. Bai, X.; Colegrove, P.; Ding, J.; Zhou, X.; Diao, C.; Bridgeman, P.; Hönnige, J.R.; Zhang, H.; Williams, S. Numerical analysis of heat transfer and fluid flow in multilayer deposition of PAW-based wire and arc additive manufacturing. *Int. J. Heat Mass Transf.* **2018**, *124*, 504–516. [CrossRef]
9. Aucott, L.; Huang, D.; Dong, H.B.; Wen, S.W.; Marsden, J.A.; Rack, A.; Cocks, A.C.F. Initiation and growth kinetics of solidification cracking during welding of steel. *Sci. Rep.* **2017**, *7*, 40255. [CrossRef]
10. Aucott, L.; Huang, D.; Dong, H.B.; Wen, S.W.; Marsden, J.; Rack, A.; Cocks, A.C.F. A Three-stage mechanistic model for solidification cracking during welding of steel. *Met. Mater. Trans. A* **2018**, *49*, 1674–1682. [CrossRef]
11. Hwang, S.-Y.; Kim, Y.; Lee, J.-H. Finite element analysis of residual stress distribution in a thick plate joined using two-pole tandem electro-gas welding. *J. Mater. Process. Technol.* **2016**, *229*, 349–360. [CrossRef]
12. Xu, G.; Pan, H.; Wang, J. Numerical analysis model of temperature profile in swing-arc narrow gap GMAW welding. *Thansactions China Weld. Inst.* **2017**, *38*, 55–60. (In Chinese)
13. Yuan, S.; Liu, W.; Li, L.; Jiang, X. Heat source model of temperature profile in swing-arc narrow gap MAG welding considering sidewall fusion. *Thansactions China Weld. Inst.* **2018**, *39*, 95–99. (In Chinese)
14. Pereira, H.A.D.S.; Rodrigues, M.C.; Firmino, J.V.L.D.C. Implementation of weave patterns by path parameterization in the simulation of welding processes by the finite element method. *Int. J. Adv. Manuf. Technol.* **2019**, *104*, 477–487. [CrossRef]
15. Wang, Z.; Oliveira, J.P.; Zeng, Z.; Bu, X.; Peng, B.; Shao, X. Laser beam oscillating welding of 5A06 aluminum alloys: Microstructure, porosity and mechanical properties. *Opt. Laser Technol.* **2019**, *111*, 58–65. [CrossRef]
16. Guo, Z.; Saunders, N.; Schillé, J.; Miodownik, A. Material properties for process simulation. *Mater. Sci. Eng. A* **2009**, *499*, 7–13. [CrossRef]
17. Saunders, N.; Guo, U.K.Z.; Li, X.; Miodownik, A.P.; Schillé, J.-P. Using JMatPro to model materials properties and behavior. *JOM* **2003**, *55*, 60–65. [CrossRef]
18. Rosenthal, D. The theory of moving sources of heat and its application of metal treatments. *Trans. ASME* **1946**, *68*, 849–866.
19. Lecoanet, A.; Ivey, D.; Henein, H. Simulation of the temperature profile during welding with COMSOL multiphysics® software using Rosenthal's approach. In Proceedings of the 2014 COMSOL, Cambridge, UK, 17 September 2014.
20. Wang, H.X.; Sun, J.S.; Wei, Y.H.; Zheng, Y.Y. Simulation of GMAW thermal process based on string heat source model. *Sci. Technol. Weld. Join.* **2005**, *10*, 511–520. [CrossRef]
21. Eagar, T.; Tsai, N. Temperature fields produced by traveling distributed heat sources. *Weld. J.* **1983**, *62*, 346–355.
22. Rouquette, S.; Guo, J.; Le Masson, P. Estimation of the parameters of a Gaussian heat source by the Levenberg–Marquardt method: Application to the electron beam welding. *Int. J. Therm. Sci.* **2007**, *46*, 128–138. [CrossRef]
23. Fu, G.; Gu, J.; Lourenco, M.I.; Duan, M.; Estefen, S.F. Parameter determination of double-ellipsoidal heat source model and its application in the multi-pass welding process. *Ships Offshore Struct.* **2014**, *10*, 204–217. [CrossRef]
24. Goldak, J.; Chakravarti, A.; Bibby, M. A new finite element model for welding heat sources. *Metall. Trans. B* **1984**, *15*, 299–305. [CrossRef]
25. Podder, D.; Mandal, N.; Das, S. Heat source modeling and analysis of submerged arc welding. *Weld. J.* **2014**, *93*, 183–191.
26. Kik, T. Computational techniques in numerical simulations of arc and laser welding processes. *Materials* **2020**, *13*, 608. [CrossRef] [PubMed]
27. Kik, T.; Górka, J. Numerical simulations of laser and hybrid S700MC T-joint welding. *Materials* **2019**, *12*, 516. [CrossRef] [PubMed]
28. Kik, T. Heat source models in numerical simulations of laser welding. *Materials* **2020**, *13*, 2653. [CrossRef] [PubMed]
29. Hashemzadeh, M.; Chen, B. Comparison between different heat sources types in thin-plate welding simulation. In *Developments in Maritime Transportation and Exploitation of Sea Resources*; Taylor & Francis: Abingdon, UK, 2013; pp. 329–335.
30. Tong, M.; Duggan, G.; Liu, J.; Xie, Y.; Dodge, M.; Aucott, L.; Dong, H.; Davidchack, R.L.; Dantzig, J.; Barrera, O.; et al. Multiscale, multiphysics numerical modeling of fusion welding with experimental characterization and validation. *JOM* **2012**, *65*, 99–106. [CrossRef]
31. Cai, Z.; Wu, S.; Lu, A.; Zhao, H.; Shi, Q. Line Gauss heat source model: An efficient approach for numerical welding simulation. *Sci. Technol. Weld. Join.* **2001**, *6*, 84–88. [CrossRef]
32. Norman, A.F.; Ducharme, R.; Mackwood, A.; Kapadia, P.; Prangnell, P.B. Application of thermal modelling to laser beam welding of aluminium alloys. *Sci. Technol. Weld. Join.* **1998**, *3*, 260–266. [CrossRef]

materials

MDPI

Article

Exploratory Data Analysis for the Evaluation of Tribological Properties of Wear-Resistant Surface Layers Modified with Rare-Earth Metals

Paweł Malinowski [1], Justyna Kasińska [2,*], Sławomir Rutkowski [3,4] and Monika Madej [2]

[1] Faculty of Foundry Engineering, AGH University of Science and Technology, 30-059 Krakow, Poland; pamalino@agh.edu.pl
[2] Department of Metal Science and Materials Technology, Kielce University of Technology, 25-314 Kielce, Poland; mmadej@tu.kielce.pl
[3] The Faculty of Mechanics and Technology, Rzeszow University of Technology, 37-450 Stalowa Wola, Poland; s.rutkowski@prz.edu.pl
[4] Multi-Branch Company T.S.A. Marcin Górski Sławomir Rutkowski, 37-450 Stalowa Wola, Poland
* Correspondence: kasinska@tu.kielce.pl; Tel.: +48-414324470

Abstract: The role of rare Earth metals in the improvement of the properties of metals and alloys has been analysed and described in multiple studies. Their effects on changes in microstructure and mechanical properties are most pronounced. This paper focuses on the beneficial effect of rare Earth metal oxides on the wear resistance of surface layers applied to castings intended for structural elements of machinery and equipment in mining and recycling. The experiment involved modifying prepared surfaces by adding CeO_2, Y_2O_3, and La_2O_3. Hardness measurements, a scratch test, and tribological tests were performed under dry and fluid friction. The maximum wear track depth and track area were measured from the surface profile. To determine correlations between the results, exploratory data analysis was employed. Heatmaps were used to illustrate strong positive and negative interactions. The addition of oxides at increasing carbon content resulted in increased hardness, lower coefficient of friction, and reduced track area and maximum track depth. Strong negative interactions between the track area and maximum track depth were found. The differences resulting from the test conditions (fluid and dry friction) were discussed. This study demonstrated the suitability of exploratory data analysis for analysing research results and confirmed the improvement of modified surface wear resistance.

Keywords: EDA—exploratory data analysis; rare Earth metal oxides; wear resistance; heatmap; correlation

Citation: Malinowski, P.; Kasińska, J.; Rutkowski, S.; Madej, M. Exploratory Data Analysis for the Evaluation of Tribological Properties of Wear-Resistant Surface Layers Modified with Rare-Earth Metals. *Materials* **2022**, *15*, 2032. https://doi.org/10.3390/ma15062032

Academic Editors: Qing Liu and Jiangshan Zhang

Received: 30 January 2022
Accepted: 7 March 2022
Published: 9 March 2022

1. Introduction

The interest in rare Earth elements observed in recent years is driven by rising application opportunities provided by new electronic technologies and nanotechnologies [1–4]. In addition to these specialized areas of the economy, rare Earth metals (REM) are still used in the metallurgical, chemical, and metal industries. Because of their physical and chemical properties, rare Earth oxides began to be used in surface engineering. Their beneficial effect of reducing secondary dendrites and volume fraction of non-metallic inclusions was noticed in the case of nickel-based [5,6] or iron-based [7] coating microstructures. Another advantage was the increased corrosion resistance and improved passivation. In their work [8], Liu et al. reported an advantageous effect of yttrium oxide Y_2O_3 on the modification of the nickel-based coating, indicating, however, that the range of its application is limited (from 0.4 to 0.6%).

By influencing changes in the microstructure of steel, alloys, and metallic layers, REM directly improve tribological properties, corrosion resistance, mechanical properties, and resistance to oxidation [9–11]. Rare Earth metal additions change the parameters of the structure, e.g., grain size, number and size of inclusions [12–15], and the mechanical.

Adding La_2O_3, Gd_2O_3, Lu_2O_3 particles increases the hardness of surface layers [16,17]. The authors of the work [18] showed that in the case of Si_3N_4 ceramics, the addition of rare Earth oxides reduced the friction coefficient and wear. Silicon nitride ceramics sintered with the addition of rare Earth oxides also constitute an important class of materials for high temperature applications. In addition to high temperature strengths, they possess good thermal shock resistance, creep resistance, and high oxidation resistance [16,19–22].

A number of tests and devices can be applied to determine tribological properties of materials and obtain a broad spectrum of information [23–25]. Using only selected parameters, e.g., a friction coefficient or linear wear, and ignoring mass loss, for example, a proper description of the tested material is impossible. In addition, the results obtained are affected by the test conditions, e.g., temperature, friction pair, motion, medium—dry friction and fluid friction [26].

Data analysis or elements of artificial intelligence, therefore, are increasingly used to correctly interpret scientific results. Exploratory data analysis was introduced to support scientific processes and use statistical methods in solving real problems. Exploratory analysis allows for determining the correlation between the results, its type and strength, as well as exclude or significantly limit the influence of the human factor [27,28].

Exploratory data analysis (EDA) may be defined as the art of looking at one or more datasets in an effort to understand the underlying structure of the data contained there [29,30].

The current paper uses the methods of exploratory data analysis to show the relationships between the selected properties of the layers. The authors also wanted to demonstrate the suitability of EDA for analysing the results of tribological tests.

2. Methodology

2.1. Surface Modification

As part of the experiment, the weld deposit was prepared using metallic powders and rare Earth oxides (CeO_2, Y_2O_3, La_2O_3) (Pol-Aura, Olsztyn, Poland). Nickel-based powders (COB-ARC, Chorzow, Poland), iron-based powders (COB-ARC, Chorzow, Poland), and powders containing chromium and tungsten carbides (COB-ARC, Chorzow, Poland) were used to produce the deposits. The weld metals were then applied on S355 steel (Hut-Trans, Katowice, Poland) using metal active gas (MAG) welding in combination with plasma arc welding (PAW). Metallographic specimens were prepared from the collected samples for microscopic observation of microstructural changes. For this purpose, a Phenom XL scanning electron microscope (Thermo Fisher Scientific /Phenom-World, Eindhoven, The Netherlands) was employed. Representative images of the pads made of metallic powder on a nickel matrix (chemical composition: 0.6% C, 3% Fe, 11% Cr, 3.8% Si, the rest Ni) are shown in Figure 1. The modification changed the morphology and arrangement of dendrites, which was reflected in the change of wear resistance.

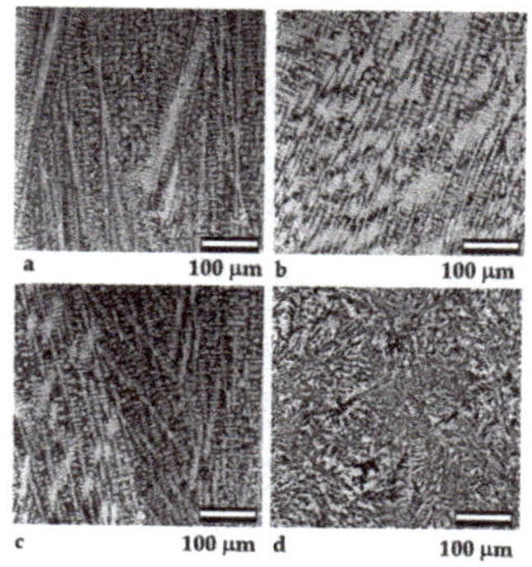

Figure 1. Microstructure of the Ni-based pad weld on the S355 steel: (**a**) non-modified and modified: (**b**) CeO_2, (**c**) La_2O_3, (**d**) Y_2O_3.

2.2. Tribological Tests

Tribological tests under dry friction and fluid friction with the addition of SiO_2 were carried out for the prepared variants, in which tribological characteristics (friction coefficient, linear wear) and wear indices (maximum wear track depth, track area) were determined.

The following tests were performed for all the welds:

- hardness test:
    - micro combi tester MCT^3 ANTON PAAR (Anton Paar, Corcelles-Cormondreche, Switzerland),
    - nominal loading force 2000 mN,
    - loading/unloading rate = 4000 mN/min,
    - Vickers indenter.
- scratch test
    - micro combi tester MCT^3 ANTON PAAR (Anton Paar, Corcelles-Cormondreche, Switzerland),
    - initial loading force 30 mN,
    - final loading force 15,000 mN
    - loading/unloading rate = 59,979.8 mN/min,
    - Rockwell indenter
    - indenter radius 100 μm.
- tribological tests under dry and fluid friction:
    - tribometer TRB3 Anton Paar (Anton Paar, Corcelles-Cormondreche, Switzerland),
    - reciprocating motion,
    - amplitude: 10 mm
    - frequency: 1 Hz
    - number of cycles: 10,000
    - friction type: dry friction/fluid friction (water solution 10% SiO_2)
    - temperature: $23 \pm 1°$
    - humidity: $50 \pm 1\%$
- tests for the maximum wear track depth and area measured from the surface profile:
    - optical profilometer: Leica DCM8 (Leica, Wetzlar, Germany)
    - the maximum track depth and the track area measured from the surface profile were taken as measures of the sample wear.

The tribological tests allowed determining wear resistance of the welds. Table 1 shows sample results from the experiments performed.

2.3. Exploratory Data Analysis

The data analysis was performed in three steps:

Acquisition

Wrangling

Exploration

In the first stage, exploratory data analysis (EDA) was carried out, which included description and visualisation of the data without assuming any initial hypotheses. The description and visualisation of the data made it possible to identify trends, patterns, missing data, outliers, etc.

Prior to exploratory data analysis, the following questions were generated:

- What are your analysis goals and outcomes?
- What tasks do you perform during analysis?
- What tools do you use?

Table 1. Compilation of sample test results.

			Tribological Tests				Hardness *			Scratch Test *	
			Dry Friction		Friction with Lubricant						
Sample Designation	RE Oxygen	Rare Earth Oxides % mas	Coefficient of Friction	Linear Wear (μm)	Coefficient of Friction	Linear Wear (μm)	Vickershardness (HV)	Instrumental Hardness H_{IT} (MPa)	Young's Moudulus (GPa)	Coefficient of Friction	Maximum Penetration Depth of the Indenter (μm)
1	-	0	0.06	39.16	-	-	-	-	-	0.181	−27.677
2	CeO_2	2	0.55	47.24	-	-	301.3	3192.5	207.4	0.214	−13.918
3	La_2O_3	2	0.48	33.54	-	-	-	-	-	0.193	−16,982
4	Y_2O_3	2	0.57	46.71	-	-	301.7	3196.4	194.1	0.221	−19.903
3-1	-	0	0.78	55.61	0.35	124.26	333.9	3537.6	184.1	0.158	−7.283
3-2	CeO_2	2	0.82	71.74	0.55	45.45	322.4	3415.9	171.9	0.163	−5.659
3-3	La_2O_3	2	0.54	64.1	0.6	57.14	327.7	3471.9	186.5	0.186	−9.281
3-4	Y_2O_3	2	0.79	71.07	0.18	96.74	428.2	4536.7	205	0.131	−6.557
4-1	-	0	0.67	55.8	0.5	60.97	335	3549.8	181.2	0.159	−7.46
4-2	CeO_2	2	0.38	97.73	0.62	64.03	356.7	3779.6	189.5	0.164	−9.219
4-3	La_2O_3	2	0.27	55.68	0.52	59.47	354	3750.9	184.7	0.165	−9.338
4-4	Y_2O_3	2	0.69	54.15	0.5	97.26	501.4	5312.6	214.5	0.119	−5.899

The first thing to do with any data set is to read it. This is done not only to get to know all the data collected, but also to reduce the workload during analysis. The initial data investigation is known as exploratory data analysis or EDA and it primarily focuses on visually inspecting the data. The main aim of EDA is to understand what data you have, what possible trends there are, and therefore which statistical tests will be appropriate to use [28].

In the EDA process, descriptive and visualisation analyses were performed, including data set description (number of samples, number of not e number (NaN)values), the removal of columns with a large number of empty NaNs, the insertion of missing data using strategy = mean, descriptive analysis (mean, standard deviation, min, max, median, 1st quartile, 3rd quartile), the visualisation of elements in individual classes, the identification of outliers, unsupervised learning using clustering, heat maps showing Pearson's correlations between features, and the visualisation of strong positive and negative correlations divided into 4 classes.

The following tools and libraries were used (open source):

- Python 3.10.2,
- Jupyter Notebook 6.4.5,
- Numpy 1.22.0,
- Pandas 1.4.1,
- Matplotlib 3.5.1,
- Seaborn 0.11.2,
- Scipy 1.8.0.

Explanation of the methods used:

- Inserting missing data. When the set has empty spaces, they can be filled using several strategies, such as inserting the mean, median, or the most common value.
- Identification of outliers. Outliers are data that do not follow the distribution of other data. Since they are anomalies that should not be modelled, they must be identified and removed. Outliers also have a negative effect on the Pearson's linear correlation coefficient.
- Cluster analysis. It is an unsupervised learning method that groups similar data using various algorithms (K-means, Hierarchical Clustering). This method enables cluster analysis, anomaly detection, and dimensionality reduction.
- Heatmap. A heatmap is a data visualisation technique that shows the magnitude of a phenomenon as a colour in two dimensions. Colour variation can be due to shade or intensity, giving visual clues as to how this phenomenon is clustered and how it changes in space.

In preparing the dataset for analysis, new columns were introduced, the abbreviations of which are shown in Table 2.

The set was divided into two parts:

- dry friction,
- fluid friction.

A NaN—Not a Number analysis was performed for the data set, Table 3.

Due to the small number of records (49), selected columns were removed (a large number of NaN values), i.e., Ni, Mo, Mn Co, B, W, V, WC. The remaining columns were completed with the mean values for each column using the SimpleImputer function. Tables 4 and 5 show the first five elements of both sets with the added mean values.

The descriptive information shown in Table 6 is then presented for the dry friction set, which shows the calculated values of mean, standard deviation, minimum, maximum value, and first, second, and third quartiles. Similar calculations were performed for fluid friction, as shown in Table 7.

Table 2. Description of abbreviations.

Abbrev.	Description	Unit	Test Type
cof1	Coefficient of friction (dry friction)		Tribological tests
lw1	Linear wear (dry friction)	(μm)	Tribological tests
cof2	Coefficient of friction (lubricated friction)		Tribological tests
lw2	Linear wear (lubricated friction)	(μm)	Tribological tests
vh	Vickers hardness	(HV)	Hardness
ih	Instrumental hardness	(MPa)	Hardness
ym	Young's modulus	(GPa)	Hardness
cof	Coefficient of friction		Scratch test
mpdi	Maximum penetration depth of the indenter	(μm)	Scratch test
mtd1	Maximum track depth (dry friction)	(μm)	Leica
ta1	Track area (dry friction)	(μm^2)	Leica
mwd2	Maximum track depth (lubricated friction)	(μm)	Leica
ta2	Track area (lubricated friction)	(μm^2)	Leica
dw1	Disc wear (dry friction)	(g)	Mass loss
bw1	Ball wear (dry friction)	(g)	Mass loss
dw2	Disc wear (lubricated friction)	(g)	Mass loss
bw2	Ball wear (lubricated friction)	(g)	Mass loss
type	1—without additives, 2—with an addition of CeO_2, 3—with an addition of La_2O_3 4—with an addition of Y_2O_3		

Table 3. Analysis of NaN values, before and after the removal of selected columns.

NaN Values before Removal		NaN Value after Removal	
C	0		
Si	4		
Cr	0		
Ni	21		
Mo	24		
Mn	45		
Fe	8		
Co	41		
B	25	C	0
W	20	Si	4
V	28	Cr	0
WC	41	Fe	0
cof1	0	cof1	0
lw1	0	lw1	0
cof2	4	cof2	4
lw2	4	lw2	4
vh	2	vh	2
ih	2	ih	2
ym	2	ym	2
cof	0	cof	0
mpdi	0	mpdi	0
mwd1	0	mwd1	0
wa1	0	wa1	0
mwd2	4	mwd2	4
wa2	4	wa2	4
dw1	0	dw1	0
bw1	0	bw1	0
dw2	4	dw2	4
bw2	4	bw2	4
type	0	type	0
dtype:	int64	dtype:	int64

Table 4. Five elements of the set for dry friction with the added means.

No.	C	Si	Cr	Fe	cof1	lw1	vh	ih	ym	cof	mpdi	mtd1	ta1	dw1	bw1	Type
1	0.1	2.1	18.0	62.9	0.06	39.16	598	6330	214	0.181	−27,667	63.79	52,888	0.0052	0.0003	1.0
2	0.1	2.1	18.0	62.9	0.55	47.24	301	3193	207	0.214	−13,918	61.05	48,955	0.0045	0.0001	2.0
3	0.1	2.1	18.0	62.9	0.48	33.54	598	6330	214	0.193	−16,982	47.23	42,889	0.0054	0.0000	3.0
4	0.1	2.1	18.0	62.9	0.57	46.71	301	3196	194	0.221	−19,903	59.33	48,247	0.0053	0.0000	4.0
5	0.6	3.8	11.0	3.0	0.78	55.61	334	3538	184	0.158	−7,283	37.07	28,214	0.0033	0.0001	1.0

Table 5. Five elements of the set for fluid friction with the added means.

No.	C	Si	Cr	Fe	cof2	lw1	vh	ih	ym	cof	mpdi	mtd2	ta2	dw2	bw2	Type
1	0.1	2.1	18.0	62.9	0.38	58.16	598	6330	214	0.181	−27,667	8.09	4939	0.0019	0.0008	1.0
2	0.1	2.1	18.0	62.9	0.38	58.16	301	3193	207	0.214	−13,918	8.09	4939	0.0019	0.0008	2.0
3	0.1	2.1	18.0	62.9	0.38	58.16	598	6330	214	0.193	−16,982	8.09	4939	0.0019	0.0008	3.0
4	0.1	2.1	18.0	62.9	038	58.16	301	3196	194	0.221	−19,903	8.09	4939	0.0019	0.0008	4.0
5	0.6	3.8	11.0	3.0	0.35	124.26	334	3538	184	0.158	−7.283	12.37	13,033	0.0090	0.0140	1.0

Table 6. Basic information for the set—dry friction.

	C	vh	ih	ym	cof	mtd1	ta1	dw1	bw1
count	49	49	49	49	49	49	49	49	49
mean	0.877	597	6330	214	0.138	15.60	12,675.0	0.00289	0.00024
std	0.317	187	1982	20	0.039	23.13	22,171.4	0.00338	0.00018
min	0.100	281	2980	172	0.072	0.18	8.7	0.0010	0.00000
25%	0.800	447	4733	202	0.102	1.58	177.5	0.0007	0.00010
50%	1.000	625	6579	214	0.134	3.95	1093.0	0.0017	0.00020
75%	1.000	744	7885	225	0.165	10.66	78,109.0	0.0041	0.00030
max	1.300	942	9985	257	0.221	70.05	78,109.0	0.01860	0.00080

Table 7. Basic information for the set—fluid friction.

	C	vh	ih	ym	cof	mtd2	ta2	dw2	bw2
count	49	49	49	49	49	49	49	49	49
mean	0.877	597	6330	214	0.138	8.09	4938.5	0.00194	0.00082
std	0.317	187	1982	20	0.039	4.90	6094.1	0.00184	0.00195
min	0.100	281	2980	172	0.072	0.16	4.6	0.00000	0.00000
25%	0.800	447	4733	202	0.102	5.02	912.8	0.00070	0.00030
50%	1.000	625	6579	214	0.134	7.14	3008.0	0.00130	0.00050
75%	1.000	744	7885	225	0.165	10.03	5109.0	0.00220	0.01400
max	1.300	942	9985	257	0.221	28.33	30,517.0	0.00760	0.01400

In order to demonstrate the correctness of the analyses carried out with regard to the number of tests performed with various modifiers (rare Earth oxides) and their effect on tribological properties, a class chart was prepared. For the tests conducted, the lack of balance between the individual classes may influence the real assessment. Figure 2 shows the number and percentage of the elements per class.

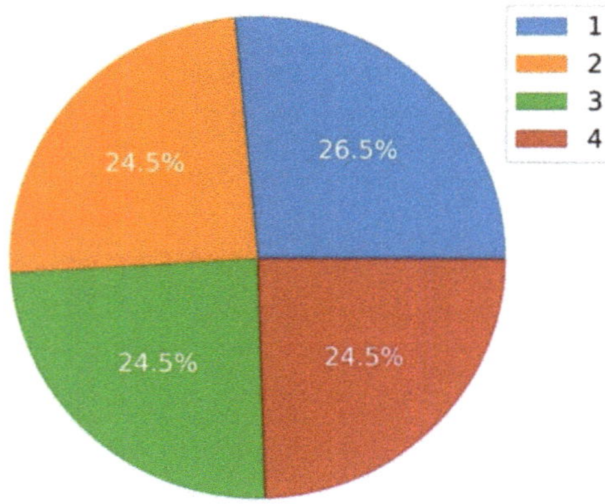

Figure 2. Elements in each class.

3. Results and Discussion

Figure 3 shows the dependence of disc wear on track area for each class.

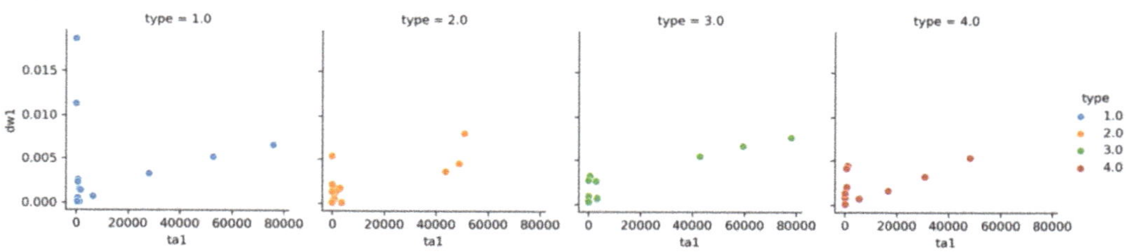

Figure 3. Dependence of dw1 on ta1 separately for each class.

Figure 4 shows the cumulative disc wear vs. track area. After removing the outliers, it can be seen that the data follow a linear dependence.

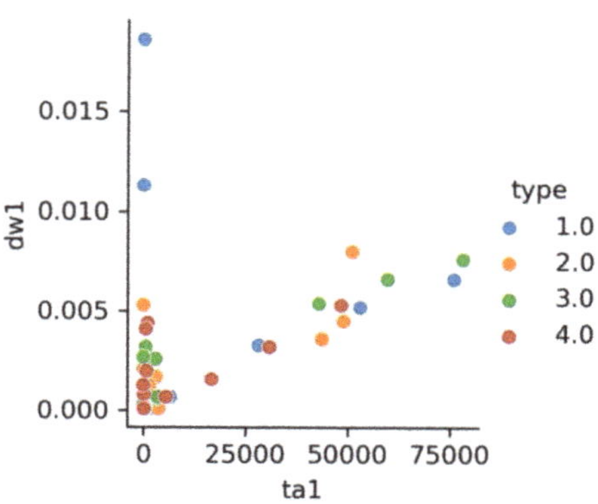

Figure 4. Dependence of dw1 on ta1 by class.

In the next step, a cluster analysis was carried out using two methods (K-means, Hierarchical). The analysis showed a high similarity between the two methods used. These are unsupervised classification methods of unsupervised learning.

Figure 5 shows the analyses for dry friction (column on the left) and fluid friction (column on the right).

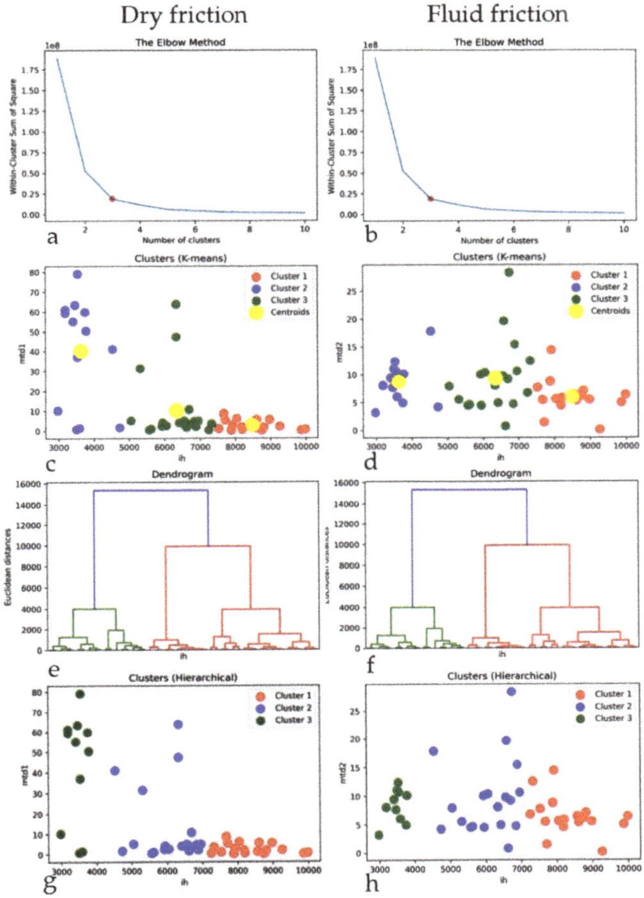

Figure 5. Cluster analysis for dry and fluid friction. Determination of optimal number of clusters using: (**a**,**b**) Elbow method; (**c**,**d**) clusters with centroids determined by K-means algorithm; (**e**,**f**) determination of optimal number of clusters for Hierarchical method using dendrograms; (**g**,**h**) clusters determined by Hierarchical algorithm.

The first row shows the optimal number of clusters determined using the so-called Elbow method. For both sets, three clusters were determined by the breakpoint. The second row shows the visualisation of the cluster analysis for both cases with the indication of the centroids for all clusters. The next line shows the dendrogram plots for dry and fluid friction on the basis of which the optimal number of clusters for the Hierarchical method was calculated.

The two heatmaps shown above are for the dry friction and fluid friction sets (Figures 6 and 7).

Pearson's correlations show that in the case of dry friction, a very strong negative correlation occurs between C-cof, C-mwd1, and C-ta1 and a very strong positive correlation occurs between vh-ih and ta1-mwd1. Pearson's correlation coefficient is a measure of linear correlation between two sets of data. In contrast, for fluid friction, a very strong negative correlation occurs between C-cof with a very strong positive correlation between ta2-mwd2. The results are summarised in Table 8. The correlation values are shown in Table 9.

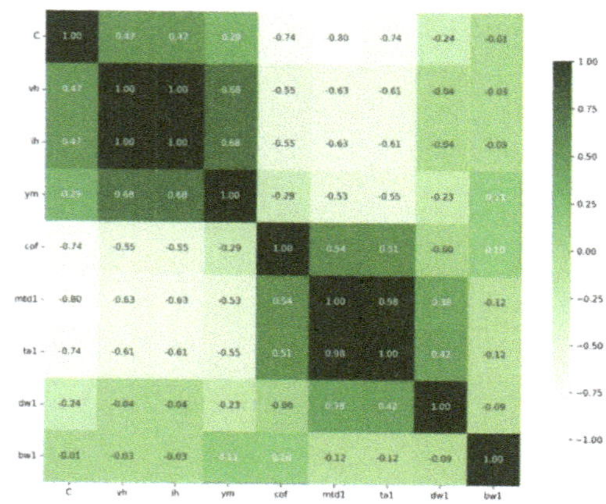

Figure 6. Heatmap for dry friction.

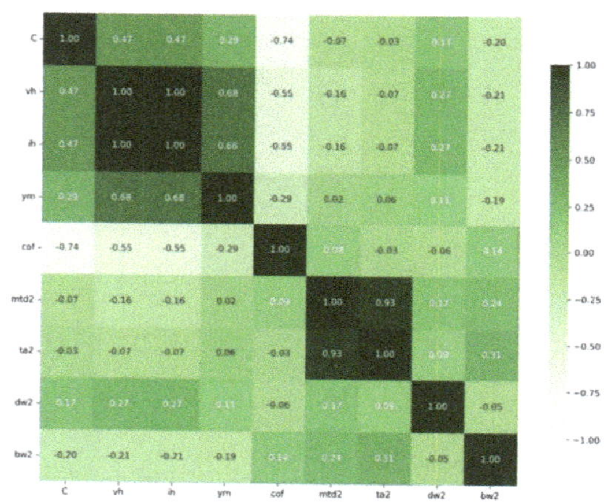

Figure 7. Heatmap for fluid friction.

Table 8. Significant correlations between the results.

Correlation	Dry Friction				Fluid Friction			
	Positive		Negative		Positive		Negative	
	Strong	V.Strong	Strong	V.Strong	Strong	V.Strong	Strong	V.Strong
	0.5–0.7	0.7–1	−0.5–0.7	−0.7–1	0.5–0.7	0.7–1	−0.5–0.7	−0.7–1
C				Cof, mwd, wa				Cof
vh	Ym	Ih	Cof, mtd, ta		Ym	Ih	Cof	
ih	Ym	Vh	Cof, mtd, wa		Ym	Vh	Cof	
ym	Vh, ih		mtd, ta		Vh, ih			
cof	mtd, ta		Vh, ih	C			Vh, ih	C
mtd	Cof	ta	Vh, ih, ym	C		ta		
ta	Cof	mtd	Vh, ih, ym	C		mtd		

Table 9. Strength of association.

Correlation	Negative	Positive
Very strong	−0.7 to −1	0.7 to 1
Strong	−0.5 to −0.7	0.5 to 0.7

Several strong correlations can be observed from the heatmaps for dry and fluid friction and for the pairwise correlations between these sets. For dry friction, the C index shows a very strong negative correlation with cof, mtd, and ta, whereas for fluid friction, only a very strong negative correlation with cof is observed. A strong negative correlation was found for ih and ym against the indices from mwd and ta for dry friction. In contrast, this correlation was not observed for fluid friction. For both sets, there is a very strong positive correlation between mtd and ta (0.98—for dry friction, 0.93—for fluid friction). Ym is strongly positively correlated with vh and simultaneously strongly negatively correlated with mtd and ta.

To confirm the correlations discussed above, they are presented in the form of relationships between the variables (Figures 8–10).

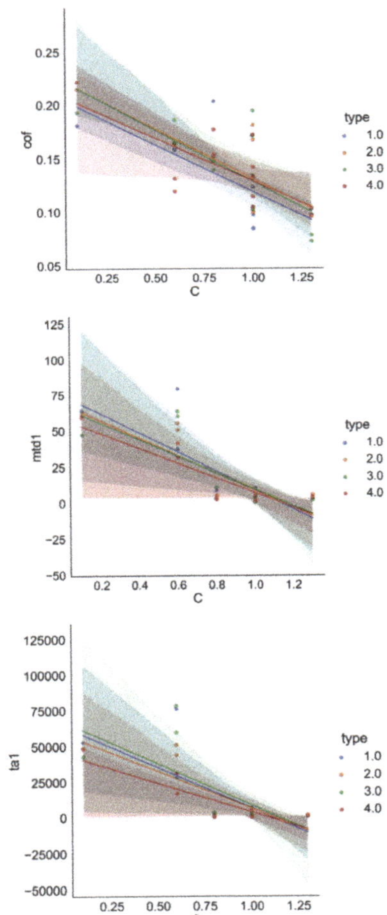

Figure 8. Relationships between the indices and carbon content.

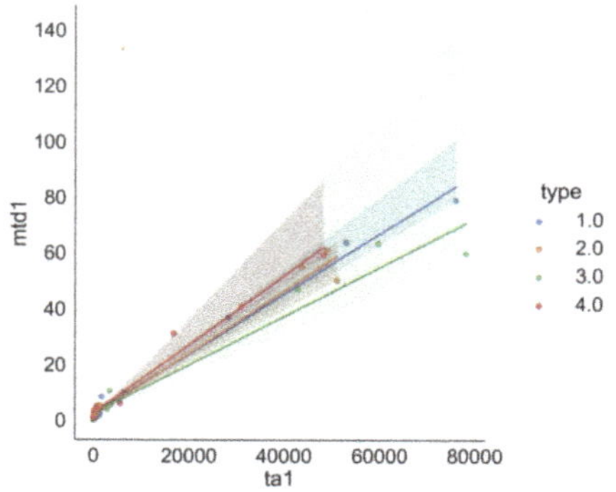

Figure 9. Maximum track depth versus track area—dry friction (positive correlation (0.98)).

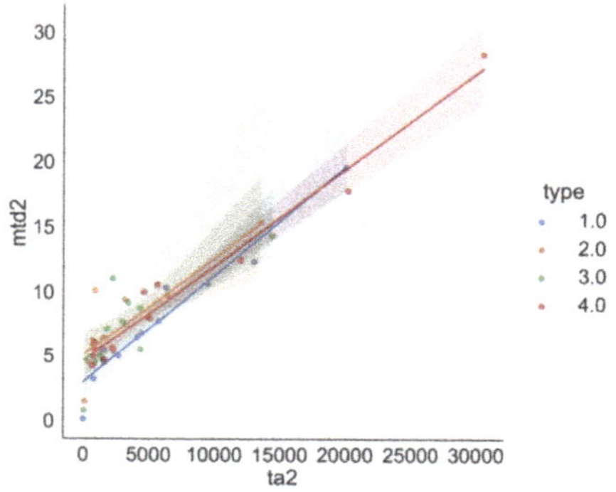

Figure 10. Maximum track depth versus track area—fluid friction (very strong positive correlation (0.93)).

The graphs for dry friction show very strong negative correlations (−0.74, −0.80, −0.74):

Figures 8–10 show the graphs in which the layers are assigned to classes, i.e., non-modified surface (class 1) and modified surfaces (classes 2, 3, 4). The differences between the individual lines corresponding to the classes are visible. Especially for class 4 (Y_2O_3), we observe a different slope of the obtained straight lines. Analysis does not reveal a quantitative but a qualitative relationship, indicating the effects of the additives on the parameters determined.

Exploratory data analysis showed dependencies between individual parameters. The proportionality of the carbon content in the welds with rare earth oxides to their mechanical properties was demonstrated. It was confirmed that its increase resulted in an increase in hardness. This is due to the formation of hard carbides in their microstructure and changes in the morphology of padding welds modified with rare Earth metals. The increase in hardness lowered the coefficients of friction "cof", which directly translated into the reduction of the surface of the track wear areas and the wear track depths. A slightly different nature of wear was demonstrated for dry friction and fluid friction. It was shown that despite the increase in hardness and the decrease in the friction coefficient, the remaining parameters in the form of the wear track area and depth were not so strongly correlated in the case of fluid friction. SiO_2 particles were used in the latter case, which changed the nature of the wear into tribocorrosion wear. The analyses confirmed both the complexity of the effect that parameters such as chemical composition or test environment have on wear processes, and the different nature of relationships between the parameters.

4. Conclusions

The authors showed that it is reasonable to use exploratory data analysis for the evaluation of the properties of RE oxide modified pad welds. In order to optimise the results, it is advisable to perform a number of additional experiments to expand the database. The expanded database will be used to predict the optimal parameters (chemical composition, percentage of rare Earth elements, padding method, etc.) while maintaining the best wear resistance. The results obtained so far indicate dependency between geometry indices and mechanical properties (friction coefficient, instrumental hardness).

Exploratory data analysis allowed determining directly proportional (positive correlations) and inversely proportional (negative correlations) dependencies. A significant dependency was observed for the "C" index, i.e., the effect of carbon on hardness, friction coefficient, and geometry indices (track area, track depth), which may result from the change in the pad weld morphology after introducing RE oxides. This issue will be addressed by the authors in their next work.

The study and analyses demonstrate that exploratory methods capture non-trivial dependencies of the obtained results.

The analysis of the results of tribological tests is the basis for the development of new techniques of forming surface layers and coatings with increased wear resistance. The wear indicators analysed in this study indicate further direction of surface modification research. The EDA results contributed to the knowledge on the influence of surface modification with rare Earth oxides on wear resistance. The study presented in this article offers a guidance to continue work on the technology of the surfacing process, selection of modifiers, and welding materials for surface layers.

Author Contributions: Conceptualization, J.K. and P.M.; methodology P.M. and J.K.; software, P.M.; validation, J.K., P.M. and M.M.; formal analysis, J.K., P.M., and M.M.; resources, S.R., J.K. and M.M.; data curation, S.R. and P.M.; writing—original draft preparation, J.K. and P.M.; writing—review and editing, J.K., P.M. and M.M.; visualization, P.M. and J.K.; supervision, J.K. and M.M. All authors have read and agreed to the published version of the manuscript.

Funding: This research received no external funding.

Institutional Review Board Statement: Not applicable.

Informed Consent Statement: Not applicable.

Data Availability Statement: Data sharing not applicable, all the data created for this study are already displayed in the article.

Conflicts of Interest: The authors declare no conflict of interest.

References

1. Naumov, A.V. Review of the world market of rare earth metals. *Russ. J. NonFerrous Met.* **2008**, *49*, 2231.
2. Lundin, R.; Wilson, J.R. Rare earth metals. *Adv. Mater. Processes* **2000**, *158*, 52–55.
3. Wang, L.; Lin, Q.; Ji, J.; Lan, D. New study concerning development of application of rare earth metals. *J. Alloys Compd.* **2005**, *408–412*, 384–386. [CrossRef]
4. Committee on Assessing the Need for a Defense Stockpile National Materials Advisory Board; Division of Engineering and Physical Science. *Managing Materials for a 21st Century Military*; The National Academies Press: Washington, DC, USA, 2008; Available online: http://www.nap.edu/catalog.php?record_id=12028 (accessed on 2 January 2022).
5. Wang, K.L.; Zhang, Q.B.; Sun, M.L.; Wei, X.G.; Zhu, Y.M. Microstructure and corrosion resistance of laser clad coatings with rare earth elements. *Corros. Sci.* **2001**, *43*, 255–267. [CrossRef]
6. Wang, X.H.; Zhang, M.; Zhou, Z.D.; Qu, S.Y. Microstructure and properties of laser clad TiC+NiCrBSi+rare earth composite coatings. *Surf. Coat. Technol.* **2002**, *161*, 195–199. [CrossRef]
7. Zhao, G.M.; Wang, K.L. Effect of La_2O_3 on corrosion resistance of laser clad ferrite-based alloy coatings. *Corros. Sci.* **2006**, *48*, 273–284. [CrossRef]
8. Liu, Q.B.; Zou, J.L.; Zheng, M.; Dong, C. Effect of Y_2O_3 con-tent on microstructure of gradient bioceramic composite coating produced by wide-band laser cladding. *J. Rare Earths* **2005**, *23*, 446.
9. Wu, C.F.; Ma, M.X.; Liu, W.J.; Zhong, M.L.; Zhang, H.J.; Zhang, W.M. Laser cladding in-situ carbide particle reinforced Fe-based composite coatings with rare earth oxide addition. *J. Rare Earths* **2009**, *27*, 997. [CrossRef]

10. Yi, W.; Zheng, C.Q.; Fan, P.; Cheng, S.H.; Li, W.; Ying, G.F. Ef-fect of rare earth on oxidation resistance of iron base flux-ing alloy spray-welding coating. *J. Alloys Compd.* **2000**, *311*, 65. [CrossRef]
11. Xing, X.G.; Han, Z.J.; Wang, H.F.; Lu, P.N. Electrochemical corrosion resistance of CeO_2-Cr/Ti coatings on 304 stainless steel via pack cementation. *J. Rare Earths* **2015**, *33*, 1122. [CrossRef]
12. Kasińska, J. Wide–ranging influence of mischmetal on properties of G17CrMo5–5 cast steel. *Metalurgija* **2015**, *54*, 135–138.
13. Kasińska, J. Influence of Rare Earth Metals on Microstructure and Mechanical Properties of G20Mn5 Cast Steel. *Arch. Foundry Eng.* **2018**, *18*, 37–42.
14. Kasińska, J.; Myszka, D. Influence of rare earths metals (REM) on the structure and selected properties of grey cast iron. *Metalurgija* **2020**, *59*, 459–462.
15. Dzioba, I.; Pala, R.; Kasińka, J. Experimental—Numerical Analysis of Stress State in Front of the Crack Tip of Modified And Unmodified G17crmo5–5 Cast Steel by Rare Earth Metals in a Brittle-Ductile Transition Region. *Arch. Metall. Mater.* **2016**, *61*, 1175–1181. [CrossRef]
16. Satet, R.L.; Hoffmann, M.J. Experimental evidence of the impact of rare-earth elements on particle growth and mechanical behaviour of silicon nitride. *J. Am. Ceram. Soc.* **2000**, *88*, 2485–2490. [CrossRef]
17. Choi, H.J.; Lee, J.G.; Kim, Y.W. High temperature strength and oxidation behaviourof hot-pressed silicon nitride-disilicate ceramics. *J. Mater. Sci.* **1997**, *32*, 1937–1942. [CrossRef]
18. Tatarkoa, T.; Kasiarováa, M.; Dusza, J.; Morgielb, J.; Sajgalíkc, P.; Hvizdoš, P. Wear resistance of hot-pressed Si_3N_4/SiC micro/nanocomposites sintered withrare-earth oxide additives. *Wear* **2010**, *269*, 867–874. [CrossRef]
19. Becher, P.F.; Painter, G.S.; Shibata, N.; Waters, S.B.; Lin, H.T. Effects of rare-earth(RE) intergranular adsorption on the phase transformation, microstructureevolution and mechanical properties in silicon nitride with RE_2O_3+MgO additives: RE = La, Gd, and Lu. *J. Am. Ceram. Soc.* **2008**, *91*, 2328–2336. [CrossRef]
20. Becher, P.F.; Shibata, N.; Painter, G.S.; Averill, F.; Van Benthem, K.; Lin, H.-T.; Waters, S.B. Observations on the influence of secondary Me oxide additives (Me = Si, Al, Mg) on the microstructural evolution and mechanical behavior of siliconnitride ceramics containing RE2O3(RE = La, Gd, Lu). *J. Am. Ceram. Soc.* **2010**, *93*, 570–580. [CrossRef]
21. Sharma, S.P.; Dwivedi, D.K.; Jain, P.K. Effect of La_2O_3 addi-tion on the microstructure, hardness and abrasive wear behavior of flame sprayed Ni based coatings. *Wear* **2009**, *267*, 853. [CrossRef]
22. Zhang, Z.Y.; Lu, X.C.; Luo, J.B. Tribological properties of rare earth oxide added Cr_3C_2-NiCr coatings. *Appl. Surf. Sci.* **2007**, *253*, 4377. [CrossRef]
23. Kalandyk, B. Microstructure and Abrasive Wear Resistance of 18Cr-4Ni-2.5Mo Cast Steel. *Arch. Foundry Eng.* **2012**, *12*, 81–84. [CrossRef]
24. Kalandyk, B.; Kasińska, J. Effects of Rare Earth Metal Addition on Wear Resistance of Chromium-Molybdenum Cast Steel. *Arch. Foundry Eng.* **2017**, *17*, 63–68.
25. Madej, M. Tribological Properties of Diamond-Like Carbon Coatings. *Adv. Mater. Res.* **2014**, *874*, 9–15. [CrossRef]
26. Styp-Rekowski, M.; Mańka, E.; Matuszewski, M.; Madej, M.; Ozimina, D. Tribological problems in shaft hoist ropes wear proces. *Ind. Lubr. Tribol.* **2015**, *67/1*, 157–164.
27. Wuest, T.; Weimer, D.; Irgens, C.; Thoben, K.-D. Machine learning in manufacturing: Advantages, challenges, and applications. *Prod. Manuf. Res.* **2016**, *4*, 23–45. [CrossRef]
28. Marian, M.; Tremmel, S. Current Trends and Applications of Machine Learning in Tribology—A Review. *Lubricants* **2021**, *9*, 86. [CrossRef]
29. Pearson, R.K. *Exploratory Data Analysis Using R*; CRC Press: Boca Raton, FL, USA, 2018.
30. Cox, V. *Translating Statistics to Make Decisions—A Guide for the Non-Statistician*; Apress: New York, NY, USA, 2017.

materials

MDPI

Article

Effect of Cerium on the Microstructure and Inclusion Evolution of C-Mn Cryogenic Vessel Steels

Liping Wu [1,2,3], Jianguo Zhi [2,3], Jiangshan Zhang [1], Bo Zhao [4] and Qing Liu [1,*]

1 State Key Laboratory of Advanced Metallurgy, University of Science and Technology Beijing, Beijing 100083, China; lipingw2021@126.com (L.W.); zjsustb@163.com (J.Z.)
2 Centre of Technology Baotou Iron and Steel Co., Baotou 014010, China; jianguoz2021@126.com
3 Inner Mongolia Enterprise Key Laboratory of Rare Earth Steel Products Research and Development, Baotou 014010, China
4 Institute of Metallurgical Technology, Iron and Steel Research Institute Beijing, Beijing 100083, China; zhao_working@163.com
* Correspondence: qliu@ustb.edu.cn; Tel.: +86-010-82375255

Abstract: The effects of Cerium (Ce) were studied on the casting slab quality, microstructure, and inclusion evolution of cryogenic vessel steel. An optical metallographic microscope, scanning electron microscope, energy dispersive spectrometer, and Thermo-calc thermodynamic software were used for characterization and analysis. The results indicated that the central segregation was significantly improved after adding Ce and reached the lowest level when the content of Ce was 0.0009 wt.%. Meanwhile, the presence of Ce reduces the size of ferrite and improves pearlite morphology. Ce also enables the modification of Al_2O_3 and MnS + $Ti_4C_2S_2$ inclusions into ellipsoid $CeAlO_3$ and spherical Ce_2O_2S + $Ti_4C_2S_2$ composite inclusions, respectively, which are easier to remove. The formed Ce_2O_2S inclusions are fine and can work as heterogeneous nucleation points to refine the microstructure of steel.

Keywords: cryogenic vessel steels; cerium; microstructure; inclusions

Citation: Wu, L.; Zhi, J.; Zhang, J.; Zhao, B.; Liu, Q. Effect of Cerium on the Microstructure and Inclusion Evolution of C-Mn Cryogenic Vessel Steels. *Materials* **2021**, *14*, 5262. https://doi.org/10.3390/ma14185262

Academic Editor: Andrea Di Schino

Received: 19 July 2021
Accepted: 9 September 2021
Published: 13 September 2021

Publisher's Note: MDPI stays neutral with regard to jurisdictional claims in published maps and institutional affiliations.

1. Introduction

Cryogenic vessel steel is an essential material for producing storage tanks and is widely used to store and transport liquefied gases. As increasing clean energy sources are developed and applied, cryogenic vessel storage equipment has been finding more applications [1–5]. Cryogenic vessel storage equipment requires stringent demands for the alloy composition, casting slab quality, and mechanical properties of the cryogenic vessel steel.

Importantly, inclusions affect the cleanliness, control, and quality of steel. During the solidification-rolling process, the morphology of inclusions is changed. Rare earth (RE) elements present a remarkable effect on modifying inclusions and in facilitating clean steel production, given their unique electronic layer structures. Earlier investigations have reported that different RE elements could purify molten steel, modify inclusion morphology, increase cryogenic toughness, and improve corrosion resistance in different steel series [6–10]. When RE is added to extra-low sulfur ($S < 0.003\%$) niobium-titanium micro-alloy steel, the deoxidation and inclusion removal rates can be increased, therefore the molten steel can be more effectively purified [11]. After adding La+Ce compounds to ship plate steel, the inclusions mainly exist in the form of spheres and approximate spheres, while the sharp-angled S-Mn-Ca phase and S-O-Mn-Ca-Al phase inclusions are reduced and the hot-rolled steel strip band structure is lightened [12]. The type of inclusions in drill steel without Ce additions are $MgAl_2O_4$ and (Ca, Mn)S. As the Ce content in drill steel reaches 0.0078% (mass fraction), the types of inclusions change to Ce-O and Ce-S and the size of the inclusions in the drill steel decrease significantly [13]. Inclusions affect the toughness, corrosion resistance, and fatigue properties of steel, and reducing their size weakens the impact of non-metallic inclusions on the properties of steel. The RE elements La, Ce, and Nd play deoxidation and desulfurization roles in molten steel, making the inclusions fine and diffuse in distribution [14]. Gao et al. found that after Ce treatment, irregular Al_2O_3 inclusions with a size of 10−15 μm are wrapped by rare earth and then gradually modified into spheroidal $CeAlO_3$, Ce_2O_3, and Ce_2O_2S inclusions with a size of ≈5 μm, distributed into interstitial free slabs [15].

It has been proposed in the literature that RE influences the microstructure of a casting slab. For example, with the addition of La into nonoriented electrical steel, the equiaxed crystal ratio increased and the columnar crystals were suppressed and refined [16]. The addition of La+Ce changed the solidification microstructure distribution of thin slabs produced by compact strip production and refined the solidification structure of 430 ferritic stainless steel, thereby improving the comprehensive properties of the material [17,18]. The addition of La had a significant effect on the distribution of arsenic by the formation of arsenic inclusions, and various RE have been widely used in the steelmaking process to improve steel quality [19].

However, there is still no systematic explanation regarding the mechanisms by which RE improves the quality of casting slabs, changes the structure morphology, and modifies the inclusions. Moreover, casting slab quality is the most important index to evaluate the finished product. The quality of a casting slab should be analyzed from a macro quality perspective, microstructure characteristics, and the evolution law of the inclusions. In this study, the characteristics of inclusions and the microstructure of C-Mn cryogenic vessel steels were comparatively studied without Ce and with the addition of different Ce content levels.

2. Experimental

The steel samples were taken from a Inner Mongolia Baotou Iron and Steel Group Company production site, located in Baotou, China. The production process uses a 240 t top and bottom combined blow basic oxygen furnace (BOF), dephosphorizing with the bottom blow, and performing automatic steelmaking with sublance 240 t ladle furnace (LF) refining, and 240 t Ruhrstahl–Heraeus (RH) refining treatment for a minimum of 25 min. The RE ferrocerium alloy (where the mass fraction of Ce is 30% and the total mass fraction of impurity elements including sulfur, oxygen, and phosphorus is lower than 0.01%, the rest being Fe) were added into the ladle at the end of vacuum treatment, and then re-pressed for 5 min. After re-pressing and soft blowing for greater than 8 min to homogenize the composition, it was ensured that the inclusion as fully floated and improved the cleanliness of the molten steel. During the continuous casting process of the slab, protective casting measures prevented an increase in nitrogen and oxygen. The casting speed was 1.2 $m \cdot min^{-1}$, and the casting slab section dimensions were 230 × 1550~2030 mm, using a 1650 mm double stand slab caster, respectively. The chemical composition of the tundish sample was analyzed by direct reading spectrum analyzer (ARL-4460, Thermo Fisher Scientific, Geneva, Switzerland), and the actual content of Ce in the steel was detected by chemical analysis (ICP-MS, Plasma Mass Spectrometer, Waltham, Massachusetts, America Perkin Elmer). Table 1 shows the chemical compositions of the experimental steel. The hot-rolled steel strip was produced on a 2250 mm rolling mill using controlled rolling and controlled cooling technology. The discharging temperature was 1190 °C, the finish rolling temperature was 840 °C, the coiling temperature was 600 °C, and the thickness of the finished product was 12 mm.

Table 1. Chemical composition (wt.%).

No.	C	Si	Mn	P	S	Ti	Als [1]	Ca	Ce
I	0.18	0.16	1.45	0.010	0.003	0.017	0.040	0.0020	0
II	0.18	0.15	1.48	0.010	0.002	0.018	0.039	0.0020	0.0006
III	0.18	0.15	1.48	0.010	0.002	0.018	0.039	0.0020	0.0009
IV	0.18	0.15	1.48	0.010	0.002	0.018	0.039	0.0020	0.0013

[1]: Als, acid-soluble aluminum.

The casting slab selected had a size of 300 mm × 1500 mm × 230 mm. After hot acid corrosion, casting slab quality was determined according to the standard diagrams for macrostructure and defects in a continuous casting slab (using YB/T 4003-2016, which is a kind of steel industry criterion in China, equal to ASTM E381-2017). In this study, we took one piece of steel at 3/4 of the width of the casting slab, with dimensional specifications of 100 mm × 100 mm × 230 mm, and we also chose three metallographic samples at 1/4 distances from the inner arc, slab center, and 1/4 distance from the outer arc, with dimensional specifications of 10 mm × 10 mm × 10 mm, following the specific sampling plan shown in Figure 1.

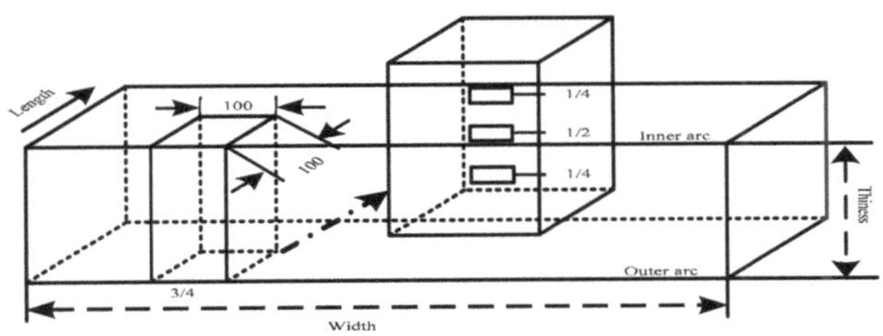

Figure 1. Casting slab sampling schematic diagram.

The sampling position of the hot-rolled finished plate corresponds to the casting slab, which is located at 3/4 of the width direction of the steel plate, and the sample size was 12 mm × 10 mm × 10 mm. The samples were mechanically ground (from #400~#1000) and then polished. Then samples were etched using 4% nital. Optical microscope (OM, Axio Observer D1 m, Jena, Germany) analysis was used to observe the microstructures. Inclusion distribution and morphology were determined using a scanning electron microscope (SEM, LEO EVO, Germany) and a field emission scanning electron microscope (FE-SEM, SIGMAN300, Carl Zeiss Company, Oberkochen, Germany). The compositions of inclusions were identified with an energy dispersive spectrometer (EDS, GENESIS, America).

In order to more intuitively show the effect of RE elements on the denaturation of other inclusions, Thermo-calc thermodynamic software was used to calculate the variation law of inclusions in C-Mn cryogenic vessel steel with the addition of Ce. Combined with the experimental results, the evolution law of various inclusions was summarized, and a dominant zone diagram of Ce inclusions was obtained.

3. Results and Discussion

3.1. Effect of Ce on the Casting Slab Quality

Casting slab results, with and without Ce, are shown in Figure 2. There are apparent differences in the observation results of the samples. It can be seen that the central segregation of the casting slabs with the Ce addition had a tendency to be ameliorated. Especially after adding 0.0009 wt.% Ce, the shape of the central segregation line almost disappeared, and the internal quality of the casting slab was excellent.

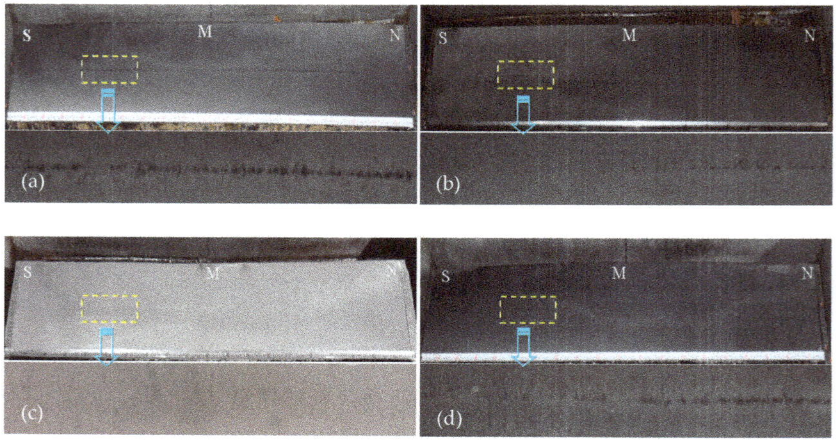

Figure 2. Casting slab quality results, (**a**) 0 wt.% Ce; (**b**) 0. 0006 wt.% RE; (**c**) 0.0009 wt.% Ce; and (**d**) 0.0013 wt.% Ce.

There are usually three types of central segregation, A represents the most serious, B is the second, and C is the lightest. Each type is divided into three levels, the smaller the number, the lighter the central segregation. The center porosity is the incompact structure formed by the final crystal shrinkage of the casting slab. The degree of porosity was judged by the size of the gap, and the casting slab quality was evaluated. Table 2 quantitatively shows the judgment results for the casting slabs. It can be seen that the central segregation of the casting slab was greatly improved after Ce treatment. The results reveal that the RE Ce plays a vitally important role in the casting slab.

Table 2. The low magnification judgment results.

Content	Central Segregation	Center Porosity
0 Ce	B 1.0	1.0
0.0006 wt.% Ce	B 1.0	1.0
0.0009 wt.% Ce	C 1.0	1.0
0.0013 wt.% Ce	B 1.0	1.0

Central segregation determines the quality of a casting slab. When high temperature stress exceeds the maximum stress that the casting slab can withstand, cracks will first form near the central segregation zone and then extend outward gradually. Cracks indirectly affect the uniformity of composition, the structure of hot-rolled finished product and the judgment level of the band structure. During the period of solidification, Ce reduces the residual sulfur and oxygen content in molten steel. High melting point RE sulfides and oxysulfides are formed which can hinder the segregation of the alloying elements, suppressing the accumulation of sulfides at the grain boundaries. Many scholars [20–22] have indicated that the formation of Ce-O-S inclusions increases the heterogeneous nuclei particles, improves the solidification structure, and reduces the center segregation of the casting slab.

Under the same casting conditions, part of the Ce is dissolved within the molten steel, reducing the segregation of C and Mn, while the other part of the Ce forms the RE inclusions. The melting point of the inclusions formed by the RE elements is also higher than that of the general metal compounds. The RE inclusions precipitate and become dispersed within the molten steel during the cooling process. The RE inclusions become nucleated particles. The outer layer will gradually become encased by the other metal inclusions, thereby optimizing the solidification structure of the molten steel and improving the quality of the casting slab. Previous researchers have indicated that adding the appropriate amounts of RE elements into weathering steel significantly improved central segregation [23].

The internal quality and macrostructure of the casting slab are closely related to the microstructure of the product. Figure 3 shows the microstructure with and without and Ce at different positions of the casting slab thickness. It can be seen that as the content of Ce increases, the proportion of pearlite increases and the proportion of ferrite decreases, however the microstructure tends to be refined. According to the reports, after the addition of mixed La and Ce elements into 0.27C-1Cr steel, the original austenite grains were refined, the grain growth rate was slowed, the nucleation rate of the phase transformation products increased, Ac_1 was unchanged, Ac_3 increased, the Ms point decreased, the pre-eutectoid ferrite or bainite precipitation period shortened, and the pearlite and bainite transformation completion time were prolonged [24].

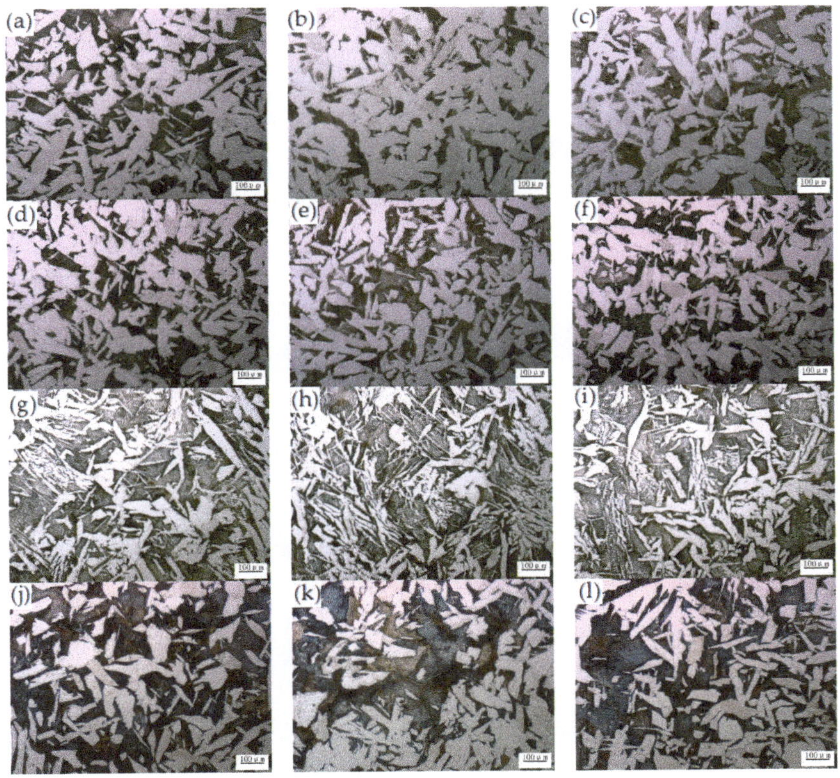

Figure 3. No Ce and different Ce microstructures of the casting slab in the direction of thickness. (**a–c**) 0 wt.% Ce, respectively represents 1/4 inner arc position, casting slab center, and outer arc 1/4; (**d–f**) 0.0006 wt.% Ce, respectively represents 1/4 inner arc position, casting slab center, and outer arc 1/4; (**g–i**) 0.0009 wt.% Ce, respectively represents 1/4 inner arc position, casting slab center, and outer arc 1/4; (**j–l**) 0.0013 wt.% Ce, respectively represents 1/4 inner arc position, casting slab center, and outer arc 1/4.

These results mean that the addition of Ce can change the structural characteristics of a casting slab. Due to a unique electronic layer structure, the edge-to-edge matching model [25] was used to calculate it. The results show that containg-Ce oxide is potential nucleant for the heterogeneous nucleation of both the δ-Fe and γ-Fe primary phases during the solidification of steels, and it promotes grain refining. Also, Ce promotes the formation of random hexagonal close packing structures during the nuclei formation process, and thus helps to reduce nucleation-free energy [26]. As shown in Figure 3, the grains will be coarsened with the increase of Ce. However, when Ce addition is 0.0009 wt.%, the pearlite and the ferrite are finer and more uniform. This is primarily due to the addition of Ce in appropriate amounts in order to form a fine RE composite inclusion. This promotes the nucleation of acicular ferrite in the austenite body, divides the austenite, and refines the grain. Also, under the same casting conditions, the Ce may be responsible for increasing the transformation temperature of austenite to pearlite. It can be concluded that the addition of the optimum content of Ce played a vital role in the size and shape of the microstructure of the casting slab. As discussed in previous work, the distribution of RE on the grain boundaries indicates a refinement to pearlite layer spacing and ferrite microstructure [27,28].

Figure 4 shows the morphology of the inclusions without and with the addition of Ce. From Figure 4a, on the metallographic samples, islands it can be seen that MnS inclusions formed near the shrinkage holes and occasionally were intermittently connected [20]. From Figure 4b, on the metallographic samples, the shape of Al_2O_3 inclusions was irregular and blocky. These inclusions deteriorate matrix continuity and are origins for the formation of cracks. For the optical microstructure of the samples, after the addition of Ce, the casting slab inclusions were not prismatic or chain-shaped but ellipsoidal. When the steel is hotly processed and cooled, due to the RE inclusions, the thermal expansion coefficient and density were close to that of molten steel [29,30], and it could thus avoid the stress around the inclusions, significantly reducing harm to the material. Figure 4 presents the inclusions in the (c), (d), and (e) steel samples after adding the Ce alloy to molten steel. The morphology of these inclusions was gradually spheroidized compared to those without RE, which greatly reduced the damage to the steel matrix.

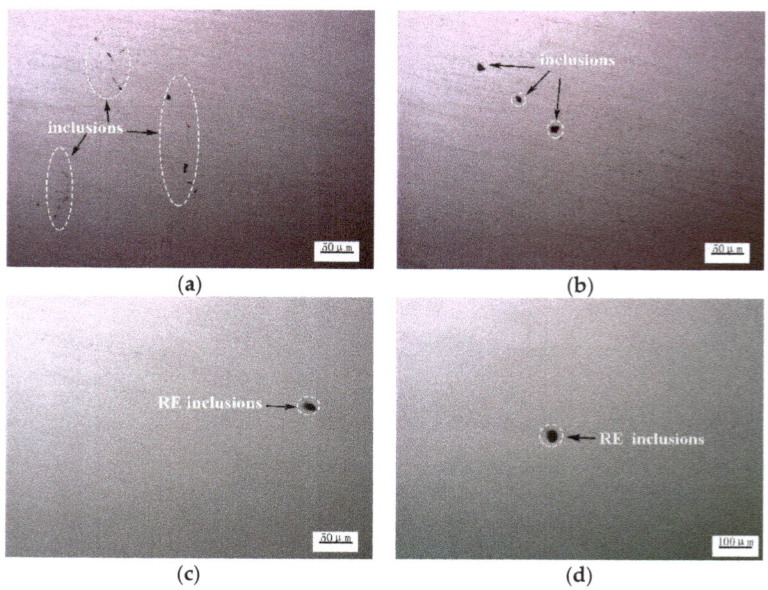

Figure 4. *Cont.*

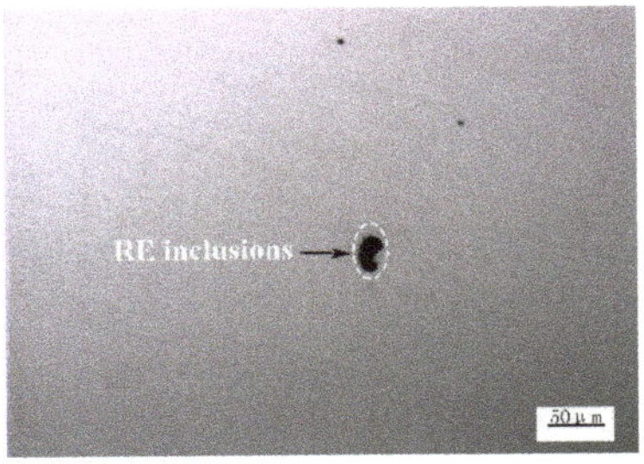

(e)

Figure 4. No Ce and different Ce typical inclusions morphology of the casting slab; (**a**) and (**b**) represent 0 wt.% Ce; (**c**) represents 0.0006 wt.% Ce; (**d**) represents 0.0009 wt.% Ce; and (**e**) represents 0.0013 wt.% Ce.

To determine the composition of the complex inclusion mentioned above clearly through further investigation, the inclusions were analyzed using SEM-EDS to determine their chemical composition, and the results are shown in Figure 5. The Al_2O_3 inclusions in the sample without Ce added were sharp-angled, while the MnS+Ti-C-S inclusions were irregular, as shown in Figure 5a,b, respectively. The addition of RE Ce modifies the Al_2O_3 inclusion morphology, forming $CeAlO_3$, which are nearly spherical inclusions. Simultaneously, the affinity of Ce to sulfur is stronger than it is to manganese [31], and thus $CeAlO_3$ or Ce_2O_2S spherical inclusions were formed instead of MnS + Ti-C-S inclusions. When an appropriate amount of Ce was added to the molten steel, the Ce in the molten steel modified the Al_2O_3 and series MnS inclusions, forming Ce complex inclusions, making their appearance approximately spherical. Some studies have indicated that the inclusions with an aspect ratio less than or equal to $\sqrt{2}$ would be treated as circular, and the other inclusions will be treated as elliptical [32]. The typical morphology and composition of inclusions with Ce additions are shown in Figure 5c–e. The irregular Al_2O_3 inclusions or MnS + Ti-C-S inclusions were wrapped by Ce, and thus gradually transformed into spheroidal Al-O-Ce-Mg-Ca, Al-O-Ce-C-Ca or Al-O-Ce-C-Ca inclusions. With the content of Ce increasing in molten steel, the inclusions were gradually modified to RE inclusions, and the size of the inclusions decreased. But when the Ce was over 0.0009 wt.%, the size of the containing-Ce inclusions shows an increased and irregular trend. It could be concluded that the appropriate addition of Ce modified the inclusions.

3.2. *Effect of Ce on Microstructure and Inclusions in Steel Strips*

Figure 6 shows the microstructures of hot-rolled steel strips with and without different Ce levels. It can be seen that the proportion of pearlite is small and the size of the structure is uneven in Figure 6a. As shown in Figure 6b–d, the proportion of pearlite is greater. The proportion of pearlite and ferrite were measured quantitatively by the Photoshop CC (Creative Cloud) image analysis tool, and the results are as shown in Figure 7. It can be found that with the increase of the Ce element, the content of pearlite shows an increasing trend, and the content of ferrite shows a decreasing trend. Meanwhile, when the Ce was 0.0009 wt.%, the presence of Ce reduced the size of ferrite, and the microstructure was uniform and fine. It can be concluded that the microstructure in the casting slab is directly inherited after hot rolling, such that more Ce is added, the better and coarser the microstructure. One possible reason is that in the C-Mn cryogenic vessel steels, Ce promotes the diffusion of C, which is conducive to the formation of pearlite, and the formation of fine Ce inclusions improves the nucleation of pearlite and finer ferrite. Zang et al. reported that the addition of Y element causes a large ferrite structure to appear in 20Cr13 martensitic stainless steel [33].

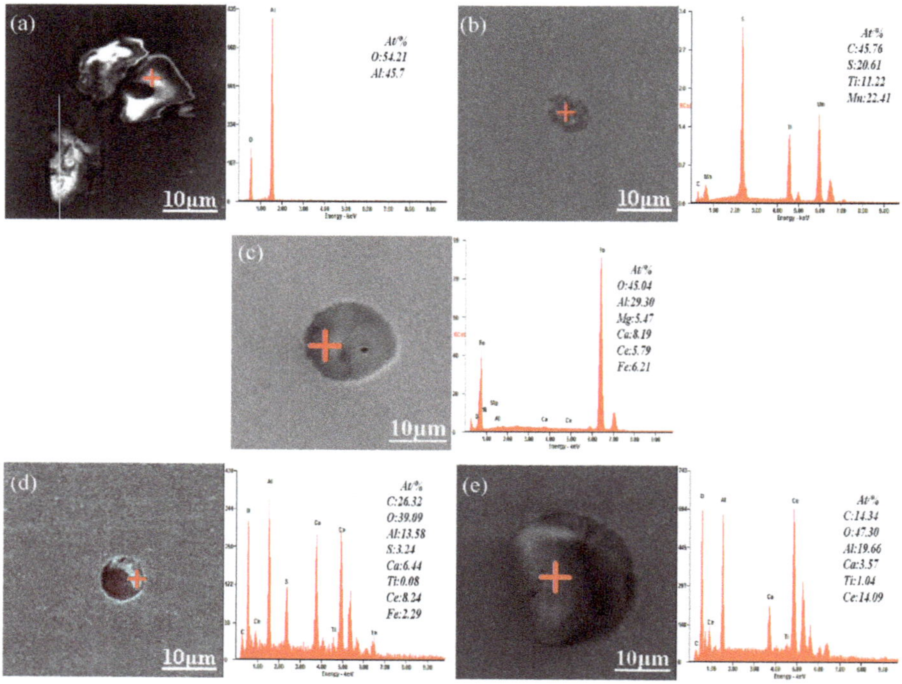

Figure 5. Inclusion morphology and spectrum results of the casting slab; (**a**) and (**b**) represent 0 wt.% Ce; (**c**) represents 0.0006 wt.% Ce; (**d**) represents 0.0009 wt.% Ce; and (**e**) represents 0.0013 wt.% Ce.

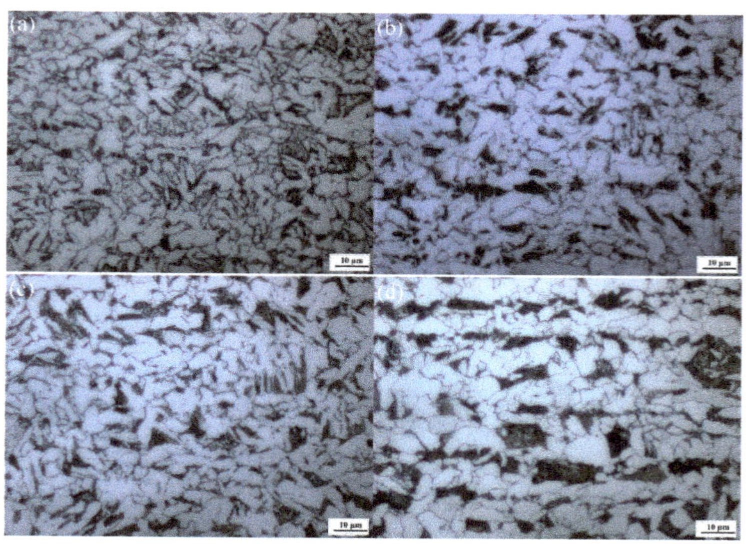

Figure 6. Microstructures of hot rolled steel strips; (**a**) 0 wt.% Ce; (**b**) 0. 0006 wt.% Ce; (**c**) 0.0009 wt.% Ce; (**d**) 0.0013 wt.% Ce.

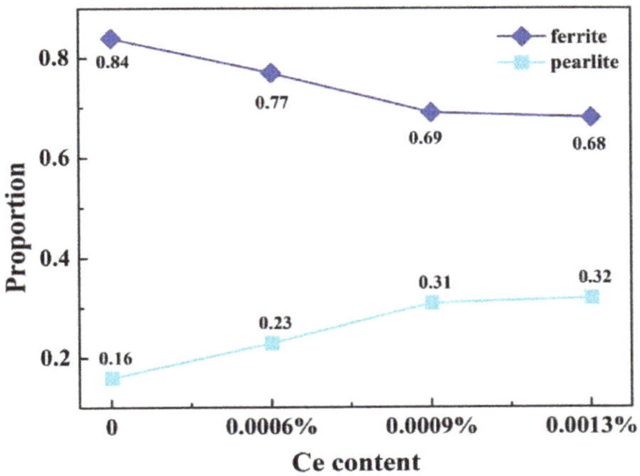

Figure 7. The statistics of ferrite and pearlite in various samples, with and without Ce content.

The above results show that Ce can be a modified inclusion in the casting slab. Figure 8 shows the typical nonmetallic inclusions after the corresponding casting slab is rolled into the finished steel strip. Figure 8a,b show the without adding Ce inclusions, the shape is irregular and rectangular; these kind of inclusions are more harmful to the steel matrix and also a sensitive source of crack propagation. Figure 8c–e show ellipsoidal RE non-metallic inclusions observed in the microstructure after adding different contents of Ce, which reduces the possibility of crack formation and propagation [34]. The size of the inclusions decreases and tends to spheroidize, and when Ce is 0.0009 wt. %, the shape of inclusions is spherical and their size is the smallest. These data indicate that the addition of an appropriate quantity of Ce played an imporptant role in the size distribution and shape deformation of the inclusions.

Combining the casting slab microstructure analysis in Figures 4 and 5, it was concluded that Ce additions modified the grain size of the studied steels containing them. Meanwhile, after adding Ce, large particles of inclusions floated up, so a typical finer spherical inclusion morphology was chosen for the surface scan analysis as shown in Figure 9, which demonstrates that the inclusions should be Ce_2O_2S. These finer RE inclusions act as heterogeneous nucleation points [35], reducing the distribution of harmful elements in the steel matrix.

3.3. Ce Inclusions Formed Thermodynamic Mechanisms

To illustrate the evolution mechanism of the inclusions, Thermo-Calc software was used to calculate the formation results of S-containing inclusions in C-Mn cryogenic vessel steel with and without the addition of RE Ce. As shown in Figures 10 and 11, it was found that after adding Ce, the contents of $Ti_4C_2S_2$ and MnS#2 type inclusions were significantly reduced, forming high melting point Ce_2O_2S and Ce_2S_3 inclusions, while MnS#2 is a composite inclusion phase dominated by MnS. One possible reason is that the RE data of the thermodynamic software is not comprehensive, which leads to certain limitations in the calculation results.

To further explore the influences of different Ce contents on the evolution of inclusions [36], the Gibbs free energy of the reactions under different Ce contents (0–100 ppm) was calculated. The values of other elements were selected from Table 1 (No.II). The results are shown in Figure 12.

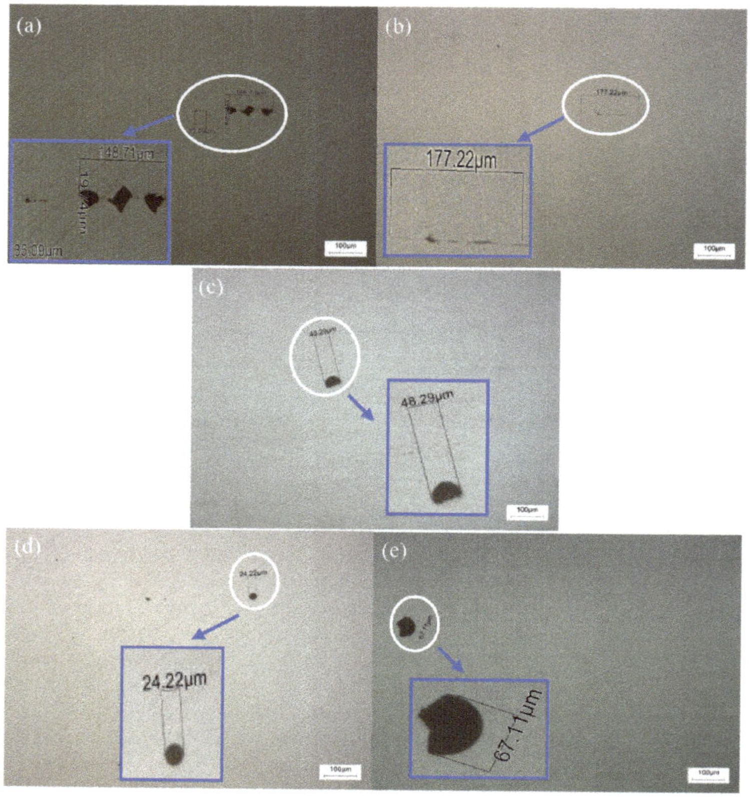

Figure 8. Non-metallic inclusion morphorgy of hot rolled steel strips; (**a**) and (**b**) represent 0 wt.% Ce; (**c**) represents 0.0006 wt.% Ce; (**d**) represents 0.0009 wt.% Ce; and (**e**) represents 0.0013 wt.% Ce.

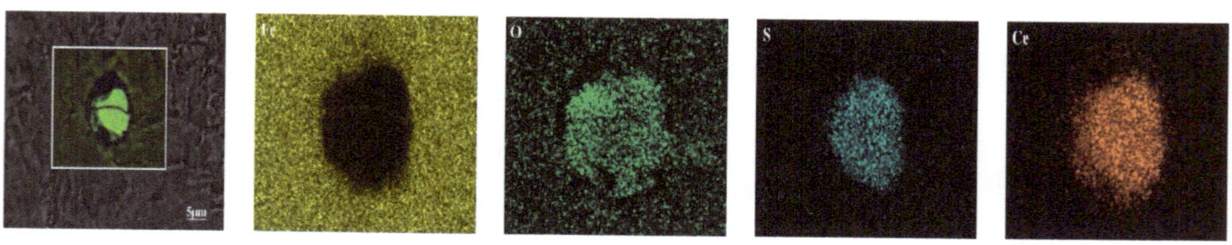

Figure 9. EDS results of the typical inclusions observed in the sample with 0.0009 wt.% Ce.

From Figure 12, the content of the RE elements has little effect on the precipitation order of the inclusions. After the initial addition of a small amount of Ce, the Gibbs free energy of each reaction was significantly reduced. However, increasing by the amount of Ce added, the change in the Gibbs free energy gradually became smaller. According to the relative size of the Gibbs free energy, it can be concluded that $CeAlO_3$ is the most stable, followed by Ce_2O_3, then Ce_2O_2S.

The formula $\Delta G = -RT \ln K$ and previous research data [37–39] were used to calculate the conditions for the formation of inclusions, and the results are shown in Table 3. The phase diagram calculation results of Ce-O-S at a temperature of 1873K are shown in Figure 13.

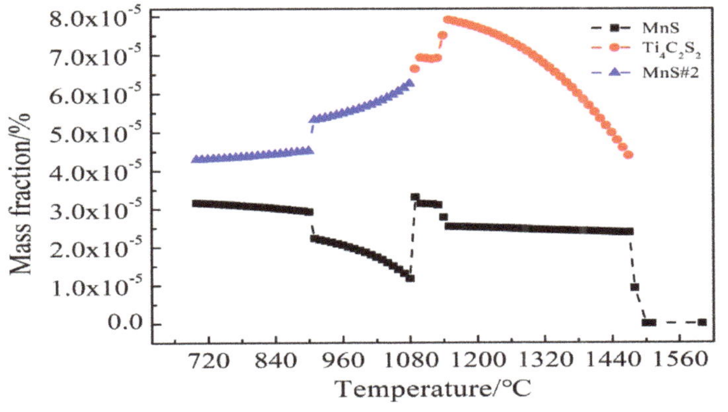

Figure 10. No Ce formation of sulfide.

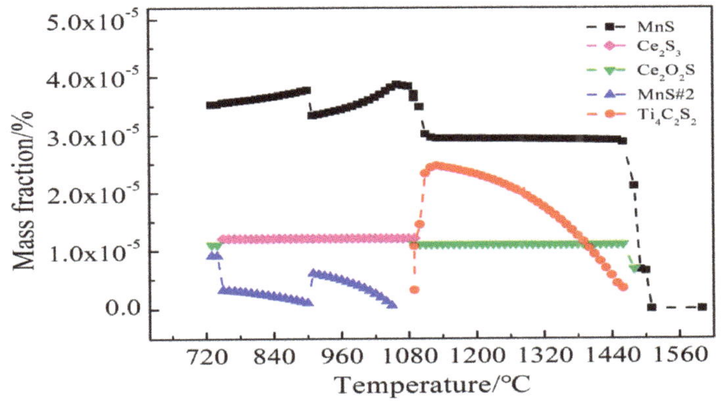

Figure 11. Various Ce formation of sulfide.

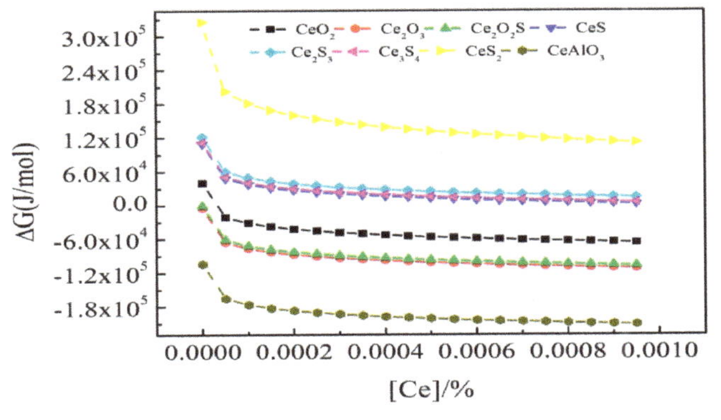

Figure 12. Relation between reaction Gibbs free energy and Ce content.

Table 3. Inclusion conversion relationship and formation conditions.

Reactions	Equilibrium Constant	The Relation between $a_{[O]}$ and $a_{[S]}$
$2CeO_2 = Ce_2O_3 + [O]$	0.12866	$a_{[O]} = 0.12866$
$Ce_2O_3 + [S] = Ce_2O_2S + [O]$	0.079	$a_{[S]} = 12.67 \times a_{[O]}$
$2CeO_2 + [S] = Ce_2O_2S + 2[O]$	0.01017	$a_{[S]} = 98.3284 \times a_{[O]}^2$
$Ce_2O_2S + [S] = 2CeS + 2[O]$	0.8505×10^{-5}	$a_{[S]} = 1.1758 \times 10^5 \times a_{[O]}^2$
$Ce_2O_2S + 2[S] = Ce_2S_3 + 2[O]$	1.452×10^{-5}	$a_{[S]} = 262.433a_{[O]}$
$Ce_2S_3 = 2CeS + [S]$	0.5857	$a_{[S]} = 0.5857$

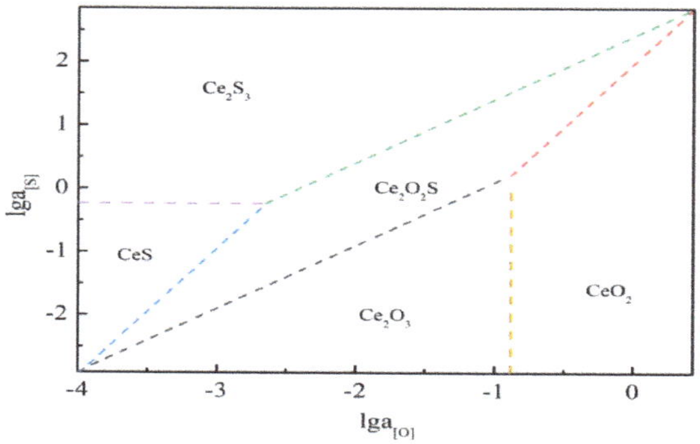

Figure 13. Ce-S-O inclusions generation diagram.

Combining Table 3 and Figure 13, the conclusions are as follows:

(1) Formation conditions of RE oxides:when $a_{[O]} > 0.232$, $a_{[S]}/a_{[O]}^2 < 98.329$ the inclusions generated are CeO_2, when $a_{[O]} < 0.232$, $a_{[S]}/a_{[O]} < 12.67$, the inclusions generated are Ce_2O_3.

(2) Formation conditions of RE oxysulfides:when $a_{[O]} < 0.232$, $12.67 < a_{[S]}/a_{[O]} < 262.433$ or $a_{[S]} < 0.5857$, $98.3284 < a_{[S]}/a_{[O]}^2 < 117580$, the inclusions generated are Ce_2O_2S.

(3) Formation conditions of RE sulfides:when $a_{[S]} < 0.5857$, $a_{[S]}/a_{[O]}^2 > 117580$, the inclusions generated are CeS, and when $a_{[S]} > 0.5857$, $a_{[S]}/a_{[O]} > 262.433$, the inclusions generated are Ce_2S_3.

3.4. Inclusion Formation Mechanisms with the Addition of Ce

RE elements have a strong mutual chemical affinity to [O] and [S] [40]. Both elements begin to aggregate and form high melting point oxides, oxysulfides and sulfides, thus having the ability to remove the deleterious types of inclusions that are formed [13,32,41]. The combination of Ce with [O] and [S] in steel has lower Gibbs free energy. It was easy to change the inclusion morphology from chains and strips to generate $CeAlO_3$, Ce_2O_2S, and Ce_2O_3 spherical inclusions [42]. The RE inclusions $CeAlO_3$ and Ce_2O_2S in molten steel were stable at 1873K, decreasing the mass fraction of dissolved oxygen in the molten steel, while Ce and S formed complex RE sulfides, resulting in a decrease in $CeAlO_3$ and increase in Ce_2O_2S [43,44]. According to the discussion in Sections 3.2 and 3.3, Ce_2O_2S inclusion is especially beneficial for heterogeneous nuclei. The evolution process of RE inclusions during solidification is shown in Figure 14.

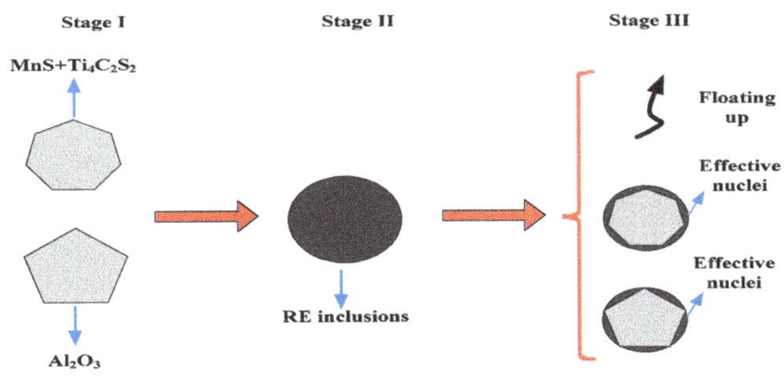

Figure 14. Schematic diagram of the evolution process of inclusions with RE-Ce alloy treatment.

Firstly, when the melting temperature is stable at 1873K, it is easy to form Al_2O_3 and $MnS+Ti_4C_2S_2$ inclusions in molten steel. As shown in Figure 13 (Stage I), the inclusions are primarily irregularly shaped and deteriorate the mechanical properties of steel [45]. The reaction equation is as follows:

$$[S] + [Mn] = MnS_{(S)} \tag{1}$$

$$4[Ti] + 2[C] + 2[S] = Ti_4C_2S_{2(S)} \tag{2}$$

$$2[Al] + 3[O] = Al_2O_{3(S)} \tag{3}$$

Secondly, when RE-Ce alloy are added to the molten steel, large-sized irregular inclusions are also transformed into small-sized spherical or ellipsoidal $CeAlO_3$ and Ce_2O_2S inclusions via Ce. The reaction equation [37,38] is as follows:

$$[Ce] + 3[O] + [Al] = CeAlO_{3(S)} \tag{4}$$

$$2[Ce] + 2[O] + [S] = Ce_2O_2S_{(S)} \tag{5}$$

$$4[Ti] + 2[C] + 2[S] = Ti_4C_2S_{2(S)} \tag{6}$$

Finally, as the above reactions proceed, part of the large-size inclusions float to the top. The other Ce inclusions generated during Stage II act as heterogeneous nucleus cores. This result suggests that RE elements can modify inclusion morphology and purify molten steel, further increasing casting slab quality control, as shown in Figure 13 (Stage III).

Based on the reasoning above, combined with Figure 2, the central segregation morphology of the casting slab almost disappears after the addition of RE Ce. According to in-situ statistical analysis, it was found that the addition of RE elements reduces the segregation of easily segregated elements such as C, S, and P, significantly weakening the central segregation defect [23,46]. This study also shows that the formed $CeAlO_3$ and Ce_2O_2S inclusions have a low degree of mismatch with the Fe matrix, increasing the effectiveness of the nucleation core and improving the solidification structure of C-Mn cryogenic vessel steels. Moreover, due to the strong surface activity of the Ce element, it is segregated in the austenite grain boundary, reducing the interface energy, weakening the driving force for nucleation, refining the structure, and improving the quality of the casting slab. After the addition of composite La and Ce, many high melting point $(Ce, La)_2O_2S$ secondary-particles can be used as an effective non-uniform nucleation core, which increases the proportion of equiaxed crystals in the solidified structure, reduces the size of the equiaxed crystals, and significantly refines the solidified structure of the alloy [47].

When RE atoms are dissolved in the matrix, a damping peak of solute atoms segregating at the grain boundary appears in the high-temperature zone. When the RE content exceeds the maximum solid solution of the matrix, it will react with other alloying elements, forming a compound [48]. Therefore, it can be concluded that the addition of 0.0009wt.% Ce reduces central segregation, accelerates the formation of needle-shaped ferrite, increases the proportion of pearlite, and modifies Al_2O_3 and $MnS+Ti_4C_2S_2$ inclusion morphology in the C-Mn cryogenic vessel steels.

4. Conclusions

In industrial production, when the RE amount is over 0.0015 wt.%, the nozzle of the continuous casting mold will appear as a nodule, which is not conducive to the efficient production of long casting times. In this study, through comparative experimental studies, it was found that when the Ce amount is 0.0009 wt.%, the central segregation, microstructure characteristics, and inclusion control of C-Mn cryogenic vessel steel benefited, which can further provide theoretical guidance for the industrial production of RE steel. Based on the above analysis, the main results can be summarized as follows:

(1) During the solidification of the RE-treated steels, the central segregation was significantly improved after adding Ce, reaching the lowest level when the content of Ce was 0.0009 wt.%.

(2) Comparing the microstructures of the casting slabs at different positions in the thickness direction after the addition of Ce, the microstructure and inclusions from the casting slab to the hot-rolled steel strip had a certain heritability, the ferrite became needle-shaped, and the proportion of pearlite increased.

(3) Experimental observation and thermodynamic calculations may indicate that Ce can modify the sharp-angled Al_2O_3 inclusions into ellipsoid $CeAlO_3$ composite inclusions, and that and the formation of $MnS+Ti_4C_2S_2$ inclusions was reduced in favor of forming $Ce_2O_2S + Ti_4C_2S_2$ spherical inclusions. Meanwhile, finer Ce_2O_2S inclusions can serve as heterogeneous nucleation points and reduce the genetic harm of inclusions.

Author Contributions: Conceptualization, L.W. and J.Z. (Jianguo Zhi); methodology, L.W. and Q.L.; software, B.Z.; validation, L.W. and B.Z.; formal analysis, L.W.; investigation, L.W. and Q.L.; data curation, L.W. and B.Z.; writing—original draft preparation, L.W. and J.Z. (Jiangshan Zhang); writing—review and editing, L.W. and Q.L.; visualization, L.W. and J.Z. (Jiangshan Zhang); supervision, Q.L.; project administration, J.Z. (Jianguo Zhi); funding acquisition, Q.L. and J.Z. (Jianguo Zhi). All authors have read and agreed to the published version of the manuscript.

Funding: This research was supported by the financial support of Local Science and Technology Development Fund Projects Guided by The Central Government (2020ZY0034) and the Inner Mongolia Natural Science Foundation (2020MS05017).

Institutional Review Board Statement: Not applicable.

Informed Consent Statement: Not applicable.

Data Availability Statement: Data sharing not applicable. No new data were created or analyzed in this study. Data sharing is not applicable to this article.

Acknowledgments: The authors are grateful for the support of the financial support of Local Science and Technology Development Fund Projects Guided by The Central Government (2020ZY0034) and the Inner Mongolia Natural Science Foundation (2020MS05017).

Conflicts of Interest: The authors declare no conflict of interest.

References

1. Chen, Z.; Pan, J.H.; Jin, T.; Hong, Z.Y.; Wu, Y.C. Estimation of fracture toughness of 16MnDR steel using Master Curve method and Charpy V-notch impact energy. *Theor. Appl. Fract. Mech.* **2018**, *96*, 443–451. [CrossRef]
2. Liu, D.F.; Yang, X.L.; Hou, L.F.; Cui, T.X.; Hu, Y.T.; Wei, Y.H. Research and Application of Ultralow Temperature 9Ni Steel for LNG Storage Tank. *J. Iron Steel Res.* **2009**, *21*, 1–5.
3. Xu, L.; Li, Z.C.; Zhang, J.; Xu, Y.S. LNG land storage and transmission in China. *Nat. Gas Ind.* **2002**, *22*, 89–91.
4. Barsouma, I.; Lawala, S.A.; Simmonsb, R.J.; Rodrigues, C.C. Failure analysis of a pressure vessel subjected to an internal blast load. *Eng. Fail. Anal.* **2018**, *91*, 354–369. [CrossRef]
5. Miao, C.J.; Zheng, J.Y.; Gao, X.Z.; Huang, Z.; Guo, A.B.; Ye, D.Y.; Ma, L. Investigation of low-cycle fatigue behavior of austenitic stainless steel for cold-stretched pressure vessels. *J. Zhejiang Univ. Sci. A* **2013**, *14*, 31–37. [CrossRef]
6. Xin, W.B.; Zhang, J.; Luo, G.P.; Wang, R.F.; Meng, Q.Y.; Song, B. Improvement of hot ductility of C-Mn Steel containing arsenic by rare earth Ce. *Metall. Res. Technol.* **2018**, *115*, 419–427. [CrossRef]

7. Garrison, W.M.; Maloney, J.L. Lanthanum additions and the toughness of ultra-high strength steels and the determination of appropriate lanthanum additions. *Mater. Sci. Eng. A* **2005**, *403*, 299–310. [CrossRef]
8. Pan, F.; Zhang, J.; Chen, H.L.; Su, Y.H.; Kuo, C.L.; Su, Y.H.; Chen, S.H.; Lin, K.J.; Hsieh, P.H. Effects of Rare Earth Metals on Steel Microstructures. *Materials* **2016**, *9*, 417. [CrossRef] [PubMed]
9. Wan, Y.; Chen, W.Q.; Wu, S.J. Effect of lanthanum content on microstructure and magnetic properties of non-oriented electrical steels. *J. Rare Earths* **2013**, *31*, 727–733. [CrossRef]
10. Liu, H.J.; Liu, H.J.; Chen, Y.Z. Effect of RE on the structure and property of steel 45. *Chin. Rare Earths* **2012**, *33*, 24–27.
11. Lin, Q.; Ye, W.; Chen, N.; Liu, Y.H.; Guo, Y.; Chen, Z.Q.; Zhu, Q.H. Effect of rare earths on extra-low sulfur microalloy steel. *J. Chin. Rare Earth Soc.* **1997**, *22*, 228–233.
12. Xiao, J.G.; Cheng, H.J.; Wang, F.M. Effects of rare earth in ship plate steel on its microstructure and low temperature toughness. *Chin. Rare Earths* **2010**, *31*, 52–57.
13. Huang, Y.; Cheng, G.G.; Xie, Y. Modification Mechanism of Cerium on the Inclusions in Drill Steel. *Acta Metall. Sin.* **2018**, *54*, 1253–1261.
14. Imashuku, S.; Wagatsuma, K. Cathodoluminescence Analysis of Nonmetallic Inclusions in Steel Deoxidized and Desulfurized by Rare-Earth Metals (La, Ce, Nd). *Metall. Mater. Trans. B* **2019**, *51*, 79–84. [CrossRef]
15. Gao, S.; Wang, M.; Guo, J.L.; Wang, H.; Zhi, J.G.; Bao, Y.P. Characterization Transformation of Inclusions Using Rare Earth Ce Treatment on Al-Killed Titanium Alloyed Interstitial Free Steel. *Steel Res. Int.* **2019**, *90*, 1900194. [CrossRef]
16. Fan, L.F.; Zhu, R.; He, J.Z.; Lu, B. Effect of Rare Earth Element La on Texture and Inclusion of Nonoriented Electrical Steel Produced by Thin Slab Casting and Rolling Process. *ISIJ Int.* **2018**, *58*, 2348–2353. [CrossRef]
17. Chang, H.; Jin, Z.L.; Ren, H.P.; Song, W.Z. Effect of rare earth on microstructure and non-metallicinclusions in thin slab producted by CSP. *J. Chin. Rare Earth Soc.* **2010**, *28*, 601–606.
18. Wang, L.M.; Zhu, G.L.; Xun, J.; Qi, G.P.; Zhu, J.X.; Chen, D.X. Study of Rare Earth Effect and Mechanism in Ferrite Stainless Steel(430). *Chin. Rare Earths* **2008**, *28*, 67–71.
19. Wang, H.P.; Jiang, S.L.; Yu, P.; Bai, B.; Sun, L.F.; Wang, Y. Distribution of Arsenic Inclusions in Rare Earth Steel Ingots. *Metals* **2020**, *10*, 146. [CrossRef]
20. Yan, Y.X.; Li, P.Z. Influence of RE addition on the central segregation in 16Mn concast slabs and the fractograph of finished products. *Iron Steel* **1986**, *21*, 13–19.
21. Tuttle, R.B. Study of Rare Earth Additions Effect on the Solidification and Properties of 4130 Steel. *JMEPG* **2019**, *28*, 6720–6727. [CrossRef]
22. Fruehan, R.J. The Free Energy of Formation of Ce_2O_2S and the Nonstoichiometry of Cerium Oxides. *Metall. Trans. B* **1979**, *10*, 143–148. [CrossRef]
23. Guo, H.H.; Song, B.; Mao, J.H.; Zhao, P. Effect of rare earth elements on macrosegregation in weather-resisting steel. *J. Univ. Sci. Technol. Beijing* **2010**, *32*, 44–49.
24. Lv, W.; Zheng, D.H.; Xu, Z.Y. Influence of Rare earth on TTT Diagram for 0.27C-1Cr Steel. *Acta Metall. Sin.* **1993**, *29*, 308–311.
25. Ji, Y.P.; Kang, L.; Song, Y.Q.; Qu, W.; Ren, H.P. Crystallographic Calculation about Heterogeneous Nucleation Potency of RE_2O_3 in Liquid Steel. *Rare Met. Mater. Eng.* **2017**, *46*, 2889–2894.
26. Gao, X.Y.; Wang, H.Y.; Xing, L.; Ma, C.N.; Ren, H.P. Evolution of local atomic structure during solidification of Fe-RE (RE = La, Ce) alloy. *J. Non-Cryst. Solids* **2020**, *542*, 120109. [CrossRef]
27. Bao, X.R.; Chen, L.; Guo, Y.P. Effect of RE on Dynamic Recrystallization and Pearlite Spacing of Heavy Rail Steel. *Iron Steel* **2012**, *47*, 76–78.
28. Jiang, J.K.; Xiao, L.F.; You, D.L.; Shu, J.S.; Ren, L.E. Effect of Rare Earth addition on Properties of Controlled Rolling Low Carbon MnMoNb Ferrite Steel. *Acta Metall. Sin.* **1987**, *23*, 165–167.
29. Wang, L.M. Application Prospects and Behavior of RE in New Generation High Strength Steels with Superior Toughness. *J. Chin. Rare Earth Soc.* **2004**, *22*, 48–53.
30. Li, X.; Jiang, Z.H.; Geng, X.; Chen, M.J.; Peng, L.Z. Evolution Mechanism of Inclusions in H13 Steel with Rare Earth Magnesium Alloy Addition. *ISIJ Int.* **2019**, *59*, 1552–1561. [CrossRef]
31. Zhang, L.F. *Non-Metallic Inclusions in Steel*; Metallurgical Industry Press: Beijing, China, 2019; Volume 26.
32. Meng, Y.; Yan, C.L.; Yang, X.; Ju, X.H. A Statistical Analysis on the Complex Inclusions in Rare Earth Element Treated Steel. *ISIJ Int.* **2020**, *60*, 534–538. [CrossRef]
33. Zang, Q.Y.; YJin, Y.F.; Zhang, T.; Yang, Y.T. Effect of yttrium addition on microstructure, mechanical and corrosion properties of 20Cr13 martensitic stainless steel. *J. Iron Steel Res. Int.* **2020**, *27*, 451–460. [CrossRef]
34. Pang, F.X.; Yin, S.C.; Li, J.J.; Wang, S.B. Study on deoxidization thermodynamics of trace rare earth element and formation mechanism of inclusions in LZ50 molten steel. *J. Taiyuan Univ. Technol.* **2011**, *42*, 646–649.
35. Wang, H.P.; Yu, P.; Jiang, S.L.; Wang, Y. Effect of Heterogeneous Nucleation on Removal of Arsenic from Molten Steel by Rare Earth Addition. *Metals* **2020**, *10*, 664. [CrossRef]
36. Wang, Y.G.; Liu, C.J. Agglomeration Characteristics of Various Inclusions in Al-killed Molten Steel Containing Rare Earth Element. *Metall. Mater. Trans. B* **2020**, *51*, 2585–2595. [CrossRef]
37. Waudby, P.E. Rare earth additions to steel. *Int. Met. Rev.* **1978**, *2*, 74–98.

38. Kwon, S.K.; Park, J.S.; Park, J.H. Influence of Refractory-Steel Interfacial Reaction on the Formation Behavior of Inclusions in Ce-containing Stainless Steel. *ISIJ Int.* **2015**, *55*, 2589–2596. [CrossRef]
39. Jiang, M.Z.; Yu, Y.C.; Li, H.; Ren, X.; Wang, S.B. Effect of Rare Earth Cerium Addition on Microstructures and Mechanical Properties of Low Carbon High Manganese Steels. *High Temp. Mater. Process.* **2017**, *36*, 145–153. [CrossRef]
40. Fan, H.Q.; Xu, W.C.; Wei, L.; Zhang, Z.H.; Liu, Y.B.; Li, Q. Relationship between La and Ce additions on microstructure and corrosion resistance of hot-dip galvanized steel. *J. Iron Steel Res. Int.* **2020**, *27*, 1108–1116. [CrossRef]
41. Wang, L.J.; Liu, Y.Q.; Wang, Q.; Chou, K.C. Evolution mechanisms of $MgO \bullet Al_2O_3$ inclusions by cerium in spring steel used in fasteners of high-speed railway. *ISIJ Int.* **2015**, *55*, 970–975. [CrossRef]
42. Wang, H.; Bao, Y.P.; Duan, C.Y.; Lu, L.; Liu, Y.; Zhang, Q. Effect of Rare Earth Ce on Deep Stamping Properties of High-Strength Interstitial-Free Steel Containing Phosphorus. *Materials* **2020**, *13*, 1473. [CrossRef]
43. An, Y.Q.; Wang, L.Z.; Li, Y.T.; Song, G.B.; Lei, S.Z.; Tan, J. Effects of Rare Earth Cerium on the Characteristics of Inclusions. *Wide Heavy Plate* **2020**, *26*, 11–17.
44. Li, Y.D.; Liu, C.J.; Zhang, T.S.; Jiang, M.F.; Peng, C. Liquid Inclusions in Heat-Resistant Steel Containing Rare Earth Elements. *Metall. Mater. Trans. B* **2017**, *48*, 956–965. [CrossRef]
45. Wang, C.; Han, Y.S.; Zhang, J.S.; Xiao, D.; Yang, J.; Chen, J.; Liu, Q. Behaviour of oxide inclusions and sulphur in 'two-stage basicity control' refining method of Si-killed spring steel. *Ironmak Steelmak* **2020**, *48*, 466–476. [CrossRef]
46. Gao, R.Z.; Chen, H.Q. The effect of rare earth elements on the solidification characteristics and crystal structure of steel. *Chin. Rare Earths* **1985**, *3*, 27–33.
47. Yu, Y.C.; Chen, W.Q.; Zheng, H.G. Effect of Cerium and Mischmetal on Solidification Structure of Fe-36Ni Invar Alloy. *J. Chin. Rare Earth Soc.* **2012**, *30*, 175–180.
48. Wang, H.Y. Effect of Rare Earth on NbC Dissolution and Precipitation Behaviors in Microalloy Steels. University of Science and Technology: Beijing, China, 2017; Volume 14.

materials

MDPI

Article

Full-Field Temperature Measurement of Stainless Steel Specimens Subjected to Uniaxial Tensile Loading at Various Strain Rates

Krzysztof Żaba [1,*], Tomasz Trzepieciński [2], Stanislav Rusz [3], Sandra Puchlerska [1] and Maciej Balcerzak [1]

[1] Department of Metal Working and Physical Metallurgy of Non-Ferrous Metals, Faculty of Non-Ferrous Metals, AGH—University of Science and Technology, al. Adama Mickiewicza 30, 30-059 Cracow, Poland; spuchler@agh.edu.pl (S.P.); maciejbalcerzak1@gmail.com (M.B.)
[2] Department of Manufacturing and Production Engineering, Faculty of Mechanical Engineering and Aeronautics, Rzeszow University of Technology, al. Powst. Warszawy 8, 35-959 Rzeszów, Poland; tomtrz@prz.edu.pl
[3] Department of Mechanical Technology, Faculty of Mechanical Engineering, VŠB—Technical University of Ostrava, 17 listopadu 15, CZ 708 33 Ostrava–Poruba, Czech Republic; stanislav.rusz@vsb.cz
* Correspondence: krzyzaba@agh.edu.pl

Abstract: This article presents a study on the effect of strain rate, specimen orientation, and plastic strain on the value and distribution of the temperature of dog-bone 1 mm-thick specimens during their deformation in uniaxial tensile tests. Full-field image correlation and infrared thermography techniques were used. A titanium-stabilised austenitic 321 stainless steel was used as test materials. The dog-bone specimens used for uniaxial tensile tests were cut along the sheet metal rolling direction and three strain rates were considered: 4×10^{-3} s^{-1}, 8×10^{-3} s^{-1} and 16×10^{-3} s^{-1}. It was found that increasing the strain rate resulted in the intensification of heat generation. High-quality regression models (Ra > 0.9) developed for the austenitic 321 steel revealed that sample orientation does not play a significant role in the heat generation when the sample is plastically deformed. It was found that at the moment of formation of a necking at the highest strain rate, the maximum sample temperature increased more than four times compared to the initial temperature. A synergistic effect of the strain hardening exponent and yield stress revealed that heat is generated more rapidly towards small values of strain hardening exponent and yield stress.

Keywords: digital image correlation; mechanical properties; stainless steel; temperature; thermovision; uniaxial tensile test

Citation: Żaba, K.; Trzepieciński, T.; Rusz, S.; Puchlerska, S.; Balcerzak, M. Full-Field Temperature Measurement of Stainless Steel Specimens Subjected to Uniaxial Tensile Loading at Various Strain Rates. *Materials* **2021**, *14*, 5259. https://doi.org/10.3390/ma14185259

Academic Editors: Qing Liu and Jiangshan Zhang

Received: 28 June 2021
Accepted: 7 September 2021
Published: 13 September 2021

1. Introduction

Metals and their alloys have a crystalline structure, which is characterised by a regular arrangement of atomic cores. Technical metals obtained by conventional metallurgical methods have a polycrystalline structure [1,2]. This means that metals are made up of grains characterised by approximately the correct crystal structure. Polycrystalline materials consist of grains with various orientations. For many polycrystalline materials, the grain orientations are random before any working (deformation) of the material is undertaken. Therefore, even if the individual grains are anisotropic, the property differences tend to average out and, overall, the material is isotropic. When a material is formed, the grains are usually distorted and elongated in one or more directions which makes the material anisotropic [3].

In metals subjected to plastic working, the most important defects of the crystal lattice are dislocations and grain boundaries. Strain hardening—hardening of a material with deformation—results from interaction and multiplication of dislocations during plastic deformation. Grain boundaries have a much greater influence on the mechanical properties of metals than dislocations [4,5]. The high energy level of grain boundaries determines the occurrence of many phenomena, such as continuous mobility of boundaries leading to grain growth and lower corrosion resistance [6–8]. Grain boundaries also have a strong influence on the ductility of metals. The degree of strain hardening of materials by grain refining is described by the Hall–Petch relationship [9,10] according to which the yield strength of the material increases with the degree of grain refinement. On the one hand, grain boundaries inhibit the free movement of dislocations during grain deformation while participating in the process of material strengthening and in the formation of the deformation texture [1]. On the other hand, the grain boundaries are the main area of permanent deformation during creep or superplastic flow.

The deformation of the material (elastic or plastic) occurs under the influence of a load applied to it. Elastic deformation disappears when the load is removed. Plastic deformation remains after the material is unloaded. There are two main mechanisms of plastic deformation: the gliding motion of dislocations and deformation twinning [11,12]. The gliding motion does not occur simultaneously in all possible planes and directions of slip. Deformation occurs gradually, starting from the sliding planes and directions prioritised in relation to the direction of the applied load. During deformation, the number of active slip planes and free dislocations that are generated during deformation increases. Further plastic deformation requires the application of greater stress in order to initiate new dislocations [13–15]. The process by which stress increases with deformation is called work hardening. Plastic deformation is also accompanied by heat generation due to the internal resistance of the material, which can be detected at the microstructure scale when loading the material [16,17].

The investigation of heat generation and dissipation during plastic deformation was reflected in scientific research. Bodelot et al. [18] applied infrared (IR) coupled full-field measurements to observe the heterogeneities of kinematic and thermal data at the grain scale of austenitic stainless steel. Specimens were subjected to a cyclic loading. It was found that the temperature was significantly affected by diffusion and this observation tends to confirm that temperature may not be used as a damage indicator. Boulanger et al. [19] have focused on the determination of heat sources from a temperature field provided by an IR camera. They separately identified the dissipative and thermo-elastic heat sources. Thermal full-field measurements from infrared thermography were used to study the Portevin-Le Chatelier effect [20]. The temperature field provided insights into the dynamics of band formation and motion.

The non-contact optical methods for the measurement of temperature displacement include video extensometers [21], laser speckle correlation [22], interferometry methods [23–25] and temperature-calibrated CCD cameras [26]. In addition to IR thermography, the other technique that allows temperature changes to be coupled with strains is digital image correlation (DIC), in which digital images of an object before and after deformation are captured using a non-contact optic and material-independent measuring instrument, and then they are subject to correlation analysis [23,24,27]. High-speed DIC allows deformation and damage mechanisms to be analysed in a quantitative manner [28]. DIC was used to obtain strain fields and to investigate the propagation of the Lüders band in steel specimens subjected to the uniaxial tensile test [29]. The strain states associated with localised necking and diffuse strains are also investigated. Results show that the cross-section of a tensile specimen must be regarded as a structure, not as an elementary volume of material which is subjected to the uniform load. The results of investigations of Hung and Voloshin [30] indicated that for the uniform tension test, DIC is a very convenient and efficient tool for the measurement of in-plane strain measurement. Feng and Xue [31] applied infrared thermography tests and DIC tests for the display of the thermal field during mechanical tensile tests of 3D printed bolts. Cholewa et al. [32] developed the method of calibrated infrared thermal cameras with a stereo-vision DIC system applicable to scales of lengths both large and small. Żaba et al. [33] used DIC and IR thermography to find the relationship between yield stress and the Taylor–Quinney coefficient and their change with the strain rate during the stretching of nickel-based superalloys. A coupled thermography and DIC system was calibrated using a series of one sided heat exposure experiments performed on loaded sandwich composites. Infrared thermography and DIC measurements are not applicable to experiments with low conductivity materials or discontinuous materials [33,34]. Maynadier, et al. [35] developed an infrared image correlation system equipped with a single IR camera to measure simultaneous thermal fields and deformation. The disadvantage of the proposed method is that this technique requires the use of a special coating for displacement measurement using the single camera. Chrysochoos et al. [34] proposed DIC and thermography measurements for the determination of the mechanical energy and heat sources involved at local scale during a heterogeneous tensile test. Thermodynamic analysis of the energy balance showed the influence of the dissipated mechanism on the stress–strain response.

Many of studies take place at a macroscopic scale with the measurement of the average temperature of the material [36,37]. Meanwhile, due to the grain structure of polycrystalline metals and the occurrence of various plastic deformation mechanisms, the temperature of the body subjected to loading is not uniform. The way to explain this phenomenon is to take full-field measurements. In this paper, the coupled IR and DIC measurements were conducted to assess the thermomechanical response of stainless steel strip samples subjected to tensile load at different strain rates. An analysis of variance (ANOVA) was used to gain information about the relationship between the strain hardening phenomenon, yield stress, sample orientation in respect of the sheet rolling direction, strain rate, and plastic strain of the specimen and temperature.

2. Materials and Methods

2.1. Material

The research material was a titanium-stabilised austenitic stainless steel, 321 (AMS 5510). Steel 321 is a steel with good formability, weldability and resistance to intergranular corrosion that is mainly used in the chemical and aviation industries. The thickness of the sheets was 1 mm. The test samples were cut from the sheet in three directions: the rolling direction (RD), at an angle of 45° to the RD and perpendicular to the RD. The dimensions of test samples (Figure 1) were in accordance with the ASTM E8/E8M–11 standard [38]. The requirements for the chemical composition of the test steel are shown in Table 1 and are in accordance SAE AMS 5510 [39].

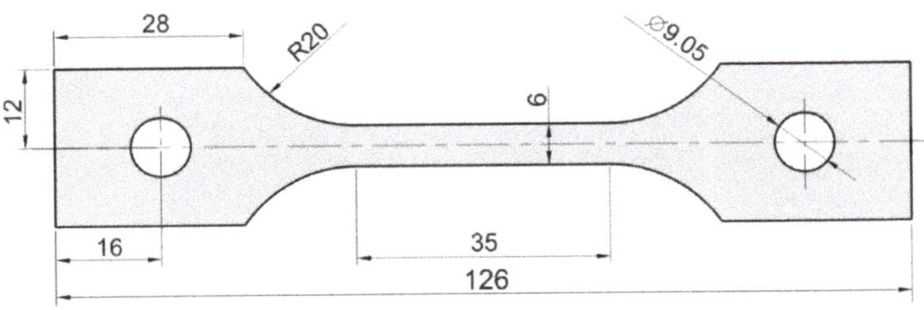

Figure 1. Dimensions (in mm) of the samples for the tensile test.

Table 1. Chemical composition of 321 steel (wt.%) [39].

C (Max.)	Si	Mo	Mn (Max.)	P + S (Max.)	Cr	Ni	Ti (Max.)	N (Max.)	Fe
0.08	0.25–1.00	0.75	2.0	0.04 P 0.03 S	17.0–19.0	9.0–12.0	0.7	0.1	balance

2.2. Uniaxial Tensile Test

Dog-bone samples (Figure 1) were stretched in a uniaxial tensile test machine Zwick/Roell Z100 (Figure 2). Three different strain rates were used in the investigations: $4 \times 10^{-3}\ s^{-1}$, $8 \times 10^{-3}\ s^{-1}$ and $16 \times 10^{-3}\ s^{-1}$. The tests were carried out at a temperature of 24 °C.

Figure 2. Test stand.

2.3. Digital Image Correlation Technique

The digital image correlation Aramis system (GOM, Braunschweig, Germany) was used to determine the character of the deformation of the sample in a non-contact manner during the stretching process of samples. The measuring system consists of two essential components. The first is a scanner, consisting of two high-resolution digital cameras (Figure 2) positioned in relation to each other in such a way as to be able to build a spatial image. The second element of the set is a computer with special software for the numerical processing of images.

Proper surface preparation of the samples is necessary for DIC analysis. First, the samples are covered with white paint, and then a random pattern of contrasting black spots is applied. The selection of the size of the spots depends on the size of the test object and the optical properties of the measuring apparatus. The measurement process consists in taking a series of images of the sample by two cameras in successive stages of loading. Correlation of images from these cameras makes it possible to determine the position of each pixel by giving them coordinates in a three-dimensional coordinate system. Then the images are divided into so-called deformation grids, each of which contains a unique pattern of dots [40]. The initial setup becomes the reference step. Due to the deformation of the sheet surface, the spots in each element of the mesh change their position in relation to each other, which makes it possible to calculate the full field deformation in relation to the reference position. However, in order to make such an analysis possible, the system divides the measurement area into fields of a fixed size called facets. Local strains are obtained from the formula:

$$\varepsilon_{eng} = \left(\lim_{l\to 0}\left(\frac{l_{in} + \Delta l_d}{l_{in}}\right) - 1\right)\cdot 100\% \tag{1}$$

where l_{in} is the initial dostance between two neighbouring facets and Δl_d is the distance increase during uniaxial tensile test.

The dimension of standard facet is 21 × 21 pixels. The second characteristic value is the distance between the centres of adjacent facets. The Aramis system offers many benefits, such as:

- a stable solution for full-field analyses of test objects of just a few millimeters up to structural components of several metres in size,
- it is a material-independent and non-contact measuring system,
- it performs high-precision measurements with a 3D measurement resolution in the sub-micrometer range,
- it is a high-resolution for point-based and full-field measurements.

2.4. Infrared (IR) Thermal Mapping

The surface temperature of the samples during the stretching was measured using a high-sensitivity IR Flir T640 camera (Flir Systems AB, Antennvägen 6, 187 66 Täby, Sweden). The measurements of temperature were correlated with the measurement of strain using the A0Aramis system. Non-invasive distance measurement was performed with an accuracy of thermographic measurements in the range of +/− 2 °C. The principal parameters of the Flir T640 camera are listed in Table 2.

Table 2. Parameters of a Flir T640 camera.

Parameter	Unit	Value
Resolution	pixel	2048 × 1536
Measuring range	°C	−40 ÷ +2000
Refresh rate	Hz	30
Thermal sensitivity	mK	30

2.5. Analysis of Variance

Quadratic multidimensional ANOVA was used to determine the relationship between material properties (strain hardening exponent, yield stress, sample orientation), process parameters (strain rate and percentage strain) and the maximum temperature appearing in the sample during the stretching process. The values of the strain hardening exponent were determined by approximating true stress-true strain curves using the well known Hollomon's power law.

ANOVA is a statistical method for examining observations that depend on one or more factors acting simultaneously. Due to the different numerical ranges of the data, they were normalised to the range [−1, +1] [41]. The *min-max* normalisation was applied by means of a linear function bringing the data to a new interval (Coded Low, Coded High).

Explanatory variables should be independent of each other. The input data were fitted with a polynomial, and the influence of individual variables on the quality of the model was checked using backward elimination. The minimum and maximum values of the input variables for samples made of 321 sheet are presented in Table 3. Intermediate values of input parameters were coded proportionally in the range [−1, +1].

Table 3. Factors and levels for analysis of variance (ANOVA) of stretching of specimens of 321 steel.

Parameter	Name	Unit	Type	Minimum	Maximum
A	Strain hardening exponent	-	Numeric	0.35	0.392
B	Yield stress	MPa	Numeric	341	367
C	Strain rate	s^{-1}	Numeric	4×10^{-3}	16×10^{-3}
D	Sample orientation	deg.	Numeric	0	90
E	Percentage strain	%	Numeric	12	117.7

The regression models that were built were subjected to significance tests. The method of backward elimination of variables was used in the analysis. It is a variable selection procedure in which all the variables are entered into the regression equation and then removed sequentially. The variable with the lowest partial correlation with the dependent variable is considered for removal first. If it meets the elimination criteria, it is removed. The basis for removing or leaving a given variable in the model is the calculation of Fisher F statistics. The independent variable with the highest probability corresponding to the Fisher parameter F is removed from the model if the probability p is sufficiently high (typically $p = 0.10$). After the first variable is removed, the next one to be removed is the one that has the smallest partial correlation with the dependent variable. The procedure exits when there are no other variables in the equation that meet the removal criteria The significance of the regression model at a level of $\alpha = 0.05$ is determined based on the variance due to the effect of a factor and the variance due to the error term.

3. Results and Discussion

3.1. Experimental Investigations

Figure 3 shows the evolution of the surface temperature of the samples cut along the sheet rolling direction of the 321 steel during its stretching with different strain rates. It is clearly visible that even within the range of proportional deformations the zone with the highest temperature is located approximately at the central point between the grippers of the testing machine. As a result of heat convection caused by internal friction of the material, an increase in the temperature of the samples was also observed in the part that is gripped where no plastic deformation occurs. Increasing the strain rate resulted in the intensification of heat generation. At the moment of the formation of a necking at the highest strain rate, the maximum sample temperature increased by more than 4 times compared to the initial temperature. Similar conclusions can be drawn for the samples cut at an angle of 45° with respect to RD and perpendicular to RD (Figure 4).

Only part of the mechanical energy is converted into heat in the deformation process. The remainder is stored in the microstructure of the material increasing the internal energy of the material. The total energy spent on deforming an elastic-plastic material is equal to the work undertaken on elastic (reversible) deformation and permanent (plastic) deformation. Moreover, the energy consumed on plastic deformation is divided into heat dissipated in the forming process and energy stored in the material [42–44]. The evolution of the microstructure during deformation depends on the type of material, the initial temperature and the loading situation [45,46]. Changing the deformation method usually leads to a partial or complete reconstruction of the dislocation systems formed in the previous stages of deformation and the formation of new systems [47,48]. The interaction of the lattice defects is related to the overlapping of their stress fields. If the overlap of stress fields caused by defects reduces the energy of the system, then with a specific activity of the respective slip systems, these defects will form configurations compliant with the principle of energy minimisation [49,50].

The plasticity margin measured as the difference between ultimate tensile stress and field stress is more pronounced in the case of 321 steel than for 17-4PH steel.

Figure 5 show the effect of percentage strain on the change in the maximum temperature of the specimens of 321 steel. These dependencies, determined by the determination coefficient R^2, show a linear trend with a high correlation value $R^2 > 0.93$. In the case of the sample made from 321 steel cut at an angle of 45° tested at a strain rate of 16×10^{-3} s^{-1}, the last point corresponds to the advanced stage of sample necking. Therefore, the R^2-value for that case is about 0.9484 (Figure 5c). In general, the higher the strain rate the steeper the trend lines. The samples made of 321 steel show anisotropic features, which are particularly visible at a strain rate of 8×10^{-3} s^{-1} (Figure 5b). Trend lines for samples cut at different angles are inclined at different angles with respect to the abscissa axis. True stress-strain curves for 321 steel determined at various strain rates are shown in Figure 6. True strain is defined as ε_t:

$$\varepsilon_t = ln\left(\frac{l}{l_0}\right) \quad (2)$$

where l and l_0 are the current and initial gauge lengths, respectively.

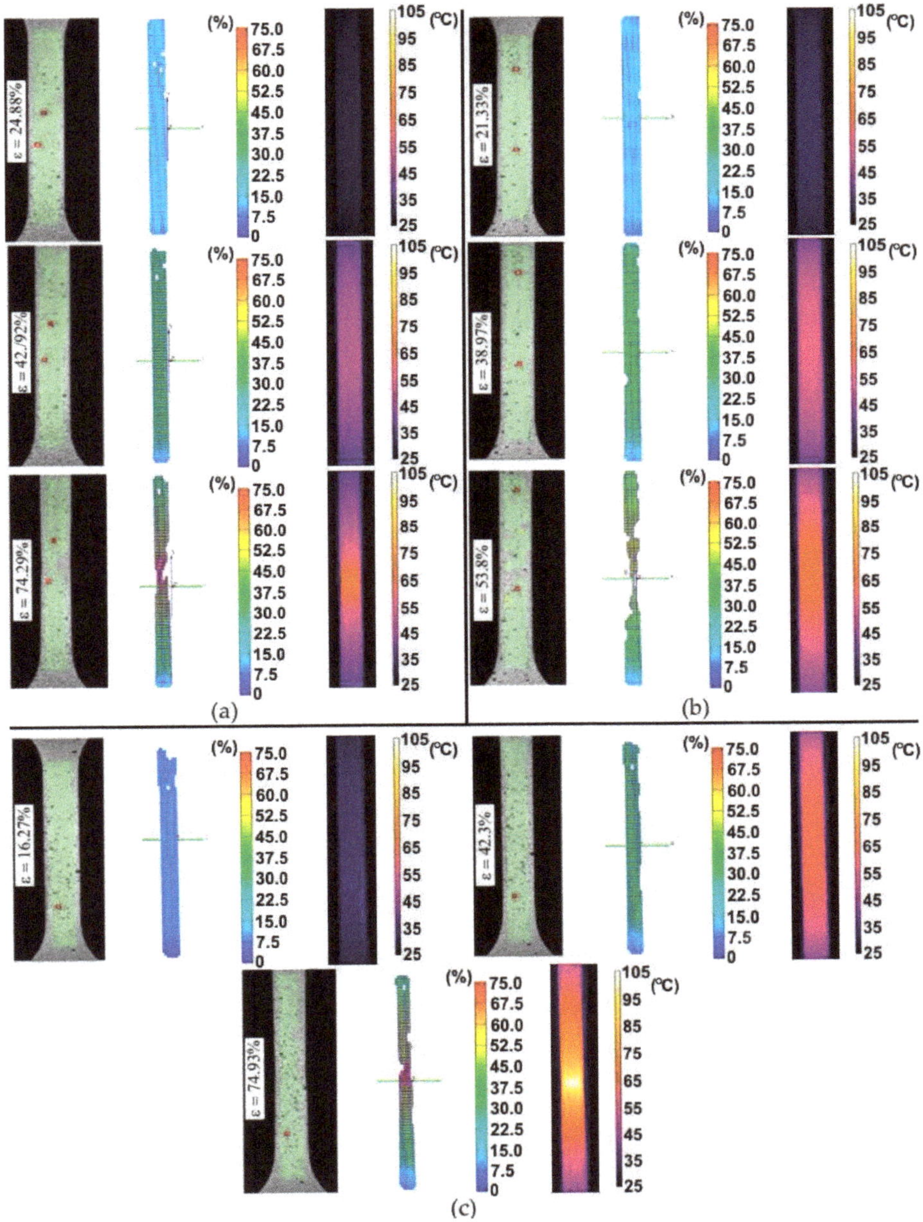

Figure 3. Infrared (IR) thermography images and digital image correlation charts for 321 steel samples cut along the sheet rolling direction, tested at the following strain rates: **(a)** 4×10^{-3} s^{-1}, **(b)** 8×10^{-3} s^{-1} and **(c)** 16×10^{-3} s^{-1}.

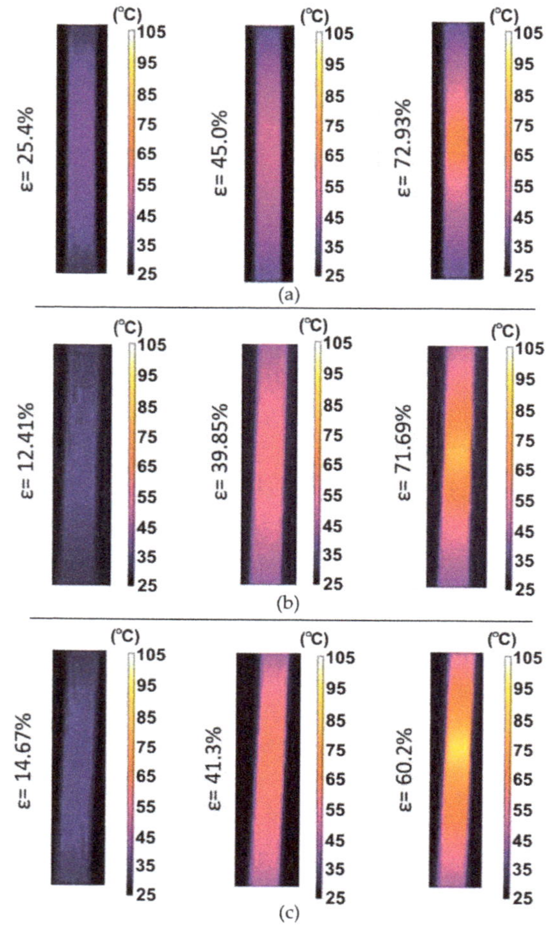

Figure 4. IR thermography images and digital image correlation charts for 321 steel samples cut perpendicular to the rolling direction (RD), tested at the following strain rates: (**a**) 4×10^{-3} s^{-1}, (**b**) 8×10^{-3} s^{-1} and (**c**) 16×10^{-3} s^{-1}.

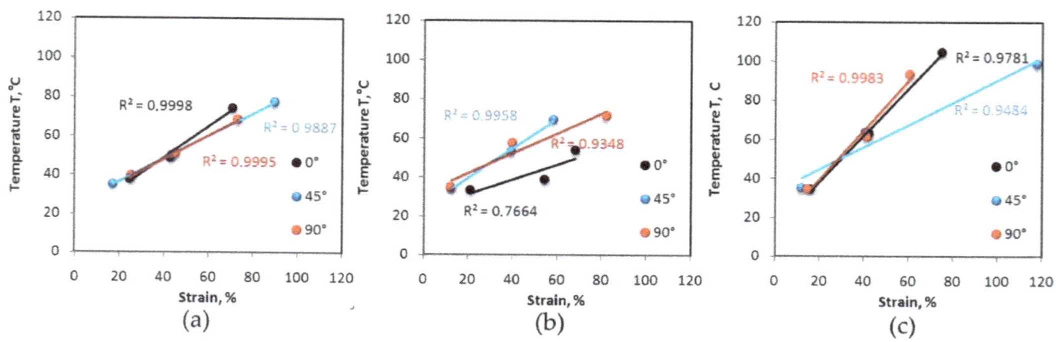

Figure 5. Influence of strain on temperature for 321 steel samples tested at the following strain rates: (**a**) 4×10^{-3} s^{-1}, (**b**) 8×10^{-3} s^{-1} and (**c**) 16×10^{-3} s^{-1}.

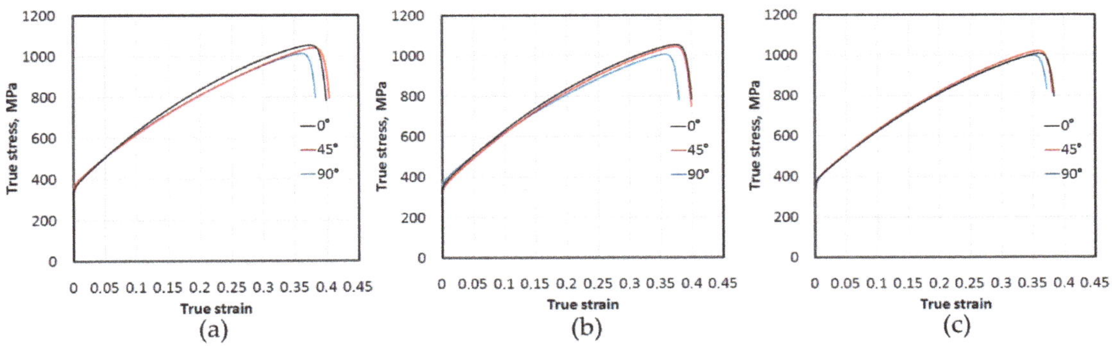

(a) (b) (c)

Figure 6. True stress-true strain curves for 321 steel determined at the following strain rates: (**a**) 4×10^{-3} s^{-1}, (**b**) 8×10^{-3} s^{-1} and (**c**) 16×10^{-3} s^{-1}.

3.2. Analysis of Variance (ANOVA) for 321 Steel Samples

A quadratic model containing only statistically significant elements was chosen to fit the experimental data. The temperature response surface for the 321 steel specimens is given by:

$$T = 76.71 - 18.56A - 10.02B - 4.19C + 4.71D + 48.61E - 4.97AB + 9.58AC + 8.27AD - 31.41AE - 12.99BC - 13.1BE - 21.35CE - 1.22DE + 13.36E^2 \quad (3)$$

The significances of the influence of the individual material parameters as well as the conditions of the stretching process and their interaction in the regression model were determined on the basis of the ANOVA. The quality of fit of the model obtained to the values measured for a given research plan (Equation (3)) was then determined on the basis of the determination coefficients R^2 and the F test. The significance of the model is confirmed by an F-value of 53.30 (Table 4). This model can, therefore, be used to predict temperature values based on material deformation. The probability that such a large F-value can result from data noise is only 0.01%.

Table 4. Results of analysis of variance for the regression model for 321 steel samples.

Source	Sum of Squares	Degrees of Freedom	Mean Square	F-Value	p-Value	Meaning
Model	8991.31	14	642.24	53.30	<0.0001	significant
A—strain hardening exponent	26.04	1	26.04	2.16	0.1723	-
B—yield stress	39.77	1	39.77	3.30	0.0993	-
C—strain rate	0.5708	1	0.5708	0.0474	0.8321	-
D—sample orientation	13.38	1	13.38	1.11	0.3167	-
E—plastic strain	989.76	1	989.76	82.14	<0.0001	-
AB	1.11	1	1.11	0.0920	0.7679	-
AC	9.23	1	9.23	0.7664	0.4019	-
AD	22.41	1	22.41	1.86	0.2025	-
AE	171.00	1	171.00	14.19	0.0037	-
BC	4.45	1	4.45	0.3696	0.5568	-
BE	36.85	1	36.85	3.06	0.1109	-
CE	758.38	1	758.38	62.94	<0.0001	-
DE	2.40	1	2.40	0.1992	0.6649	-
E^2	66.37	1	66.37	5.51	0.0409	-
Residual	120.49	10	12.05	-	-	-
Cor Total	9111.80	24	-	-	-	-

The probability values p for E, AE, CE, E^2 that are less than 0.0500 prove that these factors are statistically significant. The parameters the removal of which will not reduce the quality of the model include, inter alia, C, AB, BC, DE. Due to the satisfactory value of the determination coefficient for the model equal to $R^2 = 0.9868$ (Table 5), it was decided to leave these variables in the model. The predicted value of the coefficient of determination $R^2 = 0.8982$ is in reasonable agreement with the adjusted coefficient of determination $R^2 = 0.9683$ because the difference between their values is less than 0.2 [51]. The value of adequacy precision 27.6342 is greater than 4. The adequacy of model is, therefore, confirmed.

Table 5. Fit statistics of the regression model for 321 steel samples.

Standard Deviation	**3.47**	**R^2**	**0.9868**
Mean	54.09	Adjusted R^2	0.9683
Coefficient of variation, %	6.42	Predicted R^2	0.8982
-	-	Adequacy precision	27.6342

If the errors arising during the experimental tests are random, then the distribution of residuals in the form of deviations between the actual values and the expected values should be in accordance with the normal distribution. The distribution of residuals along a straight line (Figure 7a) shows that the studentised residuals have a normal distribution [52]. Figure 7b shows a comparison of the actual values and the predicted temperature values resulting from the introduction of independent variables to the regression model. The distribution of data around a straight line indicates that the regression model is of good quality [53,54].

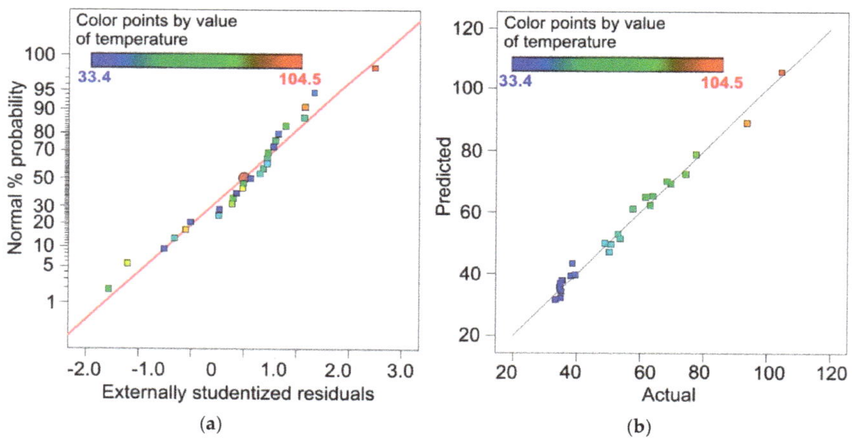

Figure 7. (**a**) Normal % probability vs. externally studentised residuals and (**b**) predicted vs. actual values of temperature plots for 321 steel samples.

The relationship between the dependent and independent variables was analysed using 2D response surface plots. Figures 8 and 9 show the interactive effect of two selected parameters on the value of the maximum temperature in the sample during the stretching test when the other input parameters are constant [55–57]. Figure 8a indicates the synergistic role of the strain hardening exponent (SHE) and yield stress (YS). The heat is generated faster towards the low values of both the SHE and YS. Figure 8b indicates that the SHE plays a more significantly influential role in heat generation. However, the strain rate (SR) is a significant factor, but only when considering small values of SHE. For small values of SHE, the sample orientation in interaction with the SHE does not significantly affect the temperature value (Figure 8c). However, if the SHE increases then the difference between the temperature for 0° and 90° orientations increases to around 30 °C. For small values of SHE, starting from a plastic strain (PS) of about 80%, the regression model overestimates the temperature values (Figure 8d). However, the trend observed in the experimental studies is maintained, i.e., with an increase in PS for a given SHE value, the temperature value increases.

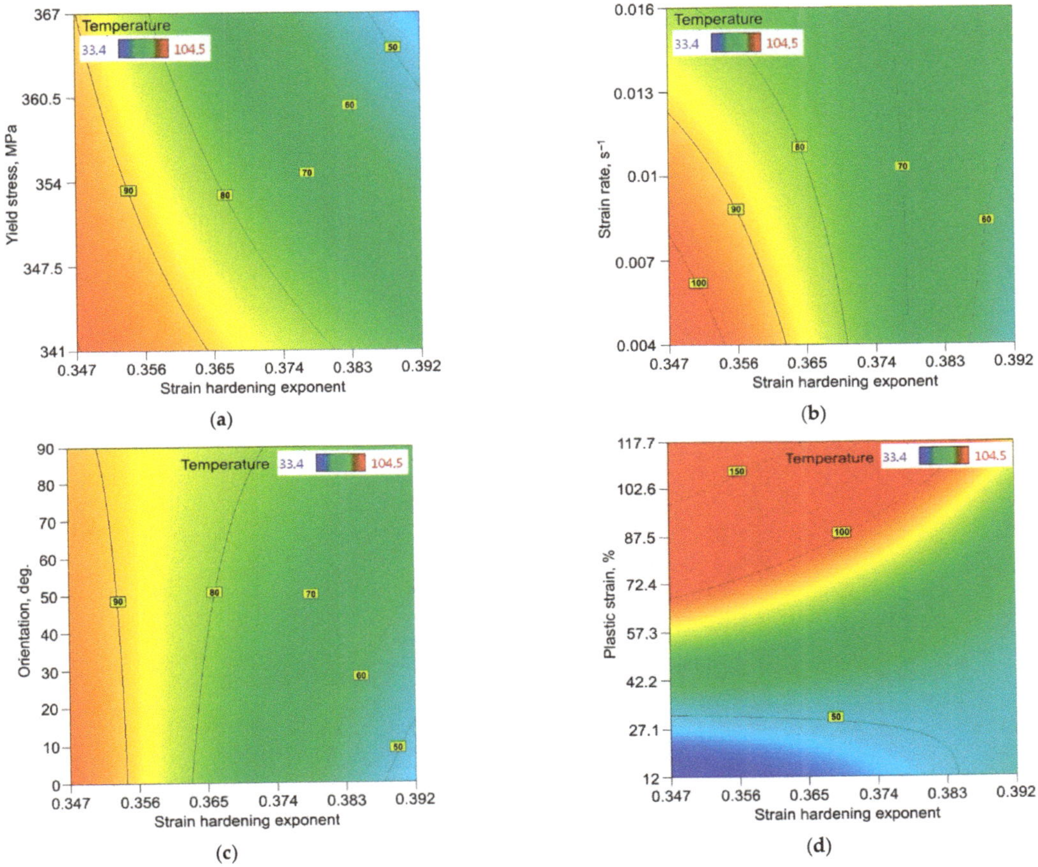

Figure 8. Response surface plots presenting the interaction between the strain hardening exponent and (**a**) yield stress, (**b**) strain rate, (**c**) sample orientation and (**d**) plastic strain, affecting the value of the maximum temperature of the surface of the 321 steel sample.

The synergistic effect of both SR and YS is more complicated (Figure 9a). Similarly, in the case of sample orientation and YS (Figure 9b), it is only for the lowest value of YS that strain rate and orientation significantly affect the value of temperature. Figure 9c indicates that the orientation when testing samples with different SR values plays the following role: for orientation 0°, the temperature is low and increases with change in the orientation to 45° and 90°. The effect of PS and SR shown in Figure 9d is very similar to the effect of PS and SHE on temperature (Figure 8d). A strong correlation between PS and temperature is also confirmed in Figure 9e,f by the high isotherm concentration. Specimens cut at 90° heat faster during deformation than specimens cut at 0° (Figure 9e). The reverse effect is observed in Figure 9f: when the YS of specimens is small, the material heats up faster.

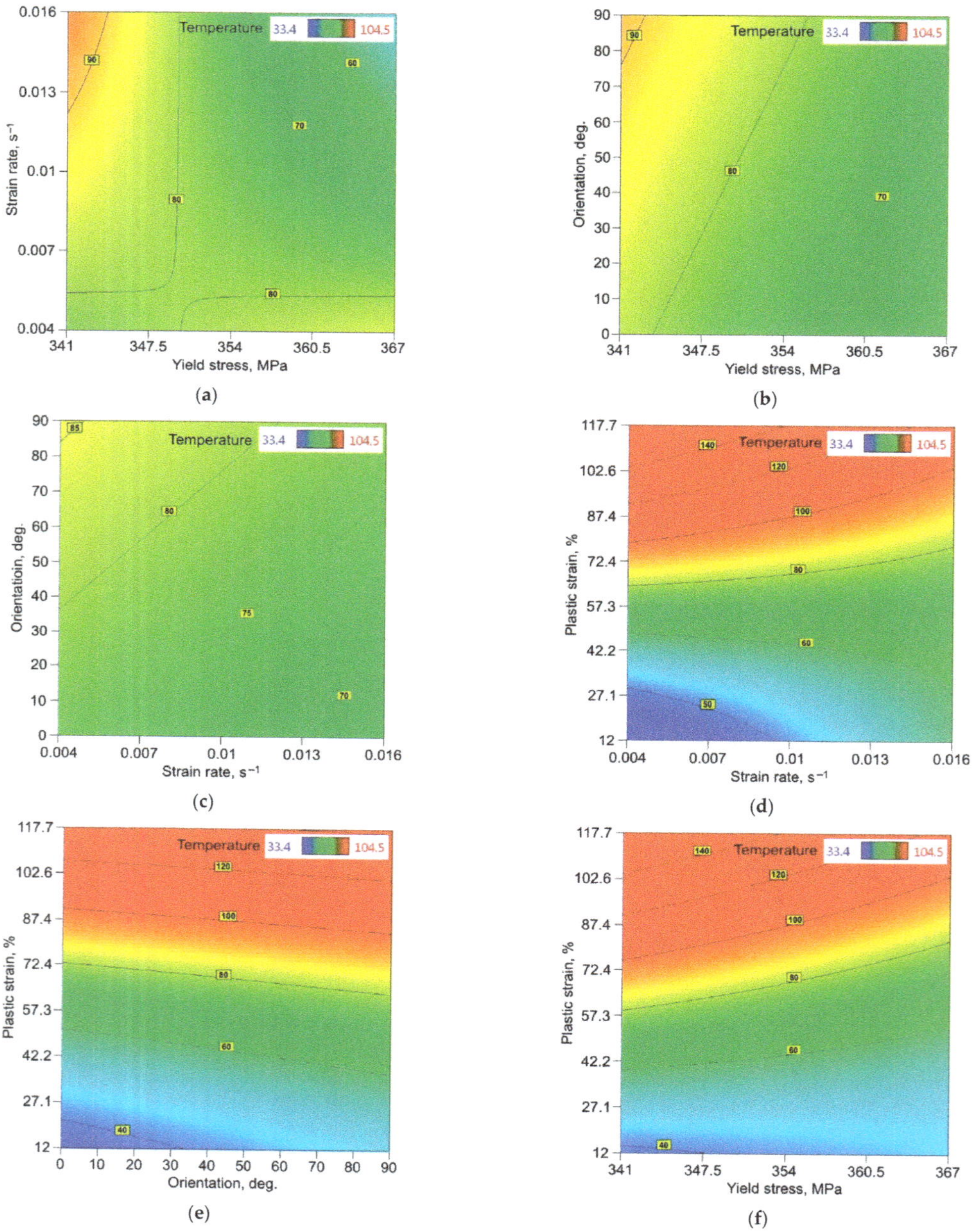

Figure 9. Response surface plots presenting the interaction between (**a**) strain rate (SR) and yield stress (YS), (**b**) sample orientation and YS, (**c**) sample orientation and SR, (**d**) plastic strain (PS) and SR, (**e**) PS and sample orientation and (**f**) PS and YS, affecting the value of maximum temperature of 321 steel sample surface.

4. Conclusions

In this paper, full-field image correlation and infrared thermography techniques were used to analyse the effect of material properties and strain rate on heat generation in stretched dog-bone specimens of titanium-stabilised austenitic stainless steel. Based on the experimental results and analysis of variance, the following conclusions can be drawn:

- The heat during stretching of the samples was not generated proportionally over the entire length of the samples. In the range of proportional deformation (before necking formation), the zone with the highest temperature was located approximately centrally along the distance between the grippers of the testing machine.
- At the moment of formation of a necking at the highest strain rate, the maximum sample temperature increased more than four times compared to the initial temperature. This phenomenon was observed for all sample orientations analysed.
- The character of the effect of plastic strain on the maximum temperature generated in the specimens was close to a linear trend in the range of strain rates analysed.
- A synergistic effect of the SHE and YS revealed that heat was generated more rapidly towards small values of SHE and YS.
- For small values of SHE, the sample orientation in interaction with the SHE did not significantly affect the temperature value.
- For all the orientations analysed, the temperature increased in a similar way with increasing PS.

Author Contributions: Conceptualisation, K.Ż.; methodology, K.Ż., S.P., S.R. and T.T.; software, K.Ż.; S.P. and M.B.; validation, K.Ż. and T.T.; investigation, K.Ż., S.P., S.R., M.B. and T.T.; data curation, K.Ż. and T.T.; visualisation, K.Ż. and T.T.; writing—original draft preparation, T.T.; writing—review and editing, K.Ż. and T.T. All authors have read and agreed to the published version of the manuscript.

Funding: This research received no external funding.

Institutional Review Board Statement: Not applicable.

Informed Consent Statement: Not applicable.

Data Availability Statement: The data presented in this study are available on request from the corresponding author.

Conflicts of Interest: The authors declare no conflict of interest.

References

1. Pater, Z.; Samołyk, G. *Podstawy Technologii Obróbki Plastycznej Metali*; Politechnika Lubelska: Lublin, Poland, 2013.
2. Di Gioacchino, F.; Edwards, T.E.J.; Wells, G.N.; Clegg, W.J. A new mechanism of strain transfer in polycrystals. *Sci. Rep.* **2020**, *10*, 10082. [CrossRef]
3. Physics on Nondestructive Evaluation. Iowa State University. Available online: https://www.nde-ed.org/Physics/Materials/Structure/anisotropy.xhtml (accessed on 19 August 2021).
4. Yang, G.; Park, S.J. Deformation of single crystals, polycrystalline materials, and thin films: A review. *Materials* **2019**, *12*, 2003. [CrossRef]
5. Venkataraman, A.; Sangid, M.D. *Deformation Mechanisms at Grain Boundaries in Polycrystals*; Final Report to U.S. Department of Energy, Office of Basic Energy Sciences, Division of Materials Sciences and Engineering under Award#DE-SC001428; Purdue University: West Lafayette, IN, USA, 2019.
6. Luo, H.; Sheng, H.; Zhang, H.; Wang, F.; Fan, J.; Du, J.; Liu, J.P.; Szlufarska, I. Plasticity without dislocations in a polycrystalline intermetallic. *Nat. Commun.* **2019**, *10*, 3587. [CrossRef] [PubMed]
7. Bahrami, A.; Taheri, P. A study on the failure of AISI 304 stainless steel tubes in a gas heater unit. *Metals* **2019**, *9*, 969. [CrossRef]
8. Bansod, A.V.; Patil, A.P.; Moon, A.P.; Khobragade, N.N. Intergranular corrosion behavior of low-nickel and 304 austenitic stainless steels. *J. Mater. Eng. Perform.* **2016**, *25*, 3615–3626. [CrossRef]
9. Hall, E.O. The deformation and ageing of mild steel: III discussion of results. *Proc. Phys. Soc.* **1951**, *64*, 747–753. [CrossRef]
10. Petch, N.J. The cleavage strength of polycrystals. *J. Iron Steel Inst.* **1953**, *174*, 25–28.
11. Roa, J.J.; Wheeler, J.M.; Trifonov, T.; Fargas, G.; Mateo, A.; Michler, J.; Jiménez-Piqué, E. Deformation of polycrystalline TRIP stainless steel micropillars. *Mater. Sci. Eng. A* **2015**, *647*, 51–57. [CrossRef]
12. Di Gioacchino, F.; da Fonseca, J.Q. An experimental study of the polycrystalline plasticity of austenitic stainless steel. *Int. J. Plast.* **2015**, *74*, 92–109. [CrossRef]

13. Cheong, K.S.; Busso, E.P. Discrete dislocation density modelling of single phase FCC polycrystal aggregates. *Acta Mater.* **2004**, *52*, 5665–5675. [CrossRef]
14. Ichiyanagi, K.; Takagi, S.; Kawai, N.; Fukaya, R.; Nozawa, S.; Nakamura, K.G.; Liss, K.D.; Kimura, M.; Adachi, S.I. Microstructural deformation process of shock-compressed polycrystalline aluminum. *Sci. Rep.* **2019**, *9*, 7604. [CrossRef]
15. Korsunsky, A.M. Crystal plasticity and hardening: A dislocation dynamics study. *Procedia Eng.* **2009**, *1*, 241–244.
16. Farren, W.S.; Taylor, G.I. The heat developped during plastic extension of metal. *Proc. R. Soc. A* **1925**, *107*, 422–451.
17. Taylor, G.I.; Quinney, H. The latent energy remaining in a metal after cold working. *Proc. R. Soc. A* **1933**, *143*, 307–326.
18. Bodelot, L.; Sabatier, L.; Charkaluk, E.; Dufrenoy, P. Optical and infrared coupled full-field measurements at a mesoscopic scale. In Proceedings of the 9th International Conference on Quantitative InfraRed Thermography, Kraków, Poland, 2–5 July 2008.
19. Boulanger, T.; Chrysochoos, A.; Mabru, C.; Galtier, A. Calorimetric analysis of dissipative and thermoelastic effects associated with the fatigue behavior of steels. *Int. J. Fatigue* **2004**, *26*, 221–229. [CrossRef]
20. Ait-Amokhtar, H.; Fressengeas, C.; Boudrahem, S. The dynamics of Portevin-Le Chatelier bands in an Al-Mg alloy from infrared thermography. *Mater. Sci. Eng. A* **2008**, *488*, 540–546. [CrossRef]
21. Völkl, R.; Fischer, B. Mechanical testing of ultra-high temperature alloys. *Exp. Mech.* **2004**, *44*, 121–127. [CrossRef]
22. Anwander, M.; Zagar, B.G.; Weiss, B.; Weiss, H. Noncontacting strain measurements at high temperatures by the digital laser speckle technique. *Exp. Mech.* **2000**, *40*, 98–105. [CrossRef]
23. Pan, B.; Wu, D.; Gao, J. High-temperature strain measurement using active imaging digital image correlation and infrared radiation heating. *J. Strain. Anal. Eng. Des.* **2013**, *49*, 224–232. [CrossRef]
24. Pan, B.; Wu, D.; Wang, Z.; Xia, Y. High-temperature digital image correlation method for full-field deformation measurement at 1200 °C. *Meas. Sci. Technol.* **2011**, *22*, 015701. [CrossRef]
25. Sharpe, W.N. A high-frequency high-temperature optical strain/displacement gage. *Exp. Mech.* **2009**, *50*, 227–237. [CrossRef]
26. Orteu, J.J.; Rotrou, Y.; Sentenac, T.; Robert, L. An innovative method for 3-D shape, strain and temperature full-field measurement using a single type of camera: Principle and preliminary results. *Exp. Mech.* **2007**, *48*, 163–179. [CrossRef]
27. Tung, S.H.; Sui, C.H. Application of digital-image-correlation techniques in analysing cracked cylindrical pipes. *Sadhana* **2010**, *35*, 557–567. [CrossRef]
28. Hild, F.; Bouterf, A.; Forquin, P.; Roux, S. On the use of digital image correlation for the analysis of the dynamic behavior of materials. In *The Micro-World Observed by Ultra High-Speed Cameras*; Tsuji, K., Ed.; Springer: Cham, Switzerland, 2018; pp. 185–206.
29. Wattrisse, B.; Chrysochoos, A.; Muracciole, J.-M.; Nemoz-Gaillard, M. Kinematic manifestations of localisation phenomena in steels by digital image correlation. *Eur. J. Mech. A Solids* **2001**, *20*, 189–211. [CrossRef]
30. Hung, P.C.; Voloshin, A.S. In-plane strain measurement by Digital Image Correlation. *J. Braz. Soc. Mech. Sci. Eng.* **2003**, *25*, 215–221. [CrossRef]
31. Feng, X.; Xue, F. Characterization of 3D printed bolts based on digital image correlation and infrared thermography. *Mater. Des.* **2020**, *191*, 108641. [CrossRef]
32. Cholewa, N.; Summers, P.T.; Feih, S.; Mouritz, A.P.; Lattimer, B.Y.; Case, S.W. A technique for coupled thermomechanical response measurement using infrared Thermography and Digital Image Correlation (TDIC). *Exp. Mech.* **2016**, *56*, 145–164. [CrossRef]
33. Żaba, K.; Trzepieciński, T.; Puchlerska, S.; Noga, P.; Balcerzak, M. Coupled thermomechanical response measurement of deformation of nickel-based superalloys using full-field digital image correlation and infrared thermography. *Materials* **2021**, *14*, 2163. [CrossRef] [PubMed]
34. Chrysochoos, A.; Huon, V.; Jourdan, F.; Muracciole, J.M.; Peyrux, R.; Wattrisse, B. Use of full-field digital image correlation and infrared thermography measurements for the thermomechanical analysis of material behaviour. *Strain* **2010**, *46*, 117–130. [CrossRef]
35. Maynadier, A.; Poncelet, M.; Lavernhe-Taillard, K.; Roux, S. One-shot measurement of thermal and kinematic fields: Infrared image correlation (IRIC). *Exp. Mech.* **2011**, *52*, 241–255. [CrossRef]
36. Curà, F.; Curti, G.; Sesana, R. A new iteration method for the thermographic determination of fatigue limit in steels. *Int. J. Fatigue* **2005**, *27*, 453–459. [CrossRef]
37. Doudard, C.; Calloch, S.; Cugy, P.; Galtier, A.; Hild, F. A probabilistic two-scale model for high-cycle fatigue life predictions. *Fatigue Fract. Eng. Mater. Struct.* **2005**, *28*, 279–288. [CrossRef]
38. ASTM International. ASTM International. ASTM E8/E8M-11. Standard test methods for tension testing of metallic materials. In *ASTM Book of Standards*; ASTM International: West Conshohocken, PA, USA, 2011.
39. SAE International. *AMS 5604-Steel, Corrosion and Heat Resistant, Sheet, Strip, and Plate 18Cr-10.5Ni-0.40Ti (SAE 30321) Solution Heat Treated*; SAE International: Warrendale, PA, USA, 1991.
40. Zielecki, W.; Kubit, A.; Święch, Ł. Analiza Doświadczalna Odkształcenia Elementów Klejonych w Próbie Statycznego Oddzierania. *Pomiary Autom. Robot.* **2013**, *2*, 71–78.
41. Trzepieciński, T.; Szpunar, M.; Kaščák, Ľ. Modeling of friction phenomena of TI-6AL-4V sheets based on backward elimination regression and multi-layer artificial neural networks. *Materials* **2021**, *14*, 2570. [CrossRef] [PubMed]
42. Brennhaugen, D.D.E.; Georgarakis, K.; Yokoyama, Y.; Nakayama, K.S.; Arnberg, L.; Aune, R.E. Probing heat generation during tensile plastic deformation of a bulk metallic glass at cryogenic temperature. *Sci. Rep.* **2018**, *8*, 16317. [CrossRef]
43. Kostina, A.; Iziumowa, A.; Plekhov, O. energy dissipation and storage in iron under plastic deformation (experimental study and numerical simulation). *Frat. Ed. Integrità Strutt.* **2014**, *27*, 28–37. [CrossRef]

44. Rusinek, A.; Klepaczko, J.R. Experiments on heat generated during plastic deformation and stored energy for TRIP steels. *Mater. Des.* **2009**, *30*, 35–48. [CrossRef]
45. Klitschke, S.; Trondl, A.; Huberth, F.; Liewald, M. Adiabatic heating under various loading situations and strain rates for advanced high-strength steels. *IOP Conf. Ser. Mater. Sci. Eng.* **2018**, *418*, 012123. [CrossRef]
46. Sada, N.B. The plastic depth of heat treatment steel alloy (AISI 01) due to torsion test. *J. Appl. Sci.* **2005**, *5*, 1196–1199. [CrossRef]
47. Shen, Y.F.; Li, X.X.; Sun, X.; Wang, Y.D.; Zuo, L. Twinning and martensite in a 304 austenitic stainless steel. *Mater. Sci. Eng. A* **2012**, *552*, 514–522.
48. Song, G.S.; Ji, K.S.; Song, H.W.; Zhang, S.H. Microstructure transformation and twinning mechanism of 304 stainless steel tube during hydraulic bulging. *Mater. Res. Express* **2019**, *6*, 1265h9. [CrossRef]
49. Rubtsov, V.E.; Kolubaev, A.V. Effect of heat generation due to plastic deformation on behavior of surface-layer material during sliding. *J. Frict. Wear* **2009**, *30*, 324. [CrossRef]
50. Maj, M. Wpływ Kierunku Wstępnego Odkształcenia na Proces Magazynowania Energii w Polikryształach. Ph.D. Thesis, Instytut Podstawowych Problemów Techniki Polskiej Akademii Nauk, Warsaw, Poland, 2007.
51. Montgomery, D.C. *Design and Analysis of Experiments*, 8th ed.; John Wiley & Sons Inc.: Hoboken, NJ, USA, 2013.
52. Sen, A.; Srivastava, M. *Regression Analysis Theory, Methods, and Applications*; Springer: New York, NY, USA, 1990.
53. Montgomery, D.C.; Peck, E.A.; Vining, G.G. *Introduction to Linear Regression Analysis*, 5th ed.; John Wiley & Sons Inc.: Hoboken, NJ, USA, 2012.
54. Alexopoulos, E.C. Introduction to multivariate regression analysis. *Hippokratia* **2010**, *14*, 23–28. [PubMed]
55. Chauhan, S.R.; Dass, K. Dry sliding wear behaviour of titanium (grade 5) alloy by using response surface methodology. *Adv. Tribol.* **2013**, *2013*, 272106. [CrossRef]
56. Khalatbari, H. Investigation of Formability of Material in Incremental Sheet Metal Forming Process. Master's Thesis, Eastern Mediterranean University, Farmagusta, Cyprus, 2012.
57. Venugopal, V.; Kumar, K.J.; Muralidharan, S.; Parasuraman, S.; Raj, P.V.; Kumar, K.V. Optimization and in-vivo evaluation of isradipine nanoparticles using Box-Behnken design surface response methodology. *OpenNano* **2016**, *1*, 1–15. [CrossRef]

materials

MDPI

Article

Study of Interaction Mechanism between Positrons and Ag Clusters in Dilute Al–Ag Alloys at Low Temperature

Xiaoshuang Liu [1,2], Peng Zhang [1], Baoyi Wang [1,2], Xingzhong Cao [1,2], Shuoxue Jin [1] and Runsheng Yu [1,*]

1 Institute of High Energy Physics, Chinese Academy of Sciences, Beijing 100049, China; liuxs@ihep.ac.cn (X.L.); zhangpeng@ihep.ac.cn (P.Z.); wangboy@ihep.ac.cn (B.W.); caoxzh@ihep.ac.cn (X.C.); jinshuoxue@ihep.ac.cn (S.J.)

2 School of Nuclear Science and Technology, Chinese Academy of Sciences, Beijing 100039, China

* Correspondence: yursh@ihep.ac.cn; Tel.: +86-010-88236990

Abstract: The microstructural evolution of dilute Al–Ag alloys in its early aging stage and at low temperatures ranging from 15 K to 300 K was studied by the combined use of Positron annihilation lifetime spectroscopy (PALS), high resolution transmission electron microscopy (HRTEM), and positron annihilation Coincidence Doppler broadening (CDB) techniques. It is shown that at low temperatures below 200 K, an Ag–vacancy complex is formed in the quenched alloy, and above 200 K, it decomposes into Ag clusters and monovacancies. Experimental and calculation results indicate that Ag clusters in Al–Ag alloys can act as shallow trapping sites, and the positron trapping rate is considerably enhanced by a decreasing measurement temperature.

Keywords: Al–Ag alloys; Ag clusters; positron annihilation; monovacancies; shallow trapping

Citation: Liu, X.; Zhang, P.; Wang, B.; Cao, X.; Jin, S.; Yu, R. Study of Interaction Mechanism between Positrons and Ag Clusters in Dilute Al–Ag Alloys at Low Temperature. *Materials* **2021**, *14*, 1451. https://doi.org/10.3390/ma14061451

Academic Editor: Elena Pereloma

Received: 9 February 2021
Accepted: 12 March 2021
Published: 16 March 2021

1. Introduction

Al–Ag alloys are important for the study of the fundamental processes of decomposition in supersaturated alloys, owing to the large difference in the atomic numbers of Al and Ag (13 versus 47) [1]. Although Al–Ag alloys have limited structural applications, Ag is a commonly used microalloying element in aluminum alloys. Adding a small amount (from 0.1 at.% to 0.5 at.%) of Ag into aluminum-based alloys can improve their mechanical properties, thermal stability, and stress corrosion cracking resistance [2,3]. This effect is widely seen in aluminum alloys, particularly in the high-strength alloys for advanced aerospace and defense applications, including Al–Cu–Mg and Al–Zn–Mg-based alloys [4–6]. However, most studies on Al–Ag alloys have focused on the relatively high Ag concentration in Al–Ag alloys. In 1942, Guinier [7] found that the Ag atoms in Al–Ag alloys first form small clusters, then evolve into Guinier-Perston (GP) zones and further larger precipitates in the early stage of the aging process. Guinier-Preston (GP) zones are the early-stage solute enriched regions in aluminum alloys [8,9]. Nagai et al. [10] claimed that the chemical composition of Ag-rich nanoclusters is $AlAg_3$ in aged Al-2.0 at.% Ag from experimental and calculated two-dimensional angular correlation of positron annihilation radiation (2D-ACAR) anisotropy results. In a recent study, B. Zou et al. [11] reported the precipitation process of Ag nanoclusters in three Al–Ag alloys with different Ag contents (Al-1 wt.%Ag, Al-5 wt.%Ag, and Al-15 wt.%Ag). However, in the above studies, little attention was paid to the Al–Ag alloys with a very low Ag concentration (less than 0.5 wt.%). Moreover, the mechanism of formation and migration of Ag clusters in Al–Ag alloys at the early stage of aging is still unclear. The main reason for this is most probably that the Ag clusters that form at the early stage of aging in dilute Al–Ag alloys are too small to be detected by conventional methods. In particular, unlike most other aluminum alloys, GP zones in the Al–Ag system form immediately after quenching [12,13], which makes it difficult to detect the precipitation of Ag clusters in the early stage of aging.

Positron annihilation spectroscopy (PAS) is a highly sensitive technique for studying subatomic scale defects, e.g., 10^{-5}~10^{-6} density defects, in metals and semiconductors [14–16] If a vacancy exists in the material, it will easily capture positrons, since the vacancy does not contain positively charged nuclei. In fact, there are two typical sites in which positrons can be trapped [17,18]. One is the deep positron trapping sites, such as vacancies and voids—since the positron affinity of these defects is less than −1 eV [19], positrons encounter a deep potential well from which it is thereby difficult to escape; the other type is the shallow positron trapping sites—e.g., dislocations. The positron affinities of these defects are greater than −100 meV [20]. Hidalgo et al. [21] studied positron annihilation in deformed copper and found that dislocations can also trap positrons, and the positron trapping rate increased when the temperature was below 77 K. In addition to the two above-mentioned typical types of positron trapping sites, in recent years, some clusters have been shown to be able to act as shallow positron trapping sites; e.g., in 2002, Huis et.al [22] found that lithium nanoclusters are able to trap positrons in MgO, and the trapping ability is enhanced at low temperatures. Indeed, the reason for positron trapping lies mainly in the different chemical elements. A table of positron affinities for most elements can be found in the work of Puska et al. [23]. The values are −4.41 and −5.36 eV for aluminum and silver, respectively. This indicated that the Ag atom has a weak but extended attractive potential to positrons in Al–Ag alloys. Zou et al. [11] found that the Ag clusters in Al–Ag alloys can act as positron trapping centers and that the confinement of positrons is enhanced with the increase in the size of Ag clusters. Moreover, Hugenschmidt et al. [24] found that coincidence doppler broadening (CDB) is sensitive to the detection of small precipitates in aluminum, and even a 0.1 nm thin Sn layer embedded underneath a 200 nm Al coating leads to a significant and unambiguous fingerprint in the CDB spectrum. Nevertheless, studies on the positron trapping mechanism in dilute Al–Ag alloys have not been reported in literature. Moreover, it is very important to investigate the formation and evolution of Ag clusters in Al–Ag alloys.

The main purpose of this work is to investigate the microstructural evolution of Al–Ag alloys in the early aging stage by means of positron annihilation spectroscopy to examine whether or not the minor Ag clusters can act as positron trapping centers and to study the temperature dependence of positron annihilation behavior. In addition to positron annihilation lifetime spectroscopy (PALS) and coincidence doppler broadening (CDB) measurements, we also used high resolution transmission electron microscopy (HRTEM) to investigate the Ag clusters at the early aging stage in Al–Ag alloys.

2. Materials and Methods

2.1. Materials and Preparation

The Al-0.2 wt.%Ag alloys were prepared from aluminum (99.999% purity) and silver (99.999% purity) using a high-frequency induction furnace in a vacuum environment. After melting, the specimens were solution treated at 823 K for 24 h in a vacuum environment, quenched by ice water as quickly as possible (~15 ms), thinned to 1 mm, and punched into 10 × 10 mm^2 square sheets. Pure aluminum and sliver samples were also prepared as reference specimens. All the square sheet samples were electrochemically polished to a mirror-like surface using 25% $HClO_4$ (Beijing Institute of Chemical Reagents, Beijing, China) and 75% C_2H_6O (Beijing Institute of Chemical Reagents, Beijing, China) polishing solution at room temperature (300 K).

Al-0.2 wt.%Ag alloys were then heated at 823 K for 30 min, followed by quenching into ice water. Then, the specimens were naturally aged for 24 h at room temperature. The specimens were measured PLAS and DBS at temperatures of 15–300 K for about 12 h. After that, the CDB spectra were measured for about 12 h.

2.2. Characterization Methods

PALS measurements were carried out using a fast–slow coincidence system with a time resolution of 196 ps (full width at a half maximum). A ^{22}Na positron source with an intensity of 10 μCi was encapsulated by a Kapton foil (made by Nilaco Corporation, Tokyo, Japan). The thickness of the Kapton foil was 7.5 microns. The total number counts were 2.0×10^6 for each lifetime spectrum. The count rate of the PALS was about 100 cps. The data from the experiment were analyzed using the LT9 program [25]. After deducting the components (12%) of the positrons annihilated in Kapton foils, one or two lifetime components were identified in our experimental spectra. At 15–200 K, the positron annihilation lifetime could only be deconvoluted into one lifetime component. Above 200 K, the positron annihilation lifetime was decomposed into two components.

Doppler broadening spectroscopy (DBS) was performed using a single high-purity Ge detector (GEM35P4-76-PL, ORTEC, Oak Ridge, TN, United States) with an energy resolution of 1.3 keV at 511.0 keV. The total number of counts for each single detector DBS spectrum was also 2.0×10^6. The S and W parameters were obtained from these DBS results. The S parameter was defined as the ratio area in 510.55–511.45 keV to the total area of the 511 keV photo peak, reflecting the ratio of the annihilation of positrons with the low-momentum electrons. The W parameter is defined as the ratio area in 514.6–521.2 keV and 500.8–507.4 keV to the total area, reflecting the ratio of annihilation of positrons with the high-momentum electrons. CDB measurements were carried out using two high-purity Ge detectors, with the efficiency of 35%. The counting rate was about 100 cps, and each spectrum took 12 h to ensure no less than 10^6 counts in each CDB spectrum. To obtain supplementary information about the Al–Ag sample microstructure, a TecnaiF30 high resolution transmission electron microscopy (HRTEM, FEI, Hillsboro, OR, United States) operating at 300 kV was utilized.

3. Results and Discussion

Figure 1 shows a typical HRTEM image of Al–Ag alloy sample after aging at room temperature for 24 h. The inset in Figure 1 shows the Fast Fourier Transform (FFT) of the HRTEM image. The HRTEM presented almost prefect crystallographic image. There were no defects or clusters that could be observed in the HRTEM image. The reason may be that the atomic size difference between Al (1.43 Å) and Ag (1.44 Å) is negligible. Thus, it was difficult to distinguish clusters using HRTEM. Moreover, the concentration of Ag clusters in Al-0.2 wt.%Ag alloys is too low (less than 2×10^{-4}). The HRTEM was not sufficiently sensitive to detect such a low content of clusters. It is thus understandable that most previous studies focused on samples with a high Ag content after aging at high temperature [26], where the sizes of GP zones in the studied samples were more than 10 nm. However, as proved by the present study, it is difficult to detect Ag cluster in Al–Ag alloys with very low Ag content after aging at room temperature, even with HRTEM. The FFT of the HRTEM image however has streaks which may indicates the presence of some very fine linear defects or clusters. To make further clarification, positron annihilation spectroscopy is thereafter applied for characterization of the nanostructure in dilute Al–Ag alloys.

The positron annihilation lifetime of quenched Al-0.2% Ag alloys at different temperatures ranging from 15 K to room temperature was obtained by LT9 analysis. At 15–200 K, the positron annihilation lifetime could only be deconvoluted into one lifetime component τ_m about 150 ps after deducting the source components. We measured the positron annihilation value of pure Al at room temperature. The observed value was 150 ps, which was very close to that in the Ag–Vacancy complex (146 ps) [27]. In addition, the positron annihilation lifetime in the vacancy is 231 ps [28], which is much larger than the positron annihilation lifetime at 15–200 K. It is worth noting that the positron annihilation lifetime at 15–200 K is very close to that in aluminum and Ag–V complexes. This provides evidence that the V–Ag complexes were formed after being quenched, which is consistent with the conclusion of Nagai et al. [27] who showed that quenched-in vacancies are bound to Ag atoms in Al–Cu–Mg–Ag alloys with very low Ag contents. Furthermore, Shasha Zhang et al. [29] reported that the Ag atoms have a strong attraction interaction with the vacancy in Al–Ag alloys. That may be the origin of the formation of Ag–V complexes in Al–Ag alloys. In 1988, F. Maury et al. [30] conducted electrical resistivity measurements on dilute Al–Ag alloys after electron irradiation. They reported that the Ag atoms in the dilute Al–Ag alloys are easily trapped by vacancies to form Ag–V complexes when the temperature is below 180 K, and Ag–V complexes can be recovered above 180 K. This coincides rather well with our current positron annihilation measurement results.

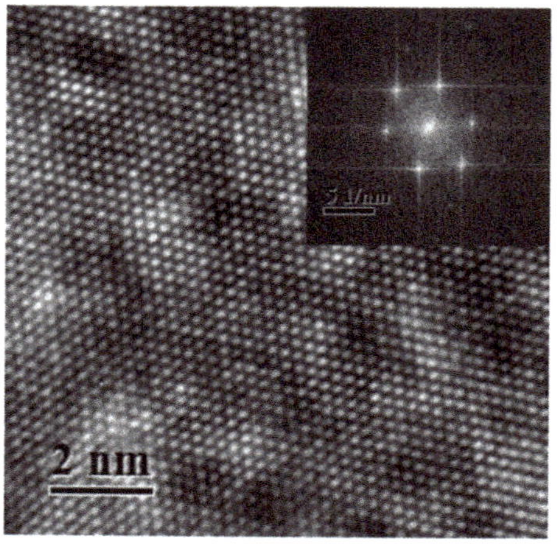

Figure 1. High resolution transmission electron microscopy (HRTEM) image of the nanostructure of Al-0.2 wt.%Ag sample aged at room temperature after quenching. The inset view is the Fast Fourier Transform (FFT) of the HRTEM image.

As shown in Figure 2, the positron annihilation lifetimes in Al–Ag alloys changed from one component of ~150 ps to two components of 118 ps and 195 ps over 200 K as the temperature increased. Most probably, in our studied dilute Al–Ag alloys, Ag–V complex formed as they were quenched, and after 200 K, the Ag–V complex decomposed into Ag clusters and vacancies. With Ag atoms gradually separated from the binding of the vacancies, the size of the open space near Ag–V complexes increased, resulting in the appearance of a long positron annihilation lifetime of 195 ps (τ_2). Indeed, above 200 K, the short lifetime (τ_1) of around 118 ps was close to the positron annihilation lifetime of Ag clusters, and the long lifetime of 195 ps was close to the positron annihilation lifetime of monovacancies [31]. This provides evidence that the Ag–V complexes are decomposed into monovacancies and Ag clusters at 200 K. Thereafter, part of the injected positrons annihilated in the vacancies, while others annihilated on Ag clusters. Furthermore, as can be seen from Figure 2, the relative intensity (I_1 and I_2) for τ_1 and τ_2 gradually decreased from 52.7 to 44 and increased from 47.3% to 56%, respectively, as the temperature increased from 200 K to 300 K. That indicates that fewer positrons annihilated with the electrons around Ag clusters and more positrons annihilated with the monovacancies as the temperature increased. Thus, the ability of Ag clusters to capture positrons is enhanced at a lower temperature.

Single detector DBS results are shown in Figure 3. It is evident that the S increased from 0.494 to 0.509 as the temperature increased from 15 K to 200 K. The W showed an opposite trend, decreasing from 0.112 to 0.097. This suggests the same conclusion as that of the PALS results; i.e., Ag atoms gradually separated from the Ag–V complex and resulted in an increase in the size of the open space near Ag–V complexes at 200–300 K. It is worth noting that the S value started to increase slowly after 200 K. In particular, from 250 K, the S value basically remained stable. This indicates that almost all Ag–V complexes decomposed into Ag clusters and monovacancies until 250 K.

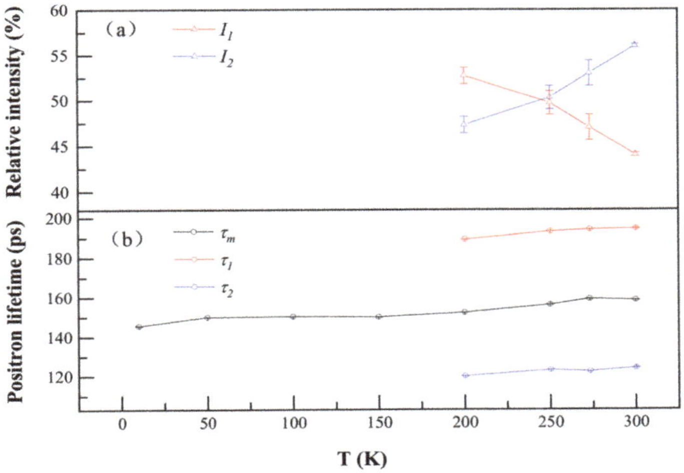

Figure 2. Positron annihilation lifetimes (**a**) and their relative intensities (**b**) of Al-0.2 wt.%Ag as a function of temperature. τ_1, τ_2 and τ_m represent a short lifetime, long lifetime and mean lifetime in the positron annihilation spectrum. I_1 and I_2 represent the relative intensity of the short lifetime and long lifetime.

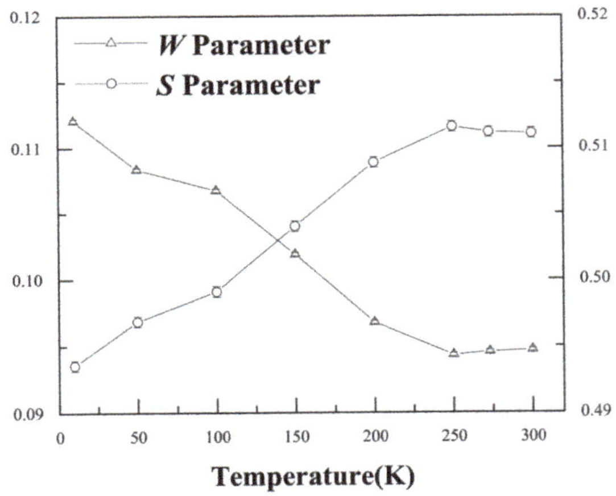

Figure 3. *S* and *W* values for the Al-0.2% Ag alloys at different measurement temperatures.

Figure 4 shows the DBS results with the *S*-*W* plot of Al–Ag alloys at different temperatures. Clearly, the *S*-*W* characteristic points are in a straight line, and the *S*-*W* characteristic points move from Ag towards Al with increasing temperature. This indicates that the surrounding elements in which positrons were trapped and annihilated gradually changed from Ag clusters (surrounding Ag–V complexes) to aluminum (surrounding monovacancies) as the temperature increased. This is in consistent with the above-mentioned conclusion from the positron annihilation lifetime results.

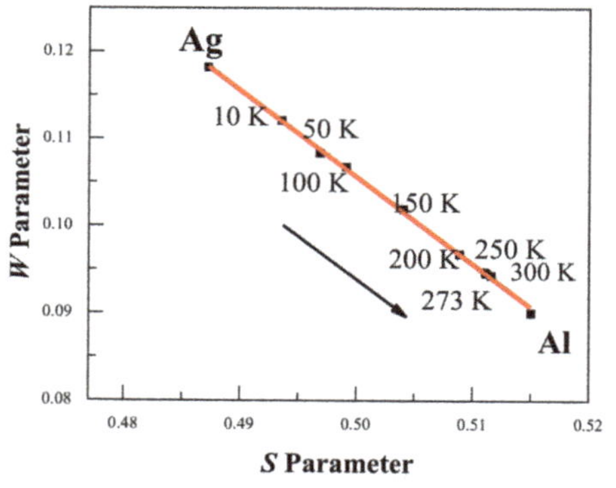

Figure 4. *S*-*W* plot for the Al-0.2 wt.%Ag alloys at different temperatures. The arrow shows the increase in the measurement temperature, and the movement of the *S*-*W* characteristic points from Ag towards Al. The red line is presented as a visual aid.

Figure 5 shows the ratio curves for CDB spectra of Al-0.2% Ag at different temperatures compared to pure Al; the ratio curve for pure Ag measured at room temperature is also added for comparison. A CDB ratio curve can reflect the momentum distribution of core electrons of certain elements annihilated with positrons, due to the significantly enhanced signal-to-noise ratio of at least 10^5:1 for a typical CDB spectrum. As shown in Figure 5, there is a narrow peak at around 11×10^{-3} m_0c in the CDB ratio curve of pure Ag, which is consistent with the experimental results of Y. Nagai [32] and B. Zou [11]. The curves in Figure 5 show a bump at 11×10^{-3} m_0c at low temperatures, which disappears at high temperatures. In particular, at 200 K, the characteristic peak at 14×10^{-3} m_0c moves to 15×10^{-3} m_0c. Combined with the positron annihilation lifetime results, this gives evidence that Ag atoms gradually become free from binding with a vacancy and the Ag–V complexes decomposed to Ag clusters and monovacancies at 200 K. When the temperature reached 300 K, however, the characteristic peak for Ag clusters in Al-0.2% Ag was hardly observable; there was only a very small ratio fluctuation around 1.0.

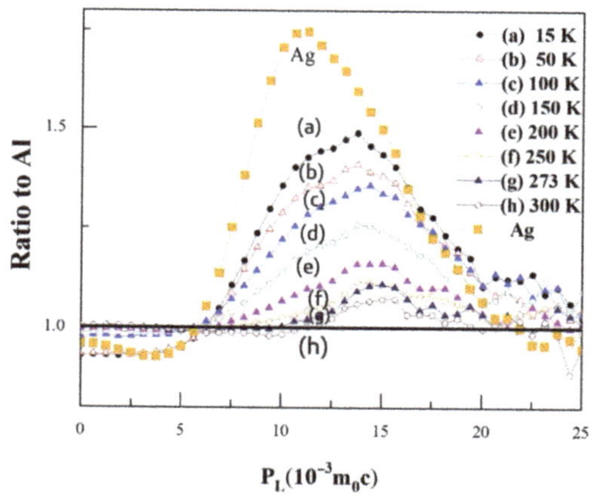

Figure 5. Coincidence doppler broadening (CDB) ratio curves for Al-0.2 wt.%Ag at different temperatures. The statistical errors are omitted for clarity of representation.

When the temperature increased above 200 K, the value of characteristic peak gradually decreased with the increase in temperature. That is, above 200 K, more positrons annihilated with low-momentum electrons around monovacancies and fewer positrons were trapped and annihilated with high-momentum inner shell 4d electrons of Ag clusters. In other words, the Ag clusters can trap positrons, but the positron trapping ability of Ag clusters decreases with the increase in temperature. At room temperature, the Ag clusters could barely trap positrons any more. Based on such temperature-dependent behavior, we may deduce that the Ag clusters can act as shallow positron trapping sites to trap positrons, and the trapping ability increases with the decrease in temperature. The conclusion is consistent with the results of B. Zou et al. [11].

Puska et al. [23] have reported that a minimum radius of embedded clusters is necessary for positron trapping to occur. There is one bound state if the radius of the precipitate is larger than the critical radius, $r_c = \frac{5.80a_0}{\sqrt{\Delta A}}$, where ΔA is the positron affinity difference between the two kinds of metals, and a_0 is the Bohr radius. For the Al–Ag alloy, the critical radius is about 5.9 a_0 with A_+ (Al) = −4.41 eV and A_+ (Ag) = −5.36 eV. This means that precipitates should contain at least ~6 Ag atoms.

We attempted to quantitatively describe the positron trapping mechanism in Al–Ag alloys at varied temperatures utilizing the three state trapping model [20,33]. At 15–200 K, there were two positron trapping states in the Al–Ag alloy: free annihilation (bulk state) and deep trapping (Ag–V complex). At 200–300 K, there were three positron states in the Al–Ag alloy: free annihilation (bulk state), shallow trapping (Ag clusters), and deep trapping (vacancies). For spherical defects, the positron trapping rate and detrapping rate capture rate can be written as [19,34]

$$\kappa(T) = r_0 C e^{-\gamma T} \tag{1}$$

$$\delta(T) = \frac{m}{4\hbar^2}\pi^{-\frac{1}{2}}C^{-1}\kappa(T)(k_B T)^{\frac{3}{2}}E_b^{-\frac{1}{2}}e^{-\frac{E_b}{k_B T}} \tag{2}$$

where γ is the trapping rate of positrons by the trapping sites at 0 K and r_0 is a constant representing the specific trapping rate of such trapping sites. C means the concentration of trapping sites, E_b means the binding energy of the trapping sites and positrons. The above equations show that the trapping rate is proportional to the concentration of trapping sites (either shallow or deep trapping). In contrast, the detrapping rate is independent of the trapping site concentration but related to the binding energy.

Using the above equations, we calculated the temperature dependence of the positron detrapping rate as a function of temperature for a range of binding energies from 0.01 eV to 1 eV. The calculated results are shown in Figure 6. It can be seen that the higher the binding energy, the smaller the detrapping rate. When the binding energy reaches 1 eV, the detrapping rate remains at around 0. This suggests that the higher the binding energies, the harder it is for the positrons to escape from the trapping sites. As regards the temperature dependence, it is clear that an increased temperature facilities the detrapping of positrons from trapping sites for all the binding energies except E_b = 1 eV. It was reported that the positron binding energy for Ag clusters is lower than 40 meV [11]; therefore, Ag clusters could only act as shallow trapping defects for positrons, and the low temperature enhancement of the effect of Ag on positron trapping is thereby understandable.

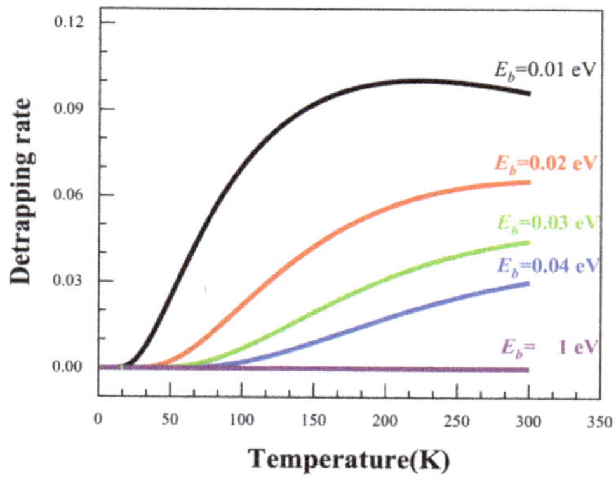

Figure 6. The temperature dependence of the positron detrapping rate at different binding energies.

4. Conclusions

In summary, we conducted a positron annihilation spectroscopy study on the early stage of precipitation in dilute Al–Ag alloys. The interaction mechanism between positrons and Ag clusters as well as monovacancies in Al–Ag alloys at varying temperatures were extensively studied. Several conclusions can be drawn. Firstly, a monovacancy in quenched Al–Ag alloys can easily combine with Ag atoms to form Ag–V complexes at 15–200 K, and within this temperature range, the Ag–V complex effect is predominant for positron annihilation in the quenched Al-0.2% Ag alloy. Secondly, above 200 K, Ag atoms gradually escape from binding with vacancies and migrate to form Ag clusters. Meanwhile, Ag clusters can act as shallow trapping defects, and the positron detrapping from Ag clusters is enhanced at higher temperatures. Last but not least, the addition of a small amount (0.2 wt.%) of Ag has a great influence on the positron trapping mechanism; in other words, low-temperature positron annihilation spectroscopy is an important method to investigate the microstructural evolution of defects and precipitates in the early aging stage of dilute alloys.

Author Contributions: Conceptualization, X.L.; methodology, P.Z. and B.W.; software, X.C. and S.J.; formal analysis, X.L.; investigation, X.L.; resources, S.J.; data curation, X.L.; writing—original draft preparation, X.L.; writing—review and editing, B.W., X.C., S.J. and R.Y.; project administration, X.C. and R.Y.; funding acquisition, X.C. and R.Y. All authors have read and agreed to the published version of the manuscript.

Funding: This research was funded by the National Natural Science Foundation of China, Nos. 11675188, 11875055 and Beijing Natural Science Foundation, Nos. 11705212.

Institutional Review Board Statement: Not applicable.

Informed Consent Statement: Not applicable.

Data Availability Statement: Data sharing not applicable, all the data created for this study are already displayed in the article.

Conflicts of Interest: We declare that we have no financial and personal relationships with other people or organizations that can inappropriately influence our work, and there is no professional or other personal interest of any nature or kind in any product, service and/or company that could be construed as influencing the position presented in, or the review of, the manuscript.

References

1. Zhang, Z.Z.; Bourgeois, L.; Rosalie, J.M.; Medhekar, N.V. The bi-layered precipitate phase zeta in the Al-Ag alloy system. *Acta Mater.* **2017**, *132*, 525–537. [CrossRef]
2. Auld, J.H.; Vietz, J.T.; Polmear, I.J. T-phase Precipitation induced by the Addition of Silver to an Aluminium–Copper–Magnesium Alloy. *Nature* **1966**, *209*, 703–704. [CrossRef]
3. Polmear, I.J. A Trace Element Effect in Alloys based on the Aluminium-Zinc-Magnesium System. *Nature* **1960**, *186*, 303–304. [CrossRef]
4. Lumley, R.N.; Polmear, I.J.; Morton, A.J. Development of mechanical properties during secondary aging in aluminium alloys. *Mater. Sci. Technol.* **2005**, *21*, 1025–1032. [CrossRef]
5. Marceau, R.K.W.; Sha, G.; Ferragut, R.; Dupasquier, A.; Ringer, S.P. Solute clustering in Al-Cu-Mg alloys during the early stages of elevated temperature ageing. *Acta Mater.* **2010**, *58*, 4923–4939. [CrossRef]
6. Nakamura, J.; Matsuda, K.; Kawabata, T.; Sato, T.; Nakamura, Y.; Ikeno, S. Effect of Silver Addition on the beta '-Phase in Al-Mg-Si-Ag Alloy. *Mater. Trans.* **2010**, *51*, 310–316. [CrossRef]
7. Guinier, A. The mechanism of precipitation in aluminium-silver alloy crystals during structural hydrogenation. *C. R. Hebd. Seances Acad. Sci.* **1942**, *214*, 34–37.
8. Dubey, P.A.; Schonfeld, B.; Kostorz, G. Shape and internal structure of guinier-preston zones in Al-Ag. *Acta Metall. Mater.* **1991**, *39*. [CrossRef]
9. Baur, R.; Gerold, V. Existence of a metastable miscibility gap in aluminium-silver alloys. *Acta Metall.* **1962**, *10*, 637–645. [CrossRef]
10. Nagai, Y.; Toyama, T.; Tang, Z.; Inoue, K.; Chiba, T.; Hasegawa, M.; Hirosawa, S.; Sato, T. Interactions between Fermi surfaces and Brillouin zone boundaries and phase stability of embedded metallic nanoparticles. *Phys. Rev. B* **2009**, *79*, 21405. [CrossRef]
11. Zou, B.; Qi, N.; Liu, Z.W.; Chen, Z.Q.; Liu, H.Q.; Yi, D.Q.; Tang, Z. Enhanced positron trapping by Ag nanoclusters at low temperatures: A challenge of positron sensitivity to quantum dots. *J. Appl. Phys.* **2017**, *121*, 125103. [CrossRef]
12. Alkassab, T.; Haasen, P. Early stages of precipitation in dilute Al-Ag alloys. *Z. Met.* **1993**, *84*, 248–250.
13. Nicholson, R.B.; Nutting, J. The metallography of precipitation in an A1-16% Ag alloy. *Acta Metall.* **1961**, *9*, 332–343. [CrossRef]
14. Wang, C.; Ma, B.; Cao, X.Z.; He, S.; Han, J.B.; Wei, M.; Evans, D.G.; Duan, X. Bridge-type interface optimization on a dual-semiconductor heterostructure toward high performance overall water splitting. *J. Mater. Chem. A* **2018**, *6*, 7871–7876. [CrossRef]
15. Ning, X.; Cao, X.Z.; Li, C.; Li, D.M.; Zhang, P.; Gong, Y.H.; Xia, R.; Wang, B.Y.; Wei, L. Modification of source contribution in PALS by simulation using Geant4 code. *Nucl. Instrum. Methods Phys. Res. Sect. B Beam Interact. Mater. Atoms.* **2017**, *397*, 75–81. [CrossRef]
16. Cao, X.Z.; Zhu, T.; Jin, S.X.; Kuang, P.; Zhang, P.; Lu, E.Y.; Gong, Y.H.; Guo, L.P.; Wang, B.Y. Detection of helium in irradiated Fe9Cr alloys by coincidence Doppler broadening of slow positron annihilation. *Appl. Phys. A Mater. Sci. Process.* **2017**, *123*, 177. [CrossRef]
17. Nagai, Y.; Takadate, K.; Tang, Z.; Ohkubo, H.; Sunaga, H.; Takizawa, H.; Hasegawa, M. Positron annihilation study of vacancy-solute complex evolution in Fe-based alloys. *Phys. Rev. B* **2003**, *67*, 224202. [CrossRef]
18. Iakubov, I.T.; Pogosov, V.V. Positron binding energy in monovacancy of metals and work functions in stabilized jellium model. In *Positron Annihilation Icpa-10, Pts 1 and 2*; He, Y.J., Cao, B.S., Jean, Y.C., Eds.; Transtec Publications Ltd.: Zurich-Uetikon, Switzerland, 1995; Volume 175, pp. 169–172. [CrossRef]
19. Smedskjaer, L.C.; Manninen, M.; Fluss, M.J. An alternative interpretation of positron-annihilation in dislocations. *J. Phys. F Metal Phys.* **1980**, *10*, 2237–2249. [CrossRef]
20. Hidalgo, C.; Linderoth, S.; Dediego, N. Positron-trapping mechanism at dislocations in Zn. *Phys. Rev. B* **1987**, *36*, 6740–6745. [CrossRef] [PubMed]
21. Riceevans, P.; Hlaing, T.; Chaglar, I. Evidence for Temperature Dependence of Positron Trapping Rate in Plastically Deformed Copper. *Phys. Rev. Lett.* **1976**, *37*, 1415–1418. [CrossRef]
22. Van Huis, M.A.; van Veen, A.; Schut, H.; Falub, C.V.; Eijt, S.W.H.; Mijnarends, P.E.; Kuriplach, J. Positron confinement in embedded lithium nanoclusters. *Phys. Rev. B* **2002**, *65*, 085416. [CrossRef]
23. Puska, M.J.; Lanki, P.; Nieminen, R.M. Positron affinities for elemental metals. *J. Phys. Condes. Matter* **1989**, *1*, 6081–6093. [CrossRef]
24. Hugenschmidt, C.; Pikart, P.; Stadlbauer, M.; Schreckenbach, K. High elemental selectivity to Sn submonolayers embedded in Al using positron annihilation spectroscopy. *Phys. Rev. B* **2008**, *77*. [CrossRef]
25. Kansy, J. Microcomputer program for analysis of positron annihilation lifetime spectra. *Nucl. Instrum. Methods Phys. Res. Sect. A Accel. Spectrometers Detect. Assoc. Equip.* **1996**, *374*, 235–244. [CrossRef]
26. Alexander, K.B.; Legoues, F.K.; Aaronson, H.I.; Laughlin, D.E. Faceting of GP zones in an Al-Ag alloy. *Acta Metall.* **1984**, *32*, 2241–2249. [CrossRef]
27. Nagai, Y.; Honma, T.; Tang, Z.; Hono, K.; Hasegawa, M. Coincidence Doppler broadening study of the pre-precipitation stage of an Al-Cu-Mg-Ag alloy. *Philos. Mag. A Phys. Condens. Matter Struct. Defects Mech. Prop.* **2002**, *82*, 1559–1572. [CrossRef]
28. Raju, P.; Gupta, R.; Siegel, W. Annihilation of a positron in a vacancy in aluminum. *Phys. Rev. B* **1980**, *22*, 4572–4589. [CrossRef]
29. Zhang, S.S.; Yao, Z.J.; Kong, X.S.; Chen, L.; Qin, J.Y. First-principles calculations of solute-vacancy interactions in aluminum. *Chin. Physics. B* **2020**, *29*. [CrossRef]

30. Maury, F.; Lucasson, A.; Lucasson, P.; Dimitrov, C. Mixed-interstitial recovery in AlTi and AlAg dilute alloys. *J. Phys. F Metal Phys.* **1988**, *18*, 657–675. [CrossRef]
31. Folegati, P.; Makkonen, I.; Ferragut, R.; Puska, M.J. Analysis of electron-positron momentum spectra of metallic alloys as supported by first-principles calculations. *Phys. Rev. B* **2007**, *75*, 054201. [CrossRef]
32. Nagai, Y.; Tang, Z.; Ohkubo, H.; Takadate, K.; Hasegawa, M. Elemental analysis of positron affinitive site in materials by coincidence Doppler broadening spectroscopy. *Radiat. Phys. Chem.* **2003**, *68*, 381–386. [CrossRef]
33. Seeger, A. Study of defects in crystals by positron-annihilation. *Appl. Phys.* **1974**, *4*, 183–199. [CrossRef]
34. Brandt, W.; Paulin, R. Positron diffusion in solids. *Phys. Rev. B Solid State* **1972**, *5*, 2430–2435. [CrossRef]

materials

MDPI

Article

Integrating a Top-Gas Recycling and CO_2 Electrolysis Process for H_2-Rich Gas Injection and Reduce CO_2 Emissions from an Ironmaking Blast Furnace

Yichao Hu [1], Yinxuan Qiu [1], Jian Chen [2], Liangyuan Hao [3], Thomas Edward Rufford [1], Victor Rudolph [1] and Geoff Wang [1,*]

1 School of Chemical Engineering, The University of Queensland, St. Lucia 4072, Australia; h.yichao@uq.edu.au (Y.H.); yinxuan.qiu@uq.edu.au (Y.Q.); t.rufford@uq.edu.au (T.E.R.); v.rudolph@uq.edu.au (V.R.)
2 College of Mechanical and Electrical Engineering, Central South University, Changsha 410083, China; chenj@csu.edu.cn
3 The Strategy Research Institute, HBIS Group Co., Ltd., Shijiazhuang 050023, China; haoliangyuan@hbisco.com
* Correspondence: gxwang@uq.edu.au

Abstract: Introducing CO_2 electrochemical conversion technology to the iron-making blast furnace not only reduces CO_2 emissions, but also produces H_2 as a byproduct that can be used as an auxiliary reductant to further decrease carbon consumption and emissions. With adequate H_2 supply to the blast furnace, the injection of H_2 is limited because of the disadvantageous thermodynamic characteristics of the H_2 reduction reaction in the blast furnace. This paper presents thermodynamic analysis of H_2 behaviour at different stages with the thermal requirement consideration of an iron-making blast furnace. The effect of injecting CO_2 lean top gas and CO_2 conversion products H_2–CO gas through the raceway and/or shaft tuyeres are investigated under different operating conditions. H_2 utilisation efficiency and corresponding injection volume are studied by considering different reduction stages. The relationship between H_2 injection and coke rate is established. Injecting 7.9–10.9 m^3/tHM of H_2 saved 1 kg/tHM coke rate, depending on injection position. Compared with the traditional blast furnace, injecting 80 m^3/tHM of H_2 with a medium oxygen enrichment rate (9%) and integrating CO_2 capture and conversion reduces CO_2 emissions from 534 to 278 m^3/tHM. However, increasing the hydrogen injection amount causes this iron-making process to consume more energy than a traditional blast furnace does.

Keywords: blast furnace; hydrogen injection; gas utilisation efficiency; energy consumption; CO_2 emission

Citation: Hu, Y.; Qiu, Y.; Chen, J.; Hao, L.; Rufford, T.E.; Rudolph, V.; Wang, G. Integrating a Top-Gas Recycling and CO_2 Electrolysis Process for H_2-Rich Gas Injection and Reduce CO_2 Emissions from an Ironmaking Blast Furnace. *Materials* **2022**, *15*, 2008. https://doi.org/10.3390/ma15062008

Academic Editor: Miguel Ángel Sanjuán

Received: 31 January 2022
Accepted: 6 March 2022
Published: 8 March 2022

1. Introduction

Traditional blast furnace (BF) iron making relies on carbon and contributes to over 70% of CO_2 emissions in the iron and steel industry [1]. In a blast furnace, coke is converted into a high-temperature CO gas and performs an exothermic reaction with iron ores, resulting in a large amount of CO and CO_2 leaving the furnace with top gas. Typically, every tonne of hot metal (tHM) produced from a traditional BF requires about 500 kg/tHM carbon and generates around 1.2 tonnes of CO_2 emissions [2,3]. Hence, there were various attempts for a clean iron-making process to reduce CO_2 emissions [4–7]. One of the approaches is using alternative reductants produced from renewable energies to replace carbon. BF operation with hydrogen as an auxiliary reducing agent was extensively investigated because of its specific advantages over CO [8–12]. Compared to CO reduction that generates CO_2, reducing iron ores by hydrogen only forms water vapor. Kinetically, hydrogen enables a higher gas flow rate, and a faster reduction in iron ores and productivity than only CO does [13,14]. The higher thermal conductivity of hydrogen helps in heat transfer efficiency between solid and gas phases [15]. In addition, the hydrogen reduction in iron ores suppresses the strong endothermic direct reduction in iron ores. However, hydrogen reduction is thermally more disadvantageous than CO reduction. Due to this endothermic reduction in iron ores by hydrogen, hydrogen addition changes the energy supply of the BF, and it is only useful to a certain extent [16–18].

Few previous experiments and mathematical models have investigated the maximal or optimal hydrogen injection to the BF, and results are controversial and need further investigation. A thermogravimetric experiment showed the efficiency of hydrogen on reduction rate is neglected when its content is lower than 5%, and H_2 content should be 5% to 15% at reduction temperatures between 700 and 1000 °C [13,19]. Wang et al. performed a pulverisation experiment at 900 °C with 70% N_2 and found that the reduction degree of burdens was more than 90% when H_2 content was higher than 20% [20]. On the basis of reduction experiments, Lyu et al. reported that the appropriate H_2 content lies between 5% to 10% in terms of the reduction rate, gas utilisation, and reasonable distribution of the energy in the BF in the CO and H_2 mixture [17]. Nogami et al. used a multiphase fluid dynamic model to simulate the effect of hydrogen injection with 2.5% oxygen enrichment [8]. They demonstrated that the coke rate decreases linearly, and the maximal hydrogen injection can reach 43.7%.

The application of hydrogen in large-scale BF iron-making processes is limited due to its supply in terms of cost, availability, storage, and transportation [21,22]. The traditional BF contains a low level of hydrogen content because it only generates from the blast air moisture and the volatiles of pulverised coal. One opportunity for BF hydrogen enrichment is injecting hydrogen-rich (so-called H_2-rich) gas or hydrogen-bearing materials from external sources, including natural gas, fuel oil, coke oven gas (COG), reformed gas, and waste plastics (C_nH_m) [23–26]. Previous studies adopted photocatalysis to produce CO and H_2 from CO_2. However, the productivity of photocatalysis is less than 1000 μmol CO/gCO_2 and 19 μmol H_2/gCO_2 [27,28]. It is challenging for photocatalysis to meet the large-scale CO_2 conversion requirement in the iron-making process. Electrochemical CO_2 conversion is one option that can recycle carbon into blast furnace gas (BFG) as CO [29]. The high energy demand is considered to be a major difficulty for electrochemical reduction in CO_2 [30]. However, it provides an added opportunity for a BF because hydrogen is coproduced in the electrolyser, which can be used for iron-ore reduction. In low-temperature electrolysis cells, CO_2 reduction is carried out in aqueous solutions, and different levels of current density can be used to produce hydrogen at various concentrations [31–33]. As an additional benefit, a pure oxygen stream is generated as another byproduct during electrolysis, which can be used directly for oxygen enrichment to the BF [34]. In this study, we use CO_2 capture and utilisation (CCU) technology to provide a reliable on-site H_2-rich gas for the BF iron-making process while avoiding CO_2 emissions. Renewable energy can be coupled to CO_2 electrolysis to achieve further emission reduction.

This study aims to use a modelling method to predict hydrogen involvement in the BF and determine the CO_2 emission reduction potential. An alternative method to determine the hydrogen utilisation efficiency is developed. First, thermodynamic and thermal balance models are introduced; then, hydrogen injection is quantitatively studied to determine the optimal injection position and volume by considering key BF performance indices. Lastly, the effect of hydrogen injection on coke consumption, gas utilisation efficiency, and the energy consumption of the iron-making system are investigated.

2. Materials and Methods

Here, we studied the effect of H_2-rich gas injection through BF tuyeres at the raceway position and/or shaft tuyeres, as shown in Figure 1. The BF is studied as the main subsystem when changing the hydrogen injection condition. The CCU and gas heating subsystem is used as a black box to provide necessary input information to BF. The BFG composition and flow rate from BF are the input for the CCU unit, and the CCU unit provides H_2-rich reducing gas as the input to the BF. The operating conditions for this BF to produce 1 tonne hot metal (1 tHM) are listed in Table 1 and were kept constant throughout simulations. The overall CO content in the gas injectant was maintained at 200 m^3/tHM. The productivity of this BF is 238 tHM per hour.

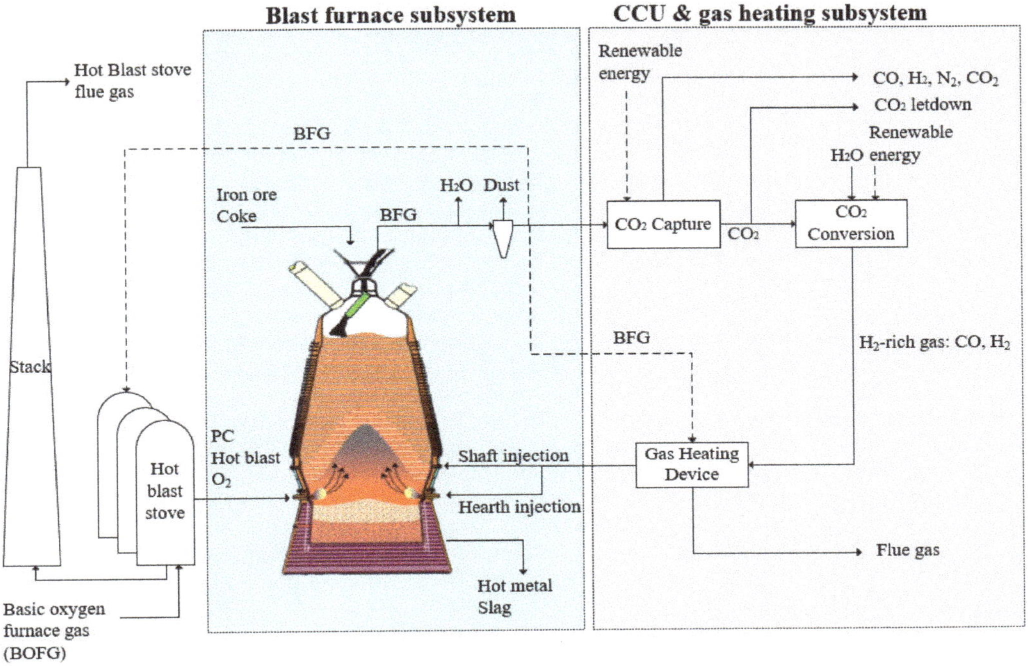

Figure 1. Blast furnace process with CO_2 capture, conversion, and H_2-rich gas injection.

Table 1. Operating conditions of the simulation.

Operating Parameters	
PCI rate (kg/tHM)	137
Blast temperature, °C	1052
Humidity of hot blast, g/m^3	12.93
Top gas temperature, °C	161
Hearth injection temperature, °C	1250
Shaft injection temperature, °C	900

As shown in Figure 1, hot oxygen-enriched blast and pulverised coal is injected through the tuyeres. The upper limit of oxygen enrichment rate for the blast was set at 14% to maintain the stable operation of the large-scale BFs. After drying and dust removal, some BFG is combusted in the hot blast stoves to heat cold blast, and in the gas heating device to provide high-temperature gas injectants. The rest of BFG enters the amine absorption CO_2 capture unit to provide a CO_2-rich stream that is processed in an electrochemical CO_2 conversion unit to produce a CO and H_2 stream containing, for example, 30% vol. H_2 and 70% vol. CO. The CO_2-lean stream from the top of the CO_2 capture unit contained a mix of CO, H_2, and N_2 that is exported to other processes in the integrated steel mill. We did not in this study consider additional separation of CO and H_2 from the N_2 in this stream for recycling back to the BF. Oxygen enrichment is required with H_2-rich gas injection to provide heat to the BF and enrich BFG for CO_2 capture [35]. The BF can take another advantage from the CCU unit, as the electrolyser produces pure oxygen in another effluent stream. Besides BFG, a small amount of the basic oxygen furnace gas (i.e., BOFG) from steel making is usually combusted as fuel to a hot blast stove. Following assumptions of the BF, CCU unit, gas heating device and hot stove are made in this study:

- Degree of indirect reduction R_i depends on the reducing gas concentration of BF bosh gas, which is estimated by an empirical equation, as shown in Equation (1) [36]:

$$R_i = 0.2777 + 0.0051 \times \%(\text{Reducing gas}) \quad (1)$$

where %(Reducing gas) is the proportion of H_2 and CO in the total amount of gas entering the bosh and shaft.

- For the amine absorption CO_2 capture unit: 30% monoethanolamine (MEA) concentration was used for CO_2 capture in this study. The capture unit recovered 90% CO_2 in BFG, and the CO_2 purity was >99%; the general thermal energy requirement for capture was assumed to be around 1000 kWh/tCO_2 (3.6 GJ/ tCO_2) [37–39]. Any additional CO_2 captured and not converted was assumed to be released as per current operation or could be sent to CO_2 storage routes, shown as CO_2 letdown in Figure 1.
- The electrochemical CO_2 conversion unit was treated in the model as a simplified input–output model. Assumptions for the material and energy balance in the CO_2 conversion unit were based on laboratory demonstration data with additional inputs from literature sources. Briefly, the model was based on multiple two-cell vapour fed electrolyser stacks with the capacity to treat 50 tCO_2 per day; further details can be found in our other report [33]. The current density of the electrolyser was altered from 2.68 V at 0 A/m^2 to 3.59 V at 1862 A/m^2 to produce the H_2-rich gas with different H_2/CO compositions.
- The electricity consumption for CO_2 conversion is proportional to the H_2 generation, which can be estimated as in Equation (2) [33]:

$$E_{conv} = 10.75V_{H_2gen} + 1282 \quad (2)$$

where E_{conv} is the power required for the CO_2 conversion unit, kWh/tHM; V_{H_2gen} is the amount of H_2 generated by the conversion unit, m^3/tHM;

- efficiency of the gas heating device was 85%;
- The hot blast stove system uses two stoves on-gas and one stove on-blast, and the efficiency of the hot blast stoves was 75%.

As indicated in Figure 1, CO_2 capture and conversion units use renewable energy to avoid their own CO_2 emissions. The type of renewable power used by the industry depends on availability and cost, such as solar power [40,41]. Besides solar power, industries can use thermal–electrical materials to recover a large amount of waste heat in an integrated steel mill to provide electricity for CCU units [42]. In addition, using the lower heating value energy to generate electricity in the steel mill and the on-site power plant can help to minimise the renewable power periodic availability problem.

To achieve the objectives of this study, two mathematical models were developed. As a reducing agent, H_2-rich gas needs to fulfil the thermodynamic requirement of the reduction reaction to capture oxygen in iron ores. H_2-rich gas also needs to provide enough heat in the shaft for keeping an effective reduction process. First, hydrogen behaviour in different parts of the BF was analysed. A thermodynamic model for hydrogen reduction was built to determine hydrogen utilisation efficiency. This model provides a guideline of the proper hydrogen injection concentration. Then, a thermal balance model was used to limit the hydrogen injection temperature and volume.

The optimal hydrogen injection amount and position were determined by increasing the reduction potential in the coke consumption and CO_2 emissions, increasing gas utilisation efficiency and lowering the energy consumption. A static mass balance model of the BF was used to calculate the above parameters.

2.1. Thermodynamic Calculations of H_2-Rich Gas Injection BF

The reduction behaviours of injected gas are discussed in different parts of the BF to determine the reducing gas utilisation.

2.1.1. Raceway

In the BF raceway, the main reactions considered in this study were carbon combustion, coke solution loss reaction between coke and CO_2, water–gas reaction between coke and moisture in the hot blast, which can be described as shown in Equations (3)–(5):

$$\text{Solution loss reaction} \quad C\ (s) + CO_2\ (g) \rightarrow 2CO\ (g) \quad \Delta H^0 = 172{,}430\ \text{kJ/kmol} \tag{3}$$

$$\text{Water–gas reaction} \quad C\ (s) + H_2O(g) \rightarrow CO\ (g) + H_2\ (g) \quad \Delta H^0 = 124{,}190\ \text{kJ/kmol} \tag{4}$$

$$\Delta G^0 = 134{,}542 - 142.28T \tag{5}$$

With excess coke existing in the BF bottom, it could be assumed that CO and H_2 combustion was negligible. This could be justified by the assumption that, if a small amount of H_2 reacts with O_2 to form water in the raceway, the generated water vapour reacts with coke and turn back to H_2. Therefore, this process can be simplified as heating the hearth gas injectant, as shown in Equation (6):

$$\Delta Q_{hearth} = \int_{25}^{T_{hearth}} V_{hearth} \cdot C_{p\_hearth} dt \tag{6}$$

where T_{hearth} is hearth gas injection temperature, °C; C_{p_hearth} denotes specific heat capacity of the gas injected to BF, kJ/m^3·°C; V_{hearth} is hearth injection volume, m^3/tHM; and ΔQ_{hearth} is the sensible heat carried by the gas injected to BF, kJ/tHM.

2.1.2. Dripping, Cohesive, and High-Temperature Zones over 1000 °C

The primary reaction in the dripping zone is a direct reaction between coke and FeO. Hot gas containing H_2 and CO that passes through the cohesive zone reacts with molten FeO or semi-molten FeO to form H_2O and CO_2. As the temperature was over 1350 °C in the dripping zone, the amount of CO_2 was negligible due to the solution loss reaction. At the high-temperature zone over 1000 °C, some reduced FeO and Fe_3O_4 were still in the solid state, and H_2 could pass through their surface. Almost all the H_2O produced by H_2 reduction rapidly participates in water–gas reaction at the presence of coke to form CO and H_2 over 1273 K (1000 °C). Therefore, the reduction reaction in this section was essentially the direct reduction in iron by coke. H_2 injected through the tuyeres at the raceway mainly catalyses direct reduction and heats the molten or semi-molten burden.

2.1.3. Shaft Zone Temperature between 800 and 1000 °C

According to the thermodynamics of iron oxide reduction and dynamics studies, H_2 has better reducing capability than that of CO above 800 °C [43,44]. At the same time, the extent of coke solution loss reaction and the water–gas reaction was less than that in the higher temperature zone. In this temperature zone, H_2 reacts with various iron oxides to generate H_2O, and the formed H_2O is not gasified into H_2 by carbon completely. Therefore, it is the primary zone to improve H_2 utilisation efficiency.

The H_2-rich reducing gas utilisation rate and volume requirement vary in different ferric oxides reduction stages. The iron oxides reduction reactions are at a nonequilibrium state in the BF. When the temperature is above 570 °C, the reduction in ferric oxides by CO and H_2 in the BF occurs in the following sequences: $1/2\ Fe_2O_3 \rightarrow 1/3\ Fe_3O_4$ (Stage I) $\rightarrow$ FeO (Stage II) $\rightarrow$ Fe (Stage III) [45]. The gas produced by the reduction in the latter stage is the reducing gas for the previous stage. Heat is gradually transferred to solid materials during the gas ascending. At the same time, part of reducing gas reacts with iron oxides and converts into CO_2 and H_2O, and finally forms top gas at around 150 to 250 °C when leaving the BF. There are 25% of the total oxygen elements removed during the reduction of Fe_3O_4

into FeO, and the remaining 75% of oxygen elements were removed in reducing FeO to Fe. Therefore, the reduction process from FeO to Fe is the key step. The required reducing gas amount is n kmol for CO, and m kmol for H_2 to produce 1 kmol iron. The value of n and m is the excess coefficient. The reduction reactions and thermodynamic parameters in Stage III for CO and H_2 are expressed as in Equations (7)–(12), and Equations (13)–(17), respectively [46]:

$$\text{Stage III}: \quad FeO\ (s) + nCO\ (g) \rightarrow Fe\ (s) + CO_2\ (g) + (n-1)CO\ (g) \tag{7}$$

$$\Delta H^0_{3CO} = -13{,}190\ kJ/kmol$$

$$\Delta G^0_{3CO} = -22{,}800 - 24.26T,\ kJ/kmol \tag{8}$$

$$\ln K_{IIICO} = -\frac{\Delta G^0_{3CO}}{RT} = -\frac{-22{,}800 - 24.26T}{RT} \tag{9}$$

$$K_{IIICO} = \frac{P_{CO_2}}{P_{CO}} = \frac{\varphi_{CO_2} \times P^0}{\varphi_{CO} \times P^0} = \frac{1}{n-1} \tag{10}$$

where K is the reaction equilibrium constant; φ denotes the fraction of gas component.

The minimal CO required for Stage III is described in Equation (11):

$$n = 1 + \frac{1}{K_{IIICO}} \tag{11}$$

The utilisation efficiency of CO in Stage III, η_{3CO}, is described in Equation (12):

$$\eta_{3CO} = \frac{\varphi_{3CO_2}}{\varphi_{3CO} + \varphi_{3CO_2}} = \frac{1}{\frac{\varphi_{3CO}}{\varphi_{3CO_2}} + 1} = \frac{1}{\frac{1}{K_{3CO}} + 1} = \frac{K_{IIICO}}{1 + K_{IIICO}} = \frac{1}{n} \tag{12}$$

$$FeO\ (s) + mH_2\ (g) \rightarrow Fe\ (s) + H_2O(g) + (m-1)H_2\ (g) \tag{13}$$

$$\Delta H^0_{3H_2} = 28{,}010\ kJ/kmol$$

$$\Delta G^0_{3H_2} = 23{,}430 - 16.16T,\ kJ/kmol \tag{14}$$

$$\ln K_{IIIH_2} = -\frac{\Delta G^0_{3H_2}}{RT} = -\frac{23{,}430 - 16.16T}{RT} \tag{15}$$

The minimal H_2 required for Stage III is calculated as in Equation (16):

$$m = 1 + \frac{1}{K_{IIIH_2}} \tag{16}$$

The utilisation efficiency of H_2 in Stage III, η_{3H_2} is described in Equation (17):

$$\eta_{3H_2} = \frac{\varphi_{3H_2O}}{\varphi_{3H_2} + \varphi_{3H_2O}} = \frac{K_{IIIH_2}}{1 + K_{IIIH_2}} = \frac{1}{m} \tag{17}$$

According to the theoretical thermochemical calculations, 50% of H_2 and CO participates in the water–gas shift reaction Equation (18) between 600 and 1400 °C [47]. Therefore, the heat consumed by the water–gas shift reaction at temperatures above 820 °C is balanced by the heat generated at the temperature below 820 °C, as calculated by Equation (19). However, H_2 promotes iron ore reduction by CO via the water gas shift reaction when the temperature is over 820 °C [48]. The CO_2 generated reacts with H_2 to reform CO, which participates in FeO reduction reaction again and improve the utilisation efficiency of CO.

$$\text{Water–gas shift reaction} \quad H_2O(g) + CO\ (g) = H_2\ (g) + CO_2\ (g) \tag{18}$$

$$\Delta H^0 = -41{,}325\ kJ/kmol$$

$$\Delta G^0 = -33{,}447 + 30.56T,\ \text{kJ/kmol} \tag{19}$$

The heat effect of FeO reduction by H_2 and CO gas is calculated by Equation (20):

$$\Delta H = \frac{X_{CO}\eta_{3CO}\Delta H^0_{3CO} + X_{H2}\eta_{3H_2}\Delta H^0_{3H_2}}{X_{CO}\eta_{3CO} + X_{H2}\eta_{3H_2}} \tag{20}$$

where X_i is the proportion of CO or H_2 in the reducing gas entering the BF shaft.

The gas produced by the reduction in Stage III is the reducing gas for Stage II. The reduction reactions and thermodynamic parameters in Stage II for CO and H_2 are expressed in Equations (21)–(28):

$$\begin{gathered}\text{Stage II}: \tfrac{1}{3}Fe_3O_4\ (s) + (n-1)CO\ (g) + CO_2\ (g) \rightarrow FeO\ (s) + \tfrac{4}{3}CO_2\ (g) + \left(n - \tfrac{4}{3}\right)CO\ (g) \\ \Delta H^0_{2CO} = 22{,}400\ \text{kJ/kmol}\end{gathered} \tag{21}$$

$$\Delta G^0_{2CO} = 35{,}380 - 40.16T,\ \text{kJ/kmol} \tag{22}$$

$$\begin{gathered}\tfrac{1}{3}Fe_3O_4\ (s) + (m-1)H_2(g) + H_2O\ (g) \rightarrow FeO\ (s) + \tfrac{4}{3}H_2O\ (g) + \left(m - \tfrac{4}{3}\right)H_2(g) \\ \Delta H^0_{2H_2} = 63{,}600\ \text{kJ/kmol}\end{gathered} \tag{23}$$

$$\Delta G^0_{2H_2} = 71{,}940 - 73.62T,\ \text{kJ/kmol} \tag{24}$$

$$\ln K_{IICO} = -\frac{\Delta G^0_{2CO}}{RT} = -\frac{35{,}380 - 40.16T}{RT} \tag{25}$$

$$\ln K_{IIH_2} = -\frac{\Delta G^0_{2H_2}}{RT} = -\frac{71{,}940 - 73.62T}{RT} \tag{26}$$

$$K_{IICO} = \frac{\varphi_{CO_2} \times P^0}{\varphi_{CO} \times P^0} = \frac{\frac{4}{3}}{n - \frac{4}{3}} \tag{27}$$

$$K_{IIH_2} = \frac{\frac{4}{3}}{m - \frac{4}{3}} \tag{28}$$

The minimal CO and H_2 volume required for Stage II is calculated as shown in Equations (29) and (30), respectively:

$$n = \frac{4}{3}\left(1 + \frac{1}{K_{IICO}}\right) \tag{29}$$

$$m = \frac{4}{3}\left(1 + \frac{1}{K_{IIH_2}}\right) \tag{30}$$

The utilisation efficiency of CO and H_2 in Stage II is calculated as shown in Equations (31) and (32), respectively:

$$\eta_{2CO} = \frac{\varphi_{2CO_2}}{\varphi_{2CO} + \varphi_{2CO_2}} = \frac{K_{IICO}}{1 + K_{2CO}} = \frac{4}{3n} \tag{31}$$

$$\eta_{2H_2} = \frac{\varphi_{2H_2O}}{\varphi_{2H_2} + \varphi_{2H_2O}} = \frac{K_{IIH_2}}{1 + K_{2H_2}} = \frac{4}{3m} \tag{32}$$

In the first stage of iron ores reduction, the transformation of Fe_2O_3 to Fe_3O_4 is very rapid due to the very high equilibrium constant of Fe_2O_3 reduction above 600 K, as shown in Equations (33) and (35):

$$\begin{gathered}\text{Stage I}: \tfrac{1}{2}Fe_2O_3\ (s) + \left(n - \tfrac{4}{3}\right)CO\ (g) + \tfrac{4}{3}CO_2\ (g) \rightarrow \tfrac{1}{3}Fe_3O_4\ (s) + \tfrac{3}{2}CO_2(g) + \left(n - \tfrac{3}{2}\right)CO(g) \\ \Delta H^0_{1CO} = -43{,}280\ \text{kJ/kmol}\end{gathered} \tag{33}$$

$$\Delta G^0_{1CO} = -52{,}131 - 41.0T,\ \text{kJ/kmol} \quad (34)$$

$$\tfrac{1}{2}Fe_2O_3\ (s) + \left(m - \tfrac{4}{3}\right)H_2\ (g) + \tfrac{4}{3}H_2O\ (g) \rightarrow \tfrac{1}{3}Fe_3O_4\ (s) + \tfrac{3}{2}H_2O(g) + \left(m - \tfrac{3}{2}\right)H_2\ (g)$$
$$\Delta H^0_{1H_2} = -21{,}810\ \text{kJ/kmol} \quad (35)$$

$$\Delta G^0_{1H_2} = -15{,}547 - 74.4T,\ \text{kJ/kmol} \quad (36)$$

The gas produced by the reduction in Stage II provides the reducing gas for Stage I. These reactions only require a low concentration of reducing gas to proceed. The minimal CO and H_2 volume required for Stage I is shown in Equations (37) and (38), respectively. With utilisation efficiency close to 100%, Fe_2O_3 reduction is an irreversible reaction.

$$n = \frac{3}{2}\left(1 + \frac{1}{K_{ICO}}\right) \quad (37)$$

$$m = \frac{3}{2}\left(1 + \frac{1}{K_{IH_2}}\right) \quad (38)$$

The utilisation efficiency of CO and H_2 in Stage I is described by Equations (39) and (40), respectively:

$$\eta_{1CO} = \frac{\varphi_{2CO_2}}{\varphi_{2CO} + \varphi_{2CO_2}} = \frac{3}{2n} \quad (39)$$

$$\eta_{1H_2} = \frac{\varphi_{2H_2O}}{\varphi_{2H_2} + \varphi_{2H_2O}} = \frac{3}{2m} \quad (40)$$

The overall gas utilisation efficiency for H_2-rich reducing gas in the BF is calculated as in Equation (41) below:

$$\eta = \frac{\varphi_{CO_2} + \varphi_{H_2O}}{\varphi_{CO} + \varphi_{CO_2} + \varphi_{H_2} + \varphi_{H_2O}} = X_{CO}\eta_{CO} + X_{H_2}\eta_{H_2} \quad (41)$$

Assuming the water generated in the Fe_2O_3 reduction is reacted with CO, in which H_2 performs only as a catalyst of CO reduction of Fe_2O_3. The water in top gas is determined by H_2 utilisation efficiency in FeO and Fe_3O_4, which was calculated as in Equation (42):

$$V_{H_2O} = \sum V_{H_2}\eta_{3H_2} + \sum V_{H_2}(1 - \eta_{3H_2})\eta_{2H_2} + \sum V_{H_2}(1 - \eta_{3H_2})(1 - \eta_{2H_2})\eta_{1H_2} \quad (42)$$

Since FeO reduction is the key step, the theoretical overall H_2 utilisation efficiency was calculated as shown in Equation (43). The highest theoretical H_2 utilisation efficiency can be obtained with the minimal H_2 requirement value on the basis of the thermodynamic requirement in Stage III, and this highest value is determined by temperature. Due to thermal restrictions and excess H_2 injected, the actual gas utilisation efficiency can only approach this theoretical value. The actual thermodynamic utilisation efficiency of H_2 is a function of the amount of H_2 introduced to the BF, as shown in Equation (43):

$$\begin{aligned}\eta_{H_2} &= \eta_{3H_2} + \left(1 - \eta_{3H_2}\right)\eta_{2H_2} + \left(1 - \eta_{3H_2}\right)\left(1 - \eta_{2H_2}\right)\eta_{1H_2} \\ &= \tfrac{1}{m} + \left(1 - \tfrac{1}{m}\right)\tfrac{4}{3m} + \left(1 - \tfrac{1}{m}\right)\left(1 - \tfrac{4}{3m}\right)\tfrac{3}{2m}\end{aligned} \quad (43)$$

As the FeO reduction is the key step in the indirect reduction process, the thermodynamic requirement of gas entering the BF shaft to produce 1 tHM is calculated as in Equations (44) and (45):

$$V_{bosh\_shaft} = \frac{1000(Fe)_{HM}(1 - R_d)}{\eta_3 \times \%(\text{Reducing gas})} \times \frac{22.4}{56} \quad (44)$$

$$R_d = 1 - R_i \quad (45)$$

where V_{bosh_shaft} is the amount of gas raised from BF bosh after direct reduction and the gas injected through the shaft tuyeres, m^3/tHM; $[Fe]_{HM}$ is the proportion of iron content in hot metal; R_d is the degree of direct reduction.

2.2. Thermal Calculations of H_2-Rich Gas Injection BF

As the heat carrier, the H_2-rich gas injected through raceway tuyeres needs to compensate for the required energy in the lower furnace and maintain the theoretical combustion temperature at a reasonable range. The gas injected through the shaft also needs to satisfy the heat requirement in the upper furnace. The energy of H_2-rich gas includes the oxidation heat release from the iron ore reduction and sensible heat. The oxidation heat release depends on gas utilisation efficiency and gas composition. The injection temperature determines the sensible heat. Thermal calculations for determining the amount of H_2-rich gas were developed by a static mass and energy model of the iron-making process.

The thermal balance for this iron-making process is developed in the lower and upper furnaces, divided by the shaft gas injection position. In this work, the lower furnace included BF raceway, dripping zone, and cohesive zone. The thermal balance of the lower furnace is shown in Equation (46) below:

$$\begin{gathered} Q_{cc} + Q_{blast} + Q_{hearth} + Q_{coke} + Q_{ore} \\ = H_{CO_2} + H_{H_2Ode} + H_{PCde} + Q_{bosh} + H_{dA} + H_{dFe} + H_S + Q_{HM} + Q_{slag} + Q_{loss\_1} \end{gathered} \quad (46)$$

where the heat income in the lower furnace includes: Q_{cc} = combustion heat of coke and pulverised coal in front of tuyeres, kJ/tHM; Q_{blast} = sensible heat of the hot blast, kJ/tHM; Q_{hearth} = sensible heat of H_2-rich gas injection to the hearth, kJ/tHM; Q_{coke} = heat of the coke brings to the lower part of BF, kJ/tHM; Q_{ore} = sensible heat of the iron ores into the lower part of the BF, kJ/tHM; and the heat expenditure includes: H_{CO2} = heat consumption of solution loss reaction due to the possible CO_2 in the hearth injection gas, kJ/tHM; H_{H_2Od} = heat consumption of water decomposition in front of tuyeres, kJ/tHM; H_{PCde} = heat consumption of pulverised coal decomposition in front of tuyeres, kJ/tHM; Q_{bosh} = heat brought to the shaft by bosh gas, kJ/tHM; H_{dA} = heat consumption by direct reduction of alloy element, kJ/tHM; H_{dFe} = heat consumption by direct reduction of FeO, kJ/tHM; H_S = heat consumption by desulphurisation, kJ/tHM; Q_{HM} = sensible heat of hot metal, kJ/tHM; Q_{slag} = sensible heat of slag, kJ/tHM; and Q_{loss_1} = heat loss in the lower furnace, kJ/tHM.

With H_2-rich gas injection to BF hearth, the raceway adiabatic flame temperature (RAFT) is calculated as in Equation (47):

$$\text{RAFT} = \frac{Q_{coke} + Q_{blast} + Q_{hearth} + Q_{cc} - Q_{H2Ode} - Q_{PCde} - H_{CO_2}}{V_{H_2}^{raceway} \times C_{H_2} + V_{CO}^{raceway} \times C_{CO} + V_{N_2}^{raceway} \times C_{N_2}} \quad (47)$$

where Q_{coke} is the heat brought to the raceway by coke, kJ/tHM; $V_i^{raceway}$ is the gas volumes of H_2, CO and N_2 in the raceway, m^3/tHM, respectively.

The energy input and output of the BF shaft can be expressed as in Equations (48) and (49), respectively:

$$Q_{shaft} + Q_{bosh} + Q_{ind} + Q_{ore\_shaft} = Q_{top} + Q_{coke} + Q_{ore} + Q_{loss\_s} \quad (48)$$

$$Q_{ind} = Q_{Fe_2O_3} + Q_{Fe_3O_4} + Q_{FeO} \quad (49)$$

where Q_{shaft} is the heat carried by H_2-rich gas injected into the shaft, kJ/tHM; Q_{ind} is the heat generation by iron oxides reduction by H_2 and CO, kJ/tHM; Q_{ore_shaft} is the sensible heat carried by iron ores entering the BF top, kJ/tHM; Q_{top}, kJ/tHM is heat loss in the shaft, kJ/tHM.

The heats provided by the reduction reactions from Fe_2O_3 to Fe_3O_4, Fe_3O_4 to FeO, and FeO to Fe by CO and H_2 are calculated as from Equations (50)–(52), respectively:

$$Q_{Fe_2O_3} = 56 \times \frac{m_{ore} \times \frac{w(Fe_2O_3)_{ore}}{160}}{3} \left(r_{ico} \times \Delta H^0_{1CO} + r_{iH_2} \times \Delta H^0_{1H_2} \right) \tag{50}$$

$$Q_{Fe_3O_4} = 56 \times \frac{2 \times m_{ore} \times \frac{w(Fe_2O_3)_{ore}}{160}}{3} \left(r_{ico} \times \Delta H^0_{2CO} + r_{iH_2} \times \Delta H^0_{2H_2} \right) \tag{51}$$

$$Q_{FeO} = 56 \times (1 - R_d) \left(\frac{2 \times m_{ore} \times w(Fe_2O_3)_{ore}}{160} + \frac{m_{ore} \times w(FeO)_{ore}}{72} \right) \left(r_{ico} \times \Delta H^0_{3CO} + r_{iH_2} \times \Delta H^0_{3H_2} \right) \tag{52}$$

where r_{ico} and r_{iH_2} are the degree of indirect reduction by CO and H_2, as shown in Equations (53) and (54), respectively:

$$r_{iH_2} = \frac{V_{H2O}}{22.4 \times 1000(Fe)_{HM}} \times 56 \tag{53}$$

$$r_{iCO} = R_i - r_{iH_2} \tag{54}$$

3. Results and Discussion

3.1. Results of the Thermodynamic Model

At a medium oxygen enrichment rate (9%), the nitrogen content in the BF shaft is calculated at around 35%. Figure 2a shows the heat effect results based on Equation (20). The FeO reduction reaction transforms from an exothermic into an endothermic process when H_2 content increases to 25% around 900 °C. A similar phenomenon shows up when the H_2 reaches 20% around 1000 °C. With more H_2 participating in the reduction at high temperatures, it causes a severe negative effect on the thermal energy supply to the BF. Therefore, the shaft gas injection temperature should not be too high to reduce its endothermic heat effect and require less preheating in the gas preheating device. The higher oxygen enrichment (less N_2 content) enables higher H_2 content in the BF, as shown in Figure 2b. It is suggested that with 20% N_2 entering the BF shaft, the H_2 content should be lower than 25% to avoid too much heat consumption by its reduction reaction.

According to Equations (11), (16), (29) and (30), the theoretical minimal H_2 and CO requirement with temperature are shown in Figure 3. From this figure, Stage III for iron ores reduction requires more H_2 when the temperature is over 625 °C. Stage III requires more CO over 650 °C. In this study, since the gas injection temperature is kept above 820 °C to ensure the high reducing capability of H_2, the thermodynamic key step for iron ores reduction is Stage III, and the other stages are proceeded with excess reducing gas. Compared with Fe_3O_4, FeO reduction requires more reducing gas to proceed, which determines the minimal amount of gaseous mixture. Fe_3O_4 and Fe_2O_3 reductions are carried out with excess reducing gas. In addition, the amount of H_2 required for FeO reduction at high temperatures is less than the amount of CO required in the reaction. Thermodynamically, injecting H_2 content to replace some amount of CO would reduce the total amount of gas mixture and reduce the fuel requirement for the gas preheating. However, CO content in the BF should be enough to meet its thermal condition.

According to Figure 4, gas utilisation efficiency for H_2 and CO was similar in Stage II when the temperature was above 820 °C, but H_2 utilisation efficiency was much higher than that for CO in Stage III. Although the utilisation efficiency of CO decreases as temperature increases, the FeO reduction reaction would still be promoted with increasing H_2 content due to the effect of the water–gas shift reaction.

On the basis of the calculation from Equation (41), gas utilisation efficiency at different H_2 contents in the reducing gas for FeO and Fe_3O_4 reduction is shown in Figure 5.

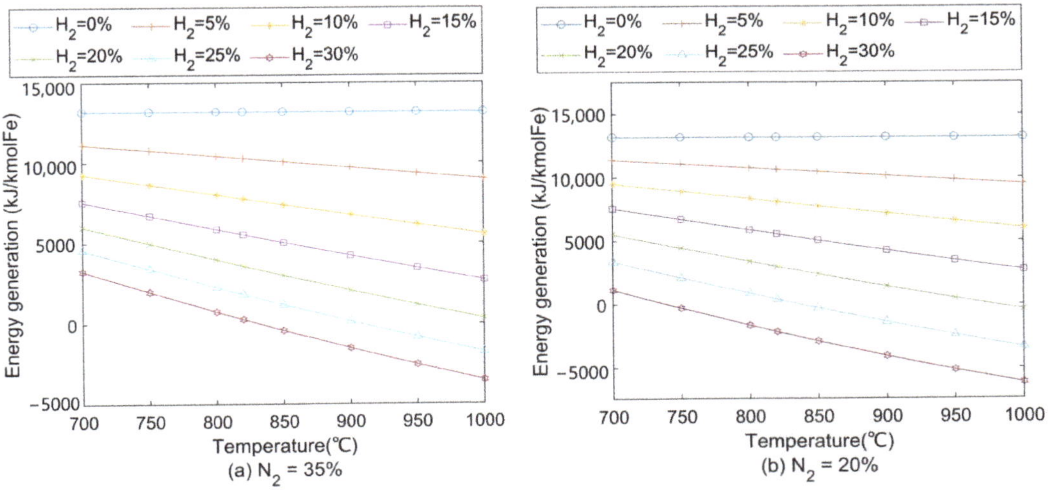

Figure 2. Heat effect of FeO reduction in H_2–CO–N_2 gas mixture at (**a**) 35% N_2 and (**b**) 20% N_2. H_2 content entering blast furnace shaft is indicated by different colours.

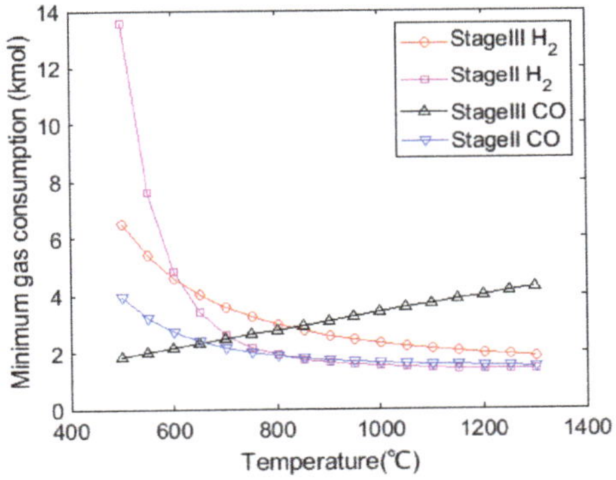

Figure 3. Variation in minimal gas consumption for iron ores reduction in pure H_2 or CO with temperature.

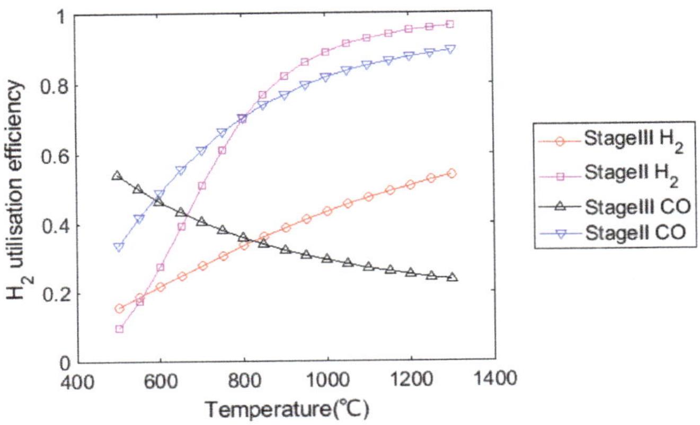

Figure 4. Variation in gas utilisation efficiency for iron ores reduction in pure H_2 or CO with temperature.

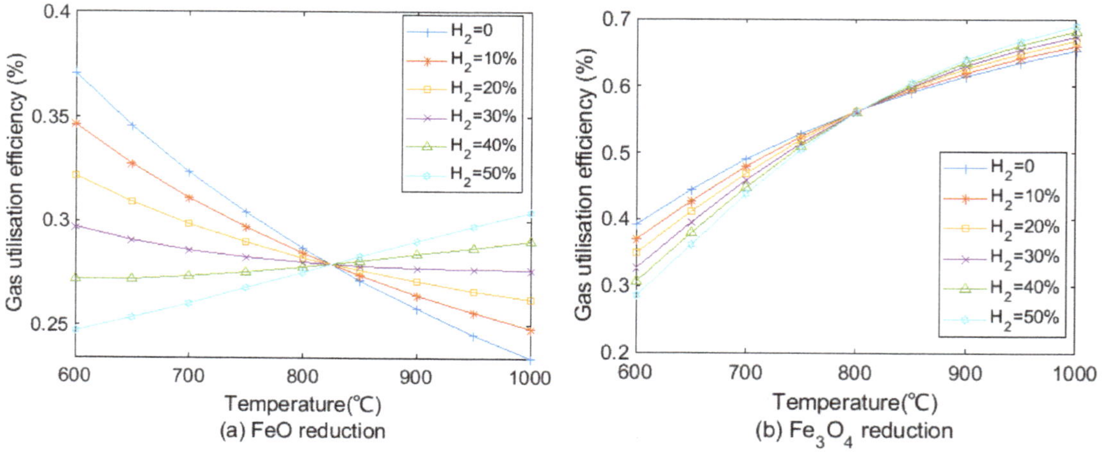

Figure 5. Gas utilisation efficiency at different H_2/CO ratio with temperature in (**a**) FeO and (**b**) Fe_3O_4 reductions.

Figure 5a presents the gas utilisation efficiency of FeO reduction. Due to the endothermic reaction of the H_2 reduction, utilisation efficiency of H_2-rich reducing gas at 1000 °C for the FeO reduction increased from 23% with no H_2 to around 30% with 50% H_2 by the equal interval. When H_2 content was less than 30% in reducing gas, gas utilisation efficiency in FeO reduction decreased with the increase in temperature. In contrast, gas utilisation efficiency rose with temperature when H_2 content is more than 40% in reducing gas. Hence, H_2 content in the reducing gas should not be too low to hinder improvement in the gas utilisation efficiency. As shown in Figure 5b, since Fe_3O_4 reductions by CO and H_2 are endothermic, gas utilisation efficiency increases with temperature. The effect of increases in H_2 content for Fe_3O_4 reduction is less than FeO reduction in terms of gas utilisation efficiency. At 1000 °C, gas utilisation efficiency increases by less than 4% when H_2 content increases from 0% to 50%. The shaft gas injection temperature should be higher than 820 °C to promote gas utilisation in FeO and Fe_3O_4 reduction, especially focusing on FeO reduction.

3.2. BF Simulation Conditions and Validation

Without gas injection, the model developed in this work can be used for a traditional BF. The measured data collected from a 2500 m^3 BF were used to validate this proposed model. The comparison of the industrial data and the model predictions is summarised in Table 2. The coke rate and top gas components were compared because they are essential measurable parameters that indicate the overall performance of an iron-making BF. The chemical composition of raw material data is shown in Tables 3–5. In general, results in this simulation show a similar trend as in practice, and the model was capable of estimating the overall iron-making process. The slight difference in top gas composition is because industrial data contain a slight amount of O_2 and CH_4 in top gas, which does not exist in this model.

Table 2. Quantitative validation of the model.

Parameter	Prediction	Industrial	Top Gas	Prediction	Industrial
Coke rate, kg/tHM	386	386	CO, %	25.1	24.9
Blast, Nm^3/tHM	1060	1089	CO_2, %	21.1	20.0
Slag rate, kg/tHM	364	373	H_2, %	1.2	0.8
Burden input, kg/tHM	1676	1676	N_2, %	50.0	53
RAFT, °C	2205	2195	R_d	0.46	-

Table 3. Chemical composition of raw material and dust, $w/w\%$.

Composition	Tfe	FeO	SiO_2	CaO	MgO	TiO_2	S	Al_2O_3
Sinter 1	55.85	9.59	5.25	10.35	2.19	0.17	0.03	2.5
Sinter 2	55.85	9.2	5.26	10.27	2.2	0.27	-	2.5
Pellet	61.92	1.46	5.31	1.45	-	1.58	-	0.98
Ti ore	40.27	-	9.07	2.05	-	10.78	-	1.69
Dust	39.80	2.58	4.05	2.04	0.83	0	-	1.31

Table 4. Chemical composition of coke and pulverised coal, $w/w\%$.

Composition	Fixed C	H_2O	FeO	CaO	SiO_2	Al_2O_3	MgO	N	O	H	S
Coke	86.06	4.50	1.6	0.39	4.49	3.69	0.35	0.30	0.21	0.33	1.80
PCI	71.15	0	0.03	1.98	8.4	7.93	0.18	0.34	3.16	2.10	0.30

Table 5. Chemical composition of hot metal, $w/w\%$.

Composition	Fe	C	Si	Mn	P	S	Ti
Hot metal	95.14	4.12	0.34	0.32	0.136	0.023	0.129

3.3. Effect of H_2 Injection on Coke Consumption Rate

The effect of injecting H_2-rich gas to hearth and/or shaft on coke rate is shown in Figure 6. The highest H_2 injection rate is limited to 160 m^3/tHM to ensure a balanced energy distribution in the BF shaft. Injecting H_2 to the BF hearth shows less coke consumption than injecting it to the shaft. This is because it provides more sensible heat to the lower part of the furnace to compensate the heat supplied by coke combustion. The relationship between H_2 injection and coke consumption rate at 9% oxygen enrichment rate is given as in Equation (55). Injecting 7.9~10.9 m^3/tHM H_2 can reduce the coke consumption rate by 1 kg/tHM, depending on the injection position.

$$K = 351X_{shaft} + 341X_{hearth} - (0.0913X_{shaft} + 0.1275X_{hearth}) \times H_{2inj} \quad (55)$$

where X_{shaft} and X_{hearth} are the proportion of H_2-rich gas injected to the shaft and hearth, respectively; and H_2 is the total H_2 injection volume to the BF, m^3/tHM.

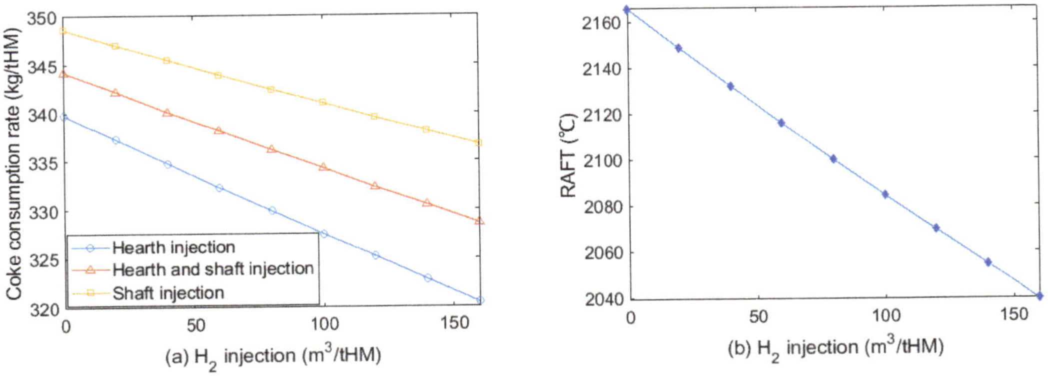

Figure 6. (**a**) Coke consumption rate at different H_2 injection volumes at different injection positions; (**b**) effect of H_2 injection to hearth on RAFT at 9% oxygen enrichment rate.

The effect of H_2 injection on RAFT at a constant oxygen enrichment rate is shown in Figure 6b. Injecting 20 m^3/tHM H_2 can reduce RAFT by around 16 °C. Increasing oxygen enrichment can achieve thermal compensation to maintain a stable RAFT and reduce coke consumption, as shown in Figure 7. Compared to the cases without thermal compensation, further reduction in coke consumption is obtained at 309 kg/tHM with 12.4% oxygen enrichment at 160 m^3/tHM H_2 injection.

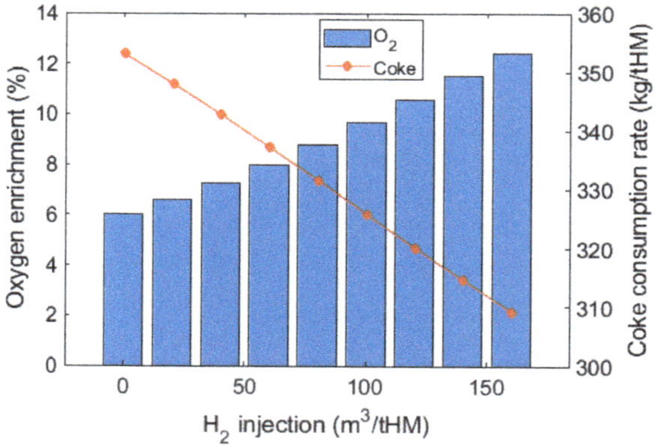

Figure 7. Coke consumption rate at different H_2 injection volumes at different oxygen enrichment rates for thermal compensation. RAFT was maintained at 2096 °C in this test.

3.4. *Effect on H_2 Utilisation Efficiency*

In the case of injecting H_2 to hearth, bosh gas composition is shown in Figure 8a. As the H_2 injection rate increased from 0 to 160 m^3/tHM at a constant CO injection rate, CO content in bosh gas drops due to less coke consumption. N_2 content also decreased since less hot blast is required for carbon combustion. There was no H_2O in the bosh gas because H_2O that formed from the iron oxide reduction by H_2 reacted with coke to generate H_2 again at high temperature. The amount of top gas and its composition is described in Figure 8b. With more H_2 injection, the moisture content in the top gas increases slightly from 2.72% to 3.12% when the H_2 injection rate reaches from 0 to 100 m^3/tHM. This is because H_2 utilisation efficiency significantly dropped from 72.6% to 22.9%, as shown in Figure 9a. CO content in top gas decreased because of less CO in the bosh gas and increased CO utilisation efficiency. CO_2 concentration was reduced by less than 2% because there was a significant drop of CO in the bosh gas and increase in CO utilisation efficiency is very limited. The H_2 injection promotes CO utilisation with the presence of the water–gas shift reaction. However, the overall effect of the water–gas shift reaction was very limited across the whole BF. The comprehensive gas utilisation efficiency gradually decreased from 43.6% to 39.0% when H_2 injection reached 160 m^3/tHM.

H_2 and CO utilisation efficiency after thermal compensation is shown in Figure 9b. Results in this simulation reflect a similar trend as in the literature results [8,49]. Oxygen enrichment increased, and the nitrogen composition in the blast decreased with H_2 injection. Hence, there was more increase in reducing gas concentration than that in Figure 9a. With less N_2 dilution and stronger indirect reduction, CO utilisation with thermal compensation was enhanced by 5% with H_2 injection from 0 to 160 m^3/tHM.

By H_2 increasing in the BF, CO utilisation increased, and H_2 utilisation efficiency decreased. One reason is that water–gas shift reaction Equation (18) would tend to proceed to the right-hand side, and both H_2 and CO_2 are generated in the upper furnace below 1000 °C. With the regeneration of H_2, FeO reduction in Equation (7) can be considered to be proceeding in two successive stages, as shown in Equations (13) and (18):

$$\text{From Equation (13)}: \ \text{FeO (s)} + \text{H}_2\ \text{(g)} \rightarrow \text{Fe (s)} + \text{H}_2\text{O(g)}$$

$$\text{From Equation (18)}: \ \text{H}_2\text{O(g)} + \ \text{CO (g)} = \text{H}_2\ \text{(g)} + \text{CO}_2\ \text{(g)}$$

$$\text{From Equation (7)}:\ \text{Overall}:\ FeO\ (s) + CO\ (g) \rightarrow Fe\ (s) + CO_2\ (g)$$

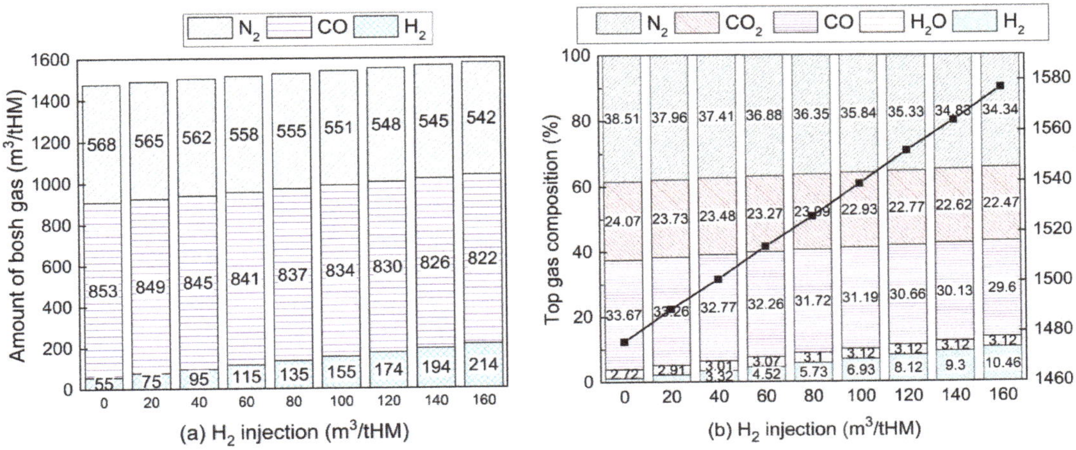

Figure 8. Variations of (**a**) bosh gas and (**b**) top gas composition with hydrogen injection. (H_2 injection at 0 m^3/tHM indicates the operating condition of a traditional BF with oxygen enrichment rate at 9%).

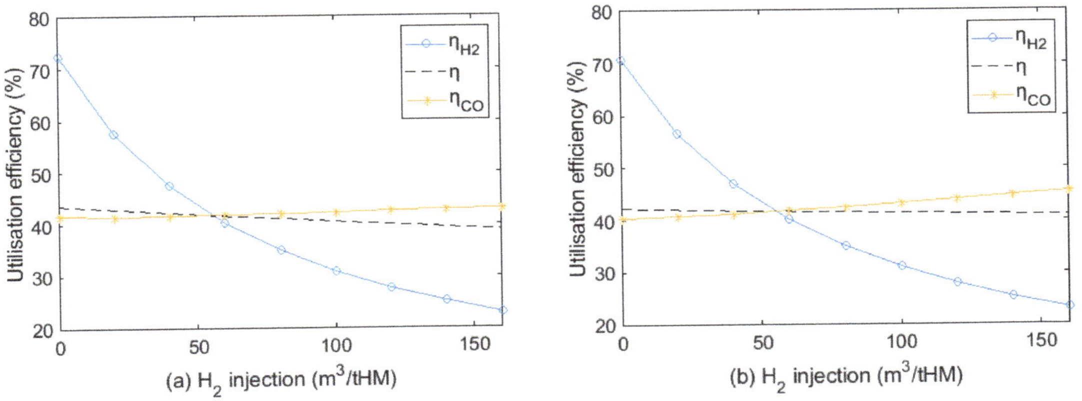

Figure 9. Variations in gas utilisation efficiency with hydrogen injection (**a**) before and (**b**) after thermal compensation.

Due to the smaller size and high diffusivity of H_2, the reaction in Equation (13) has an advantage over the reaction in Equation (7). Thus, FeO reduction by CO was promoted by increased H_2 content.

As depicted in Figure 10, with H_2 injection and thermal compensation, the degree of indirect reduction increased because H_2 reduction replaced part of the direct reduction and oxygen enrichment enhanced reducing gas atmosphere. Since the direct reduction was a huge endothermic reaction process in the lower furnace, less direct reduction reduces coke consumption. Compared to the higher H_2 injection volume, injecting 0 to 80 m^3/tHM H_2 generates more effect on the degree of indirect reduction, from 0.107 to 0.125. Further, the injection of more than 50 m^3/tHM H_2 significantly increased the indirect reduction of CO. Considering gas utilisation efficiency, H_2 injection should be around 50–80 m^3/tHM to maintain smooth BF operation and avoid too much excess H_2 in top gas.

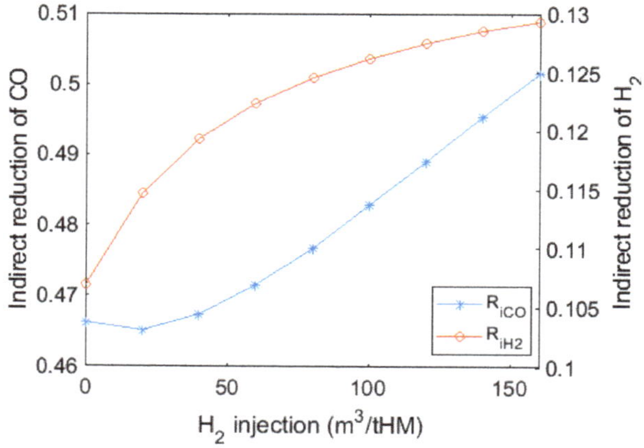

Figure 10. Variations in degree of indirect reduction by CO and H_2 with H_2 injection and thermal compensation.

3.5. *Effects on CO_2 Emissions and Energy Consumption*

In this work, the CO injected into the BF comes from the CO_2 captured in top gas, which reduces CO_2 emission compared to a traditional BF process. The emission from CCU can be negligible because it applies renewable electricity. The total emission of this iron-making process includes uncaptured CO_2, flue gases from BFG combusted in the preheater and the hot blast stoves, which is shown in Figure 11a. The main CO_2 emission reduction comes from the hot stoves. Compared with the traditional BF iron-making system that lacks CCU and gas injection, CO_2 emission dropped from 534 to around 272 m^3/tHM with 160 m^3/tHM H_2 injection in this system. When increasing H_2 content from 0 to 80 m^3/tHM, CO_2 emission only decreased by 20 m^3/tHM because more BFG was consumed to preheat the injection gas. Additionally, injecting H_2-rich gas into the shaft showed more CO_2 emission reduction capability than injecting H_2 into a hearth or both tuyeres, as shown in Figure 11b.

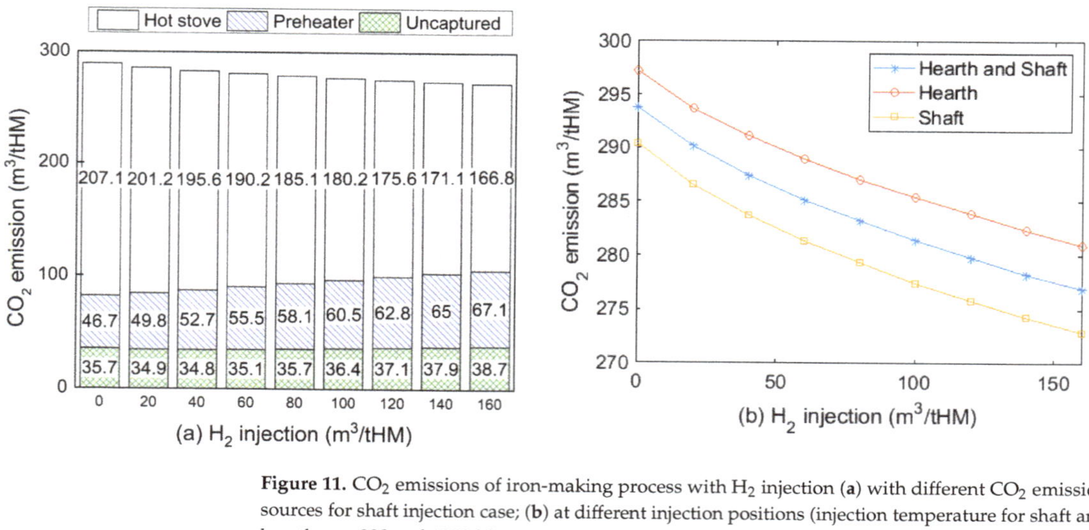

Figure 11. CO_2 emissions of iron-making process with H_2 injection (**a**) with different CO_2 emission sources for shaft injection case; (**b**) at different injection positions (injection temperature for shaft and hearth was 900 and 1250 °C, respectively).

The energy consumption of the process described in Figure 1 was calculated on the basis of static mass and energy balance, using Equation (56). The standard coal coefficient for each substance in a kilogram of coal equivalent per ton of hot metal (kgce/tHM) is used in this analysis [50].

$$E_{net} = E_{coke} + E_{PC} + E_{BOFG} + E_{blast} + E_{cap} + E_{conv} + E_{water} - E_{export} - E_{O2} \quad (56)$$

where E_{net} is the net energy consumption of the process; E_{coke} is the energy input by coke; E_{PC} is the energy input by pulverised coal; E_{BOFG} is the energy input by BOFG to the hot blast stoves; E_{blast} is the energy carried by the blast to hot blast stoves; E_{cap} is the electricity required for CO_2 capture unit; E_{conv} is the electricity required for CO_2 electrolyser, as calculated by Equation (2); E_{water} is the energy carried by the make-up water required at the humidifier for H_2 generation in the electrolsyer; E_{export} is the energy carried by the BFG that is exported to other processes in the integrated steel mill; and E_{O2} is the energy carried by the oxygen that is generated in the electrolyser and exported to other processes in the integrated steel mill.

Energy consumption results are shown in Figure 12. In a traditional BF, carbon resources from coke and pulverised coal injection provide the primary energy input, which accounts for 80% in total. When hydrogen is injected into the BF from 0 to 160 m^3/tHM, carbon only accounts for 65% to 50% of the total energy consumption. With H_2-rich gas injection from 0 to 160 m^3/tHM, net energy consumption increased from 541 to 698 kgce/tHM, which is higher than that of the traditional BF (504 kgce/tHM) because electricity required for CO_2 conversion to generate H_2 kept increasing from 158 to 369 kgce/tHM.

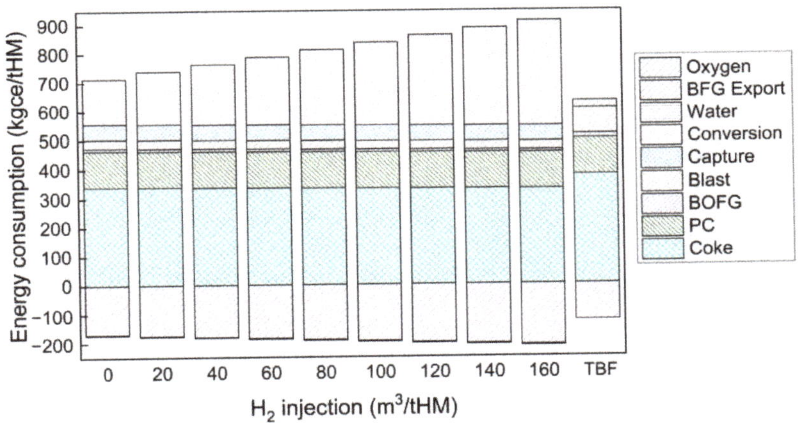

Figure 12. Energy consumption for H_2-rich gas injection BF and traditional BF process (energy carried by the make-up water needed for the CO_2 conversion and by BOFG was less than 1% of total energy consumption and thus not indicated in this figure).

In summary, increasing hydrogen injection volume can reduce coke consumption and CO_2 emissions. For coke consumption and CO_2 emission reduction, the hydrogen injection amount should be as much as possible, as long as it satisfies the energy balance in the BF. In this case, the hydrogen injection amount should be 160 m^3/tHM. However, injecting too much H_2 significantly reduces its utilisation efficiency and increases the net energy consumption of this process. Further study is recommended to develop a multiobjective optimisation model to balance these effects of hydrogen injection. In general, injecting H_2 at around 80 m^3/tHM may consume less energy and suppress CO_2 emissions under the simulation conditions.

Energy consumption to produce 1 tonne of hot metal in the case of injecting 80 m^3/tHM H_2 with 200 m^3/tHM CO is indicated in Figure 13. Compared to the traditional BF as indicated by Table 2, coke consumption decreased by 43 kg/tHM. CO_2 emission dropped from 534 m^3/tHM for a traditional BF to 278 m^3/tHM in this case (including gas heating flue gas, capture unit CO_2 letdown, and stack flue gas). However, electricity consumption in the CO_2 capture unit and electrolyser is one of the largest energy inputs. The economic impact of this CCU technology as an auxiliary system to the BF is highly recommended for future investigation.

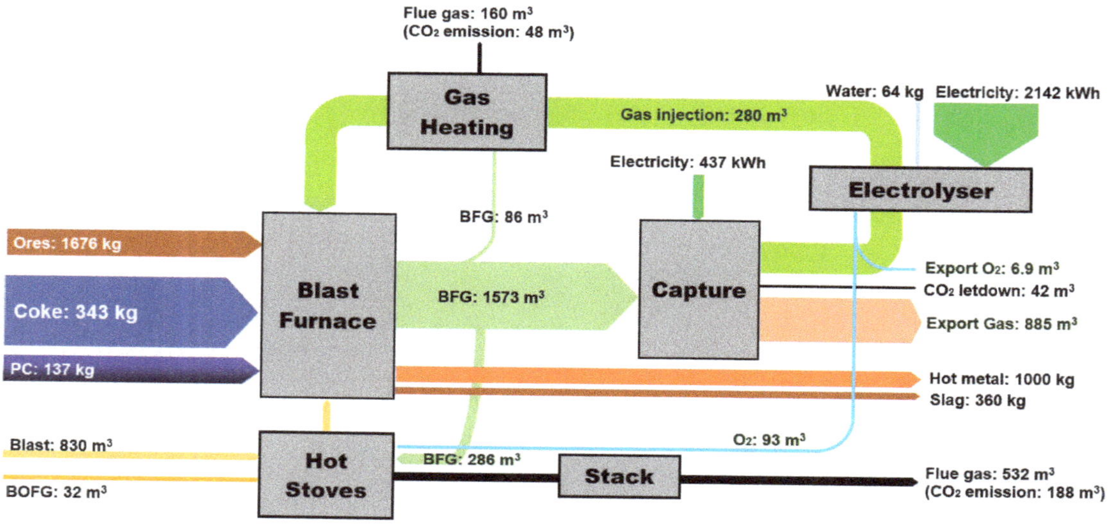

Figure 13. Energy and material consumption of the process with 80 m^3/tHM hydrogen injection. Relative flow size estimated by energy and demonstrated as bar width.

4. Conclusions

In this study, a thermodynamic model was used to determine the utilisation efficiency of H_2-rich gas by considering H_2 behaviour at different reduction stages. A static mass and energy balance model of the BF was adopted with this thermodynamic model. The effect of injecting H_2-rich gas on BF performance was determined in terms of its coke consumption, CO_2 emissions, and energy consumption. Under these simulation conditions, the major findings of this study were:

1. The desired shaft gas injection temperature should not exceed 1000 °C to suppress the endothermic FeO reduction reaction by H_2-rich gas.
2. Injecting H_2 to BF hearth has a better effect on coke rate reduction than that of injection to the shaft. The lowest H_2 consumption to save 1 kg of coke was estimated to be 7.9 m^3/tHM.
3. H_2 utilisation efficiency dropped significantly with increasing H_2 content, and the increase in CO utilisation efficiency was limited. Further research should focus on improving H_2 utilisation efficiency with a high H_2 injection rate.
4. Considering H_2 utilisation efficiency and the degree of indirect reduction by H_2 and CO, the proper H_2 injection rate should be from around 50 to 80 m^3/tHM.
5. Introducing H_2-rich gas injection can reduce CO_2 emissions of the iron-making process by up to 262 m^3/tHM compared with a traditional BF. However, injecting too much H_2 would hinder CO_2 emission reduction due to its requirement of preheating outside the BF.
6. The energy consumption of this proposed process was higher than that of the traditional BF. Although coke consumption was reduced by 43 kg/tHM more than that of the traditional BF, net energy consumption increased with the amount of injected

hydrogen due to the high electricity consumption in the CO_2 capture and electrolyser. Developing a CO_2 conversion unit with higher efficiency but less energy consumption is strongly recommended.

Author Contributions: Conceptualization, Y.H., T.E.R., V.R. and G.W.; methodology, Y.H. and G.W.; software, Y.H.; validation, Y.H. and L.H.; formal analysis, Y.H.; investigation, Y.H., Y.Q., J.C. and L.H. resources, L.H.; data curation, Y.H.; writing—original draft preparation, Y.H.; writing—review and editing, Y.H., Y.Q., J.C., L.H., T.E.R. and G.W.; visualization, Y.H.; supervision, T.E.R., V.R. and G.W.; project administration, G.W. and L.H.; funding acquisition, L.H. All authors have read and agreed to the published version of the manuscript.

Funding: This research was funded by HBIS (China) grant number ICSS2017-04.

Institutional Review Board Statement: Not applicable.

Informed Consent Statement: Not applicable.

Data Availability Statement: Data sharing not applicable. No new data were created or analysed in this study. Data sharing is not applicable to this article.

Acknowledgments: The authors gratefully acknowledge HBIS Group Company Limited for funding support and the scholarship support from the University of Queensland.

Conflicts of Interest: The authors declare no conflict of interest.

Nomenclature

Abbreviations	
BF	Blast furnace
PCI	Pulverised coal injection
CCU	CO_2 capture and conversion unit
RAFT	Raceway adiabatic flame temperature
tHM	Tonne of hot metal
Roman and Greek symbols	
E	Energy, kWh/tHM or kgce/tHM
V	Volume of gas, m^3/tHM
T	Temperature, °C
C_i	Specific heat capacity, kJ/m^3·°C
K_i	Reaction equilibrium constant of reduction stage i; i = I, II, III
G	Gibbs free energy, kJ/mol
φ	Volume fraction of gas component
η_{iCO}	Utilisation efficiency of CO in stage i, i = 1, 2, 3
η_{iH_2}	Utilisation efficiency of H_2 in stage i, i = 1, 2, 3
n	Minimal CO required for iron ores reduction, mol
m	Minimal H_2 required for iron ores reduction, mol
X_i	Volume reaction of CO or H_2 in reducing gas entering the BF shaft
η	Overall gas utilisation efficiency for H_2-rich reducing gas in the BF
%(Reducing gas)	Proportion of H_2 and CO in the total amount of gas entering the bosh and shaf
Q	Sensible heat of material, kJ/tHM
H	Enthalpy of material, kJ/tHM
R_d	Degree of direct reduction
r_i	Degree of indirect reduction
$(Fe)_{HM}$	Iron content in hot metal
ΔH^0	Enthalpy of reaction, kJ/tHM
w	Weight fraction of solid material
H_{2inj}	Total H_2 injection volume to the BF, m^3/tHM

References

1. Xu, K. Low carbon economy and iron and steel industry. *Iron Steel* **2010**, *45*, 1–12.
2. Pardo, N.; Moya, J.A. Prospective scenarios on energy efficiency and CO_2 emissions in the European Iron & Steel industry. *Energy* **2013**, *54*, 113–128. [CrossRef]
3. Babich, A.; Senk, D. *New Trends in Coal Conversion: Combustion, Gasification, Emissions, and Coking*; Woodhead Publishing: Sawson, UK, 2019; pp. 367–404.
4. Danloy, G.; Berthelemot, A.; Grant, M.; Borlée, J.; Sert, D.; Van Der Stel, J.; Jak, H.; Dimastromatteo, V.; Hallin, M.; Eklund, N.; et al. ULCOS-Pilot testing of the Low-CO_2 Blast Furnace process at the experimental BF in Luleå. *Rev. Métall.-Int. J. Metall.* **2009**, *106*, 1–8. [CrossRef]
5. Knop, K.; Hallin, M.; Burstrom, E. ULCORED SP 12 Concept for minimized CO_2 emission. *Metall. Res. Technol.* **2009**, *106*, 419–421. [CrossRef]
6. Material Economics. Industrial Transformation 2050: Pathways to Net-Zero Emisisons from EU Heavy Industry. 2019. Available online: https://materialeconomics.com/publications/industrial-transformation-2050 (accessed on 8 January 2022).
7. Chisalita, D.-A.; Petrescu, L.; Cobden, P.; van Dijk, H.E.; Cormos, A.-M.; Cormos, C.-C. Assessing the environmental impact of an integrated steel mill with post-combustion CO_2 capture and storage using the LCA methodology. *J. Clean. Prod.* **2019**, *211*, 1015–1025. [CrossRef]
8. Nogami, H.; Kashiwaya, Y.; Yamada, D. Simulation of Blast Furnace Operation with Intensive Hydrogen Injection. *ISIJ Int.* **2012**, *52*, 1523–1527. [CrossRef]
9. Usui, T.; Kawabata, H.; Ono-Nakazato, H.; Kurosaka, A. Fundamental Experiments on the H_2 Gas Injection into the Lower Part of a Blast Furnace Shaft. *ISIJ Int.* **2002**, *42*, S14–S18. [CrossRef]
10. Pinegar, H.K.; Moats, M.S.; Sohn, H.Y. Process Simulation and Economic Feasibility Analysis for a Hydrogen-Based Novel Suspension Ironmaking Technology. *Steel Res. Int.* **2011**, *82*, 951–963. [CrossRef]
11. Wagner, D.; Devisme, O.; Patisson, F.; Ablitzer, D. A laboratory study of the reduction of iron oxides by hydrogen. *arXiv* **2008**, arXiv:08032831.
12. Spreitzer, D.; Schenk, J. Reduction of Iron Oxides with Hydrogen—A Review. *Steel Res. Int.* **2019**, *90*, 1900108. [CrossRef]
13. Qie, Y.; Lyu, Q.; Li, J.; Lan, C.; Liu, X. Effect of Hydrogen Addition on Reduction Kinetics of Iron Oxides in Gas-injection BF. *ISIJ Int.* **2017**, *57*, 404–412. [CrossRef]
14. Zuo, H.-b.; Wang, C.; Dong, J.-j.; Jiao, K.-x.; Xu, R.-s. Reduction kinetics of iron oxide pellets with H_2 and CO mixtures. *Int. J. Miner. Metall. Mater.* **2015**, *22*, 688–696. [CrossRef]
15. Chu, M.; Nogami, H.; Yagi, J.-I. Numerical Analysis on Injection of Hydrogen Bearing Materials into Blast Furnace. *ISIJ Int.* **2004**, *44*, 801–808. [CrossRef]
16. Yilmaz, C.; Wendelstorf, J.; Turek, T. Modeling and simulation of hydrogen injection into a blast furnace to reduce carbon dioxide emissions. *J. Clean. Prod.* **2017**, *154*, 488–501. [CrossRef]
17. Lyu, Q.; Qie, Y.; Liu, X.; Lan, C.; Li, J.; Liu, S. Effect of hydrogen addition on reduction behavior of iron oxides in gas-injection blast furnace. *Thermochim. Acta* **2017**, *648*, 79–90. [CrossRef]
18. Watakabe, S.; Miyagawa, K.; Matsuzaki, S.; Inada, T.; Tomita, Y.; Saito, K.; Osame, M.; Sikström, P.; Ökvist, L.S.; Wikstrom, J.-O. Operation Trial of Hydrogenous Gas Injection of COURSE50 Project at an Experimental Blast Furnace. *ISIJ Int.* **2013**, *53*, 2065–2071. [CrossRef]
19. Qie, Y.; Lyu, Q.; Liu, X.; Li, J.; Lan, C.; Zhang, S.; Yan, C. Effect of Hydrogen Addition on Softening and Melting Reduction Behaviors of Ferrous Burden in Gas-Injection Blast Furnace. *Met. Mater. Trans. A* **2018**, *49*, 2622–2632. [CrossRef]
20. Wang, Y.; He, Z.-J.; Zhang, W.-l.; Zhu, H.-b.; Zhang, J.-h.; Pang, Q.-h. Reduction behavior of iron-bearing burdens in hydrogen-rich stream. *Iron Steel* **2020**, *55*, 34–40.
21. Andersson, J.; Grönkvist, S. A comparison of two hydrogen storages in a fossil-free direct reduced iron process. *Int. J. Hydrog. Energy* **2021**, *46*, 28657–28674. [CrossRef]
22. Hydrogen Council. Path to Hydrogen Competitiveness: A Cost Perspective. 2020. Available online: https://hydrogencouncil.com/en/path-to-hydrogen-competitiveness-a-cost-perspective/ (accessed on 8 January 2022).
23. Uribe-Soto, W.; Portha, J.-F.; Commenge, J.-M.; Falk, L. A review of thermochemical processes and technologies to use steelworks off-gases. *Renew. Sustain. Energy Rev.* **2017**, *74*, 809–823. [CrossRef]
24. Da Rocha, E.P.; Guilherme, V.S.; de Castro, J.A.; Sazaki, Y.; Yagi, J.-I. Analysis of synthetic natural gas injection into charcoal blast furnace. *J. Mater. Res. Technol.* **2013**, *2*, 255–262. [CrossRef]
25. Halim, K.S.A. Theoretical Approach to Change Blast Furnace Regime with Natural Gas Injection. *J. Iron Steel Res. Int.* **2013**, *20*, 40–46. [CrossRef]
26. Trinkel, V.; Kieberger, N.; Bürgler, T.; Rechberger, H.; Fellner, J. Influence of waste plastic utilisation in blast furnace on heavy metal emissions. *J. Clean. Prod.* **2015**, *94*, 312–320. [CrossRef]
27. Domingo-Tafalla, B.; Martínez-Ferrero, E.; Franco, F.; Palomares-Gil, E. Applications of Carbon Dots for the Photocatalytic and Electrocatalytic Reduction of CO_2. *Molecules* **2022**, *27*, 1081. [CrossRef]
28. Zhong, H.; Sa, R.; Lv, H.; Yang, S.; Yuan, D.; Wang, X.; Wang, R. Covalent organic framework hosting metalloporphyrin-based carbon dots for visible-light-driven selective CO_2 reduction. *Adv. Funct. Mater.* **2020**, *30*, 2002654. [CrossRef]

29. Küngas, R. Review—Electrochemical CO_2 Reduction for CO Production: Comparison of Low- and High-Temperature Electrolysis Technologies. *J. Electrochem. Soc.* **2020**, *167*, 044508. [CrossRef]
30. Parvanian, A.M.; Sadeghi, N.; Rafiee, A.; Shearer, C.J.; Jafarian, M. Application of porous materials for CO_2 reutilization: A Review. *Energies* **2022**, *15*, 63. [CrossRef]
31. Delacourt, C.; Ridgway, P.; Kerr, J.; Newman, J. Design of an Electrochemical Cell Making Syngas ($CO + H_2$) from CO_2 and H_2O reduction at Room Temperature. *J. Electrochem. Soc.* **2007**, *155*, B42. [CrossRef]
32. Lu, Q.; Jiao, F. Electrochemical CO_2 reduction: Electrocatalyst, reaction mechanism, and process engineering. *Nano Energy* **2016**, *29*, 439–456. [CrossRef]
33. Garg, S.; Li, M.; Idros, M.N.; Wu, Y.; Wang, G.; Rufford, T.E. Is Maximising Current Density Always the Optimum Strategy in Electrolyser Design for Electrochemical CO_2 Conversion to Chemicals? 2021. Available online: https://chemrxiv.org/engage/chemrxiv/article-details/61a7235f7848050092a110b4 (accessed on 8 January 2022).
34. De Ras, K.; Van de Vijver, R.; Galvita, V.V.; Marin, G.B.; Van Geem, K.M. Carbon capture and utilization in the steel industry: Challenges and opportunities for chemical engineering. *Curr. Opin. Chem. Eng.* **2019**, *26*, 81–87. [CrossRef]
35. De Castro, J.A.; Takano, C.; Yagi, J.-i. A theoretical study using the multiphase numerical simulation technique for effective use of H_2 as blast furnaces fuel. *J. Mater. Res. Technol.* **2017**, *6*, 258–270. [CrossRef]
36. Zhang, W. Fundamental Study and Process Optimization of the Oxygen Blast Furnace Ironmaking. Doctoral Dissertation, Northeastern University, Shenyang, China, 2015.
37. Kwak, N.-S.; Lee, J.H.; Lee, I.Y.; Jang, K.R.; Shim, J.-G. A study of the CO_2 capture pilot plant by amine absorption. *Energy* **2012**, *47*, 41–46. [CrossRef]
38. Bui, M.; Adjiman, C.S.; Bardow, A.; Anthony, E.J.; Boston, A.; Brown, S.; Fennell, P.S.; Fuss, S.; Galindo, A.; Hackett, L.A.; et al. Carbon capture and storage (CCS): The way forward. *Energy Environ. Sci.* **2018**, *11*, 1062–1176. [CrossRef]
39. Choi, J.; Cho, H.; Yun, S.; Jang, M.-G.; Oh, S.-Y.; Binns, M.; Kim, J.-K. Process design and optimization of MEA-based CO_2 capture processes for non-power industries. *Energy* **2019**, *185*, 971–980. [CrossRef]
40. Hari, S.R.; Balaji, C.P.; Karunamurthy, K. Carbon Capture and Storage Using Renewable Energy Sources: A Review. *IOP Conf. Ser. Earth Environ. Sci.* **2020**, *573*, 12004. [CrossRef]
41. Aresta, M.; Dibenedetto, A.; Angelini, A. The use of solar energy can enhance the conversion of carbon dioxide into energy-rich products: Stepping towards artificial photosynthesis. *Philos. Trans. R. Soc. London. Ser. A Math. Phys. Eng. Sci.* **2013**, *371*, 20120111. [CrossRef]
42. McBrien, M.; Serrenho, A.C.; Allwood, J. Potential for energy savings by heat recovery in an integrated steel supply chain. *Appl. Therm. Eng.* **2016**, *103*, 592–606. [CrossRef]
43. Mazanek, E.; Jasienska, S.; Brachucy, A.; Bryk, C. The influence of hydrogen in the gas mixture of hydrogen and CO on the dynamics of the reduction process. *Metal. Odlew.* **1982**, *8*, 53–70.
44. Piotrowski, K.; Mondal, K.; Lorethova, H.; Stonawski, L.; Szymanski, T.; Wiltowski, T. Effect of gas composition on the kinetics of iron oxide reduction in a hydrogen production process. *Int. J. Hydrog. Energy* **2005**, *30*, 1543–1554. [CrossRef]
45. Jozwiak, W.; Kaczmarek, E.; Maniecki, T.; Ignaczak, W.; Maniukiewicz, W. Reduction behavior of iron oxides in hydrogen and carbon monoxide atmospheres. *Appl. Catal. A Gen.* **2007**, *326*, 17–27. [CrossRef]
46. Wu, S.; Wang, X.; Zhang, J. *Ferrous Metallurgy*; Metallurgical Industry Press: Beijing, China, 2019; pp. 46–48.
47. Bernasowski, M. Theoretical Study of the Hydrogen Influence on Iron Oxides Reduction at the Blast Furnace Process. *Steel Res. Int.* **2014**, *85*, 670–678. [CrossRef]
48. Li, J.; Wang, P.; Zhou, L.; Cheng, M. The Reduction of Wustite with High Oxygen Enrichment and High Injection of Hydrogenous Fuel. *ISIJ Int.* **2007**, *47*, 1097–1101. [CrossRef]
49. Tang, J.; Chu, M.; Li, F.; Zhang, Z.; Tang, Y.; Liu, Z.; Yagi, J. Mathematical simulation and life cycle assessment of blast furnace operation with hydrogen injection under constant pulverized coal injection. *J. Clean. Prod.* **2021**, *278*, 123191. [CrossRef]
50. *China GB/T 2589-2020*; General Principles for Calculation of the Comprehensive Energy Consumption. Standardization Administration of China: Bijing, China, 2020.

MDPI
St. Alban-Anlage 66
4052 Basel
Switzerland
Tel. +41 61 683 77 34
Fax +41 61 302 89 18
www.mdpi.com

Materials Editorial Office
E-mail: materials@mdpi.com
www.mdpi.com/journal/materials

www.ingramcontent.com/pod-product-compliance
Lightning Source LLC
Chambersburg PA
CBHW061621220926
51850CB00005B/565